Patterns in Biology

David Harrison Ph.D.

A HALSTED PRESS BOOK

JOHN WILEY & SONS
New York

First published 1975
by Edward Arnold (Publishers) Ltd.
25 Hill Street,
London, W1X 8LL

Published in the U.S.A.
by Halsted Press, a Division
of John Wiley & Sons, Inc.
New York

Library of Congress Cataloging in Publication Data

Harrison, David.
Patterns in biology.

"A Halsted Press book."
Bibliography: p. 241
Includes index.
1. Biology. I. Title. [DNLM: 1. Biology. QH307.2

H318p]
QH308.2.H37 574 75-16300
ISBN 0-470-35555-7

Printed in the United States of America

10 9 8 7 6 5 4 3 2 1

Preface

Part of the stimulus and excitement of biology is the recognition of the relationships between the many processes that take place in living organisms. In this book these inter-relationships, or patterns, are discussed in sufficient depth to engender a feeling for their continuity and, while the content may prove challenging to many students at, say, Advanced Level, the rewards, in terms of developing a way of thinking about biological situations, are likely to be considerable.

With the explosion of interest in biology and the accelerating, apparently exponential, rate of growth of knowledge in the subject there is a real danger of a growing schism between factual biology taught in schools and the biological scene in the world at large. Fact-orientated biology has the advantage of being convenient to teach and easily examinable, but the capacity to memorize facts and reproduce them on demand is not necessarily correlated to a student's comprehension of the subject, nor to his subsequent success should he pursue the subject after school. This kind of biology tends to be static because the quantity of data that can be assimilated in a course is more or less constant and successive generations of students tend to plod through the same material. Meanwhile, the material itself becomes increasingly distant from the topics that are interesting practising biologists; school biology threatens to turn into a sort of scientific Classics.

The best solution so far to this educational dilemma has been offered by the Nuffield Foundation programmes in biology, which are based on empiricism, that is on observation and experiment. In principle the thought processes by which researchers are extending the frontiers of knowledge are also empirical and there therefore exists a real link between the work done by the most junior and the most advanced biologists.

In this book, *Patterns in Biology*, the author has drawn attention to some of the themes present in modern biology and discusses relatively recent developments in a number of fields. The topics embraced are broadly those in which the most rapid growth of knowledge is taking place. Although it is not a Nuffield-plan book in the sense that it encourages students to work in the laboratory along certain lines, it is complementary to that concept in that it seeks to explore by broadly empirical means some of the paths that modern biology is treading.

Patterns in Biology does not set out to cover all the topics listed in an Advanced Level Biology syllabus. This is quite deliberate because the author believes that the range of discussions now being aired in biology cannot effectively either be written by one person without extensive plagiarism, or be bound in a single, readily portable volume. Many topics at A Level concern areas in which the advance of knowledge is relatively slow, such as comparative anatomy and physiology. There are many excellent books now on the market covering these areas, which provide very satisfactory reading for the good A Level or University Entrance candidate. *Patterns in Biology* therefore concentrates on those topics which tend to be less well discussed in the existing literature.

1974 D.H.

Contents

General Introduction

The topics covered in this book are listed in the Contents. Their interrelated nature will be appreciated if they are visualized as being linked to form a 'paddle-shape'. This idea is illustrated on page 2, where the behaviour of molecules in cells, regulation, evolution and energy transformation in the biosphere form the blade of the paddle, and cell structure and organization form the handle. Clearly, other topics not dealt with in the book, such as the organization of tissues in the body (plant and animal anatomy) and the properties of tissues in the whole organism (physiology) can be related to the scheme and slotted into place as extensions of the handle.

The book is divided into three Sections comprising ten Chapters. The first Section describes some aspects of cell structure and discusses the roles of macromolecules in the various activities of cells, particularly those concerned in coding for and regulating the activities of cells and, through them, of the whole organism. This section concludes with a discussion of energy transformations in cells because this is ultimately related to the flow of energy through the planetary biosphere as a whole, in an ecological context.

The second and third Sections deal with biological continuity, which is the way in which the macromolecules discussed in the first Section perpetuate the characteristics of an organism during reproduction. There follows a discussion of patterns of biological discontinuity, that is, the sources of variation that arise in populations and how this in turn is related to evolution.

In the third Section the genetic basis of the evolutionary mechanism is discussed in rather more detail than might at first be thought necessary. For historical reasons, principally the publication of Darwin's Origin of Species at a time when a false view of the inheritance mechanism was widely held (discussed in greater detail in Chapter 9), genetics and evolution tended to follow different routes, the former towards an intensive exploration of the mechanics of the inheritance process and the mode of gene action, and the latter towards amassing palaeontological evidence for evolution. Although the behaviour of genes in populations was described early in this century and the role of genes as determiners of the characters upon which evolution acted was also recognized, evolution in terms of changes in the genetic make-up of populations has only been spelt out in the last two or three decades.

In view of this the author has thought it worthwhile to elaborate on some of the genetic sources of variability in populations and the ways in which genes can give rise to continuous or discontinuous kinds of variation because these are vital to the understanding of evolution. Fossil evidence only explains the evolutionary record, and then incompletely, whereas population genetics offers an explanation of the evolutionary mechanism itself; it is this facet that seems to have taken a long time to percolate through to the general area of school biology.

The Use of References

Patterns in Biology is a major departure from the usual run of school books in making fairly liberal use of references. Many of these refer to original papers where they are written in a style which in the main can be understood by younger students and in which the journals are likely to be fairly widely available. Others refer to specialized articles written particularly for this level in such periodicals as New Scientist, Science Journal (now merged with New Scientist) and Scientific American.

It is probably fair to say that the reading of original papers in scientific journals is foreign to most younger students. It will no doubt be argued that students prefer to read straightforward statements in text books, which are certainly less taxing. But such statements often represent extracts from previous text books and they in turn from text books before them, until the student is reading an oft-digested pap that is all too often a great over-simplification of the real issue.

The reading of original papers is both a stimulating and a salutary

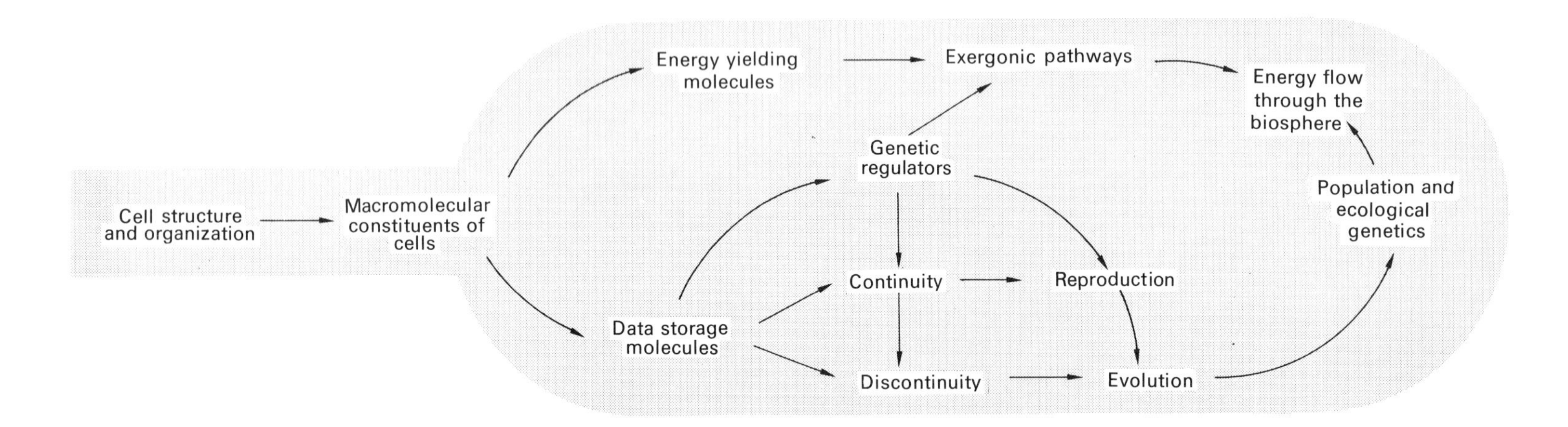

experience for students at all levels. Knowledge is not something that springs from the pages of a book, but comes from the activities and thought processes of real people in laboratories throughout the world. The author does not suggest that a student should try to read all the references in the book, but when his interest is stimulated in a particular area he should try to read some original work about it in order to gain both an insight into the way the information was gathered and a knowledge of the people involved.

References in the text are given by author, or authors, and the date of publication, for example (Beale *et al.*, 1969). In the list of references at the back of the book the paper is described under the author's name, alphabetically. This reference appears as:

BEALE, G. H., JURAN, A. and PREER, J. R. (1969). The Classes of Endosymbiont of *Paramecium aurelia*. *J. Cell Sci.*, **5**, 65.

The abbreviations of journals are given at the end of the list of references. In this case *J. Cell Sci.* is the Journal of Cell Science. The volume number (in bold type) is Volume 5 and the paper appears on page 65. The date of the volume of the journal is 1969.

For students living in university towns access to journals is relatively straightforward. It requires the exercise of some initiative in order to find out in which library of the university particular journals are kept and the eliciting of help, which is invariably generous and unending, of the university librarians. For students living outside university towns papers can often be obtained through the branches of public libraries, either in the form of photocopies of the paper itself or by asking for the particular volume of a journal which is then used as a temporary addition to the reference section of the library branch.

Section 1

The Cell and its Metabolism

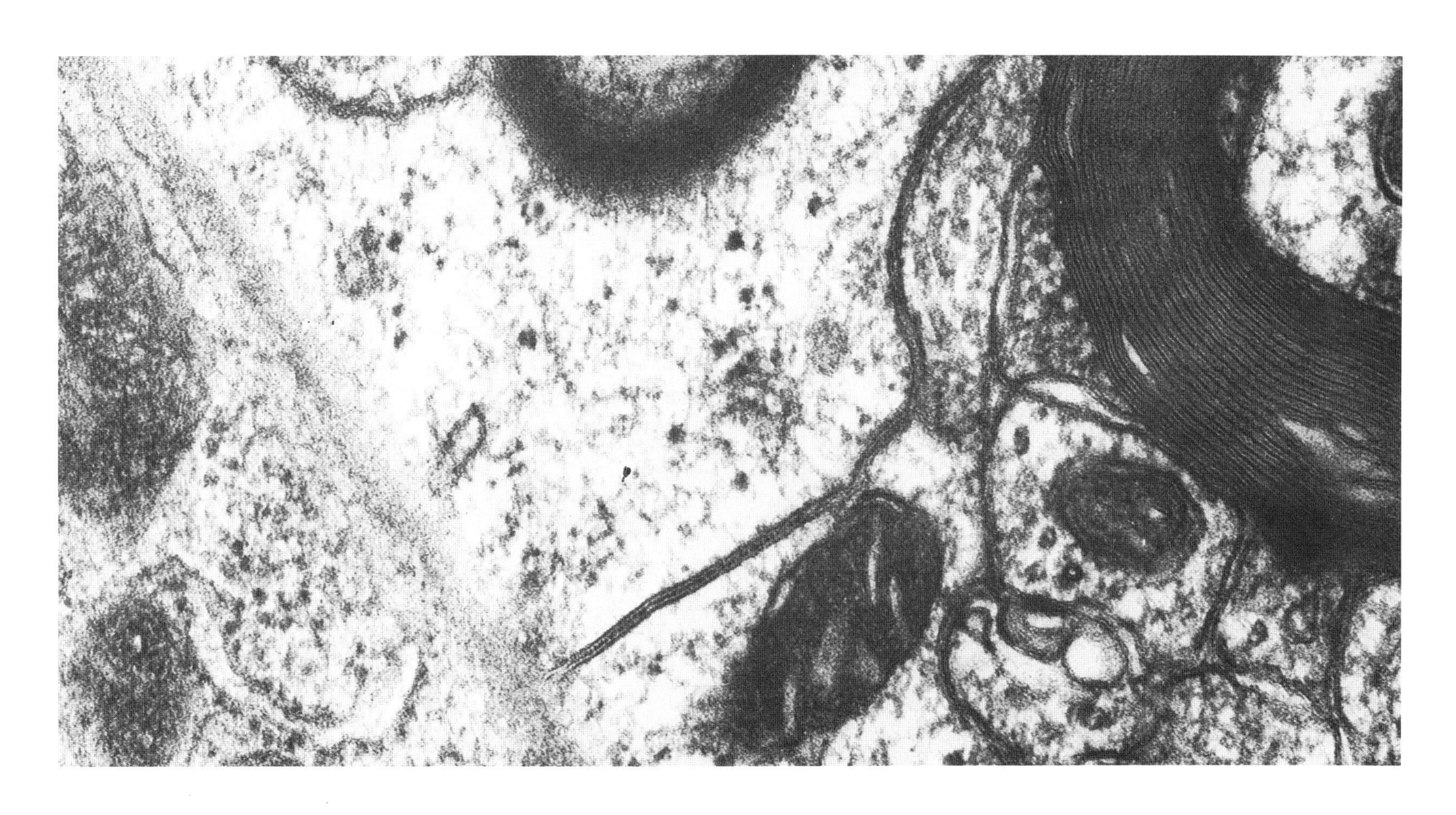

(*Photograph on preceding page*) In the formation of mammalian nerve cells the insulating myelin sheath is formed by spiral membranes of the Schwann cell. The axon cylinder (upper right) can be seen together with the Schwann cell membranes in this EM. (Courtesy of Dr. A. R. Lieberman.)

1 Cell Form and Ultrastructure

The historical view of biology has many interesting facets. In the last century the controversies that surrounded Darwin and Huxley and led to a confrontation of science and contemporary society and to clashes between various disciplines within science itself, have provided a rich source of study for the historian and sociologist. Earlier in the century the area of biology which had attracted the greatest interest was the study of cells and tissues. Germany dominated this field by virtue of her microscope technology. Never since that time has one country so commanded a fundamental area of biology.

With these points in mind there follows a brief account of the development of what has been called the *Concept of Cellularity*. Cells are then discussed in relation to neighbouring cells and to the matrix between cells, in terms both of evolution and of tissue function and differentiation. The chapter concludes with a discussion of membrane activities. The role of membranes in active transport relates to the capacity of living material to keep itself distinct from its environment (a feature which is discussed again in Chapter 10). Membranes also offer platforms in cells for protein synthesis to take place, the details of which form the principal topic of the next chapter.

1.1 The Cell Concept

In 1664 the Englishman Robert Hooke described the microscopic structure of plant tissues from a number of sources, such as cork, elder pith and carrot, and coined the term 'cells' for the box-like structures he observed (see Fig. 1.1). Other 17th century workers, notably van Leeuwenhoek in Holland, Malpighi in Italy and Grew in England, added descriptions of the cellular nature of other organisms. But the advance of knowledge in this field was limited by the slow rate of technological development of the instruments themselves. Not until the great German optical manufacturers of the early 19th century, like Carl Zeiss of Jena and Ernst Leitz of Wetzlar, began to produce achromatic microscopes of high resolution was real progress made in the study of cell structure. It is therefore not surprising that German workers became associated with the development of cell theory.

At the University of Berlin the work of Johannes Müller (Fig. 1.2), one of the most impressive figures of European science of the day, was leading to the foundation of comparative physiology as an academic discipline. His reputation and work attracted many able

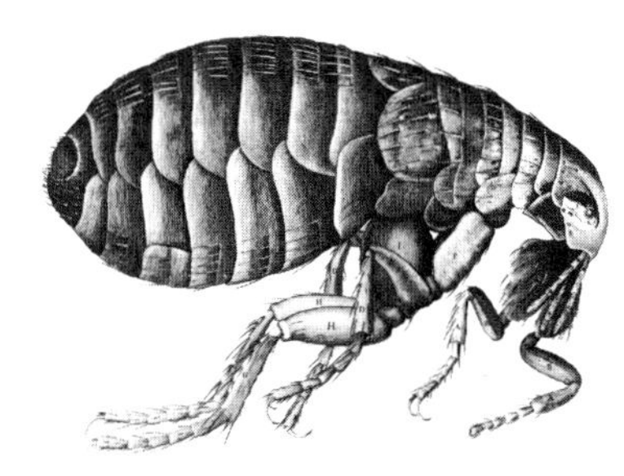

Fig. 1.1 Robert Hooke's flea, a drawing published by Martyn and Allestry in *Micrographia* (1665). Hooke built the first compound microscope and described the honeycomb-like boxes or 'cells' in a thin section of cork. (From the original *Micrographia*, in the Wellcome Institute, courtesy of the Trustees.)

intellects as co-workers and pupils. One of Müller's contemporaries was Matthias Schleiden (Fig. 1.3), a botanist. In 1838 Schleiden published a treatise* in which he demonstrated that the cell was a common element in all plant tissues. The importance of this work was recognized immediately by Müller, who proceeded to demonstrate the cellular nature, in a comparative context, of the nervous tissues of a number of vertebrates. It was Müller's pupil and assistant, Theodor Schwann (Fig. 1.4), who grasped the empirical nature of

* *Contributions to Photogenesis.*

these findings and within a year he had extended Schleiden's work to the whole animal kingdom and had, in effect, created the Cell Concept; in 1839 he published *Microscopical Researches into the Accordance in the Structure and Growth of Plants and Animals.* The timing was ripe for the publication of such an analytical and collated work and its impact on cytological studies can be compared to the impact of Mendeleef's Periodic Table (1869) on inorganic chemistry. Both works were synthetic in nature and both had the quality of predictability, such that later discoveries were seen to fit into place, and thus corroborate, their schemes.

Before the publication of Schleiden's and Schwann's works the absence of a framework had made much of the work in cytology aimless and unproductive. In the first decade of the century, for example, the French physician Bichat had attempted, with little success, to show the microscopic relationships between the organs of the human body, but it was not until the middle of the century that Kölliker and Virchow, two more pupils of the redoubtable Müller, developed tissue theory and laid the foundations of histology.

The concept of the cellular state of organisms focussed attention on the cell itself. Another pupil of Müller's, Ernst Brücke, of Vienna, by his assertion that 'the activity of all organisms is, in the ultimate analysis, the activity of the components of their tissues, the microscopic cells' (Haeckel), in effect pioneered cell physiology. In general, however, the effect of the Cell Concept was to extend a unifying quality to the many branches of biology and, notably, to give a biological basis to the medical disciplines of human anatomy, physiology and pathology.

Fig. 1.2 Johannes Müller (1801–58), founder of the great German school of cell study. (From a lithograph in the Wellcome Institute, courtesy of the Trustees.)

Fig. 1.3 Matthias Schleiden (1804–41) the German botanist who, with Schwann, founded modern cell theory. (Courtesy of the Wellcome Trustees.)

Fig. 1.4 Theodor Schwann (1810–82), the German zoologist who defined the cellular nature of nervous tissues and of the egg, and who saw the cellular relationship of plants and animals. (Courtesy of the Wellcome Trustees.)

The writings of Schleiden and Schwann, and of other 19th century workers, which collectively gave rise to the Cell Concept, can be summarized as follows:

1. Organisms are made up of microscopic cells which are distinctly organized units.
2. Within an organism cells are differentiated to form distinct cell

types, having characteristic features specific to particular tissues (Virchow, 1858).

3. A nucleus is a feature common to all cells (Brown, 1831) although in some cells, such as phloem and mammalian red blood cells, the nucleus may be lost during differentiation.

4. The living content of the cell, or *protoplasm* (Purkinje, 1839) determines the activity of that cell and thus, collectively, of the whole organism.

5. Growth is achieved by increase in cell number; new cells come into being only by the division of existing cells (Virchow, 1855). This idea, invoking the concept of an unbroken cell ancestry reaching back to the very origin of life was consolidated by Kölliker's demonstration that sperm and ova are simple cells.

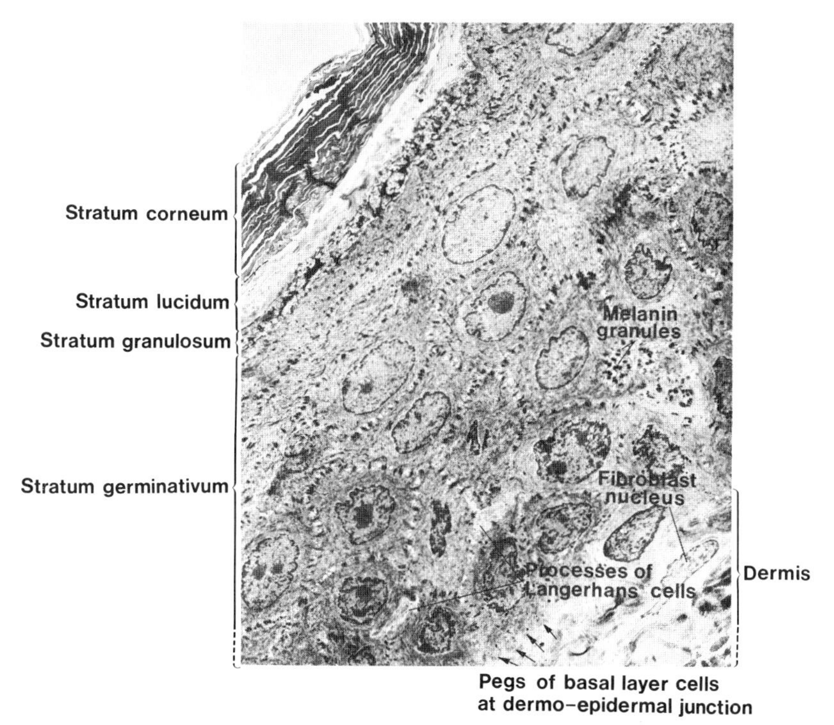

Fig. 1.5 (a) Low power electron micrograph (E.M.) of human skin. The cells of the Malpighian layer show a progressive flattening and loss of nuclear granulation towards the dead, cornified layer, the *stratum corneum*. The dark 'bridges' between the Malpighian layer cells are the desmosomes. Magnification ×1600.

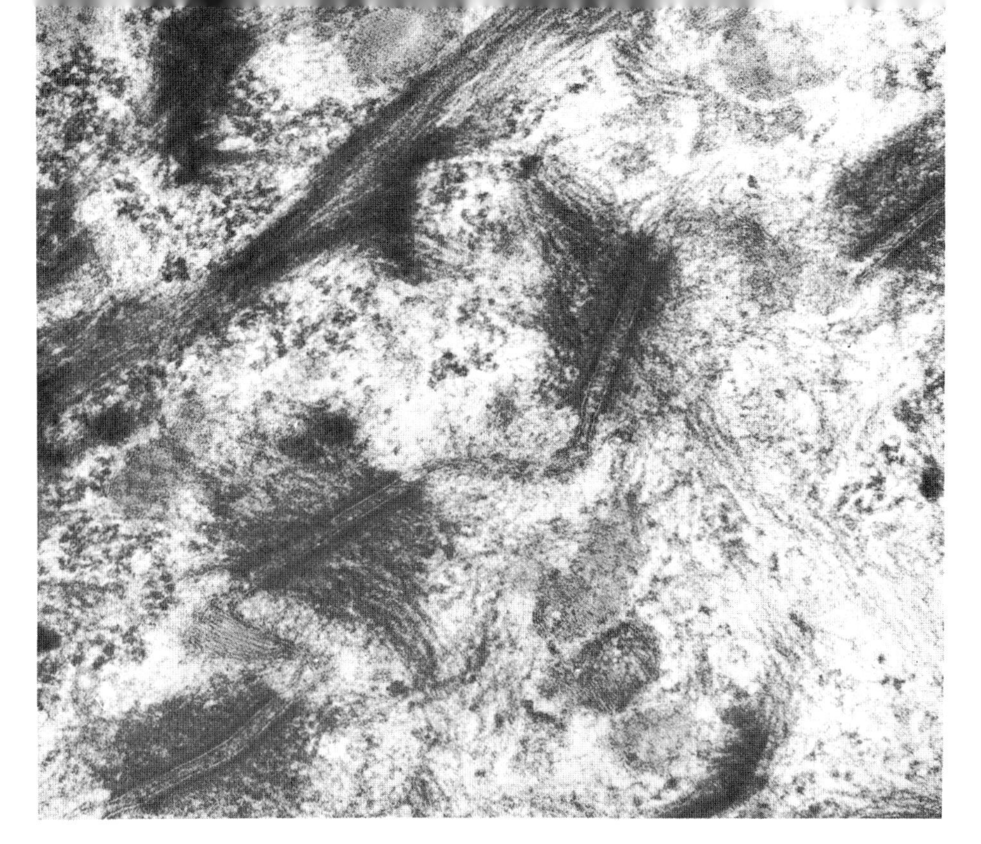

Fig. 1.5 (b) A higher power E.M. of a series of desmosomes between two skin cells. Within the desmosome the adjacent cell membranes can be seen and the filaments do not seem to pass from one cell to the other. Magnification ×23 000 (Photographs by courtesy of Dr. A. R. Lieberman.)

1.2 The Cell, Intercellular Links and the Intercellular Matrix

Any consideration of the role of cells in the function of a tissue should take account of the connections between cells and intercellular substances. In most of the tissues in which the functions are performed by the material *within* the cell (Purkinje's *protoplasm*), such as skeletal muscle, liver and epidermis, the intercellular material may be reduced to a thin cell wall outside the limiting plasma membrane of the cell, possibly visible only in the electron microscope. In such cases the intercellular material may function in cell cohesion, among other possible roles, although cell adhesion may also be a property of the **desmosomes** which have been observed at cell interfaces (Fig. 1.5). Desmosomes can be seen with the light microscope between the polyhedral cells of the epidermal *stratum spinosum* in mammals. In plants the existence of **plasmodesmata** (Fig. 1.6), which constitute protoplasmic connections between cells, has been known for some time.

Desmosomes occur most frequently in tissues which are subject to shearing strains, such as skin. At the desmosome the boundary membranes of the adjacent cells lie parallel and appear to be supported on their inner faces by supplementary plates of lipid or proteinaceous material. From these plates conspicuous filaments extend into the cytoplasm, but do not appear to pass from one cell to the other. There is no evidence that these represent 'protoplasmic bridges' between cells. Intercellular bridges have been described in animal tissues, however, where they may have a role in coordinating cell division and differentiation (Fawcett *et al.*, 1959; Davidson, 1968).

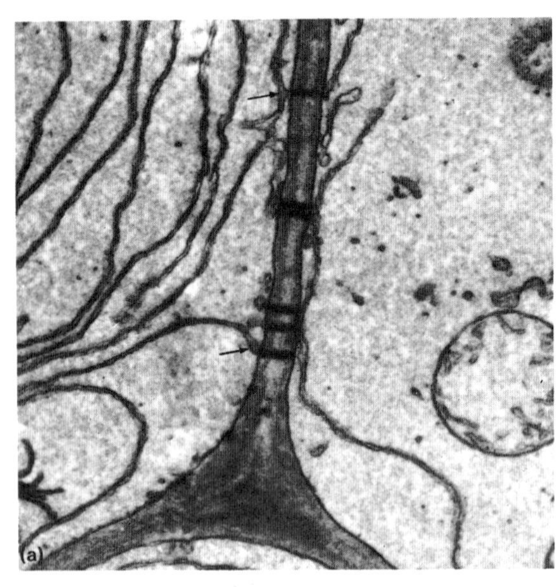

Fig. 1.6 (a) Plasmodesmata in a transverse wall between root cap cells in *Zea*. The cell walls are thin and there is little or no vacuolation. The plasmodesmata appear to be related to the endoplasmic reticulum (arrowed). (Courtesy of Dr. B. E. Juniper.)

(b) Drawing of the probable appearance of a plasmodesma showing the tube of cytoplasm between the cells and the central tubule, possibly a part of the endoplasmic reticulum.

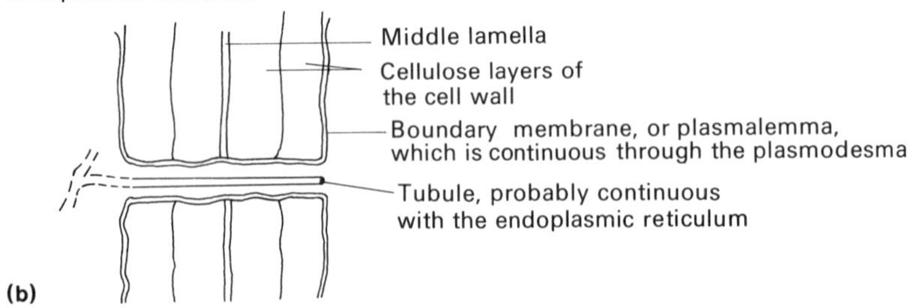

The relationship between the frequency of desmosomes and the physical stresses to which a tissue may be subjected suggests that their role is a mechanical one, although the possibility that they also have a transport or coordinating function cannot altogether be ruled out. In plants by contrast, the plasmodesmata, which extend between plant cells, act as definite protoplasmic bridges between cells and there is good evidence that they have a role in intercellular communication in respect both of solute transport and of coordination of growth and differentiation (Juniper and Barlow, 1969).

The diameter of the tube-like channel of the plasmodesma varies from 16 to 25 nm. It is lined by a unit membrane which is continuous with the boundary membranes of the adjacent cells. In the plasmodesma there is a break in the middle lamella of the cell wall and the channel is a clear passage between the cells. Under the electron microscope a filament can frequently be seen lying within the plasmodesma (Fig. 1.6b). This may be the vestigial remains of a mitotic spindle fibre, since plasmodesmata are thought to originate when the cell plate is laid down following mitotic telophase (see p. 70). Plasmodesmata are also prominent in the lamellae of pits (Fig. 1.7).

The continuity of cytoplasm between adjacent cells suggests that the plant body has a syncitial nature. Large particles, such as viruses are known to be able to pass through plasmodesmata. This raises questions, as yet unanswered, concerning possible mechanisms whereby adjacent cells linked by plasmodesmata can retain their identity, particularly when they are structurally and functionally distinct.

In the connective tissues of animals the activity of the tissue as a whole is a joint function of the cells and the intercellular matrix. In these tissues the matrix is usually chemically more complex than the thin cell wall material found in other tissues. In mammalian blood, for example, gas transport is primarily a function of the red cell, whereas the transport of other metabolites is a function of the matrix, or plasma. Foreign substances of microscopic dimensions entering the tissues of an animal may be phagocytosed, that is engulfed, by the macrophagous white cells, and at a molecular level may be agglutinated by antibodies in the γ-globulin fraction of the plasma. Clotting is a joint action of cells (platelets) and a number of soluble plasma components which give rise to fibrin.

At the other extreme, the role of some tissues is fulfilled exclusively by the matrix. The ability to withstand compression, the support capability, of cartilage and bone (Fig. 1.8) is a quality of the matrix. In these tissues the cells serve only to sustain the character of the matrix and to increase or decrease the amount of matrix present in response to the stresses which are placed upon it.

1.3 The Organization of Cells in Tissues

In evolution the appearance of the multicellular level of body organization appears to have been followed closely by cellular differentiation. In extant organisms consisting of relatively few cells, such as the free-swimming colonial, or **coenobial**, algae, which form a large part of the plankton in ponds, functional specialization frequently 'precedes' structural differentiation. Thus, in *Eudorina elegans* (Fig. 1.10), which consists of 16 to 32 biflagellate, *Chlamydomonas*-like cells embedded in the surface of a spherical mass of mucilage, no differentiation can be observed. All the cells play an active part in locomotion and all the cells divide simultaneously to form daughter colonies.

Fig. 1.7 E.M. of a section of a pit between two cells in the pea, *Pisum sativum*. The layers of cellulose in the cell walls on either side of the pit can be seen. The middle lamella of the wall continues across the pit forming the pit lamella. Magnification ×1200. (Courtesy of Dr. B. E. Juniper.)

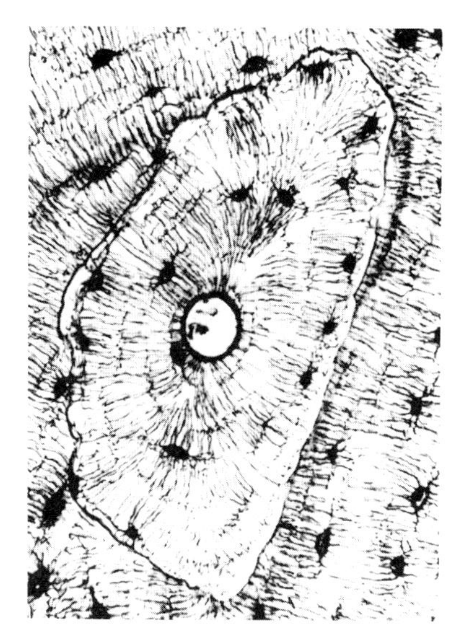

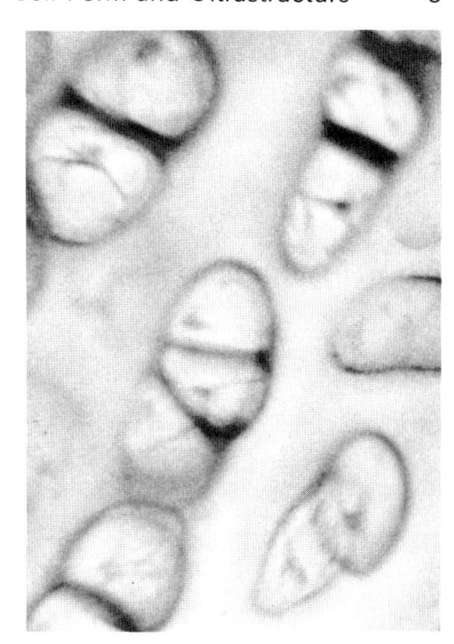

Fig. 1.8 Photomicrographs of (left) compact bone, and (right) articular cartilage. In compact bone the central space, the Haversian canal, carries blood vessels and nerves. Around this lie concentric rings of lacunae which contain the bone cells or osteocytes. Between the lacunae run the canaliculi in which lie processes of the osteocytes whose function is to maintain the calcified matrix. In cartilage the cells lie in capsules which are grouped as they arise in cell division. The matrix is relatively homogeneous. (Photograph of bone by courtesy of Gene Cox.)

Volvox (Fig. 1.11), which is a larger coenobium, or cell colony, than *Eudorina*, is a hollow sphere of up to about 20 000 cells. It shows distinct 'front-endedness' in locomotion. Some of the cells are without flagella and are larger in size; their role in the colony is to divide mitotically when growth conditions are favourable to form daughter colonies within the parent colony. Thus *Volvox* exhibits an elementary form of morphological differentiation together with functional specialization, or division of labour.

Some organisms consisting of relatively few cells exhibit a high

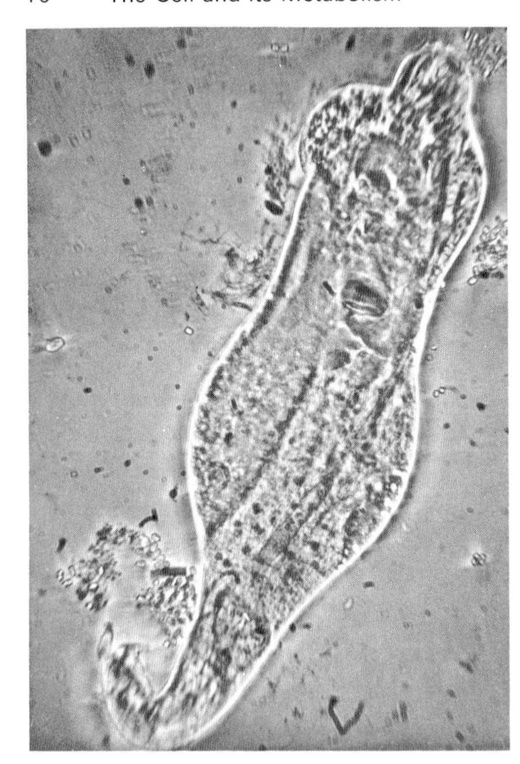

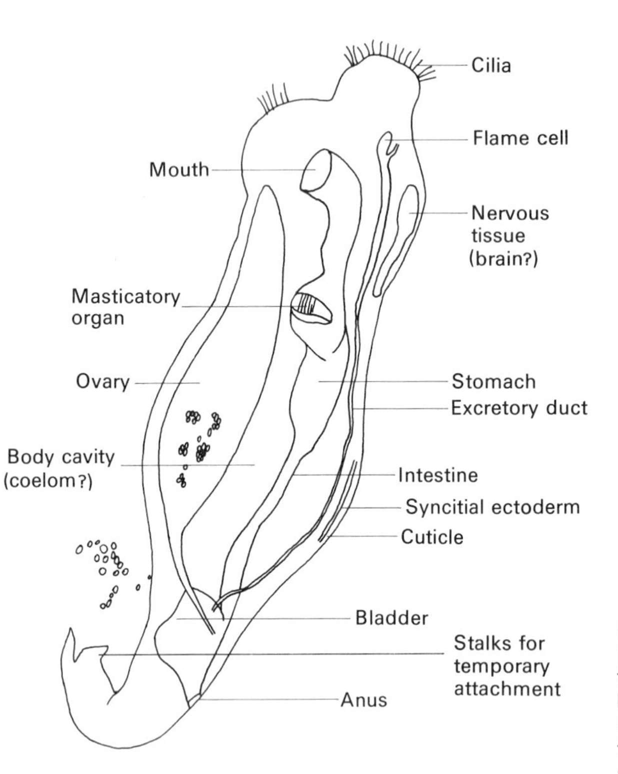

Fig. 1.9 Female rotifer, *Hydatina*. Photomicrograph (left) with a line drawing (right) to show main features. The rotifers are the smallest metazoans, being of the order of size of Protozoa. The sexes are separate, the male is about half the size of the female. The evolutionary origins and taxonomic status of the group are uncertain.

degree of morphological and physiological differentiation. The members of the Phylum Rotifera, for example, are microscopic and, indeed, smaller than some acellular Protozoa and Protophyta (Fig. 1.9). Their taxonomic position is far from clear; they possess a body cavity, so are probably more advanced than the Platyhelminthes, but it is not clear whether this cavity is a true mesodermal coelom. Some tissues, such as epidermis, appear to be syncitial, but other tissues, forming reproductive and excretory organs, nervous system and alimentary canal are highly organized.

The organization of groups of cells into tissues is of advantage to a multicellular organism since i' allows for mechanical (both structural and dynamic) and metabolic specialization. One of the features of a tissue is that the ratios of the concentrations of the enzymes in its cells differ from the enzymic ratios in other tissues. Tissues (or organs, which are aggregations of tissues) specialize in executing those particular mechanical or metabolic processes for which their enzyme pattern is best suited on behalf of all the other cells of the organism. For example, kidney cells in a mammal actively pump

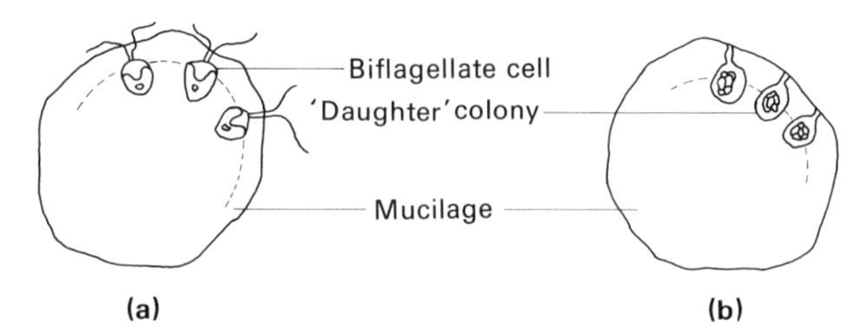

Fig. 1.10 *Eudorina elegans*. (a) Vegetative phase. (b) Reproductive phase.

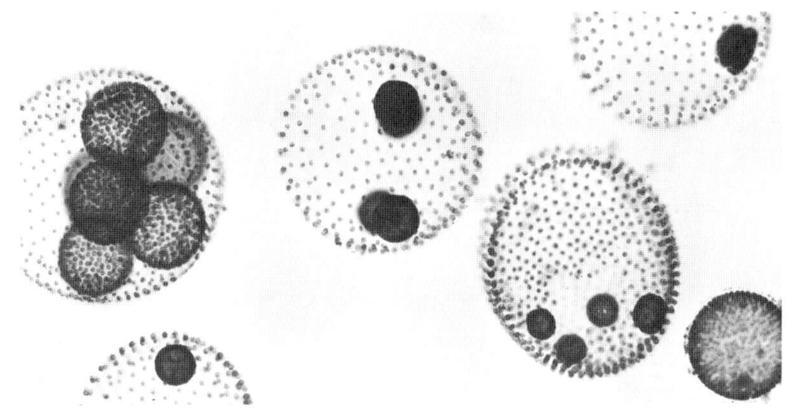

Fig. 1.11 Photomicrograph of *Volvox* showing vegetative spheres within the large colonies. (Courtesy of Gene Cox.)

blood solutes against concentration gradients to make urine, not only on behalf of themselves, but on behalf of the whole organism. In a flowering plant, palisade cells in the leaf synthesize food not only for their own use, but to provide tissue-building and energy substrates for the whole plant. Nevertheless, all the cells in an organism are genetically identical, having been formed by mitotic division from a single cell, the zygote. The cells of one organism are therefore referred to as a **clone**. Yet within this clone of cells the genes in different tissues code for different enzymic ratios. This raises the paradox of 'the differentiation process', which is dealt with elsewhere (pp. 120–123).

Two features of organization always accompany high degrees of differentiation. First, a transport system must operate to interconnect organs, in effect to make the special capabilities of each organ available to all the cells of the organism. Thus in animals the circulatory system, a vascular and anastomosing network of vessels enclosing the blood (itself a specialized tissue consisting of cells in a liquid matrix) transports oxygen from specialized gas exchange organs to tissues which are remote from the environment (Fig. 1.12a). The role of the transport system is to replenish the cellular environment, the tissue fluids, with those solutes the cells have absorbed and to remove from the cellular environment those solutes the cells have excreted. Thus the environment of the cells within tissues (Claude Bernard's *Milieu Interieur*) tends to remain constant in composition, the situation for which Cannon, in 1932, coined the term '**physio-**

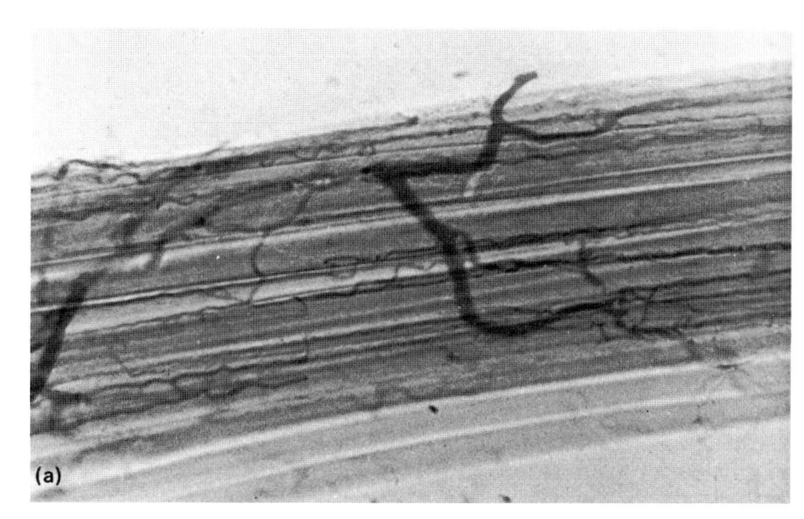

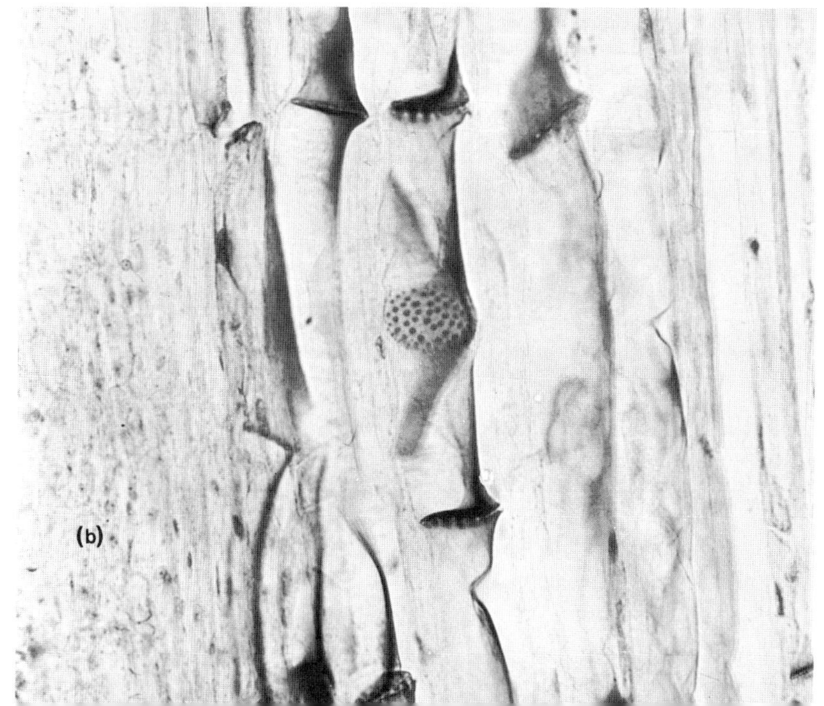

Fig. 1.12 Transport systems in animals and plants. Photomicrographs of (a) blood vessels in skeletal muscle, and (b) phloem sieve tubes in cucumber stem. (Phloem photograph courtesy of Gene Cox.)

logical homeostasis'. In plants, the products of photosynthesis are transported from the palisade tissue to growing points, to provide respiratory and growth substrates, and to storage organs, among other destinations, through a specialized transport tissue, the phloem (Fig. 1.12b).

The second feature of organization in differentiated organisms is coordination. The molecules and ions which constitute the cells of any organism are maintained at concentrations which are different from the organism's environment. One might say that the ability to maintain these differences is the definition of 'life' itself. It follows, therefore, that organisms must perceive, and respond to, physical and chemical changes in the environment, and a balanced or regulated response by the tissues which together make up the whole organism can only be achieved by a coordinating system. In plants, regulation is mainly through chemical coordinators, or hormones. In animals, regulation may also be hormonal, as in the regulation of tissue response to stress conditions by adrenalin. In addition, the autonomic nervous system of higher animals acts as a fine regulator of tissue interactions.

1.4 Cell Size

In their respective discourses leading to the Cell Concept, both Schleiden and Schwann recorded the great variation in the sizes of cells. The yolk of a fowl egg, which is a single cell, is about 4×10^4 times the diameter of a coccal bacterium. At the same time the sizes of cells in a particular tissue are relatively uniform, throughout the animal kingdom in particular. Thus the liver cells from a mouse and an elephant do not bear the same size relationships as the animals themselves.

Cell size may be limited by several factors. First, the limit of mass should be considered. In animal tissues the shape of cells and their spatial arrangement are both of importance to tissue function; they are used diagnostically by the histologist. Increase in cell size beyond a certain mass would cause loss of shape because the limiting plasma membrane, which is probably the main determiner of shape in many kinds of animal tissue, increases in mass (and therefore strength) at a lower rate than does the cytoplasm.

Second, cell size is limited by the surface area across which the cell exchanges metabolites with its environment. In approximately spherical cells the surface area of the cell is a squared function, whereas the volume is a cubed function, of the radius of the cell. Thus doubling the radius of a cell increases its volume eight times

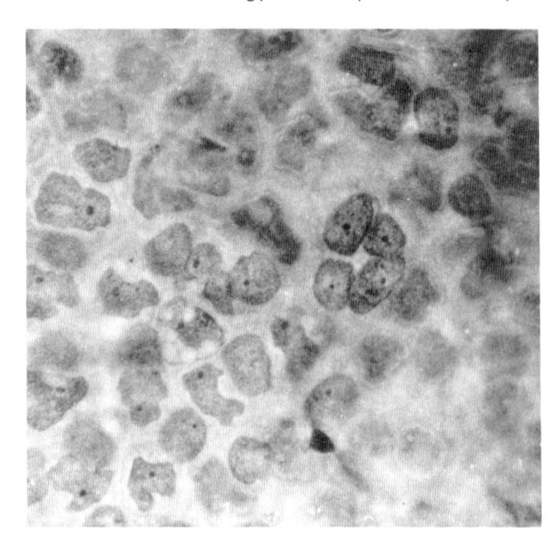
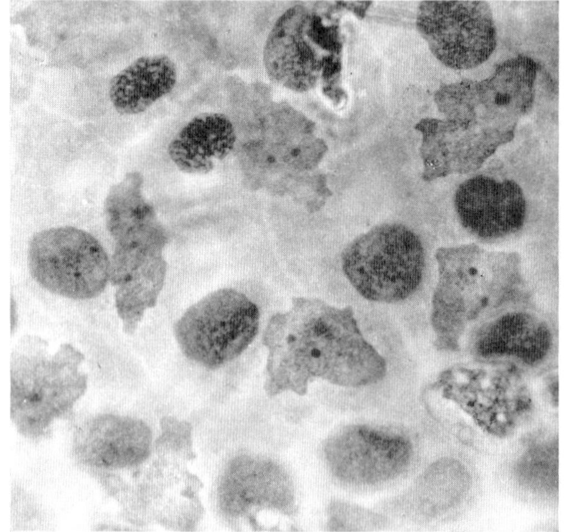
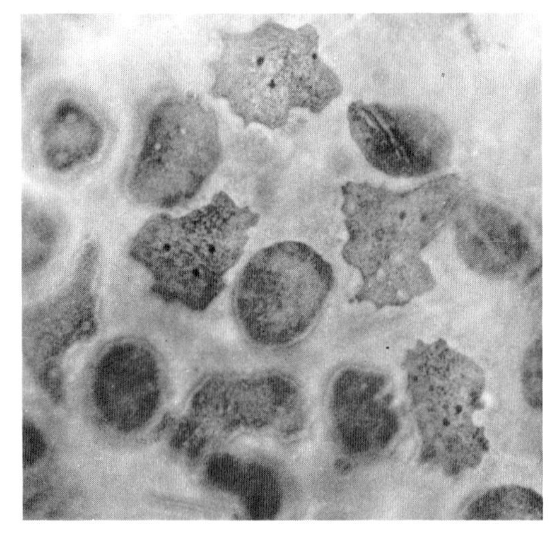

Fig. 1.13 Haploid (left), diploid (centre) and triploid (right) epidermal cells of the tadpole at the same magnification, showing that cell size is related to genetic complement. (From *Animal Growth and Development* by Professor D. R. Newth, in the series Studies in Biology. Courtesy of the author.)

but its surface area by only four times. Cylindrical cells, on the other hand, can increase in length and retain about the same ratio of volume to surface area. So, even actively respiring cells, which must exchange metabolites with extracellular fluids at a high rate, can attain lengths of several millimetres (for example, skeletal muscle) or even one or two metres (nerve cell processes).

Third, diffusion is effective only over short distances. As a cell increases in size and the inner parts of its cytoplasm become further removed from the cell surface, diffusion gradients become longer and therefore 'less steep'. The inner parts of the cytoplasm, therefore, must tolerate lower concentrations of metabolic substrates and higher concentrations of metabolic products than those parts of the cytoplasm nearer the surface. This problem could be overcome by movement of the cytoplasm, a sort of 'inner circulation' or protoplasmic streaming, such as has been observed in some plant cells, but this would require an expenditure of energy. Thus, in multicellular animals which have extracellular, circulating tissue fluids the cells are relatively small. Where a large size is unavoidable, as in some nerve cells, cytoplasmic movements of a streaming nature have been reported.

Fourth, the size of a cell may be limited by factors arising from the relationship between the nucleus and the cytoplasm. The role of the nucleus in the cell's activities is known to extend beyond that of a mere storehouse of information governing protein synthesis in the cell (p. 39). B. Commoner (1964) has drawn attention to the likely roles of different forms of DNA acting either as instructors of protein (enzyme) synthesis, in the sense of the 'Mendelian gene', or as regulators of cytoplasmic metabolism in a quantitative manner. Such a mechanism, if it is to be capable of regulation, necessitates feedback control; the instructions issued by the nucleus must be subject to monitoring according to the status of the cytoplasm at any given time, by some mechanism not understood at present.

Clearly, if the two forms of DNA (p. 160) and the cytoplasm act together as an integrated metabolic unit the volume of the cytoplasm is critical. This would suggest an explanation for the observations of Blakeslee (1934) that the cells of polyploid plants, that is plants in which the cells have multiples of two, three or more times the normal chromosome complement, are larger than the comparable diploid cells of the same species. The same situation exists in animal cells (Fig. 1.13). Plant cells are usually larger than animal cells. The shape of the plant cell (the first limiting factor described above) is maintained by an extra-cytoplasmic cellulose wall. In general, cell walls are not convoluted, or interdigitating like most animal cells. Also, the larger, differentiated plant cells are frequently vacuolated (Fig. 1.14). The role of the vacuole could be interpreted in terms of the other factors described above; the presence of a vacuole in a cell means that the volume of metabolically active cytoplasm is considerably less than the volume of the cell, and in terms of diffusion gradients it is the remote cell centre that is occupied by the vacuole.

1.5 The Cell Membranes

Boundary membranes maintain and regulate the cell's internal composition by retaining or actively transporting large molecules, although some small molecules and ions appear to enter and leave cells freely. Internally, membranes form an **endoplasmic reti-**

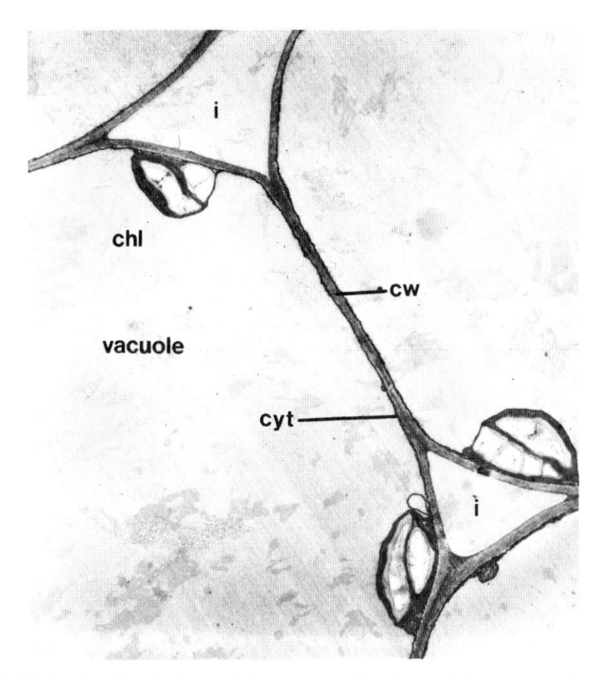

Fig. 1.14 E.M. of cells of the petiolar cortex of celery, *Apium graviolus,* showing the extent of vacuolation in a differentiated parenchymatous cell. The cytoplasm (cyt) forms a thin, dense layer along the cell walls (cw) and round the chloroplast (chl). Intercellular spaces (i) are not, of course, bounded by cytoplasm. (Courtesy of Guy Cox.)

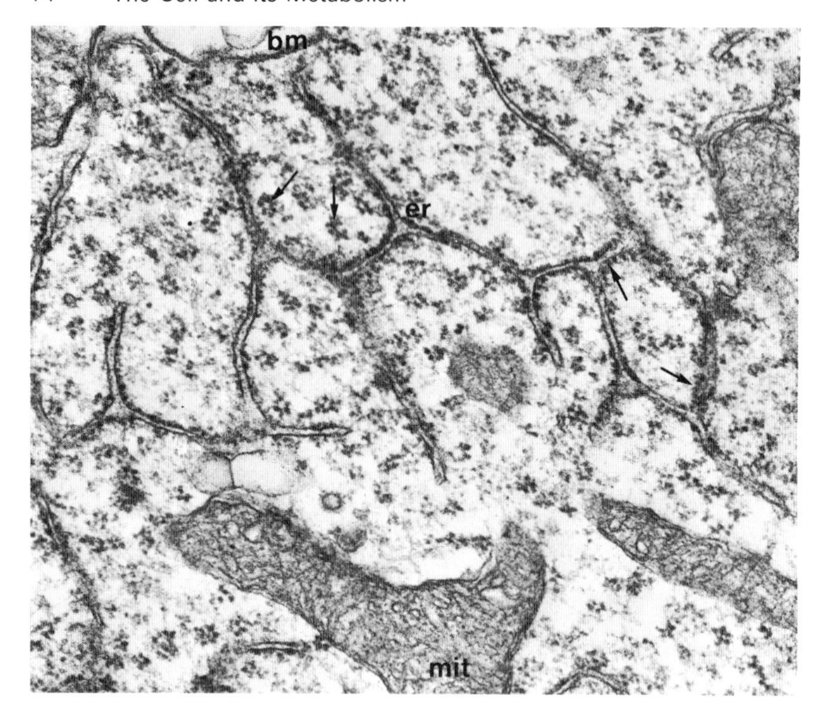

Fig. 1.15 E.M. of part of the surface of a nerve cell from the rat. It shows the anastomosing membranes of the endoplasmic reticulum (er) and the boundary membrane of the cell (bm). Ribosomes (arrowed) can be seen in clusters or polysomes in the cytoplasm and attached to the inner face of the endoplasmic reticulum. (Courtesy of Dr. A. R. Lieberman.)

culum (Fig. 1.15), which is probably continuous with the boundary membrane and the nuclear membrane in most cells, and to which are attached the ribosomes (p. 37). The outer membranes of cell inclusions such as mitochondria, lysosomes and chloroplasts have a structure similar to that of the plasma membranes. It has been suggested, from measurements of electron micrographs, ultracentrifugation, and enzymic studies, that the membrane is composed of a double layer of phospholipid with the hydrocarbon, or non-polar, chains turned inwards and bounded by a protein complex possibly containing mucoproteins and enzymes concerned with transport across the membrane. The commonest phospholipid in membranes is **lecithin** which appears to be associated with the steroid compound **cholesterol**. On the basis of diffusion studies it has been postulated that pores of about 0.7 nm diameter are present in the membrane (Fig. 1.16a).

Other configurations have been proposed for the structure of biological membranes. The enzyme phospholipase, which hydrolyses phospholipids, has been found to attack the lipids in intact membranes. This suggests that the lipid layer is not entirely coated with protein and electron microscopy has subsequently shown that the lipase digestion occurred in patches (Ottolenghi and Bowman, 1956). In such a membrane alternate 'blocks' of lipid and protein might be expected to be present (Fig. 1.16b).

The exchange of substances between the cell and its environment takes place across the boundary membrane in a number of ways. **Phagocytosis** is used as a method of feeding in Protozoa, notably the Rhizopoda, and as a method of ingesting foreign particles and cell debris by white blood cells. In Ciliates phagocytosis is a localized phenomenon associated with a differentiated organelle, the gullet.

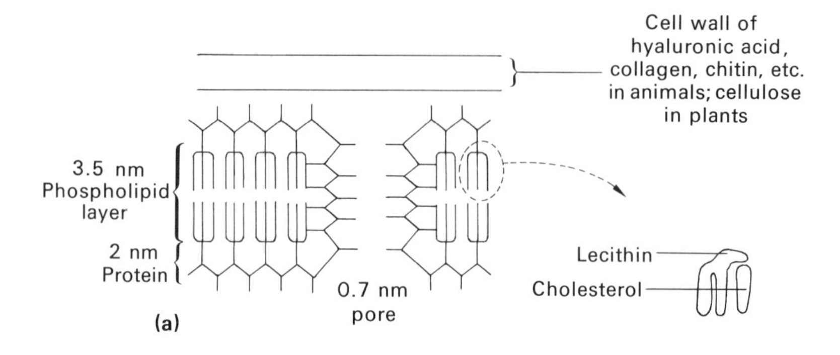

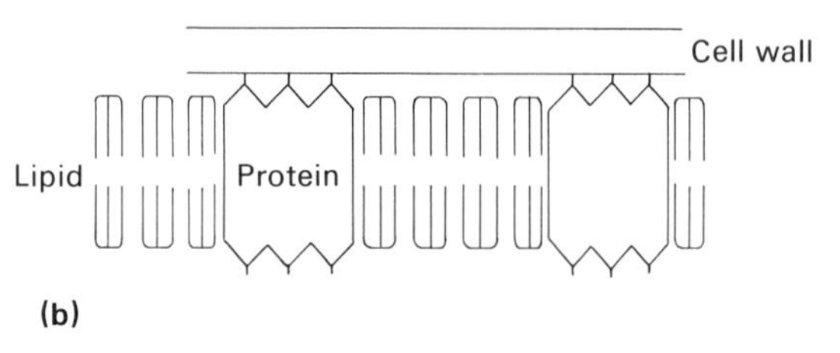

Fig. 1.16 Two postulated structures for the unit membrane. One (a) comprises a phospholipid layer bounded by protein with pores at intervals, and the other (b) consists of alternate blocks of phospholipid and protein.

Phagocytosis demonstrates that the membrane is not a static structure. When a cell engulfs a portion of its environmental fluid, containing particles in the case of free-living unicels, the boundary membrane invaginates and forms a food vacuole whose membrane is a portion of inverted boundary membrane. As digestion of the vacuolar contents proceeds the membrane becomes progressively smaller and eventually dissolves into its constituent molecules, which are presumably then available to reform boundary membrane (Figs. 1.17, 1.18 and 1.19).

Pinocytosis (Fig. 1.20) can be seen as a modification of phagocytosis, although it occurs on a much smaller scale. It concerns the ingestion of fluids and usually is associated with stationary or slow moving cells. It was first observed in cells in tissue culture by W. H. Lewis (1931) but was later discovered in many kinds of cells such as amoebae (Holter and Marshall, 1954) and meristematic cells of *Elodea* (Buvat, 1958).

Secretion (Fig. 1.21) by the cell is the reverse of phagocytosis.

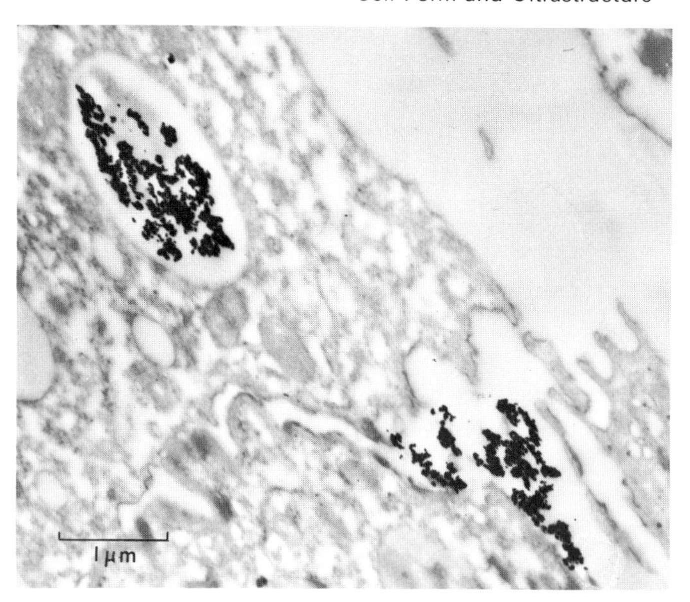

Fig. 1.18 In this electron micrograph gold particles can be seen in the medium between two cells (bottom right) and inside a food vacuole in one of the cells (top left), showing that particles can enter cells.

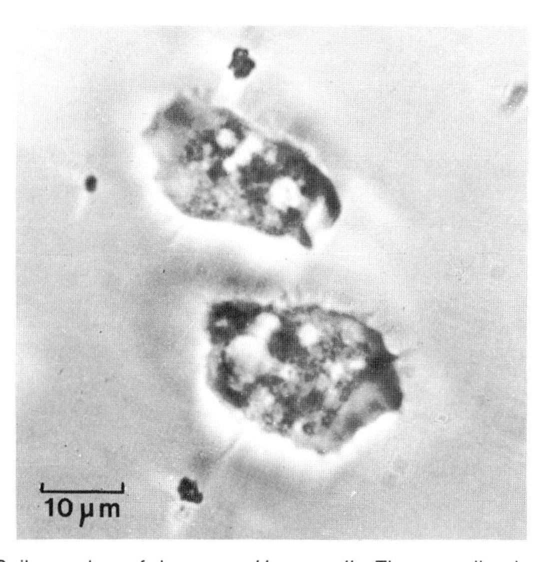

Fig. 1.17 Soil amoebae of the genus *Hartmanella*. These small animals live in the water film round soil particles and as they move along bacteria stick to their surface and collect in a clump at the posterior end. Pseudopodia are there produced which pinch off the bacteria into food vacuoles. In this photograph the amoebae are growing in sterile nutrient medium but they feed in essentially the same way and the pincer-like pseudopodia can be seen in the upper animal.

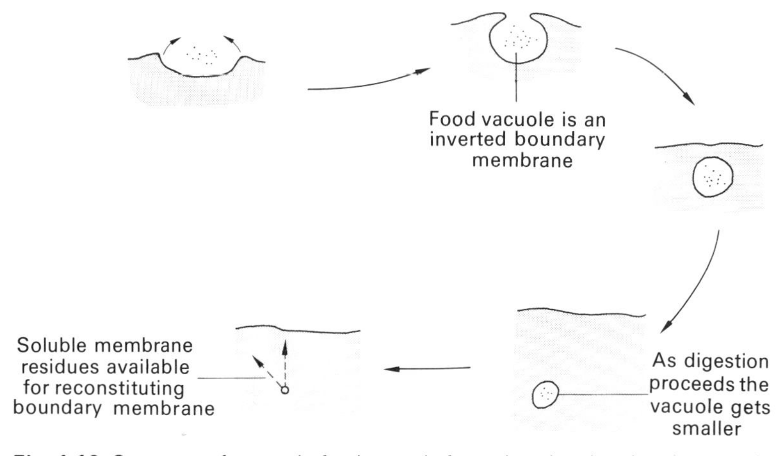

Fig. 1.19 Sequence of events in food vacuole formation showing that the vacuolar membrane is a temporary structure and is formed by inversion of the boundary membrane.

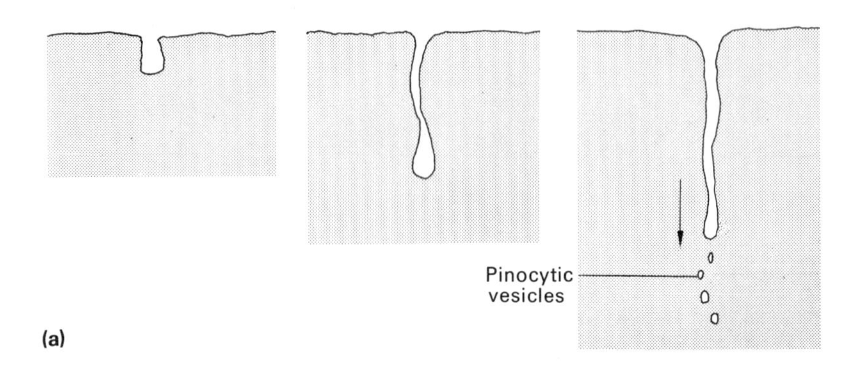

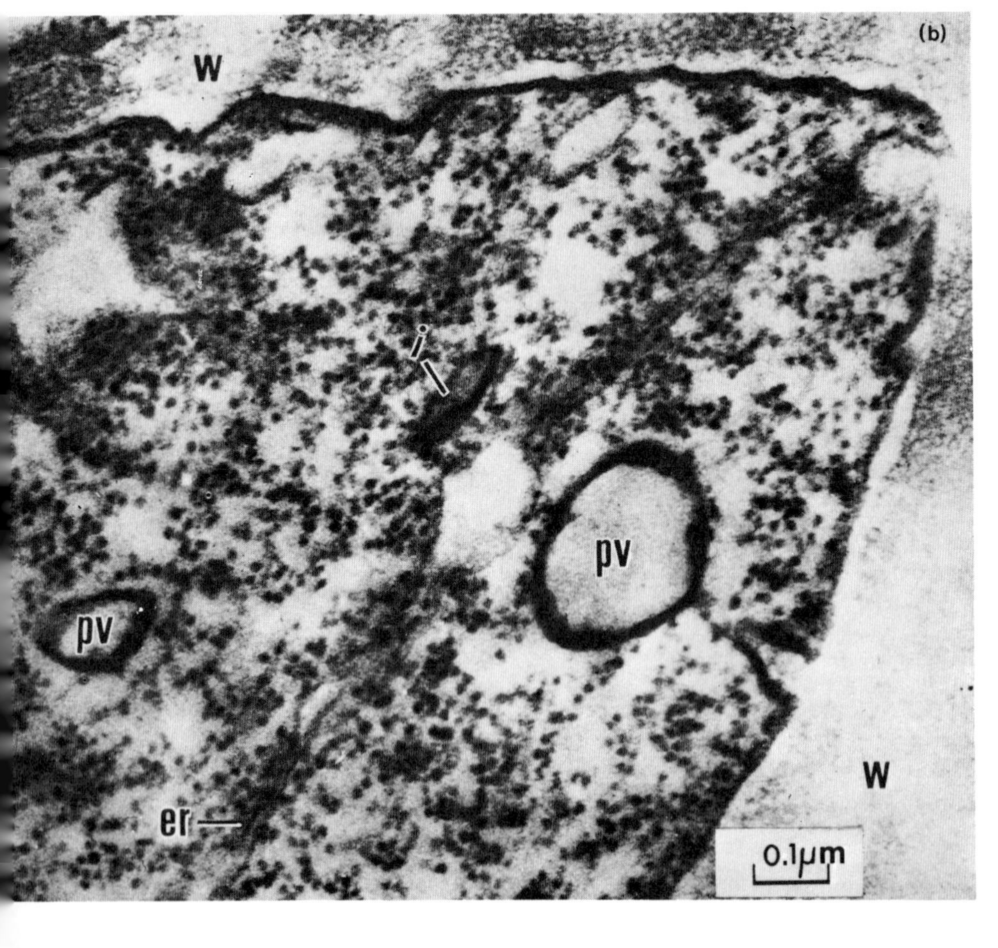

Fig. 1.20 (a) Diagram of pinocytosis. (b) E.M. of part of a meristematic cell of *Elodea* showing pinocytic vesicles (pv), the larger of which has a tubular connection to the plasmalemma adjacent to the cell wall (w). (Buvat, 1958, from Coult, *Molecules and Cells*, courtesy of Longmans.) (c) E.M. of part of an endothelial cell lining a capillary, showing abundant pinocytic vesicles probably concerned in transport of solutes across the cell. The unit membrane structure of the vesicles is clear (arrowed). The section skims across the nucleus tangentially and shows pores in the nuclear membrane (nm). Magnification × 90 000. (E.M. by Dr. A. R. Lieberman, reproduced by courtesy of Springer-Verlag, Heidelberg.)

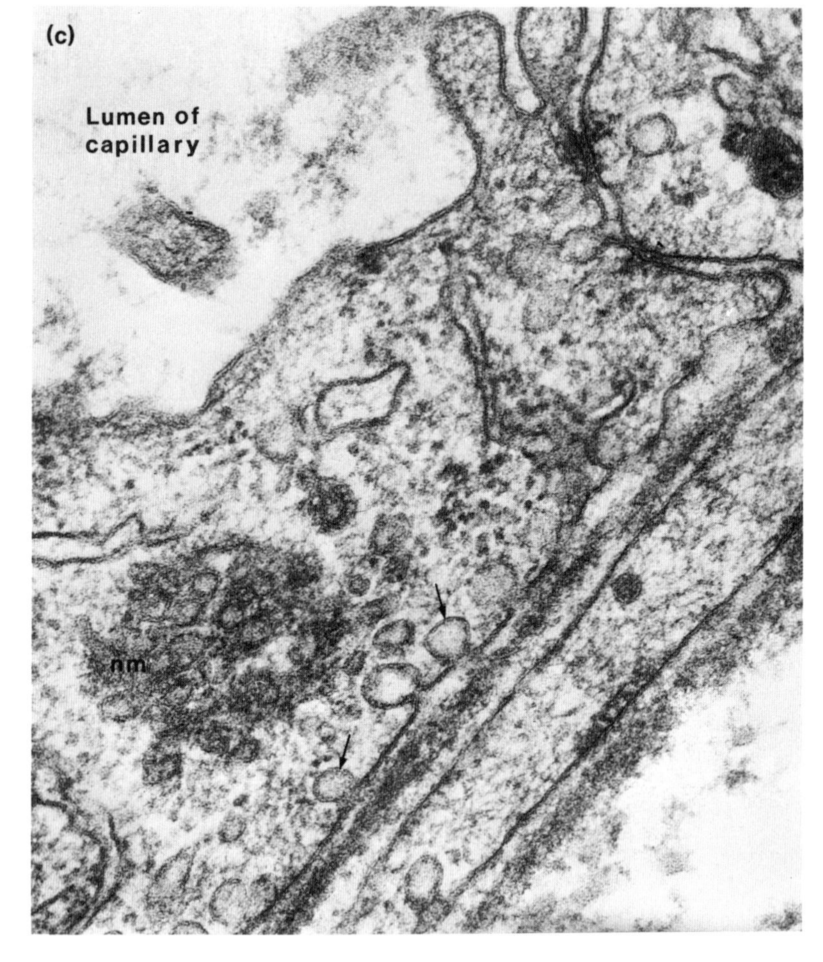

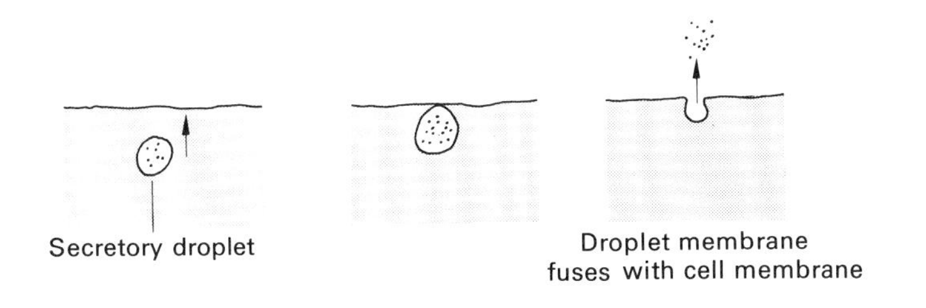

Fig. 1.21 Secretion of particles through cell membrane.

Secretory droplets in the cell (see Golgi, p. 20) move to the cell membrane and merge with it (Caro and Palade, 1964).

A combined mechanism could account for the active transport of substances across cell layers, comprising invagination and vesicle formation on one side of the cell and discharge at the opposite side (Fig. 1.22). This system might account for transport across renal tubules and gut epithelia (Paul, 1965).

It seems likely that the uniform nature of the membranes surrounding most cell organelles enables them to interact freely. Thus lysozymes, vesicles containing digestive enzymes, probably formed by an internal 'pinching off' of portions of the Golgi apparatus or endoplasmic reticulum, can fuse readily with food vacuoles formed by phagocytosis and bring about internal digestion. Similarly, the membranes of discharged contractile vacuoles of protozoa freely fuse with the boundary membrane or pellicle.

The rules governing the movement of solutes across plasma membranes are complex and appear to indicate a variety of mechanisms. Small ions which are highly charged gather a greater 'shell' of water molecules making them 'larger' than unhydrated, but intrinsically bigger ions, therefore they pass through membrane pores less

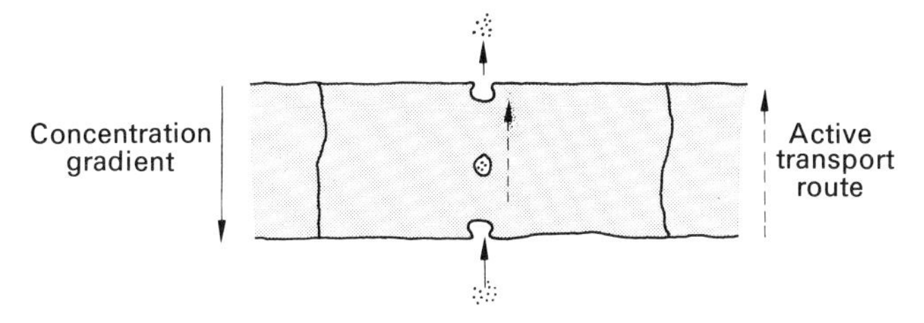

Fig. 1.22 Possible mechanism for the active transport of substances across cell layers, involving entry into the cell by pinocytosis and subsequent exit by secretion.

readily. Generally, the more soluble a substance is in lipid solvents, the more easily it passes through the membrane, as might be expected from the membrane's structure. Carbohydrates and proteins pass across cell membranes at rates often at variance with those expected from their molecular dimensions and concentration gradients (Davson and Danielli, 1943), which would seem to indicate special mechanism for their transport. Mitchell (1963) has suggested a 'portage' mechanism for the transport of substances across membranes in which the molecule fits into a special configuration on one side of the membrane and is then carried, or 'translocated' across the membrane and is finally ejected on the other side.

A mechanism postulated by Hokin and Hokin (1959) to account for the excretion of NaCl into the ducts of the 'salt-glands' of sea birds invokes the use of the phospholipids in membranes in an elegant, ATP-linked system (Fig. 1.23). The phospholipid, phosphatidic acid, collects Na^+ from the cell's cytoplasm and moves to the outside of the membrane. Here, the enzyme **phosphatidic acid phosphatase** removes both the Na^+, which leaves the cell, and the

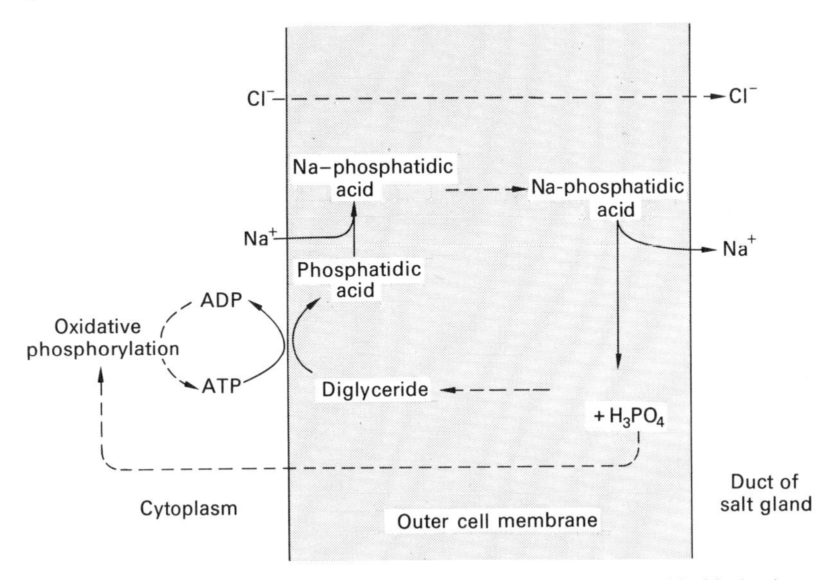

Fig. 1.23 Possible mechanism for the active transport of sodium chloride in the salt-glands of sea birds. Sodium ion is transported actively by an ATP-driven phosphatidic acid pump and chloride ion moves passively along the resultant electrical gradient. (After Hokin, L. E., and Hokin, M. R. (1959), *J. Gen. Physiol.*, **45**, 152.)

phosphate, which returns to the cytoplasm. The resulting diglyceride returns and is then phosphorylated by ATP to reconstitute the phosphatidic acid. This proposal assumes that Cl^- ion follows the Na^+ passively.

An analogous system is to be found in the estuarine marsh grass *Spartina* spp. The roots lie in brackish water and presumably absorb NaCl relatively passively. The tissues are maintained in a hypotonic state, however, by the excretion of NaCl crystals (Fig. 1.24) from special glands, or hydathodes, lying in rows between the ridges of the leaf (Skelding and Winterbotham, 1939).

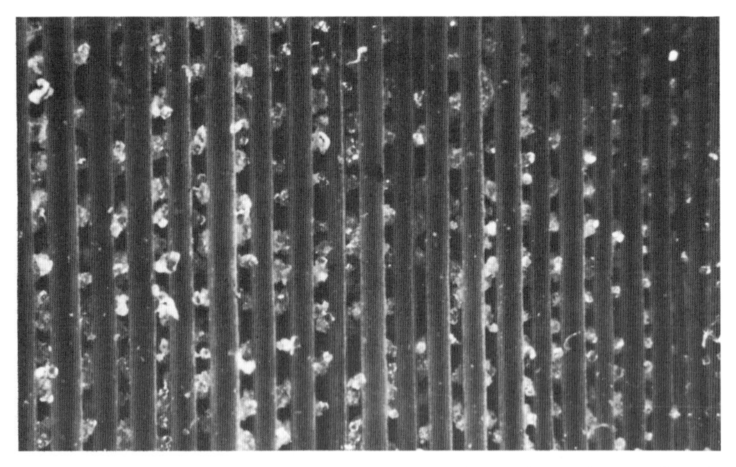

Fig. 1.24 Photomicrograph of the adaxial surface of the salt marsh grass *Spartina* × *townsendii*, showing salt crystal exudates from the hydathodes lying between the ridges. (Courtesy of Dr. C. J. Marchant.)

The movements of ions across membranes present particular problems; both the routes taken by ions and their mode of transport are imperfectly understood. Perhaps the most conspicuous ions in cells are sodium and potassium. In most animal cells these ions are maintained at concentrations of approximately 5 mM and 150 mM respectively, which is almost the reverse of their concentration in the usual surrounding fluids, such as mammalian interstitial fluids (Fig. 1.25).

Plant cells also pump ions, usually to accumulate them, against gradients. Steward (1959), by measuring ionic concentrations in cell sap of the marine alga *Valonia ventriculosa*, showed a situation analogous to that in animal tissues (Fig. 1.26).

	Interstitial fluid (mM)	Intracellular fluid (mM)
Na^+	145	12
K^+	4	155
Cl^-	120	4
Other anions	35	165
Potential	0	−90 mV

Fig. 1.25 Table of approximate concentrations of some ions in the interstitial fluid, which bathes cells, and in the cytoplasm of the cells themselves, in man. (After Woodbury, 1960.)

Suspending cells in solutions whose ionic concentrations resemble their intracellular fluid cuts down oxygen consumption by about 50%, from which it can be concluded that a large part of the cell's energy requirement is concerned with pumping ions against concentration and/or electrical gradients.

One reason for ionic pumping lies in the maintenance of electrical potentials across cell membranes. The conduction of a nerve impulse makes use of this effect and involves the movement of a wave of 'membrane depolarization' along the membrane of the nerve cell body and its processes. The propagation of such a wave depends on action currents at the wave front which further depolarize the membrane (Fig. 1.27).

Marsden (1970) has pointed out that one of the significant features of an actively maintained ionic differential is that the ions concerned 'leak' back across the membrane and in doing so may carry other substances with them. In this way glucose molecules, 'hitched' to Na^+ ions, may be taken up in the mammalian intestine. In a similar way the secretion of amylase into the salivary glands of the rat has been shown to be dependent upon simultaneous K^+ ion leakage from the secretory cells.

	Sea water (mM)	Intracellular sap (mM)
Na^+	500	43
K^+	12	590
Cl^-	580	628

Fig. 1.26 Table of approximate ionic concentrations in sea water and in the cell vacuoles of the marine alga *Valonia ventriculosa* (After Steward, 1959.)

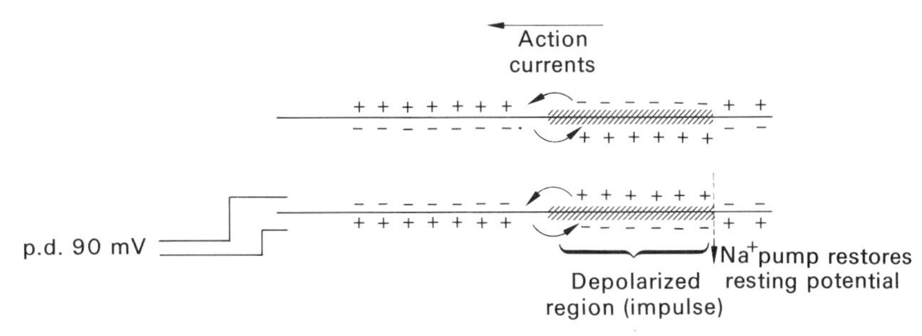

Fig. 1.27 A nerve impulse is a region of membrane 'depolarization' which allows the sudden influx of Na^+ ions and which is propagated by the resultant action currents that are generated by the ionic differences at the leading edge of the depolarized region. Before a second impulse can pass along the fibre the resting potential must be restored by the ionic pumps.

Golgi Apparatus

The presence of this structure was in dispute for many years until it was confirmed by electron microscope studies. It consists of 'smooth' membranes, that is, devoid of ribosomes. It usually lies close to the nucleus (there is some evidence that it may be formed from, or in conjunction with, the nuclear membrane) and is continuous with the membranes of the endoplasmic reticulum and the outer membrane of the nucleus. Golgi membranes are present in both animal and plant cells, although they tend to be more prominent in the former (Fig. 1.28).

Just as the granular endoplasmic reticulum tends to be most conspicuous in cells which are actively synthesizing proteins (p. 37), so Golgi membranes are most prominent in actively secreting cells, such as in endocrine and exocrine glands. Caro and Palade (1964) have provided evidence that protein synthesized at ribosomes attached to the endoplasmic reticulum passes into the Golgi apparatus which 'buds' to produce vesicles (Fig. 1.29). In the pancreatic cells which these workers examined the vesicles were zymogen granules, containing the precursors of digestive enzymes, which migrated to the cell surface where they were excreted. Such a mechanism would conform to the concept of membrane mobility discussed previously in this chapter (Fig. 1.30).

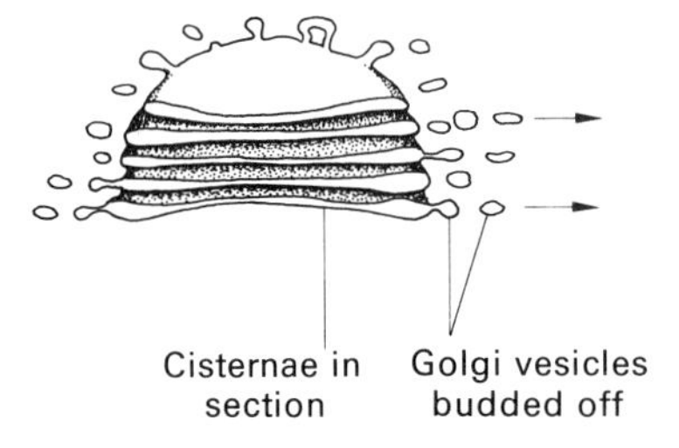

Fig. 1.28 The Golgi apparatus consists of a stack of cisternae similar to the membranes of the endoplasmic reticulum but devoid of ribosomes. From the edge of the cisternae are budded off the Golgi vesicles.

Blobel and Sabatini (1970) have examined the relationship between ribosomes and the membranes of the endoplasmic reticulum in cells treated with proteolytic enzymes. Their results suggest that those ribosomes which are attached to the endoplasmic membranes eject the proteins which they synthesize into the cisternae (Fig. 1.31). This theory neatly complements the Caro–Palade pathway and one might conclude therefore that proteins made on membrane-attached ribosomes are destined to be excreted from the cell, probably via the Golgi apparatus (Fig. 1.32), whereas those proteins synthesized on ribosomes which float free in the cytoplasm are destined to remain in the cell (see The Ribosome's Role in Polypeptide Synthesis, p. 37).

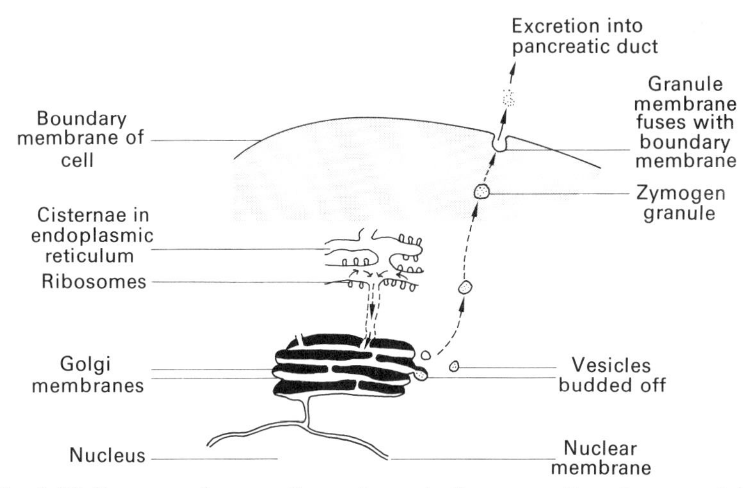

Fig. 1.29 Summary of a secretion pathway in the pancreatic cell proposed by L. G. Caro and G. E. Palade. (*J. Cell Biol.*, **20**, 473.)

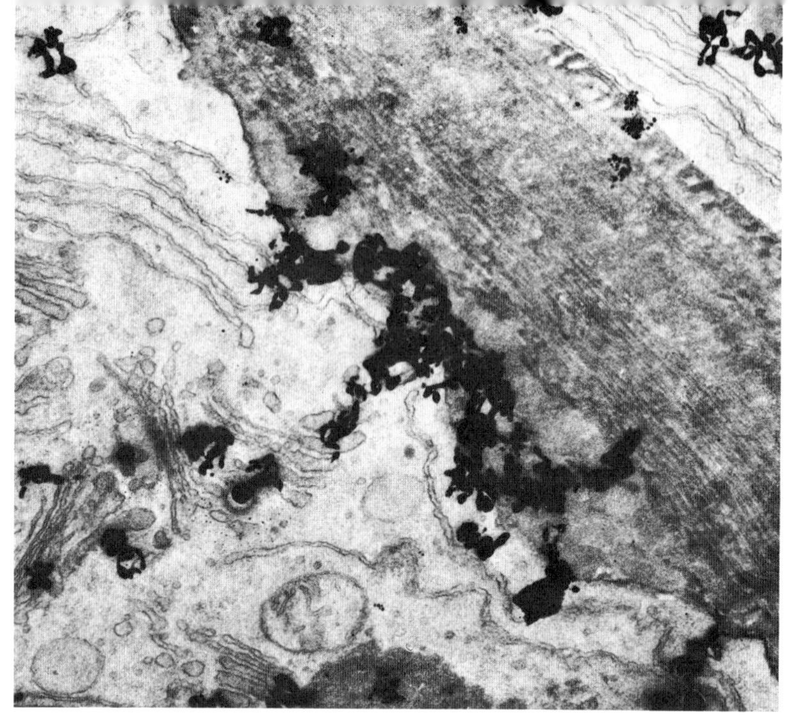

Fig. 1.30 Evidence for the role of Golgi in the secretion of carbohydrate for plant cell wall construction has been provided by Northcote and Pickett-Heaps. This combined electron micrograph and autoradiograph shows radioactive sources (black granules) in the Golgi bodies and vesicles and accumulating outside the boundary membrane in the cell wall. Before fixation the living cells had been 'fed' glucose labelled with tritium. (Courtesy of Dr. D. H. Northcote.)

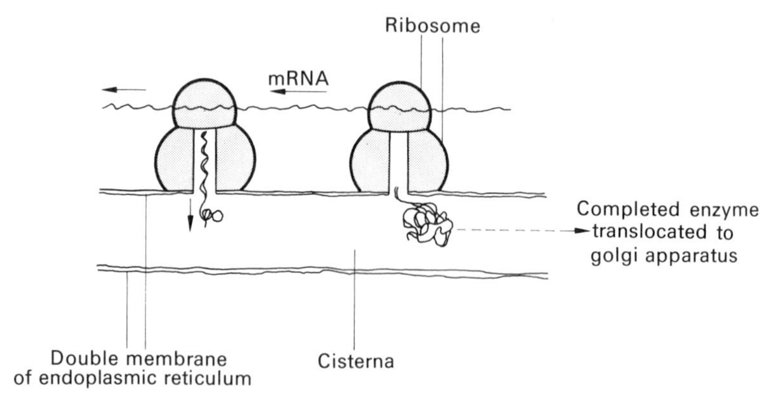

Fig. 1.31 Diagram to show the possible role of the ribosomes that are bound to the membranes of the endoplasmic reticulum. Polypeptides made on these ribosomes may pass into the cisternae and to the outside of the cell via the Golgi bodies.

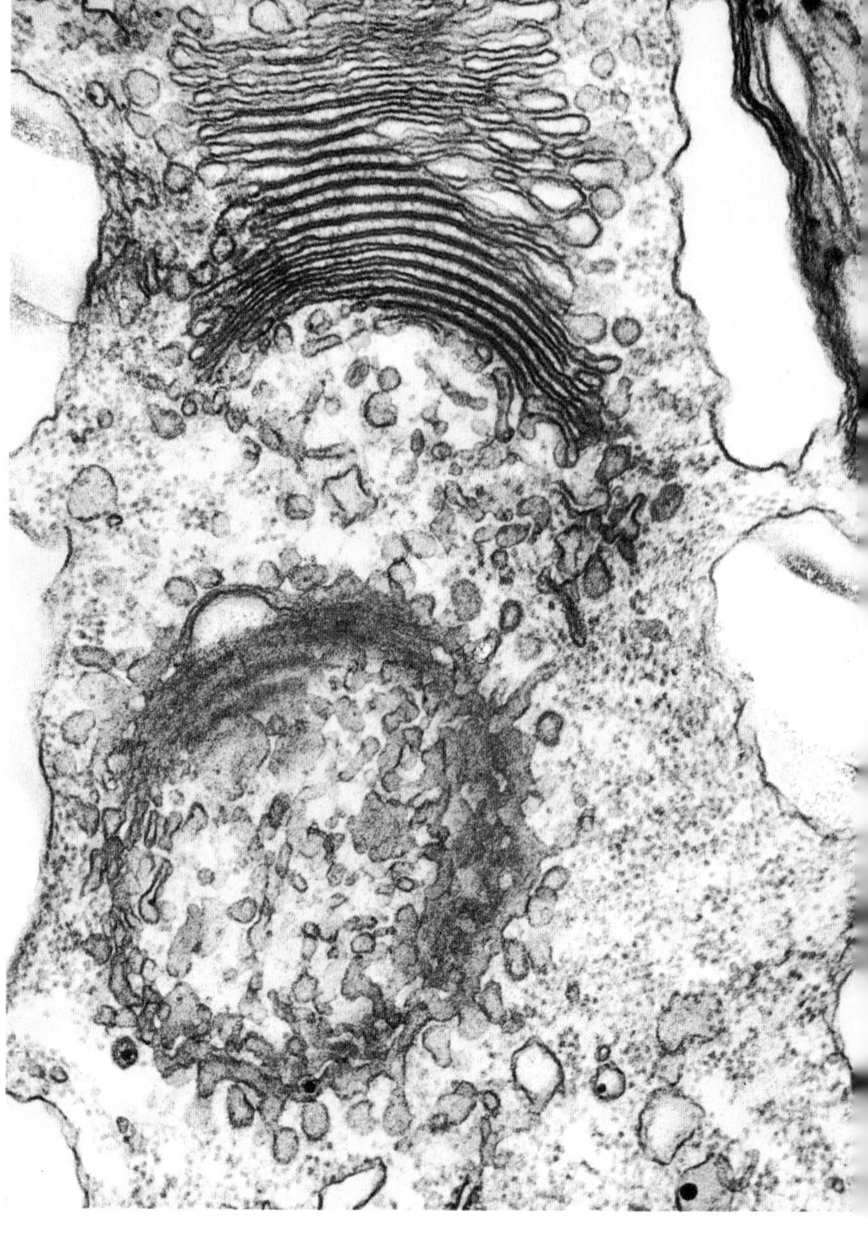

Fig. 1.32 This classic E.M. is a section of part of the flagellate *Euglena gracilis.* The section passes through two Golgi bodies, one in section and the other in a planar view. At the sides can be seen grana in the chloroplasts; the large, clear bodies are paramylon, the carbohydrate storage material equivalent to starch in higher plants. (Courtesy of Dr. G. F. Leedale.)

The Nuclear Membranes

It is probably true to say that the nuclear membranes are among the least understood structures in the cell and until recently relatively little attention has been focussed on their function. The boundary of the nucleus appears to consist of two layers of lipoprotein membrane about 20 nm apart. The outer layer is continuous with the endoplasmic reticulum and may occasionally bear ribosomes; the inner layer is devoid of ribosomes (Figs. 1.33, 1.34 and 1.35). There are a number of anomalies concerning the structure and function of these membranes. For example, the nuclear envelope is particularly tough and is able to withstand manipulations, such as occur in nuclear transplantation experiments, which readily disrupt boundary and endoplasmic membranes. At the same time, the nuclear membranes are capable of prompt collapse and reformation during meiosis and mitosis (see pp. 84 and 70).

There is also evidence that large 'annuli', of the order of 100 nm in diameter, are present in the envelope, allowing continuity of the nucleoplasm and cytoplasm. These may be pores, although recent observations by Kashnig and Kasper (1970) seem to indicate that they are regions where the double membrane of the nucleus is fused into a single iayer, which has a definite, organized structure. Nevertheless, these annuli have been shown to admit relatively large colloidal gold particles into the nucleus while at the same time exercising a selective or regulatory role on solutes passing through them. Their chief role would seem to be to form 'doors' through which messenger RNA passes from the nucleus to the sites of protein synthesis in the cytoplasm (p. 39).

Kashnig and Kasper have been able to isolate the two membranes which constitute the nuclear envelope by ultracentrifugation of isolated, disrupted nuclei from rat liver cells. They were able to show that the outer membrane had enzymic characteristics in common with the membranes of the endoplasmic reticulum, whereas the inner membrane had unique enzymic and chemical features. These findings have relevance to the results of two other American workers, Alfert and Dal (1969), who have demonstrated that DNA synthesis is related not to the volume of the nucleus, as might be expected, but to its **surface area**. They have suggested that, in the interphase nucleus, the chromosomes are attached to the inner membrane of the nuclear envelope, perhaps in much the same way as the bacterial chromosome is known to be attached to the boundary membrane of the bacterial cell (p. 150).

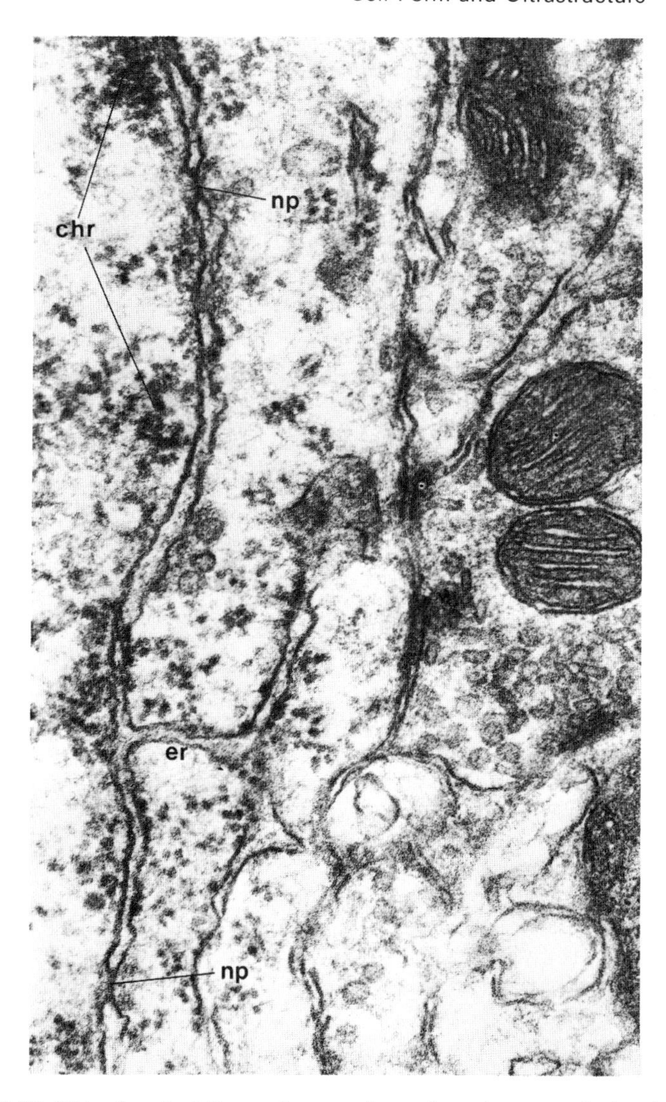

Fig. 1.33 E.M. of part of the nuclear envelope of a rat nerve cell showing the continuity between the outer membrane of the envelope and the endoplasmic reticulum (er). Also shown are chromatin masses (chr), clusters of granules associated with the inner membrane, and nuclear 'pores' or annuli where the membranes of the nuclear envelope come into close connection (np). Magnification ×60 000. (Courtesy of Dr. A. R. Lieberman.)

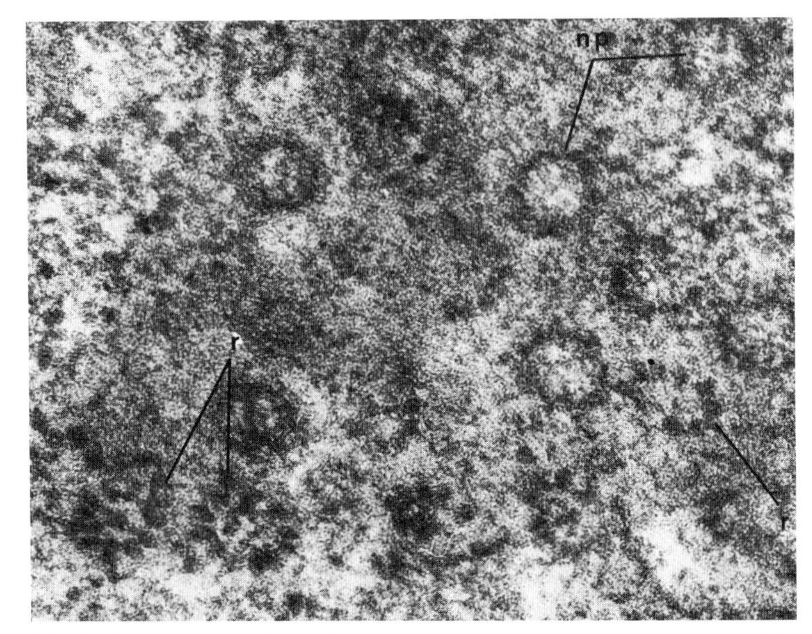

Fig. 1.34 E.M. of a rat liver cell nucleus in which the section passes tangentially across the nuclear membrane. The nuclear 'pores', about 100 nm in diameter, can be seen in face view, as can be some ribosome clusters attached to the outer membrane (r). Magnification ×120 000. (Courtesy of Dr. A. R. Lieberman.)

The nuclear membrane may have other functions. Those organisms which possess a nuclear membrane are termed **eukaryotic** and comprise all multicellular plants and animals. Bacteria, having no nuclear membrane, are termed **prokaryotic**. It would appear, therefore, that the evolutionary origin of the nuclear membrane antedates the divergence of plants and animals. This approximately coincides with the origins of barriers against endosymbiosis (p. 235). In this sense the primitive eukaryotic organisms may have evolved the nuclear membrane to protect the genome of the cell from those enzymes in the cytoplasm, notably nucleases, which hydrolyse ingested organic matter.

The Nucleolus

Inside the nuclear envelope there may be one or more bodies termed **nucleoli**. In living cells, under the light microscope, they are conspicuous owing to their highly refractive nature (see Fig. 1.36), and in fixed cells they stain heavily with basophilic dyes such as cresyl violet. There is good evidence that the nucleolus is the site of synthesis of ribosomal RNA (see p. 37 and Errera *et al.*, 1961), and possibly of the construction of the ribosomes themselves (Birnsteil *et al.*, 1963). Other forms of RNA have also been shown to be present in the nucleolus and its role may prove to be considerably more complex than was once thought (Figs. 1.37 and 1.38). Vacuoles are also present in the nucleolus. They appear to behave in a manner similar to the contractile vacuoles of fresh water protozoans, such as amoeba and *Euglena*, but it is unlikely that they would have an osmoregulatory role and their function is not clear.

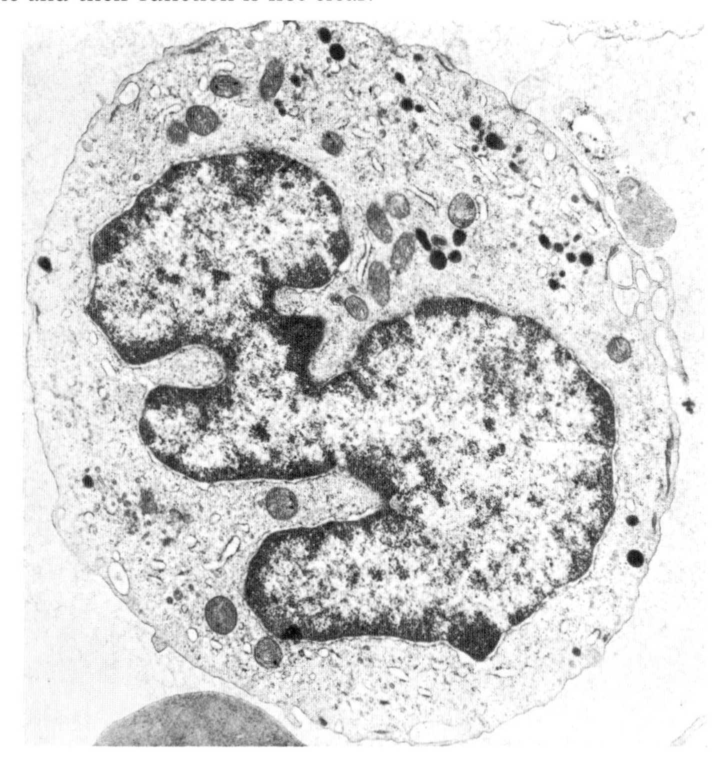

Fig. 1.35 Low power E.M. of a whole monocyte (one of the white blood cells). The cytoplasm shows relatively few mitochondria and only short strands of endoplasmic reticulum. Pinocytic vesicles formed by the 'looping' of short pseudopodia can be seen. The nucleus is characteristically deeply invaginated in these cells and chromatin granules can be seen clustered on the inside of the nuclear membrane. Magnification ×10 000. (Courtesy of Mr. Alan Ross.)

The nucleolus appears to be organized by specific loci on chromosomes. In 1934, Dr. Barbara McClintock of the Cornell genetics team, demonstrated the relationship between a particular chromosome element and the development of nucleoli in the maize plant. More recently, Brown and Gurdon (1964) have shown that mutant forms of the toad *Xenopus laevis* are unable to produce nucleoli and, therefore, ribosomal RNA. Such anucleolar mutants die at an early stage of larval development, when the supply of ribosomes obtained through the maternal cytoplasm in the egg becomes inadequate or exhausted.

The nucleolus may have other roles to play in the physiology of the cell. The formation of spindles during cell divisions coincides with structural changes in the nucleolus, which suggests a connection (Sirlin, 1962) and there are known to be strong similarities in constitution between nucleolar protein and spindle protein (Went and Mazia, 1959). Spindle formation in mitosis is also inhibited in cells in which the nucleoli have been irradiated by micro-beams of ultraviolet light (wavelength approximately 265 nm), (Gaulden and Perry, 1958).

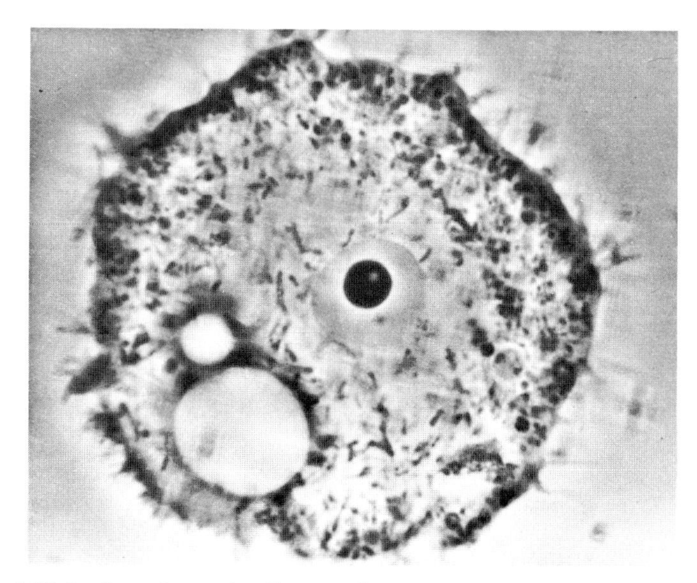

Fig. 1.36 In the soil amoeba *Hartmanella astronyxis* the nucleus appears as a translucent body under the light microscope. By contrast, the nucleolus is very dense and may contain one or more cavities, or nucleolar 'vacuoles'.

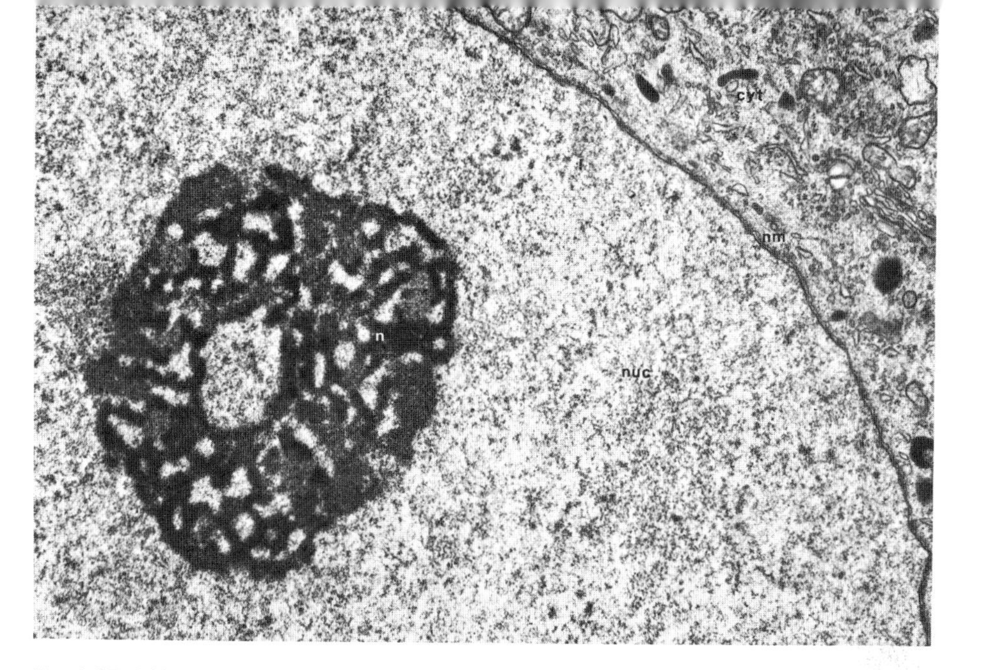

Fig. 1.37 E.M. of the nucleolus (n) and part of the nucleus (nuc) of a rat nerve cell. Using silver impregnation techniques the nucleoli of cells that are actively synthesizing proteins seem to consist of dense twisted threads. Magnification ×13 000. (Courtesy of Dr. A. R. Lieberman.)

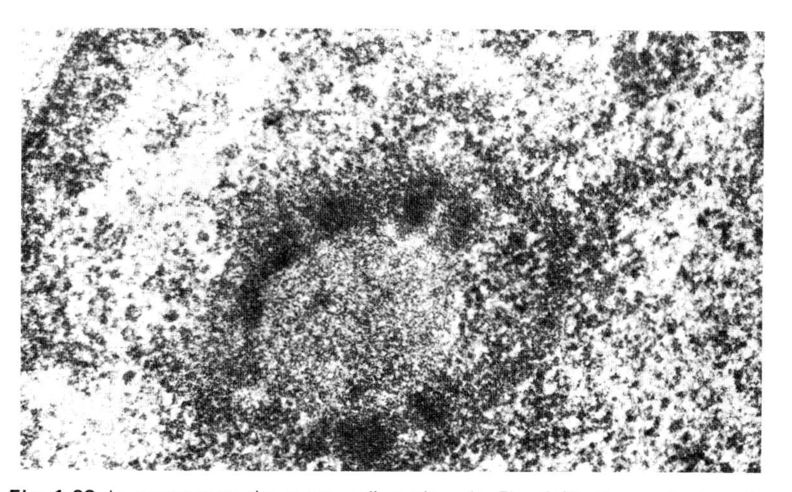

Fig. 1.38 In contrast to the nerve cell nucleus in Fig. 1.37, the nucleolus of a lymphocyte, a white blood cell which carries out relatively little protein synthesis, is more diffusely granular and less well defined. Magnification ×52 000. (Courtesy of Mr. Alan Ross.)

2 Macromolecules and their Roles in the Cell's Activities

In this chapter attention is concentrated on those macromolecules within the cell that play a part in coordinating and perpetuating its activities, and thereby those of the whole organism, namely the nucleic acids. These substances act indirectly through the medium of enzymes, which are proteins. The formation and structure of proteins is also discussed.

A description of viruses is included in this chapter not only because they are composed of nucleic acids and proteins but also because they represent aberrations in the patterns of protein synthesis in the cell leading, among other things, to the transformation of controlled and differentiated cells into cancerous ones. These can be thought of as uncontrolled and alien in the sense that they do not obey the body's rules of tissue differentiation and orderliness.

The chapter concludes with a brief account of the role of enzymes in the regulation of metabolism. One of the features of an organism's relation to its environment is that of **sensitivity**. In order to maintain its organic integrity any organism must be able to monitor changes in the environment and adjust its internal metabolism accordingly. Clearly this is more critical in motile animals in which rapid changes in the environment may arise as a result of locomotion. In these animals the need for quick monitoring and response has been met by the evolution of the nervous system. But sensitivity, the ability to respond to environmental stimuli in order to regulate the internal environment, is common to plants, animals and micro-organisms.

Regulation, however, occurs at several levels. In *Patterns in Biology* regulation of reproduction is discussed in Chapter 7 and the regulation of gene action in Chapter 8. Although the activities of genes determines the presence or absence of enzymes in the cell's cytoplasm, there are still finer regulators which relate the activities of these enzymes to the immediate needs of the cell (the requirement for ATP, for example, when the cell begins to do work). It is this level of control that is discussed in this chapter and regulation in the respiratory pathway is used as a case in point.

2.1 The Structure of Nucleic Acids

1. The Structure and Replication of DNA

In the previous chapter the historical development of the Cell Concept was described. The attention focussed on the cell during the second half of the 19th century, however, was not restricted to microscopical investigations. F. Meischer (1871), in a remarkable pioneering piece of cytochemistry, analysed the cell nucleus. He collected pus cells, taken from surgical dressings, and extracted the nuclei by dissolving away the cytoplasm in a variety of solvents, and by filtration and sedimentation. He obtained a sufficient quantity for chemical analysis, a considerable feat in those pre-centrifuge days.

Meischer's analysis of the nuclear contents revealed a substance containing 'nitrogen, sulphur and especially rich in phosphorus'.

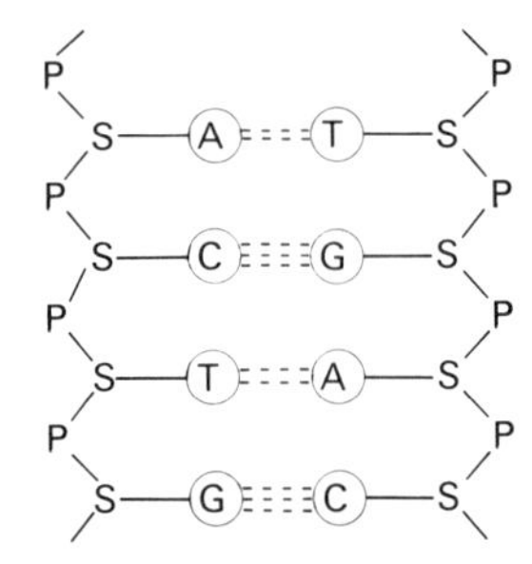

Fig. 2.1 The ladder-like concept of the structure of the DNA molecule in which the rungs are formed by pairs of bases, specified by two or three hydrogen bonds, and the sides of the ladder are formed by alternating phosphate and deoxyribose sugar units.

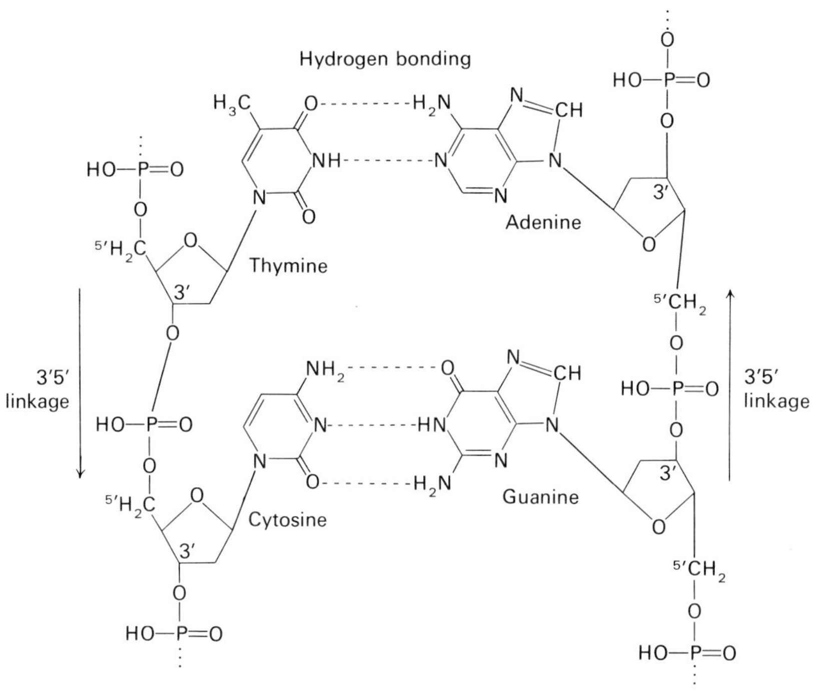

Fig. 2.2 Part of the molecular structure of the DNA molecule showing two base pairs, linked by hydrogen bonds. In the sides of the molecule, formed by alternating sugar and phosphate residues, the sugar molecules are reversed from one side to the other. Thus the 3′-5′ linkages proceed in opposite, or 'antiparallel' directions.

He discounted the possibility that this substance was a compound of lecithin and protein . . . 'It is more likely that we here have a substance *sui generis*, not comparable with any group at present known'. Meischer's substance was almost certainly nucleoprotein, the complex formed from nucleic acids, which contains the genetic information of the cell, and protein.

After the Second World War, research in the fields of biochemical genetics and particularly in the behaviour of related biological macromolecules drew attention to the nucleic acids contained in the nuclear chromatin. The work of Chargaff (1950) and Wyatt (1952) showed that certain purines and pyrimidines in nucleic acids were always present in related proportions, in particular that the amounts of adenine and thymine were approximately equal, as were the amounts of guanine and cytosine. M. Wilkins (1953) and his co-workers at King's College, London, then obtained crystals of DNA and, by X-ray crystallography, were able to show that the molecule had a regular, right-handed, helical structure. These findings, together with those of W. T. Astbury (1947) which indicated that the structure of the DNA appeared to consist of flat, aromatic bases arranged like a 'pile of pennies' with a repeat distance of 0.34 nm, provided the experimental evidence upon which the Watson–Crick hypothesis was laid.

The suggestion was made by J. D. Watson and F. H. C. Crick (1953) that the DNA molecule consisted of two helical 'ribbons', each made up of alternating phosphate and sugar groups, with the purine and pyrimidine bases lying between the ribbons, on the inside of the helix. The findings of Chargaff indicated that adenine was

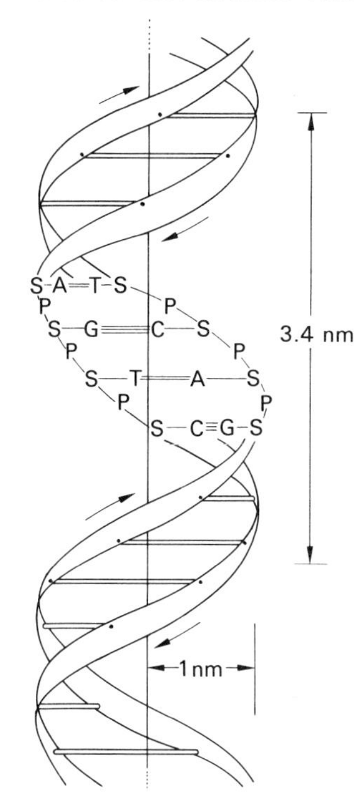

Fig. 2.3 The general form of the DNA molecule based upon that suggested by Watson and Crick in Nature (1953), **171**, 138.

always linked to thymine and that guanine was linked to cytosine. Although Watson and Crick's proposal for the detailed structure of DNA rested largely on theoretical, stereochemical arguments, subsequent work has verified their hypothesis.

The model proposed by Watson and Crick, of two chains, or ribbons, of polymerized nucleotides held together by hydrogen bonds, which are relatively weak, was particularly attractive because it indicated a mode of replication (Fig. 2.1). If the two chains were separated along the hydrogen bonds, perhaps in an 'unzipping' fashion, each half chain could act as a template for the synthesis of a complementary chain since only specific bases would pair off. As Watson and Crick concluded so succinctly in their article in Nature: '*It has not escaped our notice that the specific pairing we have postulated immediately suggests a possible copying mechanism for the genetic material.*' (See Figs. 2.2, 2.3 and 2.4.)

Size of the DNA Molecule

Owing to DNA's acidic properties the molecule is charged in solution and is usually extracted as a salt, such as sodium-DNA. It is also difficult to isolate DNA from the protein and from other nucleic acids, such as messenger RNA, which are associated with the chromatin in the nucleus. These factors present certain problems concerning the determination of molecular weights, but indirect methods, such as those using sedimentation coefficients, have shown the molecular weights of several forms of DNA to be of the order of 6×10^6.

Since a pair of nucleotides is estimated to have a molecular weight of about 660, this would suggest there are

$$\frac{6 \times 10^6}{660} = 9100 \text{ base pairs per molecule}$$

If, as Watson and Crick suggested, the base pairs are about 0.34 nm apart, this would suggest that the extended helix is, in length, approximately:

$$\frac{9100 \times 0.34}{10^3} \ \mu\text{m} = 3.1 \ \mu\text{m}$$

It is possible, however, that these calculations underestimate the molecular weight owing to fragmentation of the molecules during extraction. DNA extracted by other means, from bacteriophage virus, has given molecular weights of 130×10^6 (Peacocke and Drysdale, 1965). Since the head of the bacteriophage virus is approximately 0.1 μm in diameter, this would suggest that the DNA molecule is much folded within the head. Probably the same rule applies to DNA molecules in the nuclei of higher organisms. A discussion of the relationship between the structure of DNA and the structure of the chromosome will be found in Chapter 8.

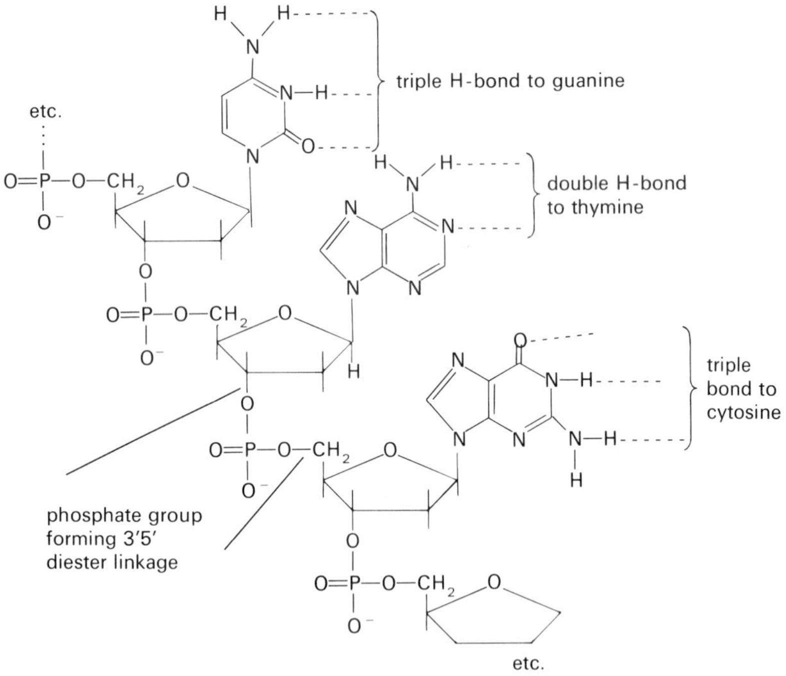

Replication of DNA and its Biological Significance

The attractiveness and the immediate acceptance of the Watson–Crick hypothesis for the structure of DNA lay not only in the manner in which it accommodated the evidence then available but in the inherent suggestion of replication. To the biologist the paradox of a chemical structure destined to carry genetic information is that it must have remarkable permanence and yet be capable of dividing and reproducing. The cells which constitute a multicellular organism represent a genotypically identical **clone** formed by mitosis from a single cell, the zygote. Some of these cells ultimately give rise to gametes, through which genes are passed on to the next generation. From our understanding of the genetic basis of speciation (p. 224) many of the genes in an organism are as old as the species to which it

Fig. 2.4 Part of a side chain of DNA involving the bases cytosine (at the top), adenine and guanine.

belongs, but most of its genes have been inherited from ancestral species. Many genes, such as those coding for enzymes involved in basic metabolic processes, for example respiratory pathways, are common to all but a few organisms and represent molecular configurations that have been in existence for possibly one or two thousand million years. Therefore, although new genes arise from time to time by the mutation of existing genes, many of them can be considered as chemical configurations of great antiquity, so any hypothesis concerning the structure of genetic molecules must incorporate the seemingly contradictory features of remarkable stability and chemical reactiveness.

Some recent experiments on the bacterium *Micrococcus radiodurans* (some strains of which are highly resistant to ultra-violet radiation) may suggest a mechanism to account for the longevity of DNA configurations. One of the consequences of u.v. damage is the formation of **dimers** in the DNA molecule (Fig. 2.5). A dimer occurs when a cyclobutane ring is formed between two pyrimidine bases, such as thymine to thymine, which are lying adjacent to one another on one strand of the DNA molecule. The effect of the dimer is to disrupt the synthesis of messenger RNA and consequently to induce a **mutation** (p. 203).

There is good evidence (Moseley, 1969) that the cell is able to recognize such damage and repair it. A number of enzyme regulated

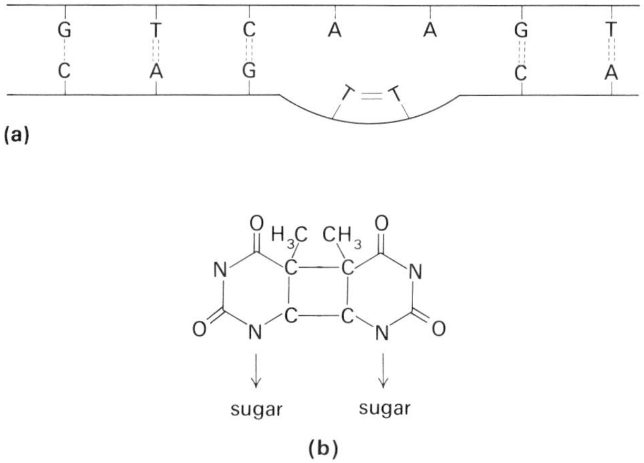

G T C A A G T
C A G C A
(a)

O H$_3$C CH$_3$ O
N C C N
C C
O N N O
sugar sugar
(b)

Fig. 2.5 (a) The thymine dimer in a DNA molecule. (b) Structure of a thymine dimer.

stages appear to be involved. After recognition of the fault, an **endonuclease** enzyme excises the offending fragment, leaving a single-stranded gap. Then **DNA polymerase** restores the break, using the complementary bases on the remaining strand as a reference. It has been suggested that the genetic disease *Xeroderma pigmentosum*, which is a dominant effect in man, represents a deficiency in the DNA repair mechanism. Persons suffering from this condition have skin which is extremely sensitive to sunlight, forming painful freckles and, in severe cases, skin cancers.

Examination of the Watson–Crick model immediately suggests that, during replication, the helix unravels and the base-pairs separate by the breaking of the hydrogen bonds which hold them together. Hydrogen bonds exist between hydrogen and only a few other atoms in the periodic table, notably nitrogen, oxygen and fluorine. In the DNA molecule hydrogen bonds occur between the adjacent N—N or N—O atoms of complementary bases (Fig. 2.6).

Hydrogen bonds are considerably weaker than the bonds which hold the polymerized nucleotides together in the sugar-phosphate, or diester linkage, groups. The hydrogen bond has an energy content of approximately 21 000 joules per mole, or 21 kJ mol^{-1}, compared with a C—O bond value of 350 kJ mol^{-1}. Thus the double helix can much more readily be separated into its two component strands than can each individual strand be broken. After separation of the strands each half strand serves as a template to which new bases can be added, as has been described.

This concept received confirmation from the work of Kornberg and his colleagues (1956) who were able to extract an enzyme, DNA-polymerase, from the bacterium *Escherichia coli*. In the presence of the bases adenine, guanine, cytosine and thymine (in the form of highly reactive phosphorylated nucleotides, that is, base + sugar + phosphates) and in a cell-free extract, this enzyme carried out the polymerization of new DNA, *but only in the presence of a DNA template, or primer* (Fig. 2.7).

Some critics of this hypothesis of replication pointed to the possibility of errors arising through faulty base pairing. But in an elegant experiment Trautner, Schwartz and Kornberg (1962) showed that the erroneous incorporation of ^{32}P—labelled guanine into an *artificial* DNA molecule, in which the primer was a polymerized adenine-thymine strand (and which therefore should take up only adenine or thymine) was less than one in 28 000 bases incorporated.

One of the most direct pieces of evidence concerning the mode of

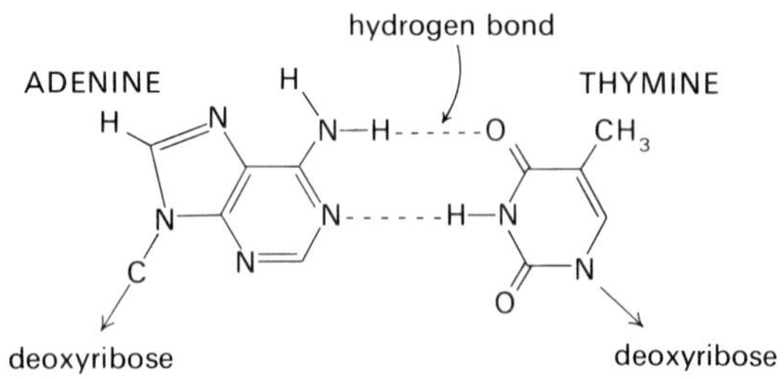

hydrogen bond

ADENINE THYMINE

deoxyribose deoxyribose

Fig. 2.6 Hydrogen bonding between the complementary bases adenine and thymine.

replication of DNA has been provided by the classic experiment of M. Meselson and F. W. Stahl (1958). These workers grew the bacterium *Escherichia coli* over several generations in a medium containing the heavy isotope of nitrogen, ^{15}N. The isotope was incorporated into the bacterial DNA which, when extracted, was shown to be heavier than the normal DNA (containing ^{14}N) when centrifuged in a density gradient made of caesium chloride solutions.

The bacteria were then transferred to a culture medium containing the normal nitrogen isotope. Thereafter, any DNA that was synthesized by the bacteria contained ^{14}N. At intervals following the transfer, samples of the bacteria were removed, DNA was extracted and its density characteristics recorded. The significant feature of this experiment was that an intermediate density was first discovered. This corresponded to a 'hybrid' DNA in which one strand contained the heavy ^{15}N and the other strand contained the light ^{14}N. Later samples, corresponding to a third generation of bacteria, were found to contain intermediate density and light density forms (Fig. 2.8).

There are, however, serious objections to the hypothesis of replication by the 'Kornberg enzyme' DNA-polymerase. Since the two strands of the DNA double helix lie in a head-to-tail manner the enzyme should be capable of working along the strands in both directions. This, it has been shown, it cannot do. The enzyme can only synthesize a new strand in the 5′ to 3′ direction and there is growing evidence that DNA-polymerase is a repair enzyme rather than a duplicating enzyme. The discovery of a mutant form of the bacterium *Escherichia coli* which is deficient in the enzyme DNA-polymerase appears to confirm this, for the strain grows well and replicates its DNA normally. The strain is, however, abnormally sensitive to ultra-violet radiation, which is known to damage DNA (de Lucia and Cairns, 1969).

The means by which DNA is replicated is therefore still not settled. That it does so by a semi-conservative principle, however, appears incontrovertible.

2. *The Structures of RNA*

Ribonucleic acid, or RNA, differs from DNA in a number of important ways. First, it is a relatively impermanent molecule, in the sense of biological continuity, and although RNA is known to play an important role in the way in which genes are expressed it does not itself constitute the genetic information. An exception to this rule occurs in most plant viruses and a few animal viruses in which DNA is absent and RNA forms the genetic data store.

The second difference between RNA and DNA is structural. As the name suggests, the sugar residues which constitute part of the spiral side chains of the RNA molecule are of ribose, whereas they are of deoxyribose in the DNA molecule. In the ribose molecule the second carbon ($C_{2'}$) carries an —OH group, for which is substituted an —H group in the deoxyribose form. This does not affect the manner in which the nucleotides are polymerized; in both cases the diester linkage is formed between the third and fifth carbons of adjacent pentose molecules, see Fig. 2.9.

A further structural difference between RNA and DNA lies in the complement of the bases. The pyrimidine base, thymine, occurs only rarely in RNA. It is replaced by **uracil**, which forms a complementary base pair with adenine. RNA analysis also reveals a number of other bases which are not present in DNA, such as methylated adenine and guanine, pseudo-uracil and inosine, although these

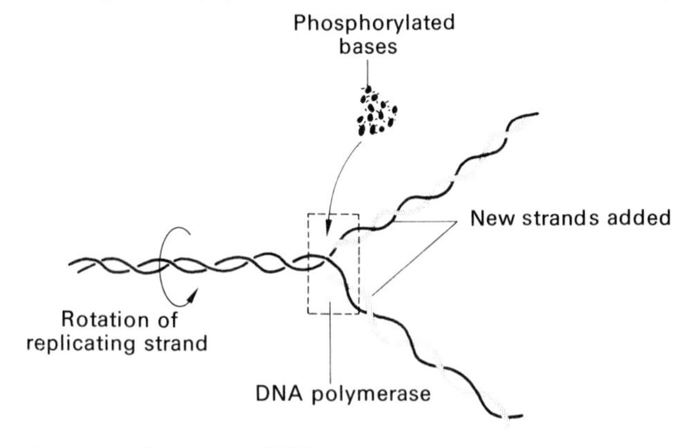

Fig. 2.7 Polymerization of new DNA.

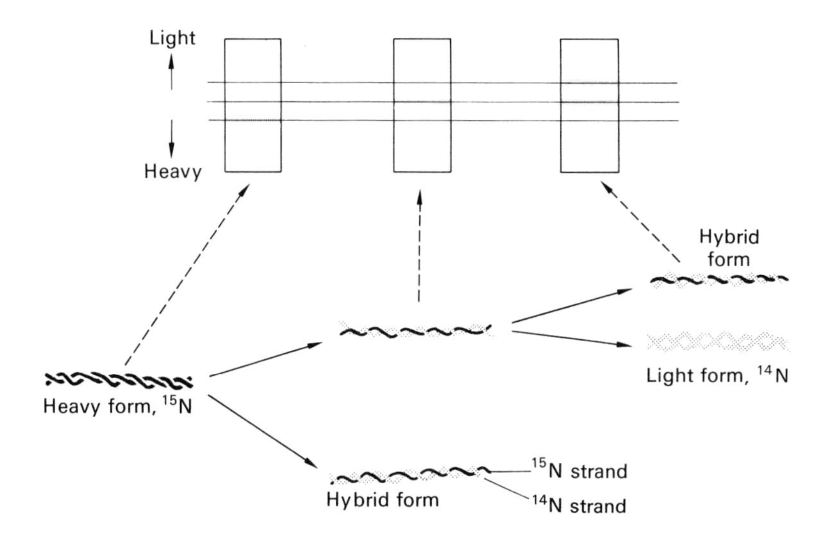

Fig. 2.8 Summary of the classical experiment of Meselson and Stahl (1958) which provided direct evidence for the semi-conservative mode of replication of DNA.

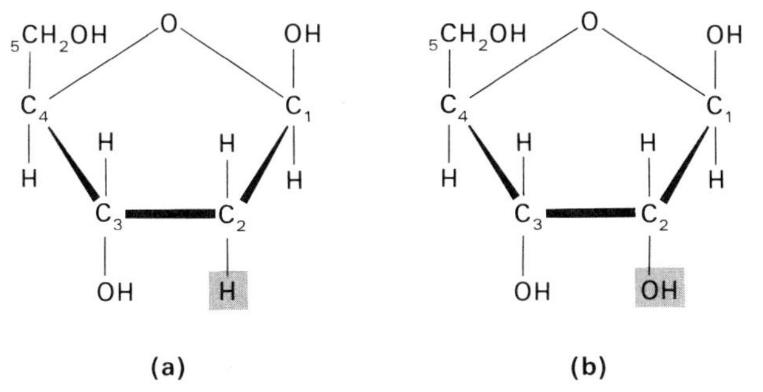

Fig. 2.9 Diester linkage in (a) deoxyribose, and (b) ribose.

usually form less than 4% of the total base complement (Smith and Dunn, 1959).

The third important difference between RNA and DNA occurs in RNA's greater versatility of form. Although single-stranded DNA has reportedly been extracted from a bacteriophage virus (Sinsheimer, 1962), in general it seems to be confined to the double helical form. RNA, on the other hand, can occur in cells in either double or single-stranded forms and as either large molecules or relatively smaller ones. The double helical, macromolecular form, which is essentially the same as DNA and has molecular weights of the order of one to ten million, is found in **ribosomal** (or **nucleolar**) RNA, designated rRNA, and in the RNA which has been mentioned as constituting the genetic information in some viruses. Unlike DNA, however, there is no accurate quantitative complementation between adenine and uracil and between guanine and cytosine, which suggests that the macromolecular form is incompletely double helical. Possibly parts of the molecule consist of single strands folded back on themselves.

RNA is also found in a less stable, single-stranded form as **messenger** RNA, or mRNA. A third form of RNA, known variously as **amino acid transfer** RNA, tRNA, or **soluble** RNA has a molecular weight of the order of 25 000, which corresponds to a single strand of approximately 75 nucleotides.

Considerable attention has recently been focussed on the structure of tRNA. The sequence of bases in a number of kinds of tRNA molecules have been worked out by Holley and his co-workers (1965). This evidence, together with that obtained by X-ray diffraction analysis, which revealed that parts of the tRNA molecule were likely to have a double helical structure (Spencer *et al.*, 1962), has led to a number of hypotheses concerning the structure of transfer RNA. One of the schemes proposed by Holley is of a single strand of RNA doubled back on itself to form a 'clover-leaf' shape, which allows for base pairing, and a helical form, along the four arms (Fig. 2.10a). It is likely that the side arms are folded back so that the whole molecule has an approximately cylindrical shape (Fig. 2.10b) (Lake and Beeman, 1968). This could be resolved by further X-ray diffraction studies but the difficulties of extracting pure forms of tRNA, and thus obtaining crystals of sufficient size for analysis, have so far proved too great.

Other kinds of RNA are likely to emerge as more detailed analyses of cell function occur. One of the difficulties in extracting pure DNA from nuclei is that fractions usually contain firmly attached RNA.

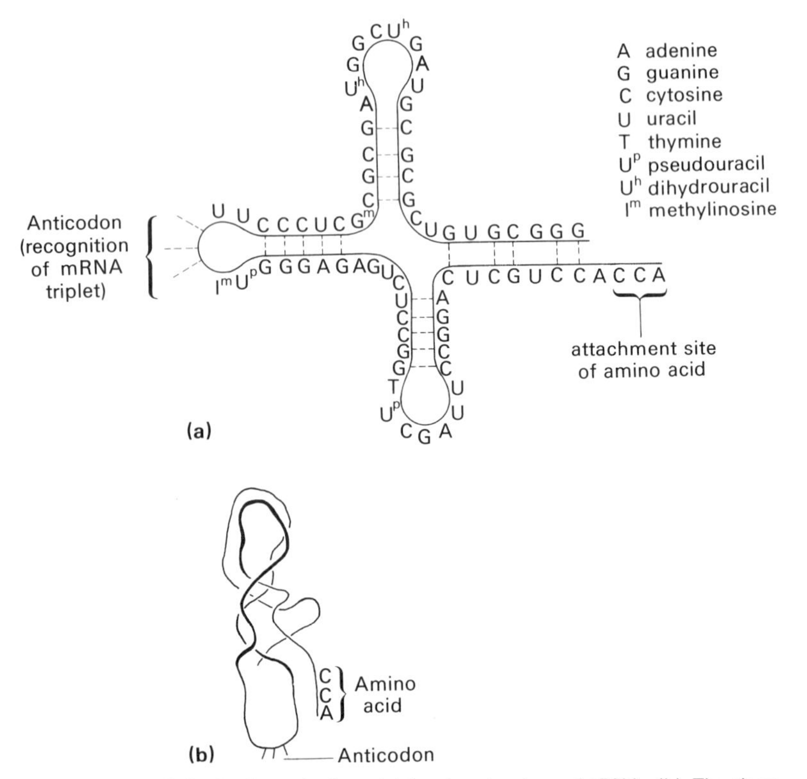

Fig. 2.10 (a) Holley's clover-leaf model for the structure of tRNA. (b) The three-dimensional view of transfer-RNA is unlikely to be like Holley's clover-leaf model. Lake and Beeman (1968) have suggested that the side arms are folded round and down to give a shape like a paper-clip.

This may not be entirely messenger RNA and the suggestion has been made that, in the nuclei of differentiated cells, there is present 'chromosomal-RNA' which has a role in the 'switching on or off' of genes as part of the differentiation process (p. 121).

2.2 The Structure of Proteins

Proteins are formed by the folding of polypeptide chains into particular configurations. The final configuration, or shape, of a protein molecule is mainly responsible for its chemical activity or its structural role. Loss of configuration usually results in loss of activity as, for example, in the thermal denaturation of enzymes. Proteins occur in structures such as membranes (p. 14), and form chemically active agents such as enzymes, hormones and antibodies. For many organisms, notably those which occupy the final stages of food chains and are termed **secondary consumers**, see Chapter 10, proteins provide a source of energy in respiratory metabolism (p. 58).

The structural unit of a protein is the amino acid. A chain of polymerized amino acids is termed a polypeptide. Amino acids have the general formula R—CH·NH_2·COOH and the twenty or so amino acids commonly occurring in proteins differ only in respect of their R—groups. In the simplest amino acid, **glycine**, the R—group is a hydrogen atom. In other amino acids the R—group may be a straight chain, as in **arginine** and **glutamine**; ring-formed, as in **tyrosine** and **phenylalanine**; or sulphur containing, as in **cysteine** and **methionine**. Most amino acids are optically active, existing in D— or L— forms, but generally the latter in nature (Fig. 2.11).

The **primary structure** of a protein lies in the sequence of its amino acids. There are frequently some hundreds of amino acids in the polypeptide chain which forms a protein molecule and since it is known that the amino acids are linked together through their

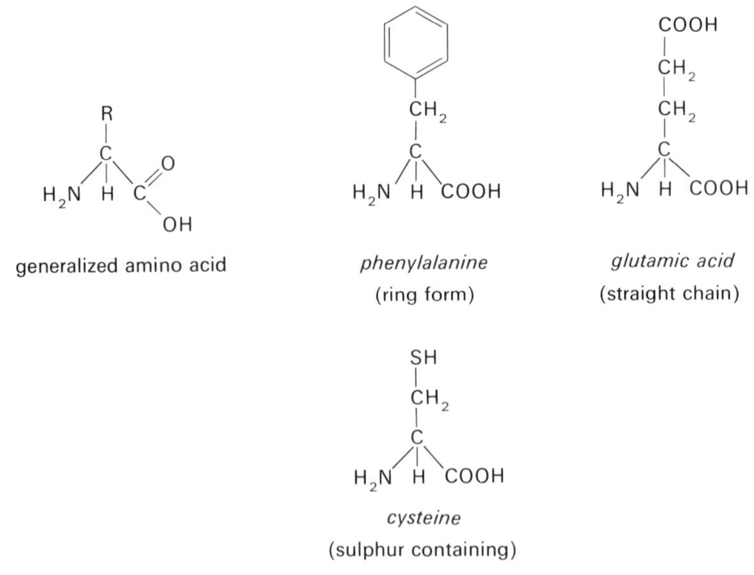

Fig. 2.11 The structure of some amino acids showing the NH_2·CH·COOH pattern common to all and the variable nature of the R—groups.

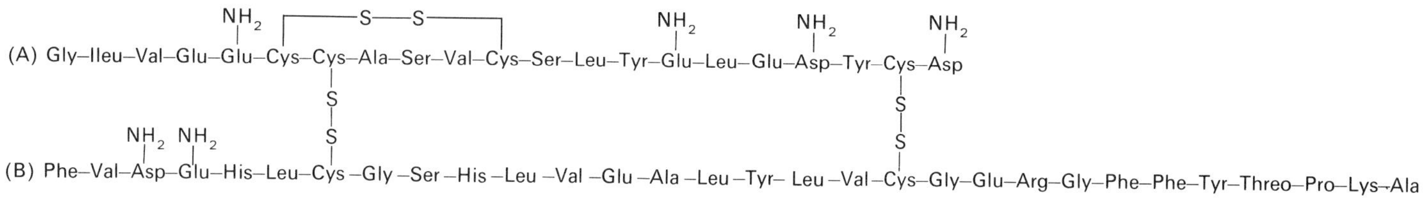

(A) Gly–Ileu–Val–Glu–Glu–Cys–Cys–Ala–Ser–Val–Cys–Ser–Leu–Tyr–Glu–Leu–Glu–Asp–Tyr–Cys–Asp

(B) Phe–Val–Asp–Glu–His–Leu–Cys–Gly–Ser–His–Leu–Val–Glu–Ala–Leu–Tyr–Leu–Val–Cys–Gly–Glu–Arg–Gly–Phe–Phe–Tyr–Threo–Pro–Lys–Ala

Fig. 2.12 The primary structure of insulin, a relatively small protein molecule consisting of two amino acid chains, termed A and B, linked by –S–S– bridges and comprising about fifty amino acids.

—COOH and —NH_2 groups, so that the R— groups stick out sideways from the chain, it can be seen that the sequence of amino acids is an important factor in determining the final shape of the molecule. The complete sequence of amino acids in a protein molecule was first determined by Sanger (1952), for the insulin molecule (Fig. 2.12).

Adjacent amino acids in a polypeptide are linked by **peptide bonds** formed by dehydration synthesis (Figs. 2.13 and 2.14).

A number of **secondary structures** of protein are known. Perhaps the commonest is the so-called **α-helix**, in which the backbone of the polypeptide, namely the repeated —NH—CO—CH— groups, form a right-handed helical structure. Each complete turn of the spiral contains about four amino acids and advances the spiral by about 0.54 nm (Pauling and Corey, 1951). Hydrogen bonds between amino acids lying on adjacent turns of the helix probably serve to stabilize the structure (Fig. 2.15).

Another kind of secondary structure is the alternate folding, or **pleating**, which has been described for a protein occurring in silk (Marsh, Corey and Pauling, 1955). This form of folded polypeptide may be relatively rare. Only a few kinds of amino acids are present in the silk protein; glycine forms about 44% of the molecule and the remaining amino acids, like glycine, have simple, small R—groups, which tend not to affect the stability of this form (Fig. 2.16).

The **tertiary structure** involves the further coiling or folding of the polypeptide, and in many cases the integration of two or more polypeptide chains,* to form the protein molecule. The protein molecule which constitutes a substantial part of hair, nails, horn and feathers, namely **keratin**, consists of a number of α-helices twisted into a secondary helix, like a rope. A similar formation, of three α-helices in a 'super helix' is believed to hold for **collagen**, which forms the principal part of white fibrous connective tissue and is present in cartilage, bone and skin (Gross, 1961). Keratin and collagen are collectively called **scleroproteins**. Their role in support tissues lies in their tertiary configurations. The primary structures of the two proteins are quite distinct. Keratin has a high proportion of the sulphur-containing amino acid **cystine**, up to 20% in human hair, whereas in collagen the principal amino acid is glycine. Proteins which adopt this extended, rope-like configuration are broadly termed **fibrous proteins** and their role is mainly structural.

Proteins which adopt an approximately spherical tertiary configuration are termed **globular proteins**. They are soluble in varying degrees and they comprise the chemically reactive proteins of metabolic systems, such as enzymes and hormones. The most detailed studies of the structure of globular proteins have been carried out by J. C. Kendrew and his team (1958, 1960, 1961). By X-ray diffraction analysis these workers have been able to construct a three-dimensional model for the molecule of myoglobin, a specialized

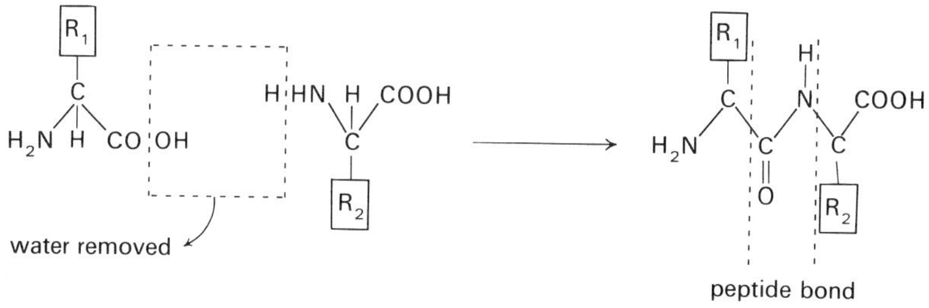

water removed

peptide bond

Fig. 2.13 The formation of a peptide bond by dehydration synthesis. In the presence of a linking mechanism, such as a ribosome (see p. 37) amino acids can be joined by the removal of a water molecule from adjacent amino and carboxyl groups.

* In some accounts of protein structure a distinction is drawn between the folding of the polypeptide chain, the **tertiary** structure, and the integration in the molecule of other polypeptides, which is termed the **quaternary structure**.

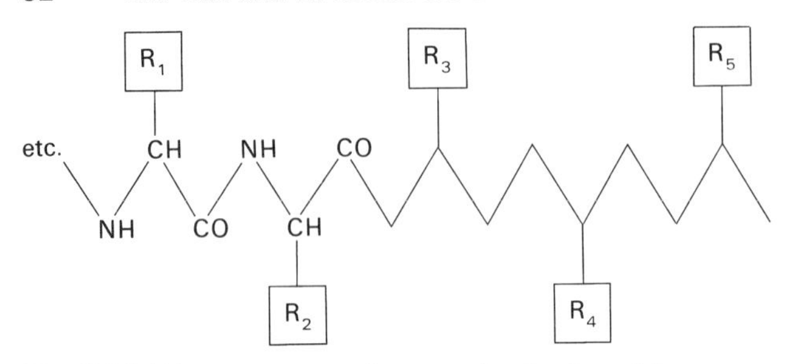

Fig. 2.14 Primary structure of a generalized polypeptide.

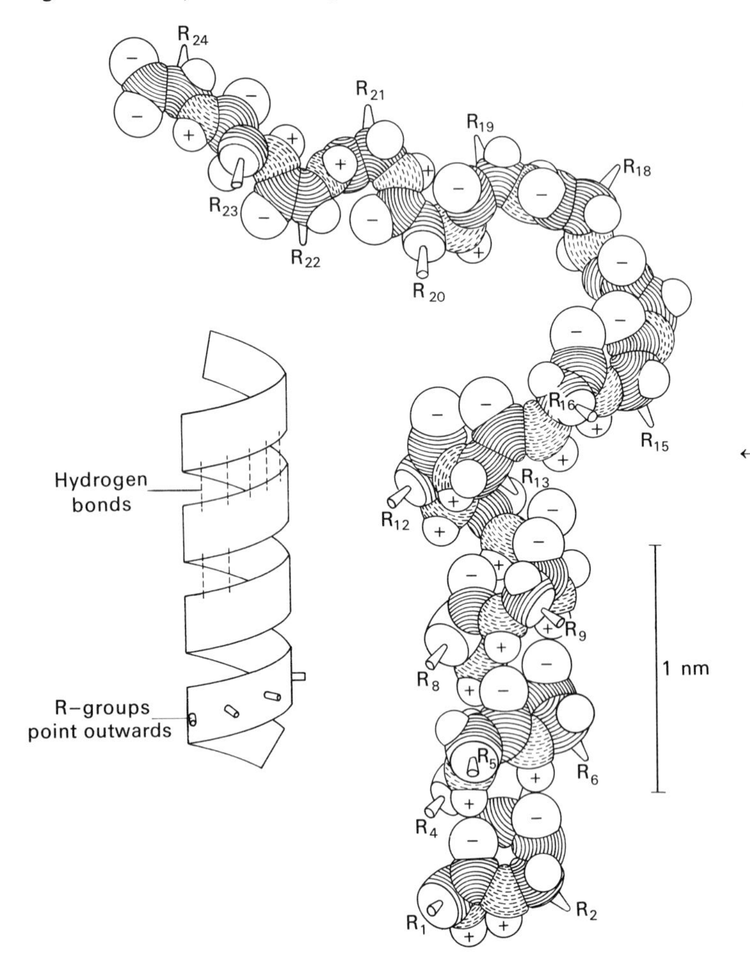

oxygen transport protein in muscle, similar to blood haemoglobin.

In this model most of the gently curving parts consist of α-helices. The configuration is stabilized by several kinds of cross-linkage between the R—groups in different regions of the chain which come to lie adjacent to one another as a result of the folding. The most common cross-linkage is the **disulphide bridge** formed between sulphur-carrying amino acids (Fig. 2.18).

Hydrogen bonds (p. 31) are also known to play an important part in maintaining tertiary configuration, even in the fibrous proteins investigated by Pauling. Although these bonds may be ten to twenty times weaker than a covalent bond, the large number of carbon, oxygen and nitrogen atoms present in the backbone of the polypeptide suggests that hydrogen bonding would be prolific. In this context it is interesting to note that Sanger's scheme for the insulin molecule (Fig. 2.12) shows only two disulphide bridges holding the two polypeptide chains together. In the tertiary configuration of insulin these disulphide bridges, together with the third bridge shown on the shorter chain, are presumably supplemented by hydrogen bonding.

A third form of bonding likely to occur is the relatively weak electrostatic bond (Fig. 2.19) formed particularly between adjacent amide and carboxyl groups (see Zwitterion Effect, p. 47).

Figure 2.20 summarizes some types of noncovalent bonds which stabilize protein structure.

← **Fig. 2.15** Secondary structure of a generalized polypeptide: 1 *The α-helix*. The drawing of the molecular model (Right) shows the polypeptide chain partially coiled into a right-hand, or clockwise, spiral and the ribbon drawn alongside (Left) shows how the helix is stabilized by hydrogen bonds and how the R-groups point outwards. (Molecular model after Coult, *Molecules and Cells*, courtesy of Longmans.)

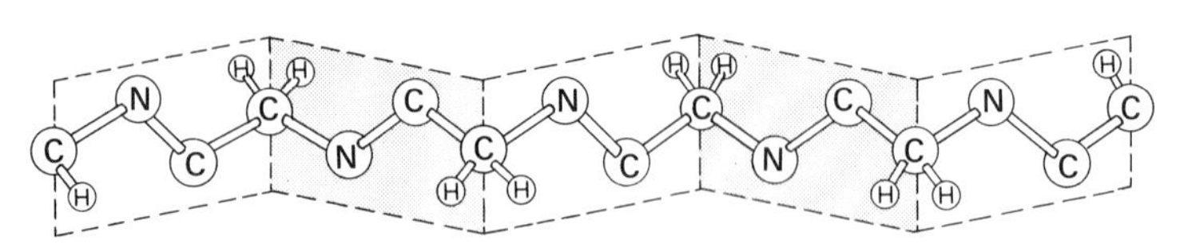

Fig. 2.16 Secondary structure of a generalized polypeptide: 2 *The pleated form*. The arrangement of amino acids in alternating planes may give a pleated form to the molecule. This is relatively uncommon but is thought to be the arrangement of protein molecules in silk fibroin. (After Paul, *Cell Biology*, courtesy of Heinemann.)

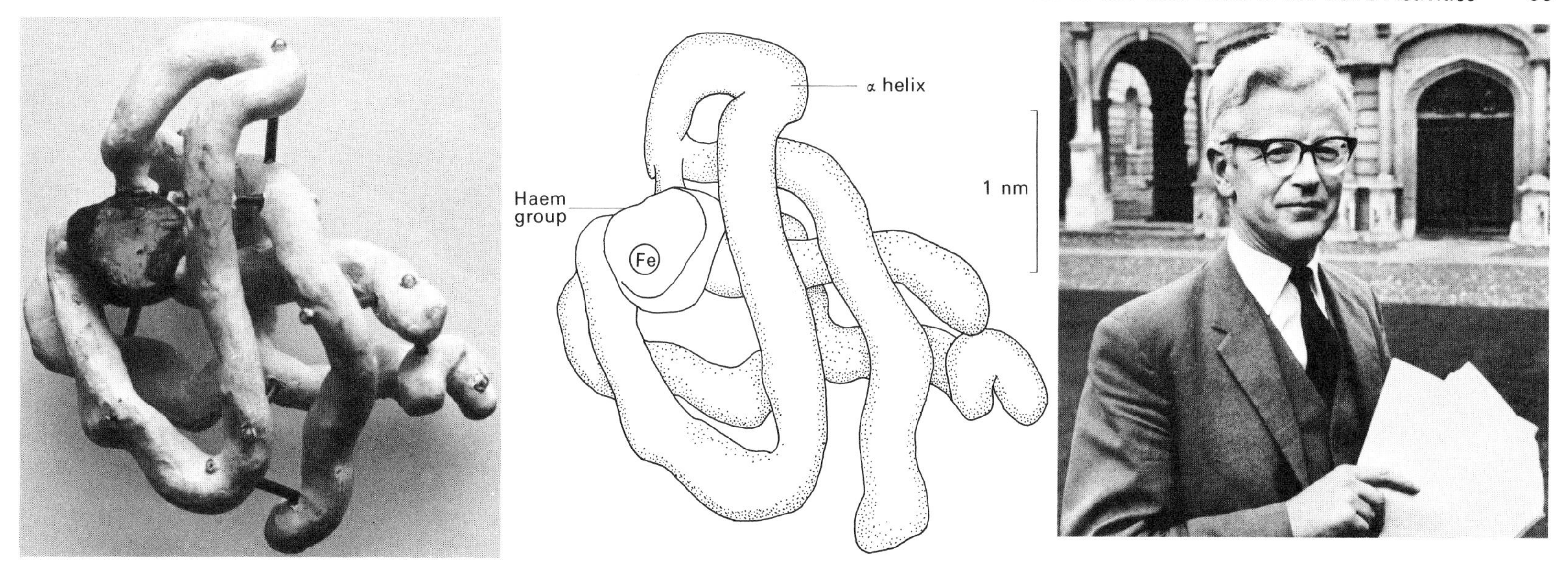

Fig. 2.17 (Left) The tertiary structure of myoglobin worked out by J. C. Kendrew is shown in this model. (Courtesy of *The Sunday Times*, and Thames and Hudson Limited.) (Centre) Drawing of the model to scale. (Right) Portrait of Dr. John Kendrew, taken at Cambridge. (Courtesy of Mr Edward Leigh and Thames and Hudson Ltd.)

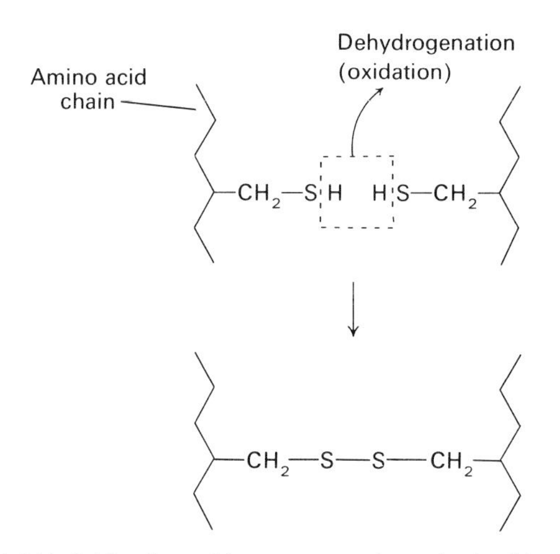

Fig. 2.18 Disulphide bridge formed between cysteine units in either adjacent parts of the same polypeptide or of different polypeptides.

It is now well established that the tertiary structure of a protein is largely determined by the sequence of amino acids which form the constituent polypeptide or polypeptides, that is, by the primary structure. Even small changes in the amino acid sequence can seriously affect the activity of the whole molecule. Ingram (1961)

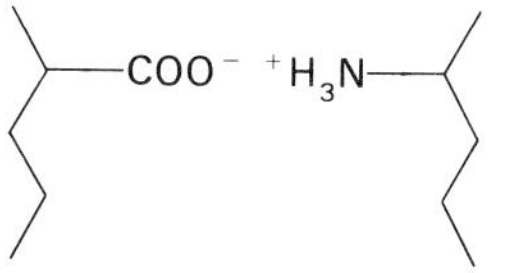

$COO^{-}\ \ {}^{+}H_3N$

Fig. 2.19 Electrostatic bond.

has shown that the haemoglobin in sickle cell anaemia differs from normal haemoglobin in respect of only one amino acid (it contains a glutamic acid unit in place of a valine unit). Also, its solubility is lower in the reduced state, which probably causes the sickling (Perutz and Mitchison, 1950). This not only seriously affects the

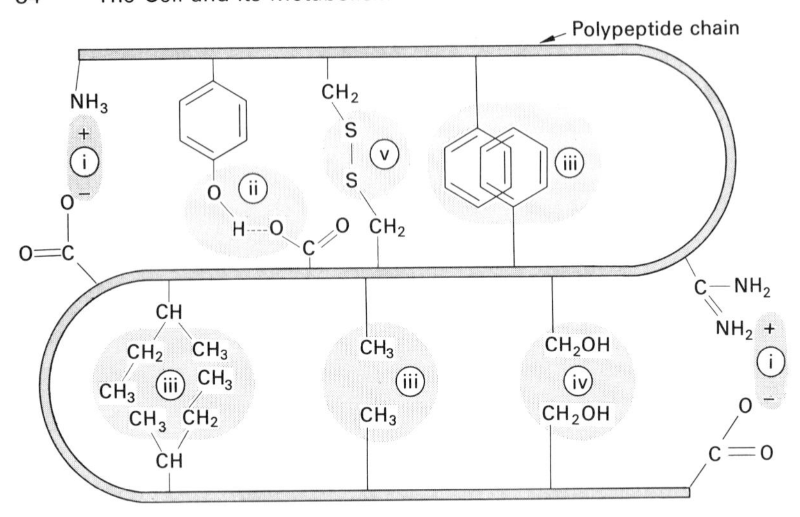

Fig. 2.20 Summary of some of the types of bonds other than covalent ones that serve to stabilize the structure of proteins. (i) Electrostatic interaction; (ii) hydrogen bonding between tyrosine residues and carboxyl groups on side chains; (iii) hydrophobic interaction of nonpolar side chains caused by the mutual repulsion of solvent; (iv) dipole-dipole interaction; (v) disulphide linkage, a covalent bond. (After Anfinsen, *The Molecular Basis of Evolution*, courtesy of John Wiley and Sons.)

oxygen-carrying capacity of the molecule but in some way renders it inedible to the malaria parasite! (See Population Genetics, p. 172.) The numerous ways in which the shape of a protein molecule can be altered, and the effect this has on the molecule's activity, is dealt with later in this chapter.

2.3 Nucleic Acid Pathways and Protein Synthesis

Ribosomes, in the cytoplasm of the cell, are known to be the sites of protein synthesis. The information, or code, which determines the sequence of amino acids in a particular polypeptide is, however, stored in the nucleus of the cell as a specific DNA base sequence (p. 35). Therefore some transport mechanism must be in operation whereby the information in the nucleus is conveyed to the ribosome. Brenner, Jacob and Meselson and co-workers (1961) have provided strong evidence that the genetic code for each polypeptide is carried from the DNA to the ribosome by an unstable, single-stranded form of RNA, for which Jacob and Monod (1961) coined the term 'messenger-RNA', or mRNA. It has also been demonstrated that messenger RNA is synthesized on a template of DNA (Geiduschek *et al.*, 1961).

Much of the evidence for the role of mRNA in protein synthesis has been obtained from studies on bacteria which have been infected with bacteriophage virus. Volkin and Astrachan (1957) showed that when virus DNA was injected into the bacterial cell a new, rapidly formed and unstable RNA appeared in the bacterial cytoplasm, which had base ratios corresponding to the **viral** DNA. In a series of elegant experiments Brenner *et al.* (1961) showed that the newly synthesized RNA which appeared when bacteria were infected with bacteriophage immediately became attached to the bacterial host ribosomes. Using heavy isotopes these workers further demonstrated that 'phage protein was synthesized on the ribosomes to which the mRNA had become attached.

In higher, **eukaryotic** organisms, in which a nuclear membrane is present, it might be supposed that this system is modified. Sibatani and co-workers (1962), however, have demonstrated the presence of an RNA fraction which has the same general properties of bacterial mRNA and which has the same base sequence as part of the DNA in the nuclei of calf thymus gland. It has similarly been demonstrated in mammalian liver nuclei (Weiss, 1960) and in yeast (Ycas and Vincent, 1960), and it is possible that the function of mRNA as a transporter of genetic information is general throughout cells.

In a field which is central to biology and which has attracted such widespread attention and research it is inevitable that several alternative theories of genetic information transport should have emerged. One such has been suggested by Bell (1969). Using the base thymine, which is not incorporated into mRNA, Bell was able to demonstrate its incorporation into newly-formed DNA in the nuclei of isolated muscle cells. He was then able to show the presence of this DNA in the cytoplasm where it was membrane-bound in the form of distinct organelles, which he termed **I-somes**. The DNA inside the I-somes he termed I-DNA, or **messenger-DNA**, and he has proposed that, in those organisms in which a nuclear membrane is present, the transport of genetic information from the genes to the cytoplasm takes place in the form of a DNA replica of the functioning genes. Once in the cytoplasm the I-DNA in the I-somes would initiate the formation of mRNA in the same manner as has been described in bacteria.

There is now good evidence, however, that I-DNA and I-somes

are artefacts which arise during cell fragmentation. Fromson and Nemer (1970) have shown that the amount of DNA associated with cytoplasmic fractions largely depends upon the form of homogenization that is used to disrupt the cells and nuclei.

Transcription, or the Formation of mRNA

Messenger RNA is a single-stranded form of RNA whose molecular weight is probably of the order of 1 to 5×10^5. It is estimated to form less than about 4% of the total cellular RNA (Peacocke and Drysdale, 1965) and is relatively unstable, having an active life of probably only a few minutes. Spirin and co-workers (1968) of Moscow, however, have isolated a form of RNA, which corresponds closely to mRNA, in a permanent, inert state, closely bonded to protein. They have coined the term **informasomes** for these mRNA/protein bodies whose structure is thought to be allied to that of some plant viruses. It has been suggested that informasomes could offer an alternative system of genetic regulation to the Jacob-Monod hypothesis (p. 136) in that, by shedding or adopting their coat, they could become active or inactive in enzyme synthesis.

Messenger RNA is almost certainly formed from only one strand of the double helix of DNA. There is good evidence that an enzyme, which has been termed **RNA-polymerase**, or **transcriptase**, moves along one strand of the DNA and constructs a mRNA molecule whose base sequence is complementary to the DNA strand. Butler (1968) suggests that the mRNA strand is taken immediately on to a ribosome particle, which may physically 'peel off' the mRNA. The direction of transcription is the same as that of DNA replication, namely in the 5′ to 3′ direction (Fig. 2.21).

The enzyme RNA-polymerase is thought to consist of two parts (Burgess *et al.*, 1969). The first, the **core enzyme**, catalyses the polymerization of mRNA from the DNA template. The second, termed the **sigma factor**, appears to indicate the precise point on

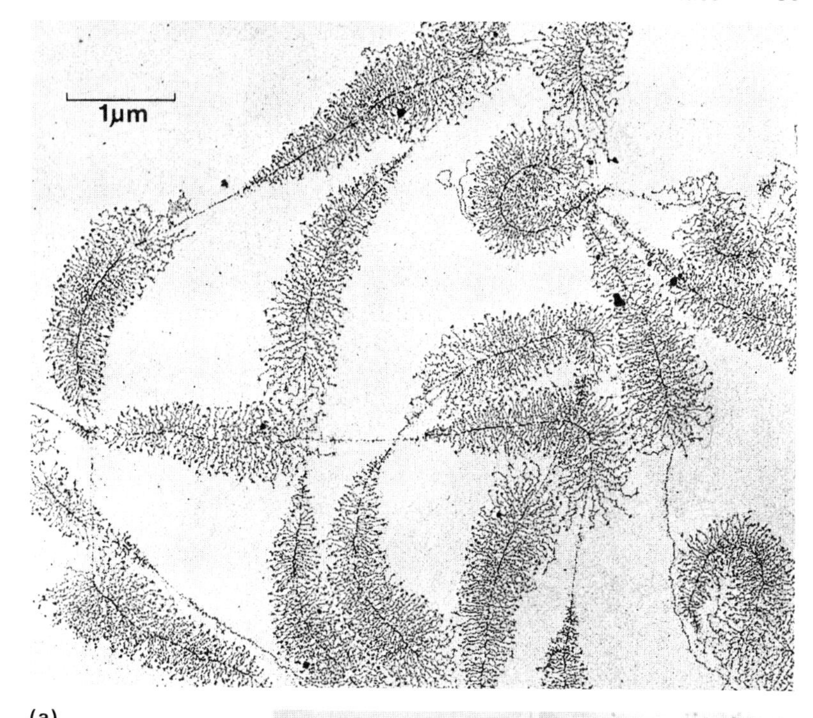

(a)

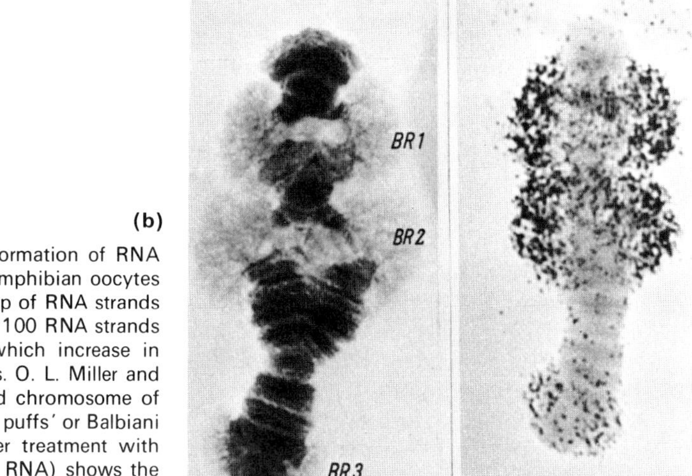

(b)

Fig. 2.21 Two pieces of direct evidence for transcription, the formation of RNA from chromosomal DNA. (a) Threads of DNA from the cells of amphibian oocytes which are actively engaged in making ribosomal RNA. Each clump of RNA strands is thought to represent a gene's output. There appear to be about 100 RNA strands (representing the activity of 100 RNA-polymerase enzymes) which increase in length as transcription proceeds along the gene. (Courtesy of Drs. O. L. Miller and Barbara Beatty.) (b) Photomicrograph of the giant salivary gland chromosome of the midge, *Chironomus*, showing three swollen regions known as 'puffs' or Balbiani rings (BR). An autoradiograph of the chromosomes taken after treatment with radioactivity-labelled uracil (which can be incorporated only into RNA) shows the active production of RNA at the Balbiani rings. (Courtesy of Dr. W. Beerman and Thames and Hudson Ltd.)

the DNA template where transcription is to begin (Fig. 2.22); the factor, in effect, determines which genes shall be transcribed into mRNA (p. 137). Other factors have been discovered that may further aid the discrimination between active and inactive genes (Roberts, 1969b).

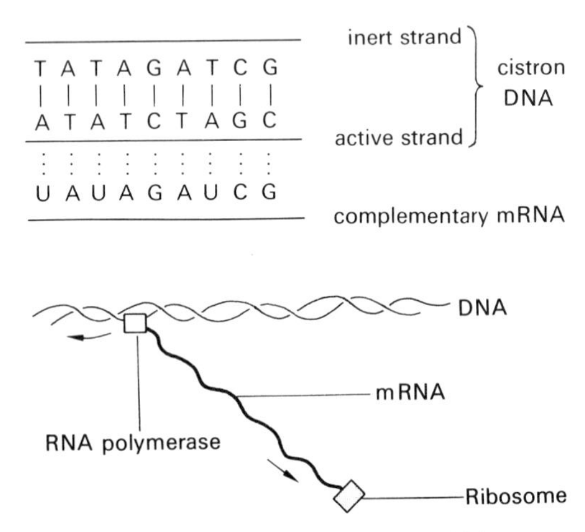

Fig. 2.22 The pattern of transcription. (After Butler, 1968.)

The Role of Transfer-RNA in Protein Synthesis

Some discussion of the structure of tRNA has been made previously. Whatever detailed structure comes to light, however, the molecule is almost certain to exhibit polarity. One end of the molecule is known to carry a recognition site, or **anticodon**, consisting of three bases. The anticodon is believed to recognize a specific base triplet, or **codon** on the mRNA molecule. Since different codons code for different amino acids (Fig. 2.23) it follows that there must be a distinct kind of tRNA molecule for each kind of amino acid.

The 'opposite' end of the tRNA molecule carries an attachment site for an amino acid. Detailed analyses of the nucleotide sequences of tRNA molecules (Holley *et al.*, 1965; Canellakis and Herbert, 1960) have shown a common terminal sequence at this attachment site, namely cytosine, cytosine, adenine and phosphate. This rules out the possibility that the amino acid acceptor site itself, by virtue of its structure, locates the particular amino acid which corresponds to the anticodon at the other end of the molecule.

The location of a particular tRNA molecule with the amino acid to which its anticodon refers is known to be performed by an **amino acid activating enzyme**. Each tRNA/amino acid complex is known to have its own kind of enzyme (Berg, 1961) which is able simultaneously to define the shape of the amino acid and to 'read' the anticodon of the tRNA.

The association of an amino acid with its tRNA molecule is known to involve two stages. In the first stage the amino acid reacts with adenosine triphosphate, or ATP. This is not a phosphorylation reaction (p. 49); instead, the diphosphate group of the ATP is displaced by the carboxyl group of the amino acid, forming an amino acid/adenosine monophosphate (AMP) complex. In the second stage the tRNA molecule binds the amino acid on to its C–C–A site and

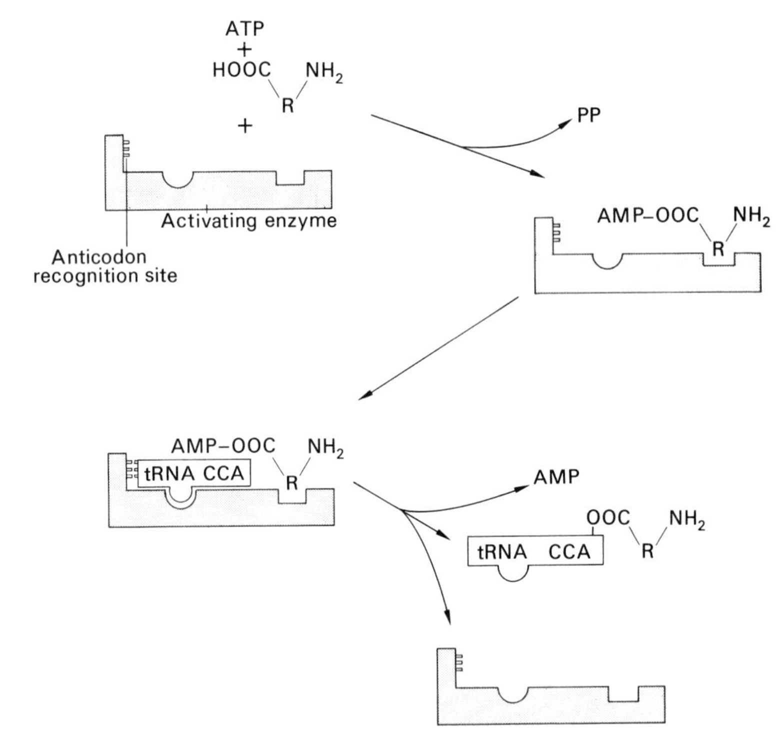

Fig. 2.23 The attachment of an amino acid molecule to its appropriate tRNA molecule involves first, the activation of the amino acid, then the linkage being formed by an activating enzyme. (After Paul, 1965.)

the AMP is released. Both stages appear to be catalysed by the activating enzyme.

The Ribosome's Role in Polypeptide Synthesis

Ribosomes are particularly conspicuous in electron micrographs of cells which are actively synthesizing proteins. In previous sections evidence has been mentioned which demonstrates that messenger-RNA, carrying coded information about the particular amino acids to be incorporated into a particular polypeptide, becomes attached to ribosomes in the cytoplasm. A ribosome is believed to become attached at one end of the mRNA strand, then to work its way along, translating each codon in turn and adding the appropriate amino acid (Fig. 2.24). When ribosomes are extracted from cells

First position	Second position U	C	A	G	Third position
U	Phe	Ser	Tyr	Cys	U
	Phe	Ser	Tyr	Cys	C
	Leu	Ser	non.	non.	A
	Leu	Ser	non.	Trp	G
C	Leu	Pro	His	Arg	U
	Leu	Pro	His	Arg	C
	Leu	Pro	Gln	Arg	A
	Leu	Pro	Gln	Arg	G
A	Ile	Thr	Asn	Ser	U
	Ile	Thr	Asn	Ser	C
	Ile	Thr	Lys	Arg	A
	Met ('capital letter')	Thr	Lys	Arg	G
G	Val	Ala	Asp	Gly	U
	Val	Ala	Asp	Gly	C
	Val	Ala	Glu	Gly	A
	Val	Ala	Glu	Gly	G

Fig. 2.24 The table of the genetic code. The first letter of a codon is given in the vertical column on the left of the table, the second letter in the horizontal row at the top, and the third letter in the vertical column on the right: 'non' represents nonsense or release codons, 'capital letter' indicates the codon for chain initiation. (After Ambrose and Easty, 1970, courtesy of Thomas Nelson Limited.)

and examined under the electron microscope they are frequently found in clusters, and it would appear that these clusters, or **poly-**

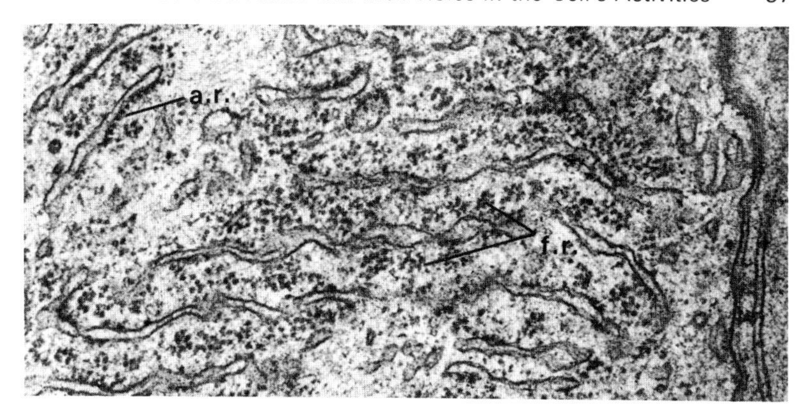

Fig. 2.25 E.M. of endoplasmic reticulum in the cytoplasm of a nerve cell from the dorsal root ganglion of a cat. Ribosomal particles can clearly be seen attached to the membranes (ar) and lying free in clusters between the cisternae (fr). Magnification × 40 000. (Courtesy of Dr. A. R. Lieberman.)

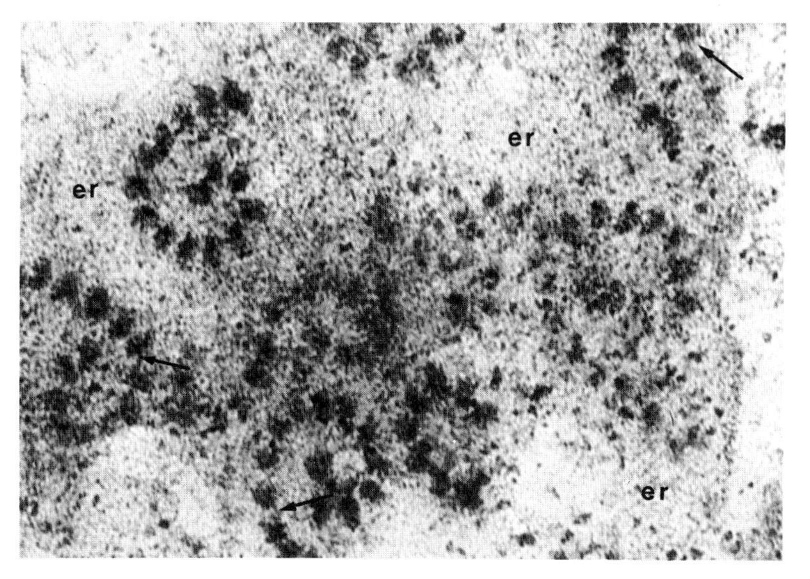

Fig. 2.26 E.M. of part of the cytoplasm of a rat nerve cell in which the section skims the surface of a membrane of the endoplasmic reticulum (er) and shows attached polysomes, each consisting of a short spiral of about 15 ribosomes. There is some evidence of interconnecting strands, possibly messenger-RNA (arrowed). Magnification × 132 000. (Courtesy of Dr. A. R. Lieberman.)

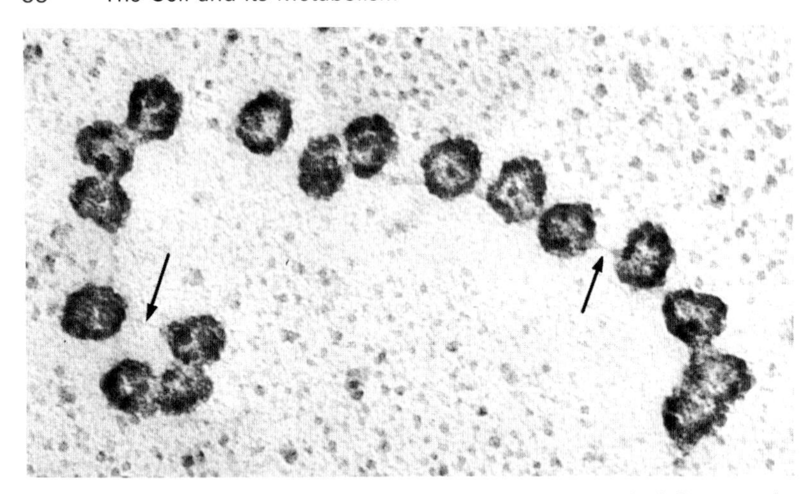

Fig. 2.27 E.M. of a polysome extracted from cells of tobacco leaf. Between the individual ribosomes there is clear evidence of a strand (arrowed), probably mRNA. (Courtesy of Dr. R. G. Milne.)

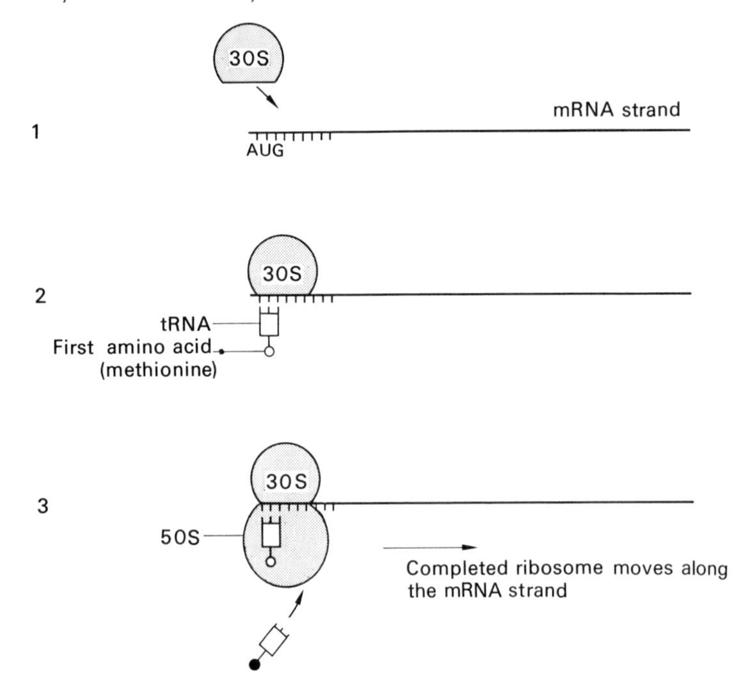

Fig. 2.28 Summary of the first stages in 'translation', the assembly of the ribosome.

somes, represent a number of ribosomes working along the same mRNA strand (Figs. 2.25, 2.26 and 2.27). The strand, however, has only a limited life and there is evidence that it survives the passage of only about twenty ribosomes, that is, one molecule of mRNA codes for the production of about twenty identical molecules of polypeptide (Levinthal *et al.*, 1962).

The role of the ribosomes is not fully understood but it is now believed that they play an **active** part both in bringing together tRNA anticodons and their respective mRNA codons, and in the

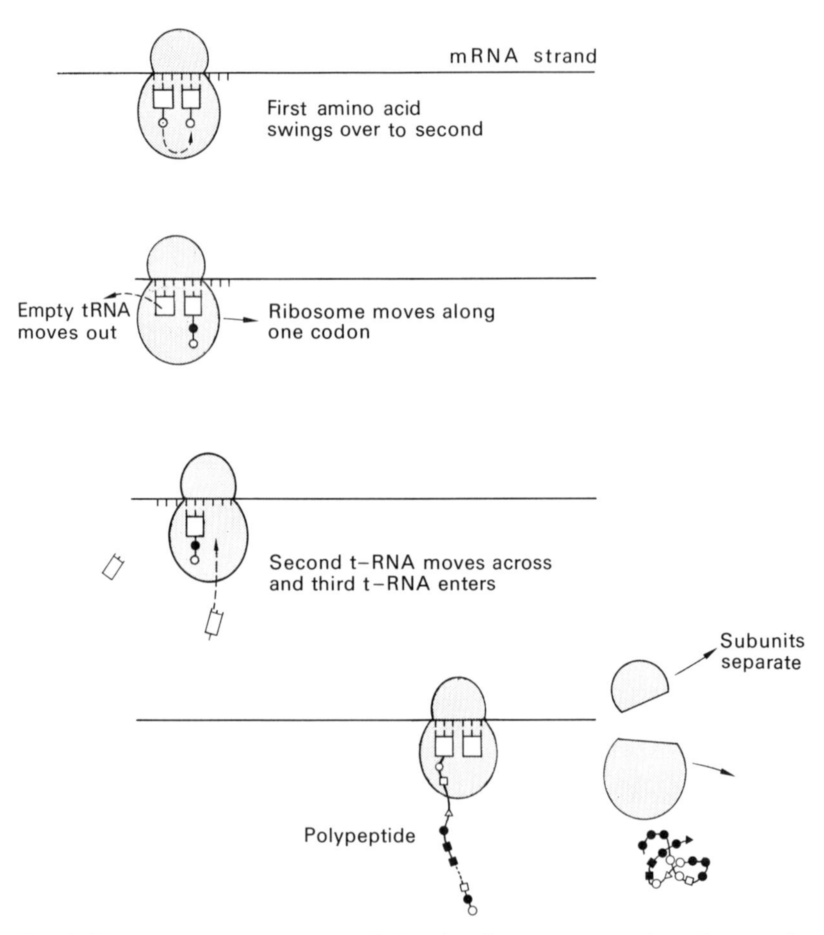

Fig. 2.29 In the later stages of translation the ribosome moves along the strand reading the codons and taking in the appropriate tRNA-amino acid combinations.

formation of a polypeptide chain by the construction of peptide links between the successive amino acids as they are added (Littlefield *et al.*, 1955).

When ribosomes are extracted from cellular debris by ultracentrifugation they tend to separate fairly readily into two parts, which have been termed 30 S and 50 S subunits. Nomura and Guthrie (1968) have provided evidence that messenger RNA is only accepted by the smaller, or 30 S subunit. It had previously been shown (Clark and Marcker, 1966) that the mRNA codon A–U–G, which codes for a tRNA molecule carrying the amino acid methionine, was always present at the 'beginning' of an mRNA sequence. It appears, then, that the combination of the A–U–G triplet and the 30 S subunit marks the beginning of a translation sequence. Nomura and Guthrie suggested that the next stage was the attachment of the methionine-tRNA. This was followed by the attachment of the 50 S subunit to form the complete ribosome (Fig. 2.28).

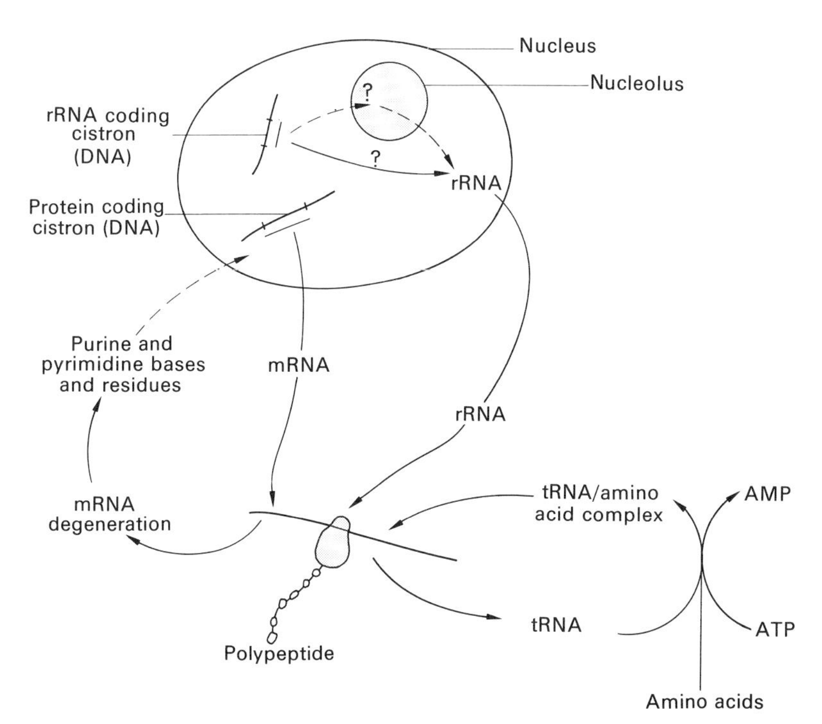

Fig. 2.30 Summary of nucleic acid pathways in protein synthesis.

In the next stage another tRNA/amino acid molecule moves into place in a second slot in the ribosome. The first amino acid is then believed to 'swing over' towards the second and form a peptide bond with it. The ribosome then moves along the mRNA strand, the 'empty' tRNA vacates its site and the second tRNA, now with two amino acids, comes to occupy the first site. A third tRNA/amino acid molecule can now enter the ribosome and the peptide forming process is repeated (Fig. 2.29).

A summary of nucleic acid pathways in protein synthesis is given in Fig. 2.30.

2.4 Viruses: A Source of Aberration in Protein Synthesis

Viruses as a source of disease have been known since the end of the last century. Extracts of diseased plant tissue which were passed through filters small enough to stop the passage of bacteria were found to transmit the disease to other plants. The agents of disease in the filtrate were termed **filterable viruses** and they were not visible under light microscopes of the highest resolution.

The invention of the electron microscope enabled the structure of viruses to be studied. Viruses vary in size from 10 to 300 nm. In general they consist of a nucleic acid core surrounded by a well organized protein **capsid**, usually composed of subunits or **capsomeres** arranged in an orderly fashion. In some viruses, such as those causing mumps and influenza, a further, less well organized coat, or **envelope**, is present. This coat can bear surface projections which may have a role in effecting the entry of the virus into the host cell (Fig. 2.31).

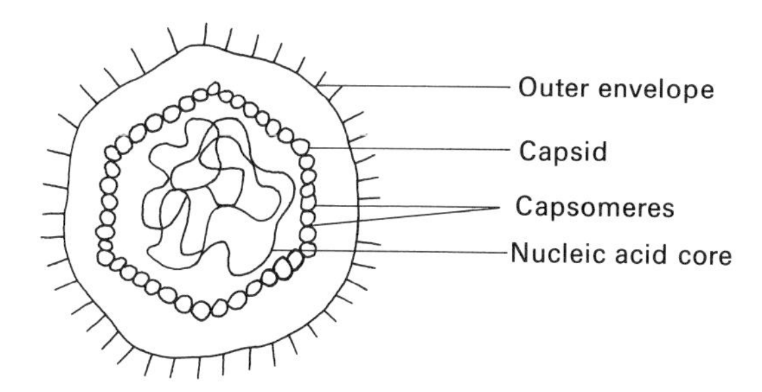

Fig. 2.31 Generalized structure of a virus of the influenza type.

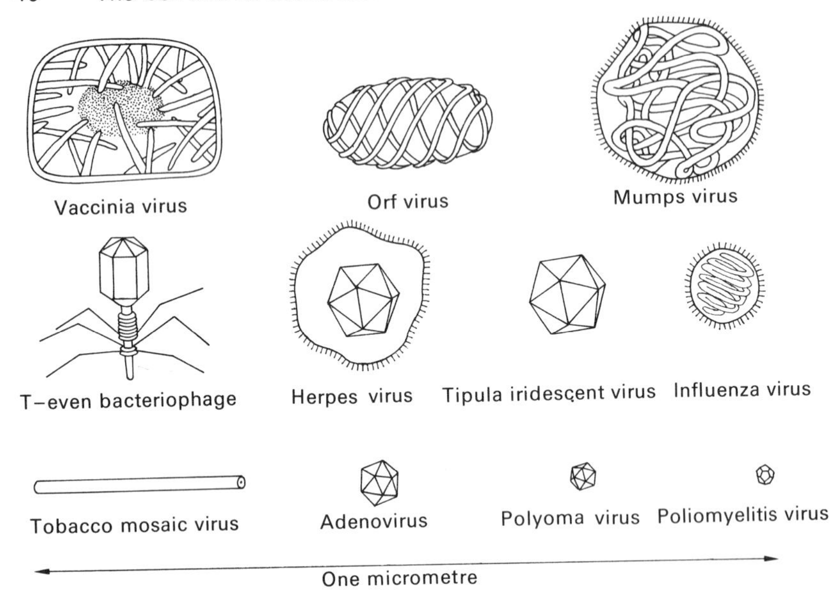

Fig. 2.32 Relative sizes of viruses referred to 1 μm scale at base of diagram. Types of symmetry shown are: *Cubic*, shown by the five polyhedral viruses, herpes, tipula iridescent, adenovirus, polyoma, and poliomyelitis; *helical*, or screw axis, shown by tobacco mosaic and the internal components of the mumps and influenza viruses; *complex symmetry* shown by vaccinia, orf, and T-even bacteriophage. (From R. W. Horne, *The Structure of Viruses*, Copyright © 1963 by Scientific American, Inc. All Rights Reserved.)

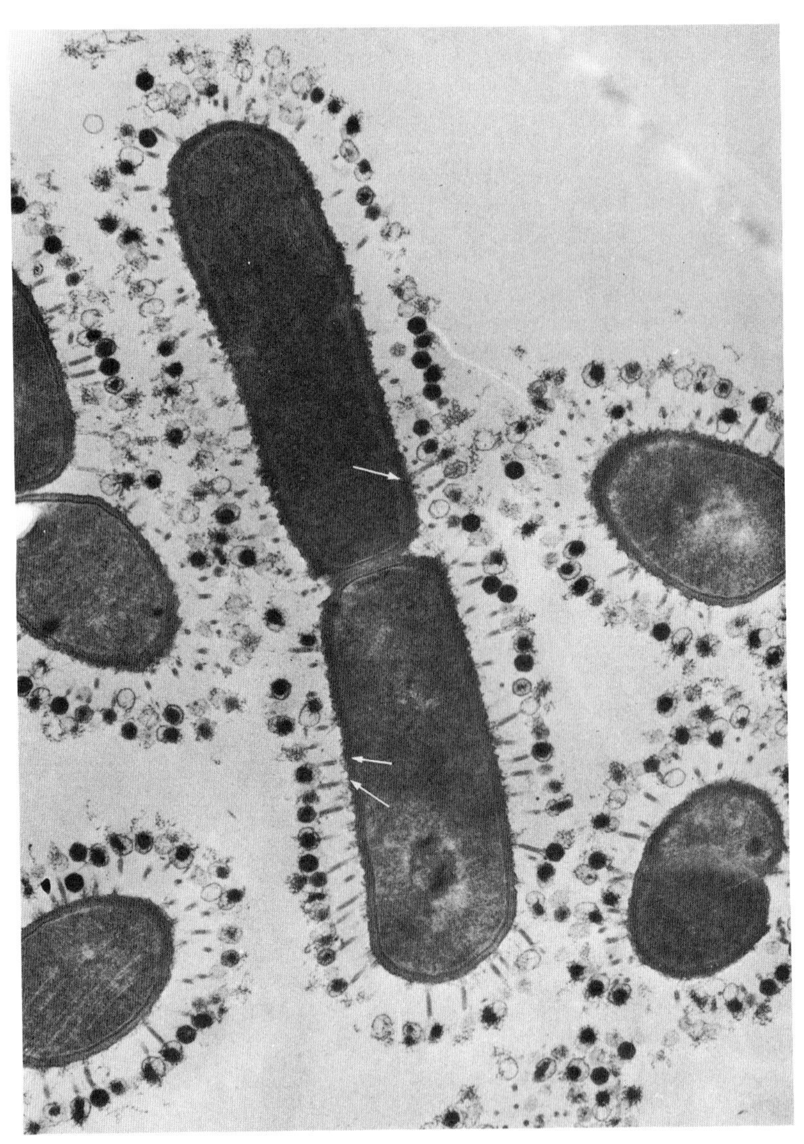

Fig. 2.33 E.M. of bacteriophage in a heavy infestation of *Bacillus subtilis* (Magnification × 36 000). Some of the polyhedral heads are empty, their contents having been injected into the bacterial cells. Base plates are visible in some places (arrowed). (Courtesy of Dr. R. G. Milne.)

Within this general framework viruses show considerable variation in structure (Figs. 2.32, 2.33, 2.34). The nucleic acid core may be DNA, as in the case of most of the viruses which invade animal cells, or RNA, as in most plant viruses. The capsids of some viruses, such as tobacco mosaic virus, are rod-shaped whereas others are polyhedral. Into the latter category fall the herpes virus, which is found in man associated with 'cold sores' round the mouth, and the Epstein–Barr virus which is found in human tumour cells in the condition known as Burkitt's or African Lymphoma.

Perhaps the most elaborate viral structure is to be found in the bacteriophage viruses (Fig. 2.33). The core material of bacteriophage is DNA, which is contained in an elongated, polyhedral head. To this head is attached a tubular tail-piece on which there are numerous appendages. When the bacteriophage attacks a bacterium the

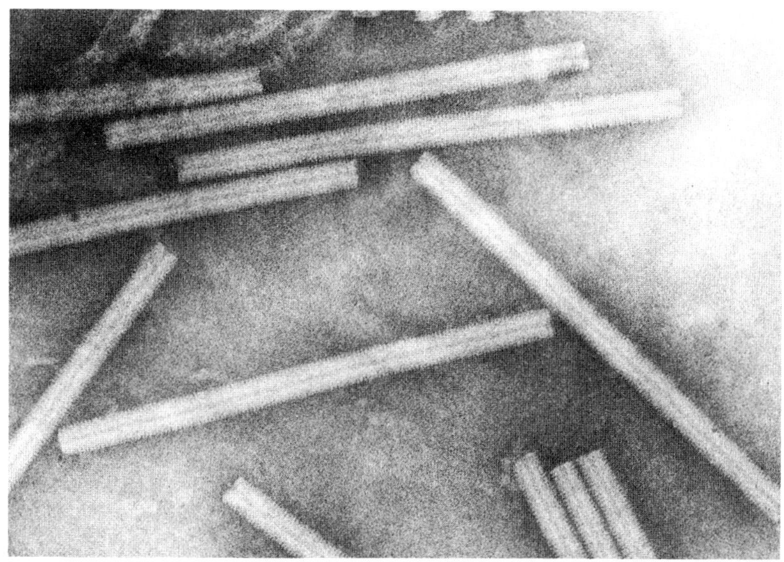

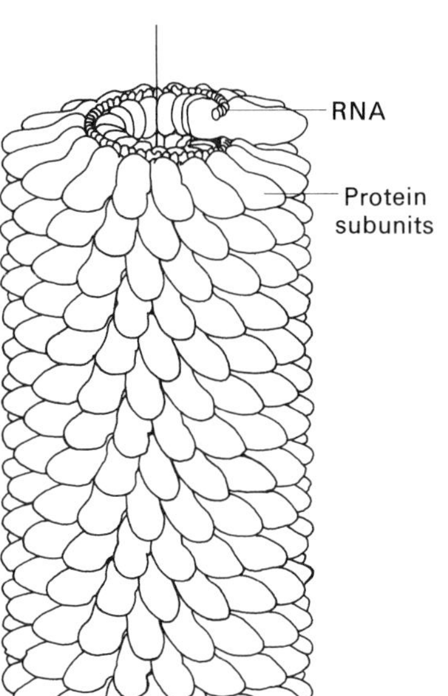

Fig. 2.34 (a) E.M. of tobacco mosaic virus. Magnification × 220 000. (Courtesy of Dr. J. T. Finch.) (b) Drawing of roughly one tenth of the length of the virus particle (after Dr. D. L. D. Casper), showing the RNA spiral backbone and the protein subunits, or capsomeres. (Evidence reviewed by Durham, 1971.)

appendages support the tail perpendicular to the cell wall and the tip of the tail digests a hole through it. A contractile sheath round the tail then contracts and the DNA is injected through the bacterial membrane like a hypodermic needle.

The method of entry of other viruses into their respective types of host cell (some viruses are species specific, some even tissue specific) is not fully understood. Dourmashkin and Tyrrell (1970) have produced good evidence that influenza virus can enter the epithelium lining the trachea (Fig. 2.36) through the cilia whose normal role is to sweep along the mucus lining the respiratory tract.

The Virus as a Parasite

The number of genes contained in the nucleic acid core of the virus is considerably fewer than the genes in a cell of a free-living organism or even in a bacterium. The virus has no cytoplasm and has no genetic capacity to code for the enzyme systems, such as those involved in energy interchange, which allow organisms to respire, excrete, divide and pump ions. In that sense viruses have no independent living existence. Outside their host cell viruses are inert, can be crystallized and show none of the properties of interchange with their environment that characterize living organisms. Yet from this crystallized state they retain their capacity for viral activity when reintroduced into cells.

The origin of viruses is obscure (Harrison, 1971) but in exhibiting the loss of genetic capacity for those features duplicated by their host they conform to the principle shown by all parasites. However, such is the extent of their genetic deficiencies they could be termed the 'ultimate parasites'.

Viral Activity in the Cell

In principle viruses operate by commandeering the metabolism of the host cell. The nucleic acid core gains entry either by injection, in the case of the bacteriophage, or possibly by the activities of the protein envelope, and instructs the host cell cytoplasm to make viruses. Since viral nucleic acids are, as a rule, either double-stranded DNA or single-stranded RNA it follows that both forms can convey genetic information sufficient to code for replication and for the synthesis of capsid and envelope proteins.

The mechanism by which the virus is able to dominate the metabolism of the cell is not understood. In some cases the host cell metabolism continues side by side with the construction of new virus but in others the host cell chemistry is wholly diverted to virus

Fig. 2.35 Blisters and discoloured areas on the leaf of tobacco are symptoms of tobacco mosaic infection.

formation.

Some viruses are lysogenic. They invade a host cell which, usually rapidly, produces a quantity of viruses in its cytoplasm. The cell membrane then ruptures, killing the cell and liberating the viruses which then invade surrounding cells. Other viruses are less virulent. Their entry into the host cell is not necessarily followed by rapid replication and cell death. It can result in passive occupation or in cell transformation. In these cases the metabolism of the host organism may influence the activity of the virus (the appearance of cold sores in times of mental stress and the appearance and disappearance of 'warts', which are probably virus induced benign tumours, are examples of this aspect of the parasite-host relationship).

An important aspect of cellular transformation is the ability of viral nucleic acid to enter, and be incorporated into, the host cell genome. In 1964 Howard Temin proposed, and produced circumstantial evidence for, a scheme whereby the RNA of some tumour-inducing viruses could produce complementary DNA (Fig. 2.37b). This DNA would then transform host cells into tumour cells by entering the host nucleus and combining with the chromosomes. This clearly contradicts what has been termed biology's 'central dogma', namely that there is a one-way instructional route in protein synthesis from DNA to mRNA to ribosomes to proteins (Fig. 2.37a and p. 39). This hypothesis of inverted transcription, or 'Teminism', while for some years regarded with scepticism, was eventually confirmed by the discovery of RNA-primed DNA polymerase specifically associated with RNA tumour viruses and by molecular hybridization techniques for matching viral RNA and portions of host cell DNA (reviewed by Allison, 1970).

The consequences of this discovery are obvious, if prospective. Since Rous isolated a virus from cancerous growths in chickens in 1911 (the Rous sarcoma virus) a wide range of animal cancers have been associated with viruses (Figs. 2.38 to 2.41). In man, leukaemia, Burkitt's lymphoma, breast cancer and cervical cancer strongly suggest a viral cause. The discovery of a possible mechanism whereby a normal cell is transformed into a cell with uncontrolled mitotic activity by insertion of genes of viral origin immediately suggests

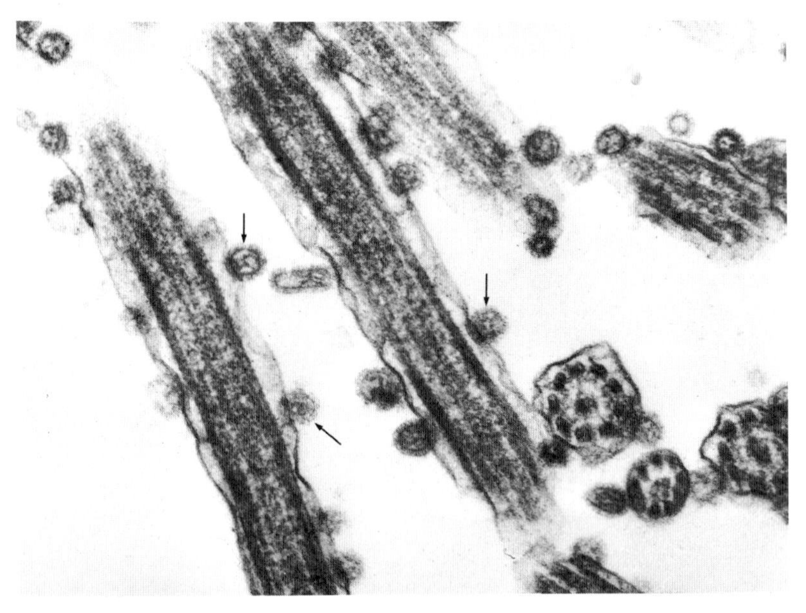

Fig. 2.36 E.M. of influenza viruses (arrowed) associated with the cilia of an epithelial cell from trachea grown in organ culture. (Courtesy of Dr. R. Dourmashkin.)

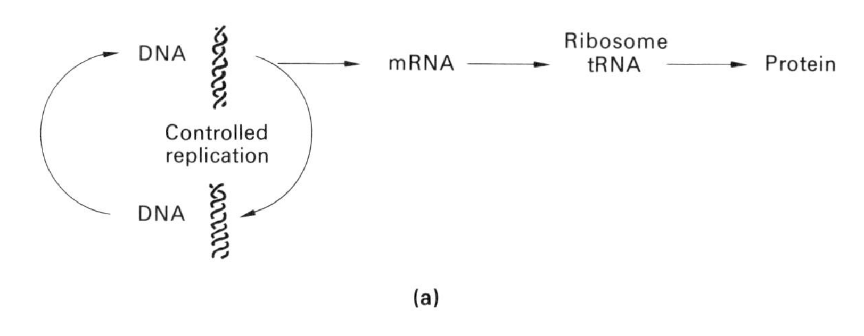

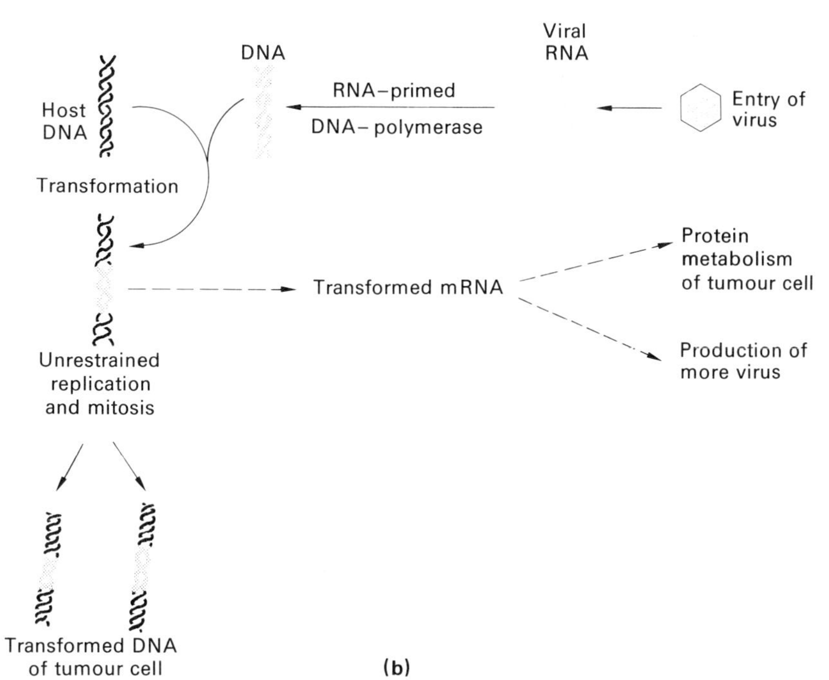

Fig. 2.37 (a) Conventional sequence in replication and protein synthesis. (b) The Temin hypothesis of inverted transcription leading to transformation.

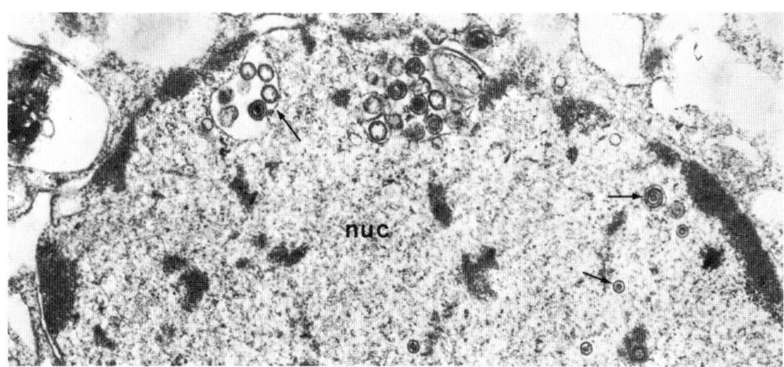

Fig. 2.38 E.M. of part of a macrophage cell from a chicken suffering from Marek's disease. Virus particles (arrowed), some with double coats, can be seen in the nucleus and in what seems to be pinocytic-like invagination of the nuclear membrane. (Courtesy of Dr. J. G. Campbell.)

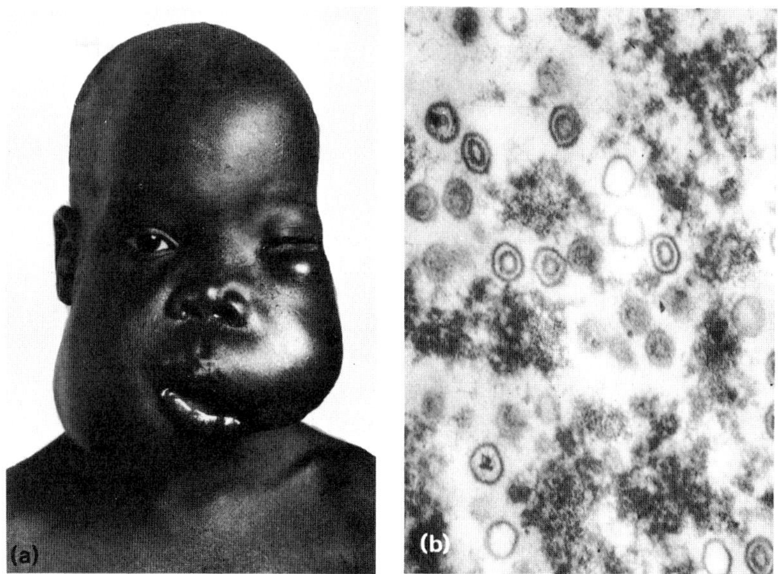

Fig. 2.39 (a) Photograph of an African boy suffering from a tumour, or neoplasm, of the jaw, a condition known as Burkitt's or African lymphoma. (Courtesy of the Wellcome Trustees.) (b) Epstein-Barr virus extracted from the cells of Burkitt's lymphoma and believed to be a causative agent in the tumour. The virus is of the Herpes type, consisting of apolyhedral capsid and an outer envelope (see Figs. 2.31 and 2.32.)

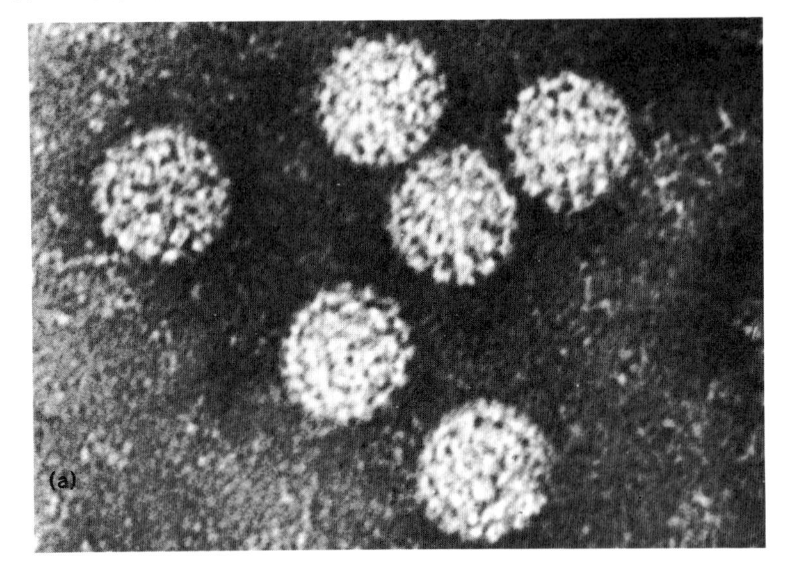

Fig. 2.40 (a) The structure of simian virus (SV-40) and (b) the induction of three tumours in a hamster injected subcutaneously with the virus. (Courtesy of Dr. A. C. Allison.)

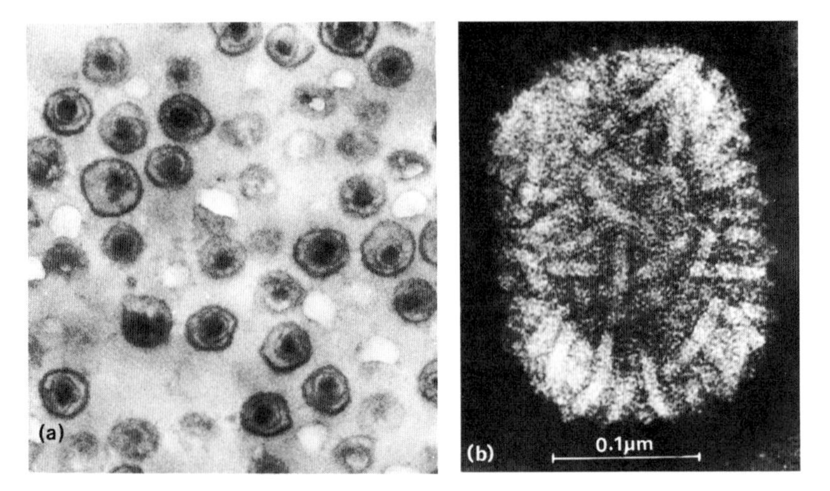

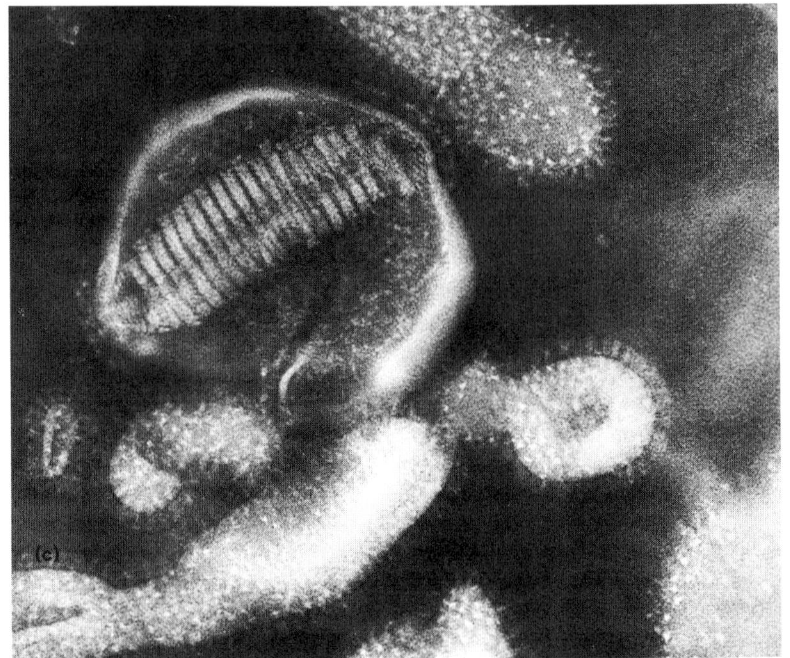

Fig. 2.41 (a) Rous sarcoma virus which is a tumour-causing virus in chickens; (b) the pox virus, vaccinia and (c) the virus of Hong-Kong 'flu. (Courtesy of Dr. R. Dourmashkin.)

chemotherapy as a possible means of blocking the specific enzyme by which the insertion is made.

2.5 The Properties of Enzymes

Almost all the biochemical changes occurring in cells are regulated by enzymes. These are complex proteins whose tertiary configurations give them a high degree of specificity of action in spite of the fact that the number of categories of biochemical action (hydrolysis, dehydrogenation, dehydration synthesis, etc.) is relatively small.

Most enzymes act catalytically, by reducing the activation energy required by the reactants, usually by forming a more reactive 'activation complex' (Fig. 2.42).

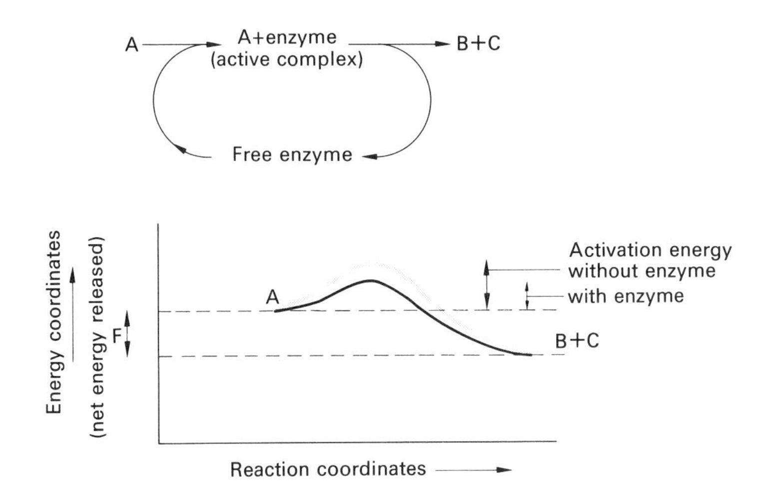

Fig. 2.42 Role of an enzyme in an exergonic reaction A→B+C.

Since the temperature of living cells is below that where organic molecules would react spontaneously, enzymes speed up, and because of their specific activity act as regulators of, rates of reaction. Most cellular enzymes appear to have a limited life and must be synthesized intermittently, or even continuously (p. 39), if the reactions with which they are concerned are to continue in the cell.

Mode of Action of Enzymes

Enzymes, proteins with molecular weights of the order of 10^4 and 10^5, are considerably larger than their substrate molecules. There are present on the enzyme molecule '**active sites**' which bring together in the correct position substrate molecules for synthesis or splitting. It is the detailed form of the active site that accounts for the high specificity of enzyme action; any alteration in the shape of the site or the molecule as a whole is likely to alter the enzyme's catalytic capability (see Regulation of Metabolism, p. 47). For example, the action of succinic dehydrogenase (which oxidizes succinic acid to fumaric acid in the Krebs Cycle) could be expressed in the form of a model (Fig. 2.43):

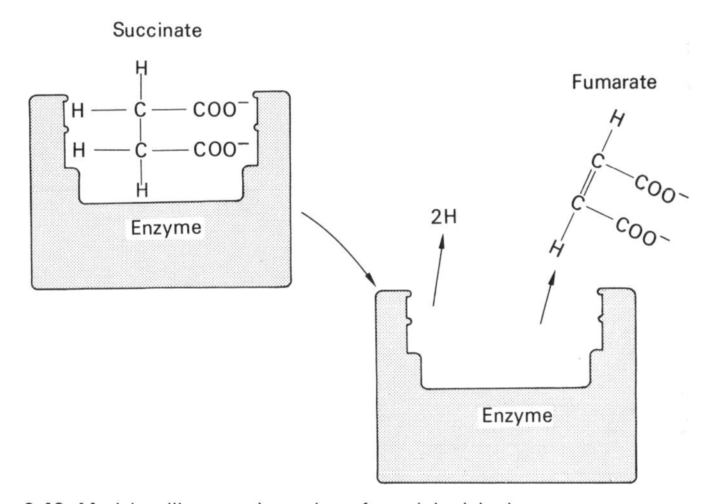

Fig. 2.43 Model to illustrate the action of succinic dehydrogenase.

Competitive Inhibition of Enzymes

If a compound which is not a substrate competes with the substrate for occupation of the active site an inert complex is formed and the activity of the enzyme is blocked. For example, considering the action of succinic dehydrogenase described above, **malonic acid** acts as a competitive inhibitor. This substance is sufficiently like succinic acid to occupy the active site of the enzyme but does not undergo dehydrogenation and is not released from the site (Fig. 2.44).

The inert complex in this case is not stable and the substrate and inhibitor compete on the basis of their respective concentrations. Thus, with a constant enzyme concentration an increase in substrate concentration would give rise to an increase in the product, fumaric

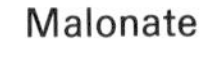

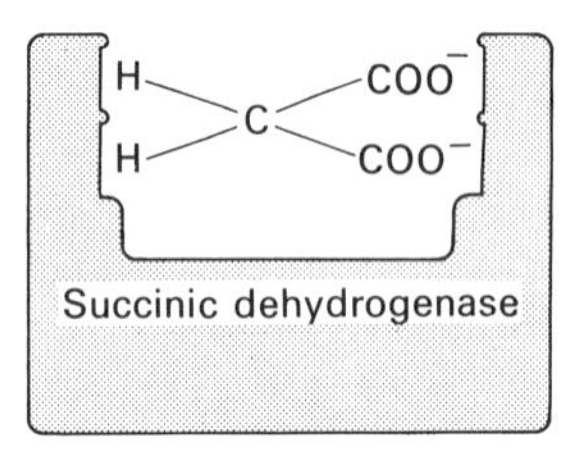

Fig. 2.44 Model to illustrate an inert complex, the inhibition of succinic dehydrogenase by malonic acid (compare Fig. 2.43).

acid. **Non-competitive inhibition** occurs when an inhibitor occupies an active site so strongly that it cannot be displaced whatever the substrate concentration. It is thought that some antibiotics act in this manner by inhibiting bacterial cell wall formation. Some nerve gases, recently developed as war tools, stop nerve synapse function by inhibiting acetylcholine esterase in a non-competitive manner.

Effects of Enzyme and Substrate Concentrations on Rate of Reaction

The rate of an enzymic reaction is directly proportional to the concentration of enzyme, providing the substrate is in excess (Fig. 2.45).

In a constant enzyme concentration increase in substrate concentration increases the rate of the reaction when the substrate is at low concentrations. At higher substrate concentrations, however, all the active sites on the enzyme are occupied and the enzyme is working at maximum 'molecular activity', so further increases in substrate concentration do not affect the reaction rate (Fig. 2.46).

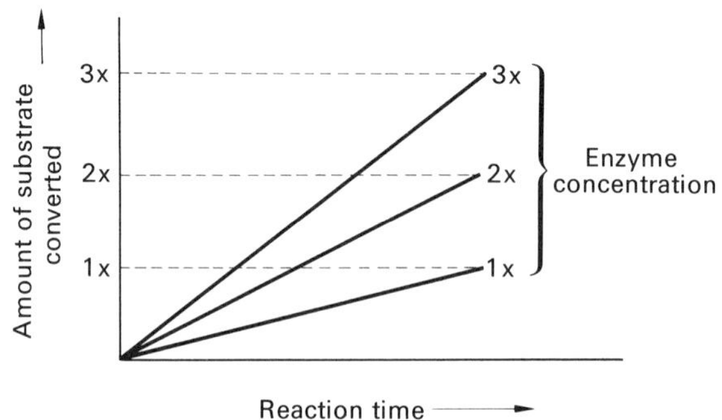

Fig. 2.45 Effect of enzyme concentration on the rate of enzymic reactions.

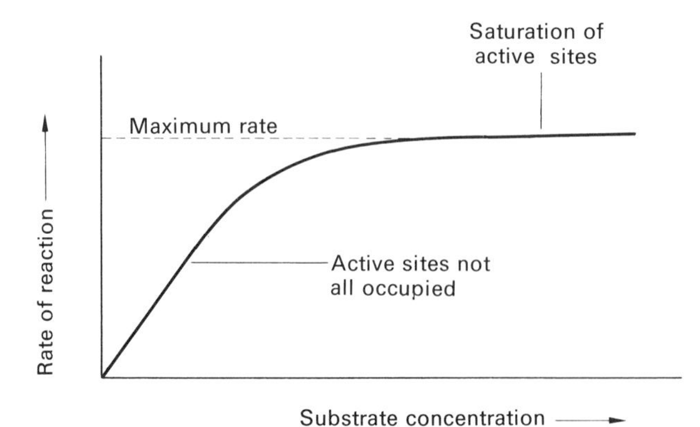

Fig. 2.46 Effect of substrate concentration on the rate of enzymic reactions.

Effect of Temperature on Enzymic Reactions

At ordinary climatic temperatures the rates of organic reactions are affected by changes in temperature in the same way as any chemical reactions. But high temperatures alter the tertiary structure of proteins; disulphide bridges (p. 33) are reduced to sulphydryl groups —SH. Consequently globular proteins denature to an elongated shape and the active sites of the enzyme are lost. Death through heat shock involves this kind of enzyme denaturation and most animals, notably birds and mammals, have physiological or behavioural methods of avoiding high body temperatures (Fig. 2.47).

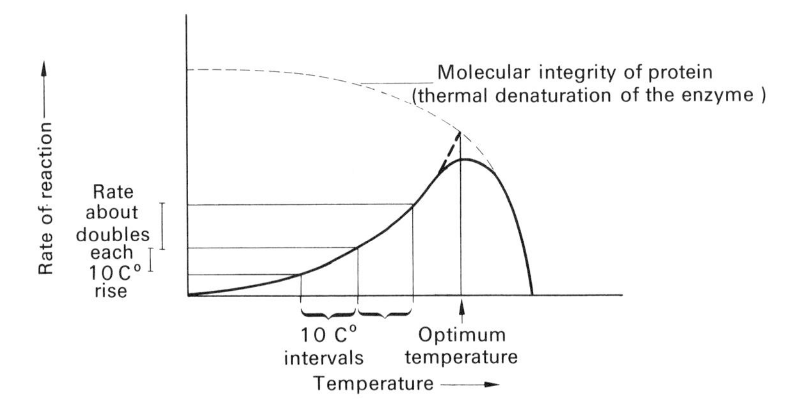

Fig. 2.47 Effect of temperature on enzymic reactions.

Effect of pH on Enzymic Reactions

The catalytic activity of an enzyme is affected by pH because amino acids show a **zwitterion** effect. A zwitterion is a polar molecule able either to accept or donate a hydrogen ion and thus become either positively or negatively charged, respectively (Fig. 2.48).

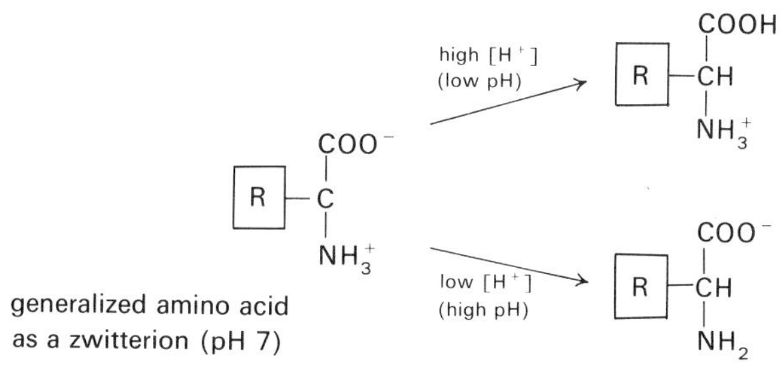

high $[H^+]$ (low pH): R—CH($COOH$)(NH_3^+)

generalized amino acid as a zwitterion (pH 7): R—C(COO^-)(NH_3^+)

low $[H^+]$ (high pH): R—CH(COO^-)(NH_2)

Fig. 2.48 Amino acid as a zwitterion.

Thus in a medium containing many hydrogen ions the appearance within the protein molecule of adjacent + charges (or − charges in a high pH medium) can distort the molecule by repulsion. In this way active sites themselves are distorted and the activity of the enzyme is altered. The graph of reaction rate against pH tends to be fairly symmetrical, as might be expected (Fig. 2.49).

Different enzymes, particularly extracellular ones, have different pH optima, since the most effective tertiary form may be promoted by more or fewer charges within the molecule.

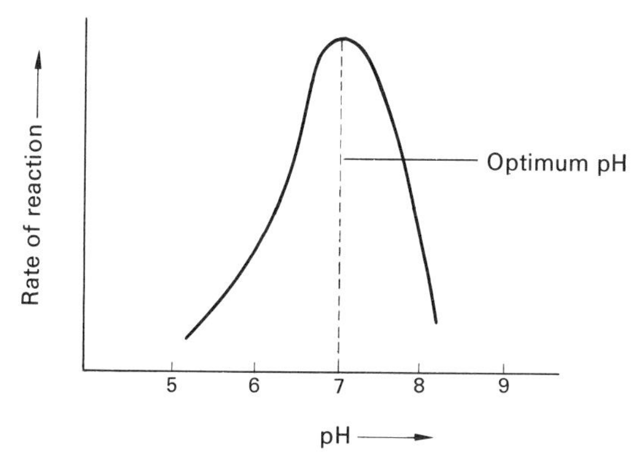

Fig. 2.49 pH/reactivity curve of urease.

Enzyme Cofactors

A number of important enzymes exist as catalytically inert proteins (apoenzymes) which are rendered active by association with a **prosthetic group** (or **coenzyme**). NAD and FAD (p. 51) are examples in which the prosthetic groups are B group vitamins. Some metallic cations such as K^+, Ca^{2+}, Mn^{2+}, act as enzyme cofactors.

Regulation of Metabolism by Enzymes

The phenomenon of **allostery**, whereby the configuration of a protein molecule is changed by the adsorption of some effector molecule on to its surface, is increasingly recognized as an important principle in biological systems. The addition of an effector to an enzyme, remote from the active site, might distort the molecule and render it either less active, or more so. In other words it could act as a regulator of the enzyme's activity.

In glycolysis the enzyme phosphofructokinase catalyses the reaction in which a second phosphate group is added to fructose phosphate:

$$\text{fructose-6-P} + \text{ATP} \longrightarrow \text{fructose-1,6-di-P} + \text{ADP}$$

and the enzyme appears to be inhibited allosterically by its own substrate, ATP, when the latter is in high concentration (Fig. 2.50).

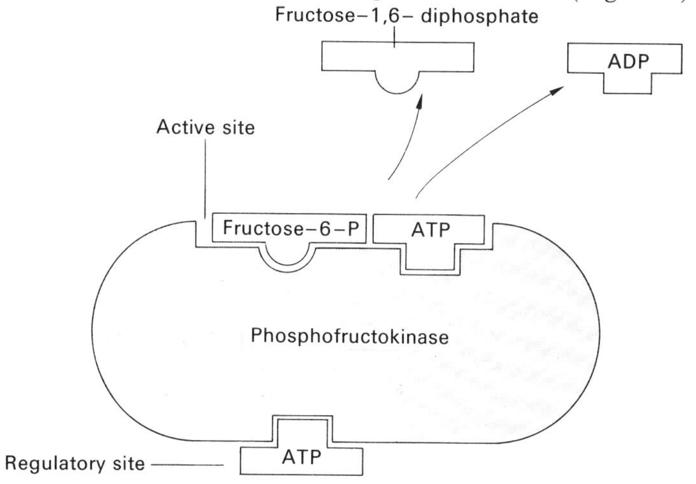

Fig. 2.50 Model of possible allosteric role of ATP in the regulation of enzyme activity.

If such a system operates in cells it seems likely that ATP would be adsorbed preferentially on to the active site. But at high concentrations of ATP the regulatory site also would be occupied, reducing the molecular activity of the enzyme and reducing the output of fructose-1,6-di-P. Thus when the cell is 'resting' a high level of ATP would depress glycolysis; conversely, when the cell is utilizing ATP to do work, the enzyme would be most active and glycolysis would proceed. (Weitzman, 1969.) (Fig. 2.51).

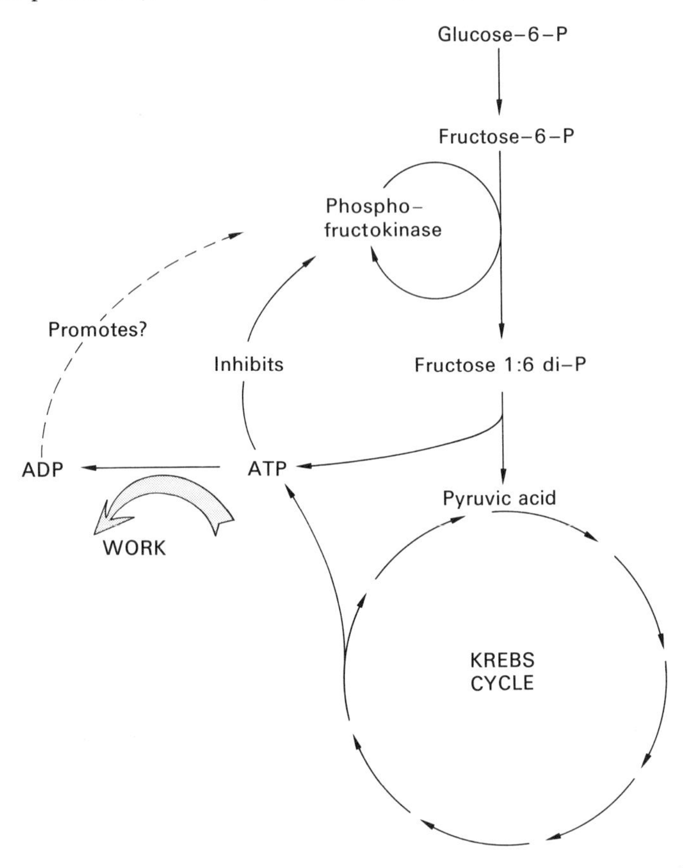

Fig. 2.51 How the allosteric effect may serve to regulate the respiratory pathway. The accumulation of ATP in the cell inhibits the enzyme phospho-fructokinase. When the cell increases its work and uses up ATP the pathway is switched on again. The switch could be a positive one if ADP allosterically promotes the activity of the enzyme.

3 Energy Interchange in Cells

In this chapter some of the chemical processes concerned with energy in the cell are discussed. In the previous chapter the roles of macromolecules as determiners of biochemical activities were described, and in this chapter some of those activities are examined in more detail. The transformation of energy from one form to another, particularly the transfer of chemical bond energy into work, is the basis of some key patterns in biology. Knowledge of the ways in which energy moves through organisms and through their cells, for example, is fundamental to an understanding of the way in which energy moves from one organism to another in the ecology of the biosphere.

This area of 'ecological energetics' is explored later in the book, in Chapter 10, where the relationships between ecology, genetics and evolution are discussed, but at this stage we must first deal with energy transfer at the cellular level.

3.1 Work and ATP in the Cell

Living cells perform work in a variety of ways. Some cells contract and execute physical work by applying a force over a given distance, as in the mechanics of locomotion. All cells pump ions and molecules against concentration gradients and do work by making molecules into larger and more orderly configurations in what are termed **endergonic** (energy-adding) syntheses.

The biochemical reactions by which cells obtain the energy to do all these forms of work may seem rather complicated. The underlying principles, however, are relatively straightforward and some knowledge of them is vital to the comprehension both of the activities of the whole organism with respect to nutrition and gas exchange and of the ways in which organisms interact and pass on energy in food chains, in what is termed the energy ecology of the biosphere (Chapter 10). In this book only the principal stages will be described; more detailed accounts occur in works devoted to biochemistry, such as Harrison (1965) and Nicholson (1970).

Based on the observations of Lipmann (1941) cells are commonly thought to obtain energy to perform chemical and mechanical work by the hydrolysis of certain high-energy or energy-rich compounds. The most common energy-rich compound in cells is the phosphorylated nucleotide **adenosine triphosphate**, or **ATP**. When the pyrophosphate bond of the terminal phosphate group of ATP is hydrolysed (Fig. 3.1) the energy yield greatly exceeds that from the hydrolysis of a simple orthophosphate bond, thus:

$$\text{ATP} + H_2O \longrightarrow \underset{\text{adenosine diphosphate}}{\text{ADP}} + H_3PO_4 \quad (\Delta F = -33.6 \text{ kJ mol}^{-1})$$

$$\text{glucose-6-phosphate} + H_2O \longrightarrow \text{glucose} + H_3PO_4 \quad (\Delta F = -14 \text{ kJ mol}^{-1})$$

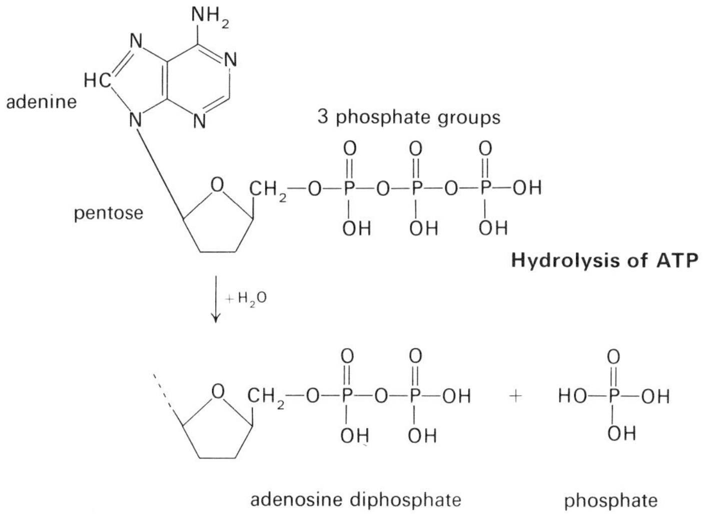

Fig. 3.1 The structure of adenosine triphosphate and its hydrolysis.

Therefore, when the cell does work ATP is used up. This ATP is then reconstituted in two ways. The first way is common to plants and animals and is termed **respiration**. The second occurs only in the chlorophyll-containing cells of green plants in the presence of light, and is called **photophosphorylation**.

It should perhaps be added that there are alternative views of the role of ATP in cellular reactions (reviewed by Banks, 1970) which hold that it has no function to provide energy by hydrolysis. In the account of energy interchange in this chapter, however, the conventional view will be adopted, that is the formation of ATP by progressive exergonic steps. In the event of Banks and her colleagues' views being substantiated by the demonstration that exergonic reactions power cell metabolism direct, without the intervention of ATP as a form of 'energy currency', the stages of metabolism outlined here will not materially be affected.

3.2 ATP Formation by Respiratory Pathways in the Cell

In respiration the energy for the restitution of ATP from ADP and inorganic phosphate comes from the progressive dehydrogenation of complex molecules containing energy-yielding bonds, notably glucose:

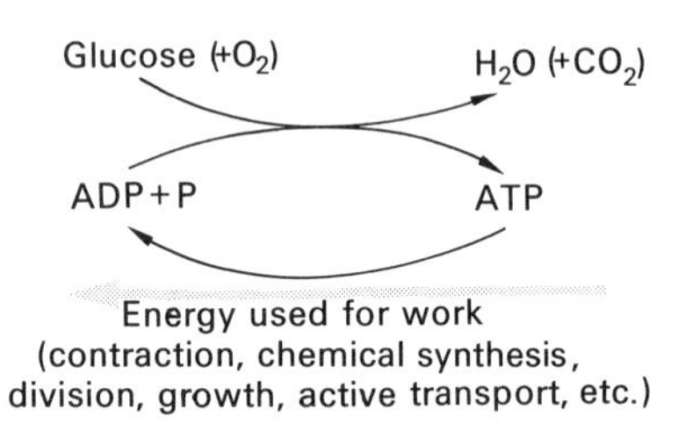

In a few bacteria and fungi this dehydrogenation sequence is incomplete (**anaerobic respiration**) terminating at lactic acid, as in the bacteria which cause milk to go sour, or at ethyl alcohol, as in the brewing yeasts. In other organisms oxygen is used as a terminal hydrogen acceptor, forming water. This enables the dehydrogenation sequence to proceed further, to CO_2 and H_2O, with consequently a greater energy yield (**aerobic respiration**). The latter process can be said to be nineteen times more efficient since the net yield aerobically is 38 moles of ATP per mole of glucose, compared with 2 moles of ATP produced anaerobically.

The respiration of the substrate glucose can be looked at in three sequences, or stages.

Stage 1 Glucose to Pyruvic Acid

This sequence of reactions can be accomplished in several ways, of which the two most important are the Embden–Meyerhof **glycolysis** and the Warburg–Dickens **pentose phosphate pathway**. There is evidence that both pathways are used extensively by plants and animals; within one organism some tissues may favour one pathway and other tissues the other.

The pentose phosphate pathway is aerobic. As the dehydrogenation proceeds the hydrogen acceptor passes the hydrogen through the cytochrome system (p. 56) where it is oxidized to water. The important features of this pathway are that it is, in part, the reverse of the dark reactions of photosynthesis and it is a means of relating the metabolism of hexoses and pentoses.

Glycolysis, on the other hand, is anaerobic because there exists in the pathway a means whereby the hydrogen acceptor can become re-oxidized by passing on its hydrogen to pyruvic acid or its derivative. Both glycolysis and the pentose phosphate pathway can be seen as converting six-carbon units (glucose) to three-carbon units (glyceraldehyde-phosphate and pyruvic acid).

The Embden–Meyerhof Pathway

The reactions in the Embden–Meyerhof pathway are relatively direct. Glucose is phosphorylated in steps, by successive ATP activity, to fructose-1,6-diphosphate. This 6-carbon molecule is unstable and readily cleaved into two 3-carbon sugars, or triose-phosphates. Two kinds of triose-phosphate are formed, glyceraldehyde-phosphate and dihydroxyacetone-phosphate, which are isomers and readily interconvertible by an isomerase enzyme.

Only glyceraldehyde-phosphate is concerned in further glycolytic stages. As it is used up it is replenished by its isomer so that both forms of the 3-carbon sugar are involved. In effect, therefore, each mole of 6-carbon sugar entering the glycolytic sequence forms two moles of glyceraldehyde-phosphate.

Glyceraldehyde-phosphate then enters several reactions during which dehydrogenation and dephosphorylation take place. The final product is pyruvic acid. In these reactions ATP is formed from ADP and inorganic phosphate. Pyruvic acid forms a junction in carbohydrate metabolism. In anaerobic respiration pyruvic acid can be reduced to form either acetaldehyde and ethyl alcohol or lactic acid.

Alternatively it can pass via acetyl co-enzyme A into the aerobic sequence of the Krebs Cycle.

Figure 3.2 shows that 2 moles of ATP are used in the early stages to phosphorylate glucose and glucose monophosphate, and 4 moles of ATP are gained in the conversion of glyceraldehyde phosphate to pyruvic acid. This gives a net gain in anaerobic respiration of 2 moles of ATP per mole of glucose.

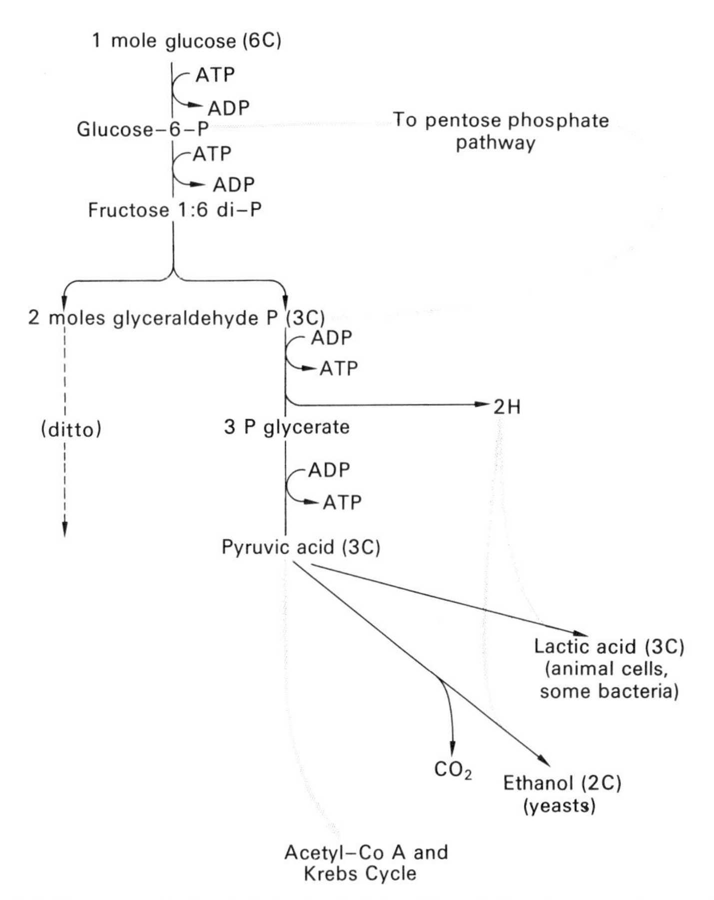

Fig. 3.2 Summary of glycolysis in the Embden-Meyerhof pathway, and anaerobic respiration (fermentation).

Dehydrogenation

The energy-yielding reactions in the respiratory pathway are those involving the removal of hydrogen. In aerobic respiration the hydrogen is passed along a chain of hydrogen acceptors, pyridine nucleotides, flavoproteins and cytochromes, before combining with the final hydrogen acceptor, oxygen, to form water.

Two widely occurring hydrogen acceptors are coenzymes incorporating B group vitamins:

NAD (nicotinamide-adenine dinucleotide) contains niacin, and
FAD (flavine-adenine dinucleotide) contains riboflavin

The structure of nicotinamide-adenine dinucleotide is shown in Fig. 3.3. Parts of the molecule will immediately be recognizable, the adenine part, the pentose sugar and the phosphate, from the nucleotides which comprise nucleic acids (Chapter 2). In addition there is a second pentose which links the nucleotide with the nicotinamide

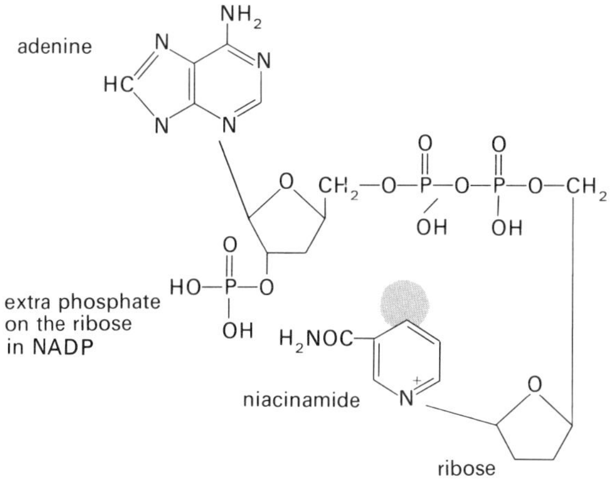

Fig 3.3 The structure of the hydrogen acceptor NAD (nicotinamide adenine dinucleotide) showing the position of the hydrogen acceptor site on the niacinamide residue (shaded) and the additional phosphate group which is present in NADP.

part. It is the ability of the pyridine ring in the latter to become reduced that confers upon the whole molecule the ability to act as a hydrogen acceptor and donor in oxidation and reduction reactions in the cell (Fig. 3.4).

In some reduction-oxidation reactions in the cell the phosphoryla-

the pyridine ring of NAD ⇌ (reduction / oxidation) reduced pyridine ring of NADH

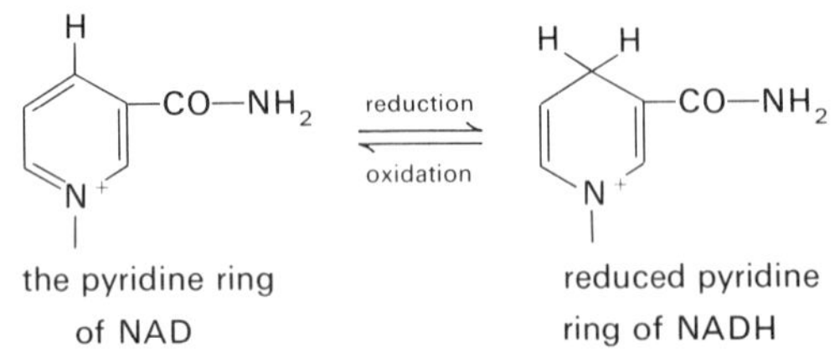

Fig. 3.4 Reduction and oxidation of the pyridine ring of NAD.

ted form of the coenzyme is used, namely nicotinamide-adenine dinucleotide phosphate, or NADP. In some books NAD and NADP may be described by their former names, DPN (diphosphopyridine nucleotide or Coenzyme I) and TPN (triphosphopyridine nucleotide or Coenzyme II).

An example of the role of NAD can be found in the pathway of anaerobic respiration or fermentation (Fig. 3.5). Dehydrogenation or the removal of hydrogen occurs in the step from glyceraldehyde-phosphate to 3-P-glycerate. The hydrogen is taken by NAD and used in the subsequent reduction of pyruvic acid.

The Pentose-Phosphate Pathway

This sequence (Fig. 3.6) provides an alternative pathway from fructose-6-phosphate to glyceraldehyde. It is considerably more complex than the Embden–Meyerhof pathway, but it is worth looking at in some detail because it is the way in which organisms provide the 5-carbon pentose sugars which have important roles in the activities of a variety of substances, such as nucleic acids and hydrogen acceptors. It also mirrors in part the route of the dark reactions of photosynthesis which will be described later in this chapter.

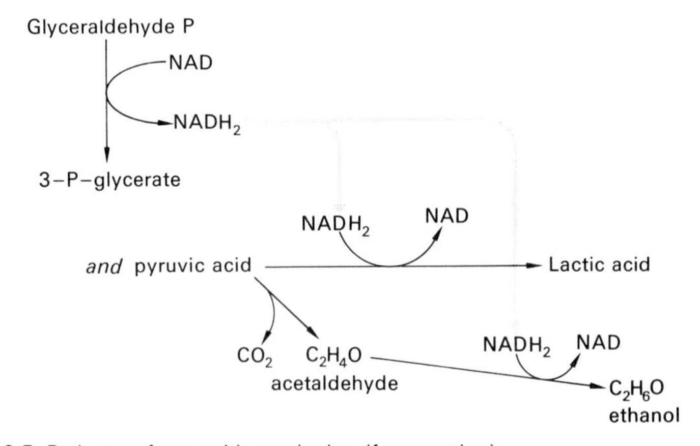

Fig. 3.5 Pathway of anaerobic respiration (fermentation).

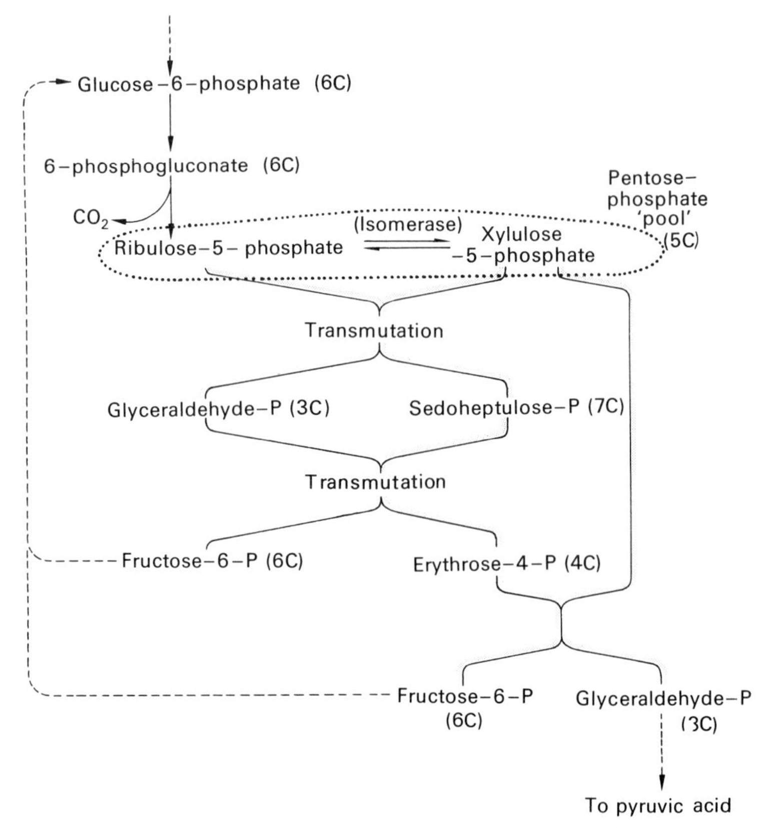

Fig. 3.6 Summary of the Pentose-Phosphate (Warburg-Dickens) Pathway. (The tinted dotted lines show that part which coincides with some of the dark reactions of photosynthesis, compare with Fig. 3.28.)

The 6-carbon sugar glucose-6-phosphate is dehydrogenated (oxidized) to 6-phosphogluconate which in turn loses one carbon in the form of CO_2 and becomes the 5-carbon sugar ribulose-5-phosphate. This sugar, or more correctly its aldopentose isomer ribose-5-

phosphate, is readily converted to another 5-carbon isomer xylulose-5-phosphate. This conversion is carried out by isomerase enzymes analogous to the hexose-isomerases which operate in the Embden–Meyerhof pathway between glyceraldehyde-phosphate and dihydroxyacetone-phosphate.

These two 5-carbon isomers interact and the 10-carbon unstable intermediate compound is transmuted to a 7-carbon sugar, sedoheptulose-phosphate, and the triose sugar glyceraldehyde-phosphate. These in turn interact and are transmuted to the 6-carbon sugar fructose-phosphate and a 4-carbon sugar erythrose-phosphate.

The erythrose-phosphate then interacts with a further molecule of xylulose-phosphate and the complex is transmuted to the familiar 6-carbon fructose-phosphate and the 3-carbon glyceraldehyde-phosphate. The glyceraldehyde-phosphate enters the glycolytic sequence and is dephosphorylated, forming ATP on its way to becoming pyruvic acid. The fructose-phosphate, together with the original molecule of fructose phosphate from the 7-carbon sugar, is reconverted to glucose phosphate for the cycle to be repeated.

This rather complicated pathway can be understood more readily by counting the carbons at each stage. At the level of ribulose-phosphate and its isomers (what might be termed the pentose-phosphate 'pool') there are involved three molecules of 5-carbon sugar, i.e., a total of 15 carbons. Through successive transmutations these emerge as two 6-carbon sugars and one 3-carbon sugar, again, a total of 15 carbons.

The pentose-phosphate 'pool' of 15 carbons is achieved by the combination of the two 6-carbon sugars that have been re-cycled plus one 6-carbon sugar which enters anew, or 18 carbons. From this, three CO_2 molecules are given off, leaving 15 carbons to enter the pool.

The pentose-phosphate pathway is summarized in Fig. 3.6. A more detailed arrow-diagram and description of the reactions is provided by Nicholson (1970).

The Role of Mitochondria in Respiration

There is now very good evidence (discussed on p. 231) that *Stages* 2 *and* 3 of respiration were made possible by an endosymbiotic association at an early stage of organic evolution. The endosymbionts which are thought to have entered primitive cells were probably bacteria-like and conferred upon the host cell the ability to respire **aerobically.** It is possible that they were the ancestors of mitochondria. A fuller discussion of the possible endosymbiotic origins of

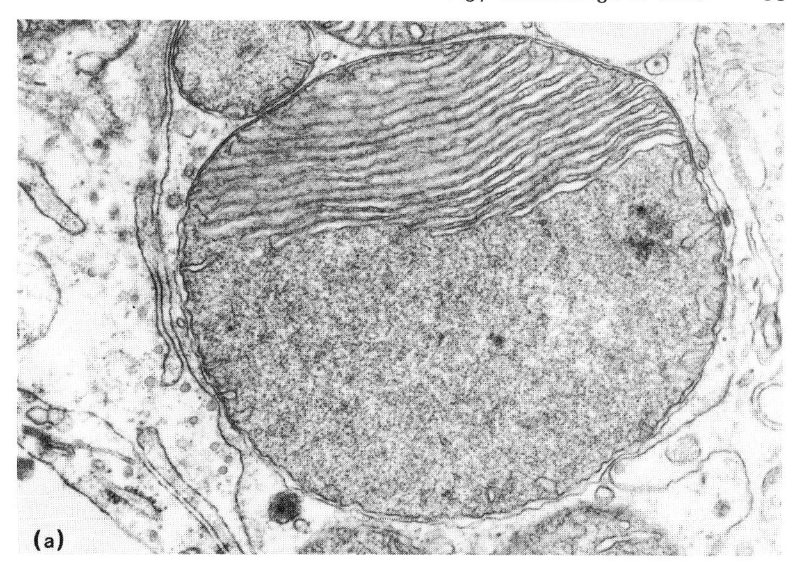

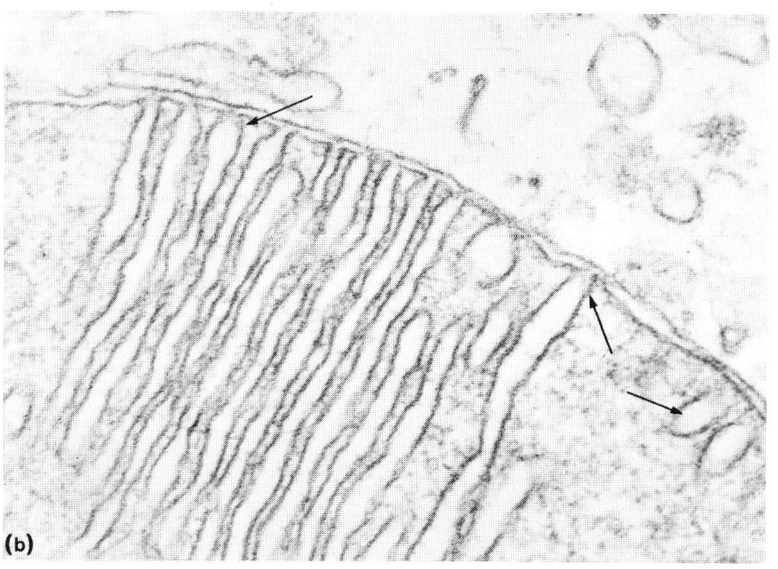

Fig. 3.7 Electron micrographs of mitochondria in kidney epithelial cells of a 12-week human foetus. Christa formation appears to be taking place by infolding of the inner mitochondrial membrane (arrowed). Magnification (a) × 30 000, (b) × 90 000. Courtesy of Mr. Alan Ross.)

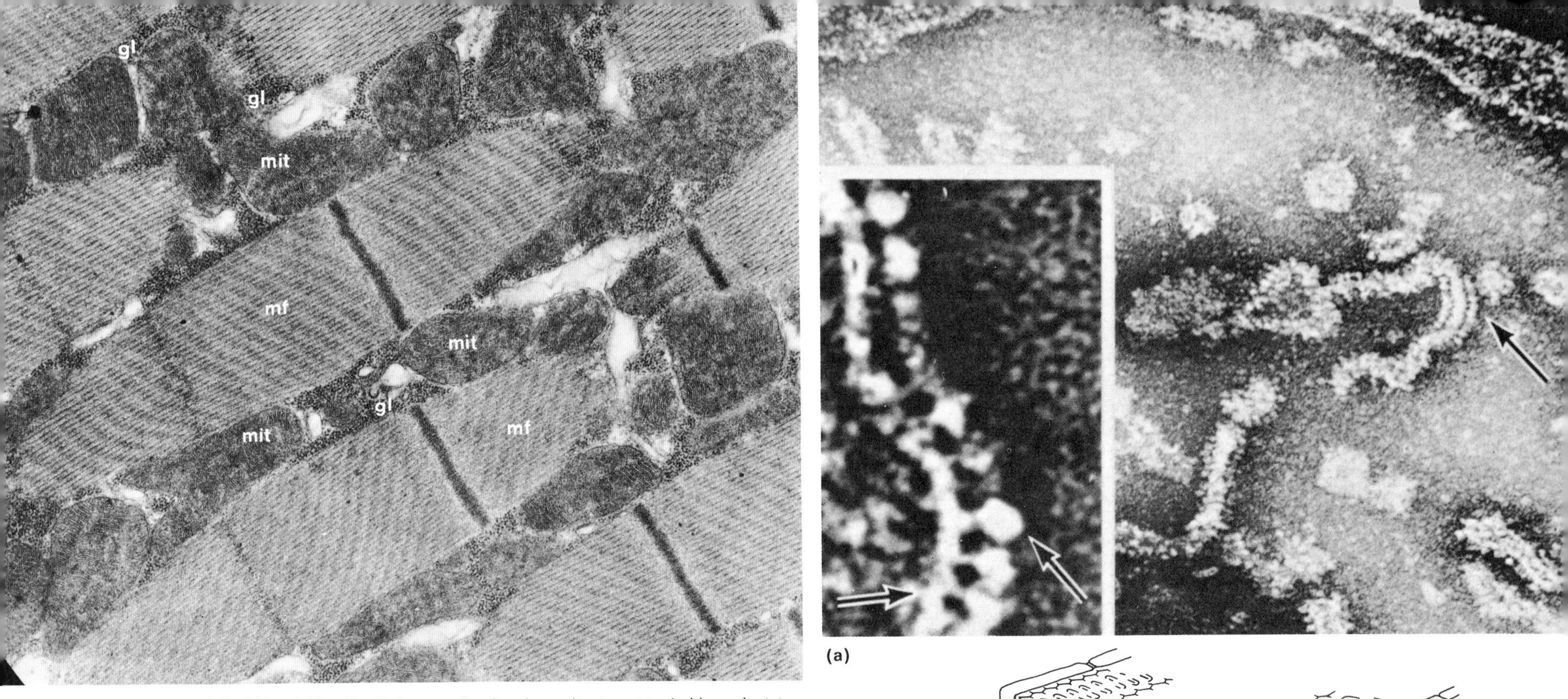

Fig. 3.8 E.M. of blowfly flight muscle showing mitochondria (mit) packed between the contractile myofibrils (mf). Glycogen granules (gl) are conspicuous. Magnification × 20 000. (Courtesy of Dr. M. A. Tribe.)

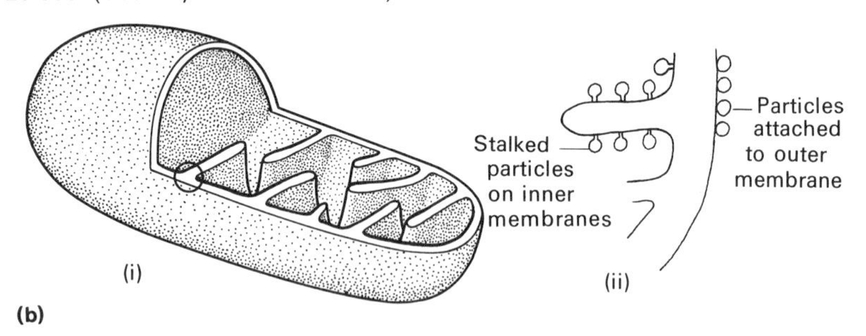

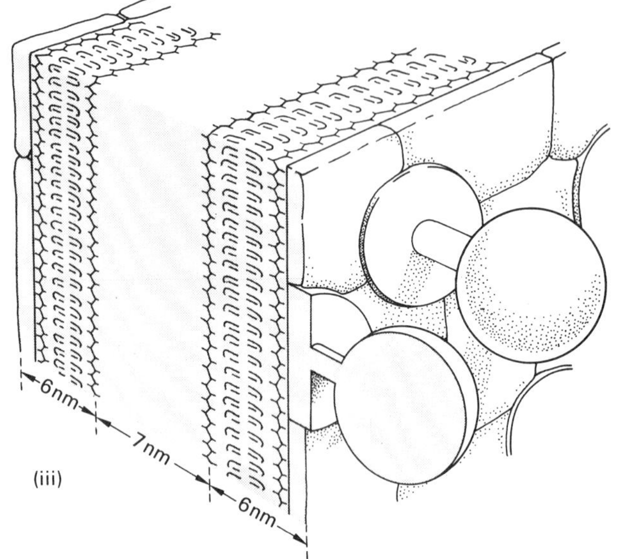

Fig. 3.9 (a) E.M. of rat heart mitochondrial membranes in a negatively stained preparation which shows stalked particles on the christae. (Courtesy of Professor H. Fernandez-Moran.) (b) Diagrammatic representation of the structure of the mitochondrion: (i) general structure showing cristae, (ii) position of elementary particles on surface of crista, and (iii) the molecular skeleton of the mitochondrion. (After Paul, 1965. Both (a) and (b) by permission of Heinemann, Limited.)

mitochondria and other cell inclusions occurs in Chapter 10.

Mitochondria occur most densely in the cytoplasm of cells that are actively working (Fig. 3.7). Individual liver cells may contain about a thousand mitochondria. In insect flight muscle the mitochondria are densely packed in alternate blocks between the contractile elements (Fig. 3.8). By contrast, cells that have a structural rather than a biochemical or dynamic role in the body, such as surface epithelia, have relatively few mitochondria.

Mitochondria were first described by Kolliker in 1850 as granules in muscle cells. Later, in 1898, Michaelis was able to relate mitochondrial function to respiration by showing that isolated mitochondria could produce colour changes in redox dyes. Although the membranous nature of mitochondria had been postulated as early as 1888 it was not until the advent of the electron microscope that their structure became clear. Mitochondria vary in shape but are essentially cylindrical. They consist of an enclosing outer membrane, and an inner membrane which is folded internally to form incomplete partitions, or **cristae** (see Fig. 3.9). Negative staining techniques have shown that the membranes of the mitochondria are covered by distinct spheres, or **particles**. These appear to be attached directly to the outer membrane, but are attached to the inner membrane by short stalks (Fernandez–Moran, 1962; Stoeckenius, 1963).

The activities of these particles are not completely defined, but it is known that the outer (stalkless) particles contain enzymes such as pyruvate oxidase which act upon the pyruvic acid produced by Stage 1, described above. In this way they introduce pyruvic acid into the sequence of reactions known variously as the Krebs Cycle (after its discoverer, Sir Hans Krebs), citric acid cycle, or tricarboxylic acid cycle.

Krebs Cycle enzymes are found inside the mitochondria, both in the matrix and associated with the membranes. The particles on the inner faces of the cristae appear to be mainly associated with the third stage of respiration, the electron transport system, or **oxidative phosphorylation** (Fig. 3.10). These particles are known to carry the cytochrome complexes which form a sequential redox system,

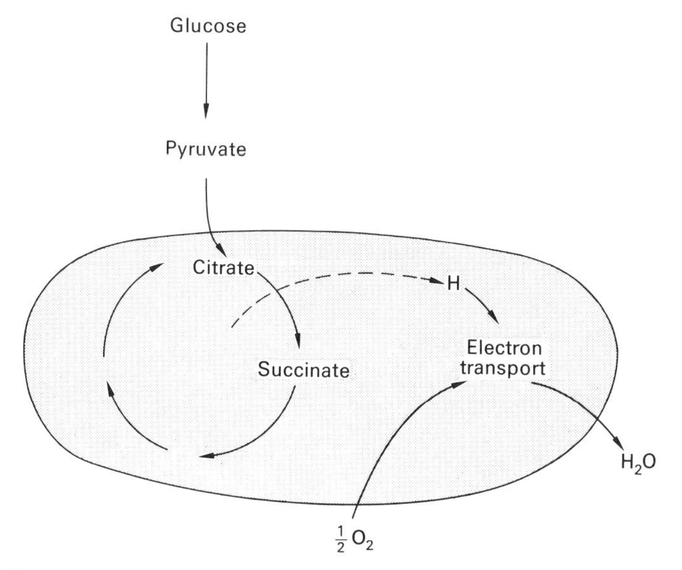

Fig. 3.10 Summary of the role of the mitochondrion in respiration.

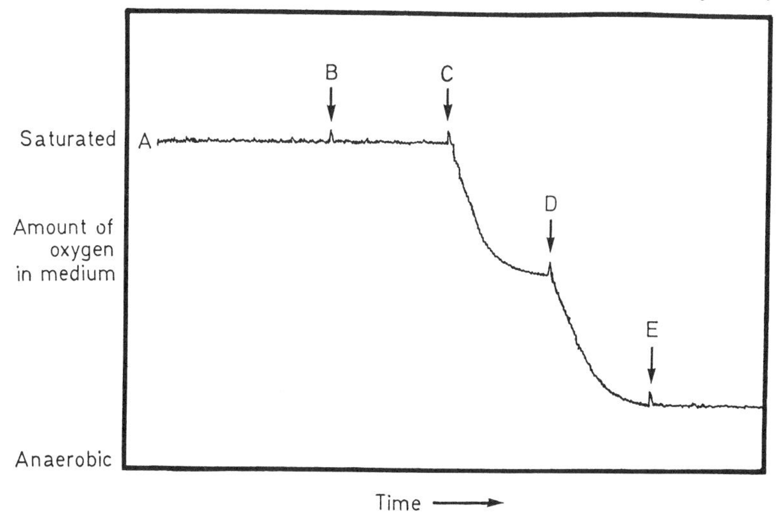

Fig. 3.11 A typical oxygen consumption trace obained with an oxygen electrode attached to a recorder. Mitochondria in the resting state have little or no oxygen consumption (A). Substrate is added at point (B). If the preparation is a good one, there should be no effect on the oxygen uptake. At point (C), a known quantity of ADP is introduced into the vessel. The trace is immediately deflected downwards, showing that oxygen is being taken up. When all the ADP has been utilized, the trace stabilizes at a new level. Further addition of ADP (D) results in an identical response. A third addition of ADP (E) elicits no response as all the substrate has now been consumed. The time taken is about five minutes, and from these results a P/O ratio (the number of molecules of ATP synthesized per atom of oxygen used) may be calculated. (From Bryant: *The Biology of Respiration*, Edward Arnold.)

the end reaction of which is the oxidation of hydrogen to form water (see Fig. 3.11, and Fowler and Richardson, 1963).

Stage 2 The Krebs Cycle

Aerobically respiring organisms obtain considerably more energy from their respiratory substrates than that released solely from glycolysis by the further dehydrogenation of pyruvic acid. The pyruvic acid is first decarboxylated and combined with coenzyme A, forming acetyl-Co A. This 2 C compound combines with the 4 C residue of the Krebs Cycle, oxaloacetic acid, to form a 6 C compound, citric acid. As this molecule is processed round the Krebs Cycle 2 molecules of CO_2 are produced leaving the 4 C residue mentioned above (Fig. 3.12).

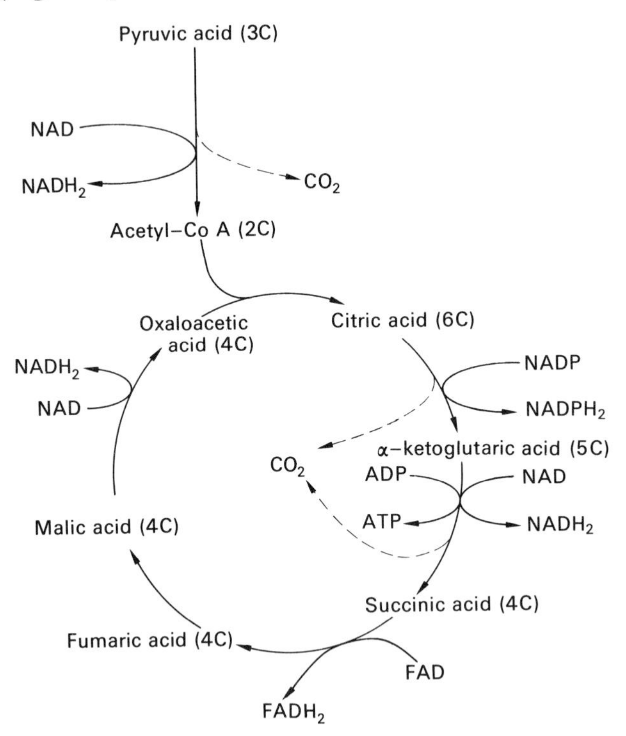

Fig 3.12 Summary of the Krebs Cycle. The principal feature of the Krebs Cycle is the progressive removal of the hydrogen (protons and electrons) from which energy is derived.

Stage 3 The Electron Transport Chain

During photosynthesis solar energy is used indirectly to provide a source of electrons which reduce carbon to form energy-rich compounds (p. 65). Cells remove the electrons from (oxidize) these compounds during the respiratory pathways outlined in Fig. 3.12 and use the energy to reform ATP by **oxidative phosphorylation**.

The removal of electrons coincides with dehydrogenation and makes use of a number of redox compounds, notably flavins, ubiquinone and cytochromes. These have different redox potentials and almost certainly act in sequence (Fig. 3.13). In living cells these compounds are associated with proteins in mitochondria, probably in the particles of the inner membrane as was described earlier.

A summary of the sources of ATP in respiration is given in Fig. 3.14.

3.3 The Energetics of Respiration

The glycolytic pathway shows a net yield of 2 moles of ATP per mole of hexose metabolized. If the organism concerned makes

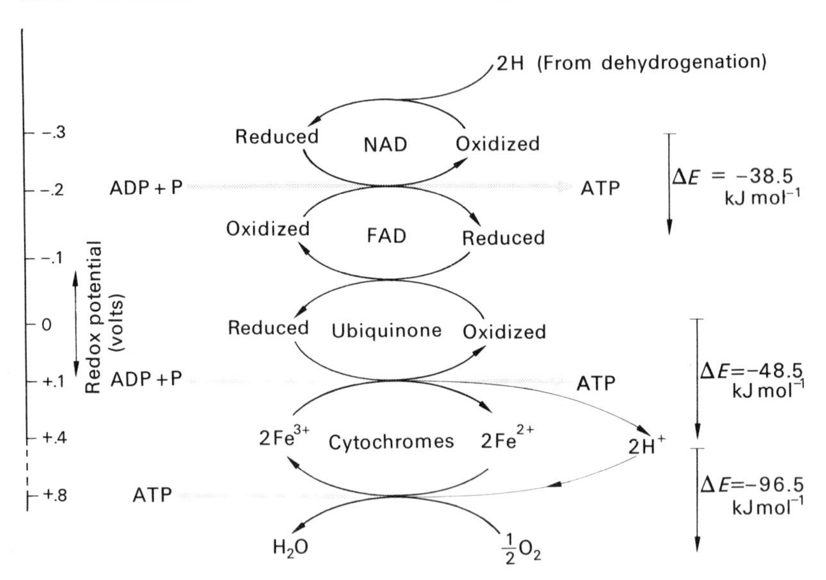

Fig. 3.13 Summary of the electron transport chain in which ATP is generated in oxidative phosphorylation. ATP can be formed when the energy released at any particular step in the redox reactions exceeds about 30 kilojoules per mole.

ethanol, the theoretical yield from thermodynamic data is 234 000 joules:

$$C_6H_{12}O_6 \longrightarrow 2C_2H_5OH + 2CO_2 \quad \Delta F = -234 \text{ kJ}$$

But the yield of 2 ATP:

$$C_6H_{12}O_6 + 2ADP + 2H_3PO_4 \longrightarrow 2C_2H_5OH + 2CO_2 + 2ATP + 2H_2O$$

corresponds to 2×-33.5 kJ available for cell functions (p. 49). One should consider, however, that the ΔF for the **formation** of ATP may be greater, say 40 kJ, owing to the relatively low concentration of reagents in the cell.

Thus the yield of 2×-33.5 kJ corresponds to an **efficiency of energy conservation** of

$$\frac{2 \times -33.5}{-234} \quad \text{or about } 29\%$$

In contrast, the theoretical free energy yield for the complete (aerobic) pathway is 2 860 000 joules:

$$C_6H_{12}O_6 + 6O_2 \longrightarrow 6CO_2 + 6H_2O \quad \Delta F = -2860 \text{ kJ}$$

As we have seen, this pathway yields 38 moles of ATP. Using the same calculation the **efficiency of energy conservation** is

$$\frac{38 \times -33.5}{-2860} \quad \text{or about } 45\%$$

The lower efficiency of fermentation can be accounted for by the difference between the energies of formation and hydrolysis, noting that two ATP must be used to produce four. The remaining energy is expended as heat.

The figure of 45% for aerobic respiration, although only an approximation, represents a high degree of efficiency when compared with mechanical engines, such as steam engines (about 10%) and petrol engines (about 30%).

That the aerobic respiratory pathways are common to plants and animals suggests that aerobic respiration appeared when life on the planet was at a primitive evolutionary level (p. 231). Its widespread

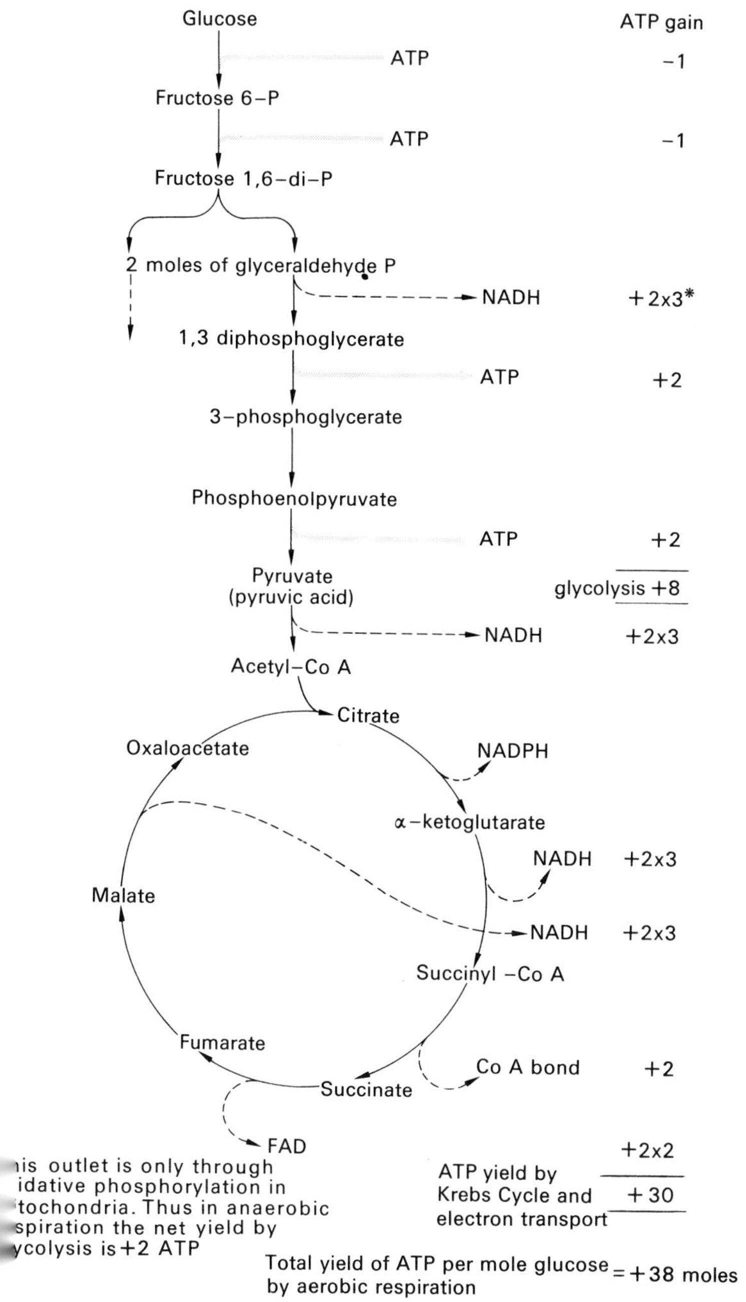

Fig. 3.14 Summary of the sources of ATP in respiration, steps in outline only.

adoption by organisms reflects its utility when there is competition for energy substrates, clearly shown by comparison of ATP yields:

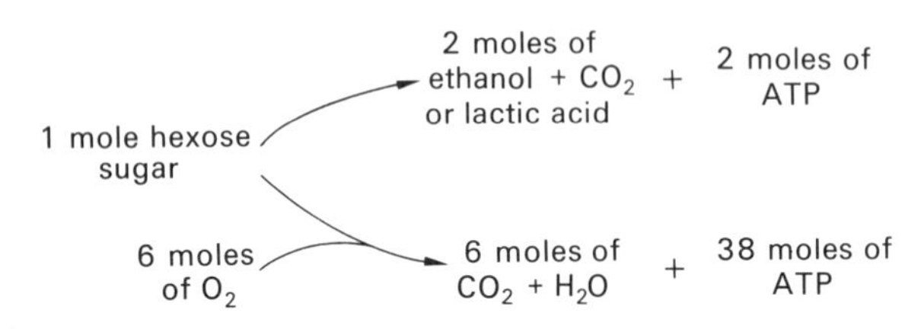

Respiratory Substrates other than Glucose

In many organisms proteins and lipids form a substantial part of their diet. Protein digestion yields amino acids which in turn, by deamination, can yield organic acids. These may enter the respiratory pathway at various points, pyruvic acid, α-ketoglutaric acid, and so on. Lipid digestion yields fatty acids and glycerol, which can enter the respiratory pathway at glyceraldehyde phosphate and acetyl-Co A respectively. By these pathways (Fig. 3.15) the solar energy originally employed in the photosynthetic process to synthesize proteins and lipids (p. 65) can be converted to the high energy pyrophosphate bonds of ATP to be used by the cell to do work.

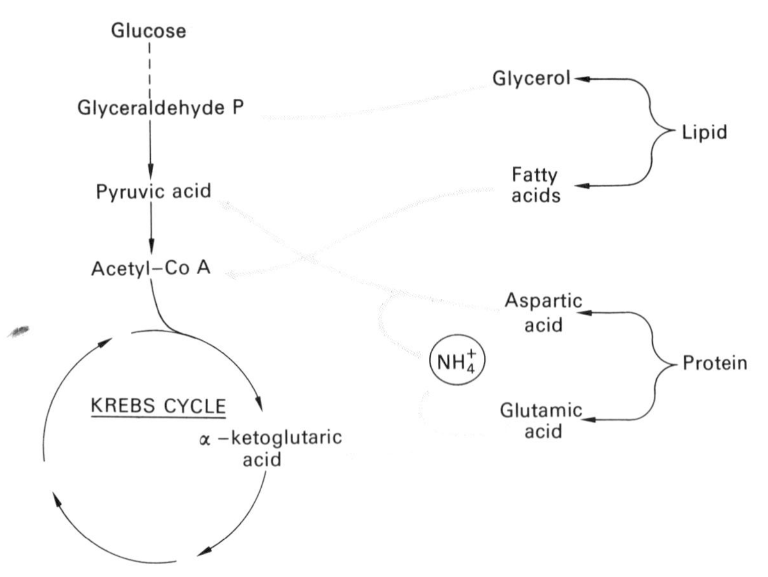

Fig. 3.15 Some entry points for substrates other than glucose.

3.4 Energy Transformation in Chloroplasts

Chloroplasts belong to a category of structures that are found only in plant cells, namely **plastids** which, when mature, are visible under the light microscope. Other forms of plastid include amyloplasts (starch grains), chromoplasts (pigment grains), proteinoplasts (protein grains) and elaioplasts (lipid grains). It is probable that all types of plastids have a common origin in undifferentiated **proplastids**, which are visible only under the electron microscope. To some extent plastids appear to be interchangeable in nature (Fig. 3.16), although only mature chloroplasts and proplastids are thought to be capable of division (Clowes and Juniper, 1968).

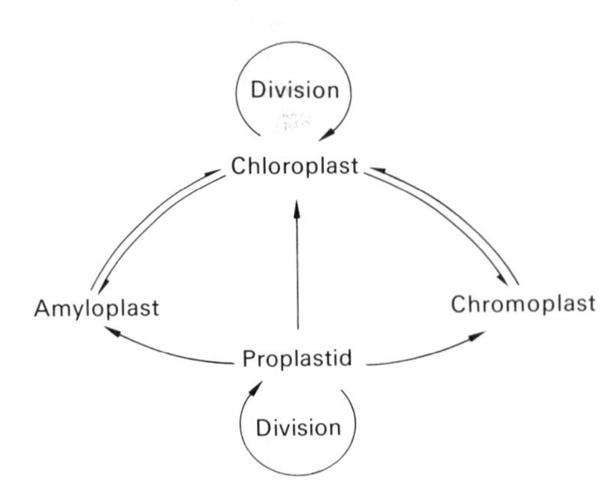

Fig. 3.16 The interchangeable qualities of some plastids. (After Clowes and Juniper, 1968.)

Some discussion of the evolutionary origin of chloroplasts occurs in Chapter 10. Chloroplasts contain DNA and a protein synthesizing mechanism which are atypical of the cells in which they are found and a primeval endosymbiotic union has been postulated to account for these observations.

Under the electron microscope sections of chloroplasts from higher plants show a series of orderly 'stacks' of membranes like piles of coins, which are termed **grana** (Fig. 3.17 and 3.19). Tubular membranes link the grana through the ground substance, or **stroma**

(Fig. 3.17), and hold them perpendicular to the surface of the chloroplast. If chloroplasts from a plant which has been photosynthesizing are examined starch can be seen between the grana.

Chlorophyll, the central pigment in the photosynthetic process, appears to be confined to the grana. The chlorophyll and other pigments such as carotenoids are thought to be sandwiched between layers of protein (Fig. 3.18). These protein layers contain enzymes

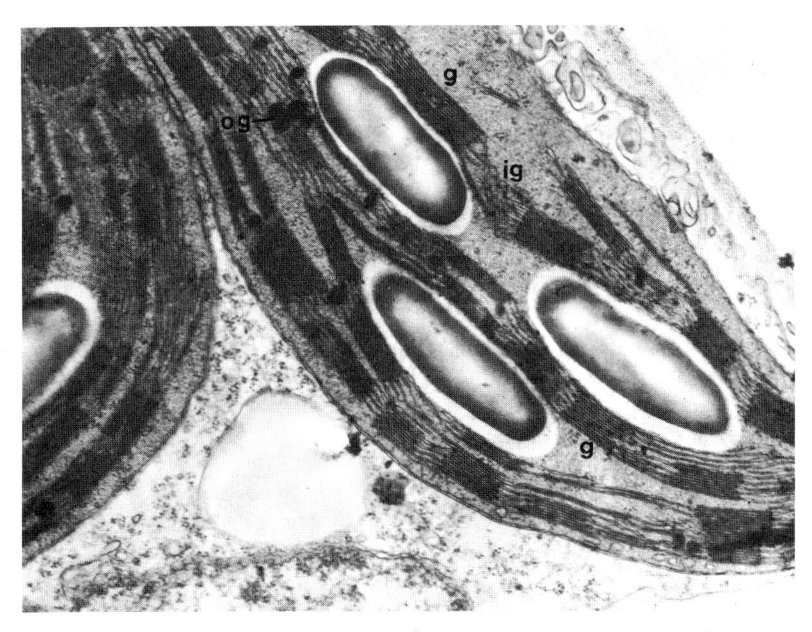

Fig. 3.17 E.M. of a section of chloroplast from the leaf of the pepper plant, containing three conspicuous starch deposits in the stroma. Membranes can be seen in close stacks or grana (g), and running more loosely between the grana (ig). When stained with osmic acid a number of electron dense spheres can be seen. These osmiophilic globules (og) occur also in other plastids but their function is unknown, although the fact that they are rich in lipids suggests they may play a part in membrane formation. (Courtesy of Dr. B. E. Juniper.)

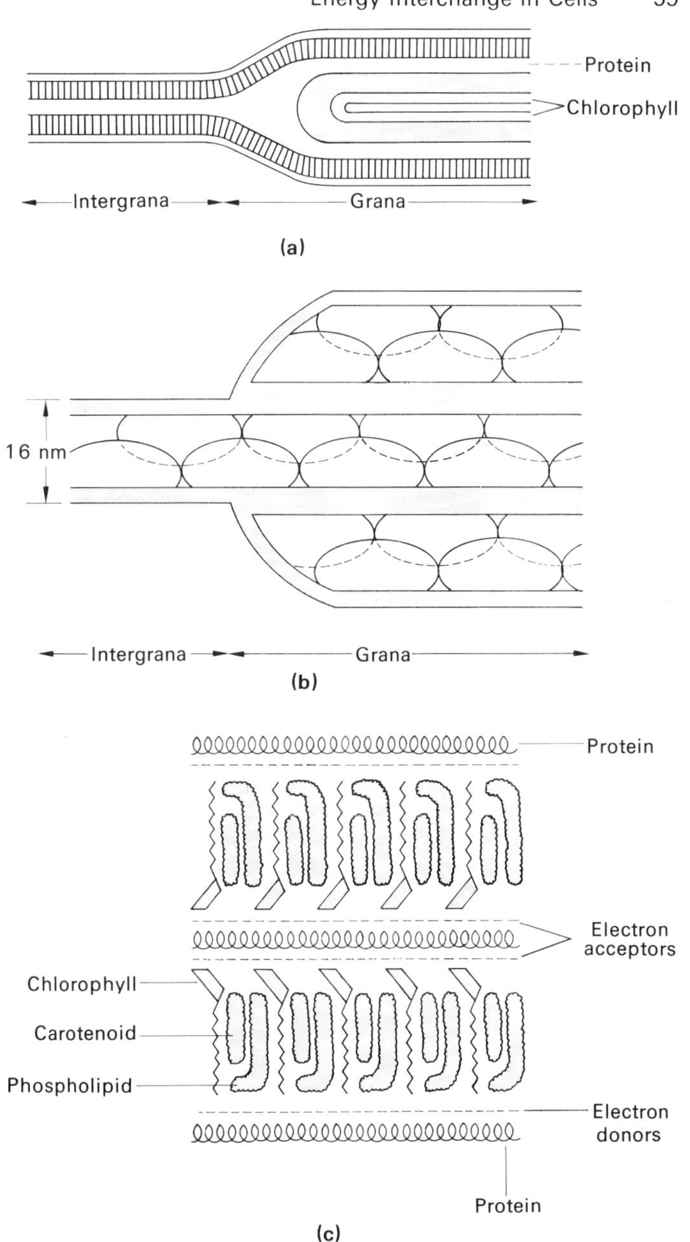

Fig. 3.18 Drawings of (a) chlorophyll location within the grana, (b) the arrangement of quantasomes attached to unit membrane and (c) the way in which chlorophyll and lipids are thought to be sandwiched in the system. (Figures (a) and (c) after J. Paul, 1965; figure (b) after R. G. Park and N. G. Pon, both courtesy of Thomas Nelson Ltd.)

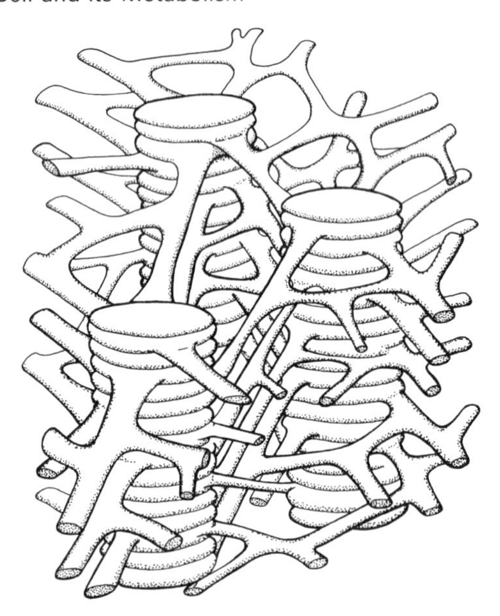

Fig. 3.19 Drawing of grana and the interconnecting tubular membranes or intergrana. (By courtesy of T. Elliot Weier.)

for catalysing the reactions concerned in starch synthesis, together with electron transport or redox substances such as are found in mitochondria (quinones, cytochromes, and so on).

When chloroplasts are disrupted and the lamellae of the grana are examined under a high power electron microscope, particles, termed **quantasomes**, can be seen (Fig. 3.20). These measure 10 to 20 nm and chlorophyll and other pigments are deposited in them. The highly organized and compartmentalized nature of quantasomes suggests a similarity to the membrane-bound particles of mitochondria. The quantasomes have been likened to 'photoelectric cells coupled to storage batteries' where light produces a flow of electrons which are stored in the accumulation of ATP (Du Praw, 1968).

The Structure of Chlorophyll

Chlorophyll has a porphyrin or **tetrapyrrole** ring structure comprising four 'five-sided' pyrrole groups. Within the ring lies a single magnesium atom and one of the groups carries a long **phytol** side chain. The tetrapyrrole form occurs widely in nature (Kamen, 1958). With an iron atom replacing the magnesium and with different side chains the same structure is found in the **haem** fraction of haemoglobin and in the **cytochromes** which are involved in electron transport in mitochondria and chloroplasts (Fig. 3.21).

The Stages of Photosynthesis

The sequence of reactions by which green plants construct their organic constituents can be divided into two series. The first is known variously as the Light Reaction or the Hill Reaction (after the Cambridge botanist R. Hill who, in 1937, demonstrated that isolated chloroplasts possessed reducing power and evolved oxygen when they were illuminated). This crucial experiment in photosynthesis is readily repeatable (Baron, 1963). The second series of reactions is known as the Dark Reactions in that light is not required directly for the pathway to proceed.

The Light Reaction

In its overall activity the chloroplast closely resembles the mitochondrion in that both use a source of high-energy electrons to syn-

Fig. 3.20 Freeze-etched electron micrograph of chloroplast membrane extracted from tobacco leaf. In this preparation the membranes are seen *en face* and the organized layer of quantasomes can clearly be seen. Magnification ×72 000. (By permission of K. H. Moor. From *Cell Biology*, courtesy of Thomas Nelson Ltd.)

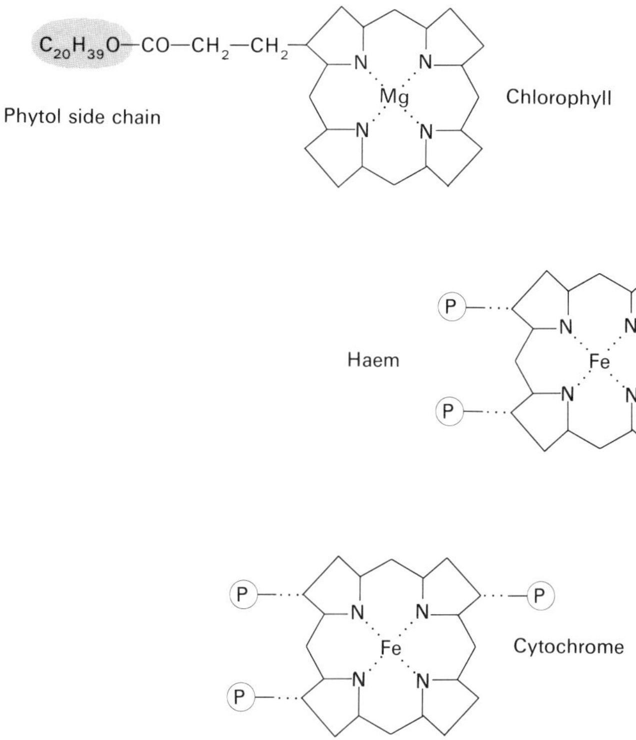

Fig. 3.21 Outline molecular diagrams of chlorophyll, haem and cytochrome to show their similar dependence upon the porphyrin or tetrapyrrole ring. Chlorophyll has a long chain phytol group attached. Haem is attached to a globulin protein in the formation of haemoglobin both at the Fe atom and at the points marked Ⓟ. In cytochrome the porphyrin is attached slightly differently to the protein.

thesize ATP using a step-wise redox sequence of quinones and cytochromes. The main difference is that the source of electrons is illuminated chlorophyll in one and respiratory substrates in the other.

There are two products of the Light Reaction. The first is **ATP** which can be used directly in cell metabolism by those cells that have chloroplasts. It is used to phosphorylate several compounds in the Dark Reaction sequence and thus enables endergonic syntheses to take place. The second product is **reduced nucleotide**. The overall principle of the photosynthetic reactions is the reduction of carbon dioxide to form carbohydrates. For this two things are required, a strong reducing agent (reduced nucleotide) and a source of energy (ATP).

In photosynthetic bacteria the nucleotide concerned is usually flavine adenine dinucleotide

$$\text{FAD} \xrightleftharpoons{\text{reduction}} \text{FADH}_2$$

but in higher plants the nucleotide is usually nicotinamide adenine dinucleotide-phosphate

$$\text{NADP} \xrightleftharpoons{\text{reduction}} \text{NADPH}$$

(see Hydrogen Acceptors, p. 51). Reduced nucleotides are then used in the dark reactions for the progressive reduction of simple sugars. These reaction stages closely resemble the reversed pathway of the Warburg-Dickens or pentose-phosphate reactions (p. 52).

In higher plants two kinds of chlorophyll are found, **chlorophylls *a*** and ***b***, which differ from one another in respect of a small side-chain substitution. The impingement of photons of light excites the chlorophyll molecule and causes electrons to be raised into higher energy orbits from which they can be 'captured' by adjacent electron acceptors. The lost electrons are replaced at a lower energy orbit from an electron donor substance. Thus the role of chlorophyll is essentially to pass electrons from a donor to an acceptor, rather like an 'electron pump'. The hydrogen acceptor then passes on the electron which is used to drive reduction-oxidation reactions.

Both forms of chlorophyll absorb light primarily at the red and blue ends of the spectrum. In a remarkably elegant experiment the German botanist T. W. Engelman, in 1881, shone a spectrum of light on to the filamentous alga *Cladophora* sealed in an air-tight cell. He was able to show that motile bacteria accumulated densely close to the filament in the red and blue illuminated zones, and concluded that these were where most oxygen was being evolved (Fig. 3.22).

In photosynthetic bacteria the light reaction is described as **cyclic** photophosphorylation because the electrons, after passage through the ATP-generating cytochrome sequence are returned to the electron donors and back to the chlorophyll. In higher plants the electrons are donated from OH^- ions in water and since they do not return, this process is termed **non-cyclic** photophosphorylation.

Non-cyclic Photophosphorylation. The stages of the light reaction in higher plants use both chlorophyll *a* and chlorophyll *b* apparently in distinct reactions, although both act by producing high-energy electrons as described (Fig. 3.23).

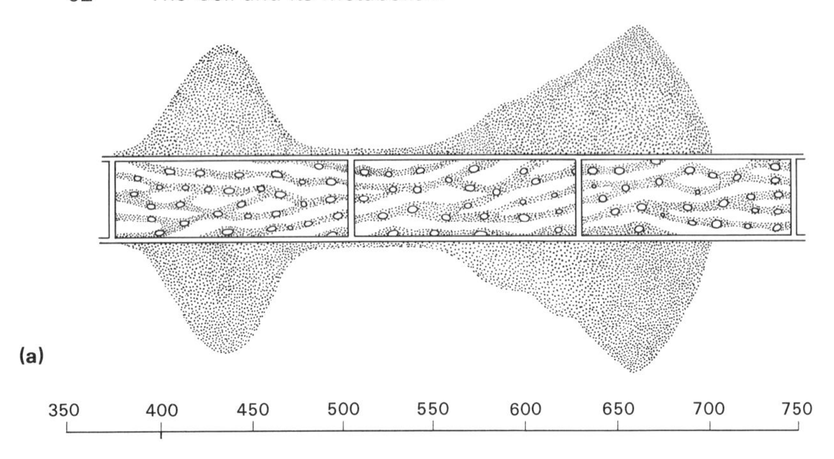

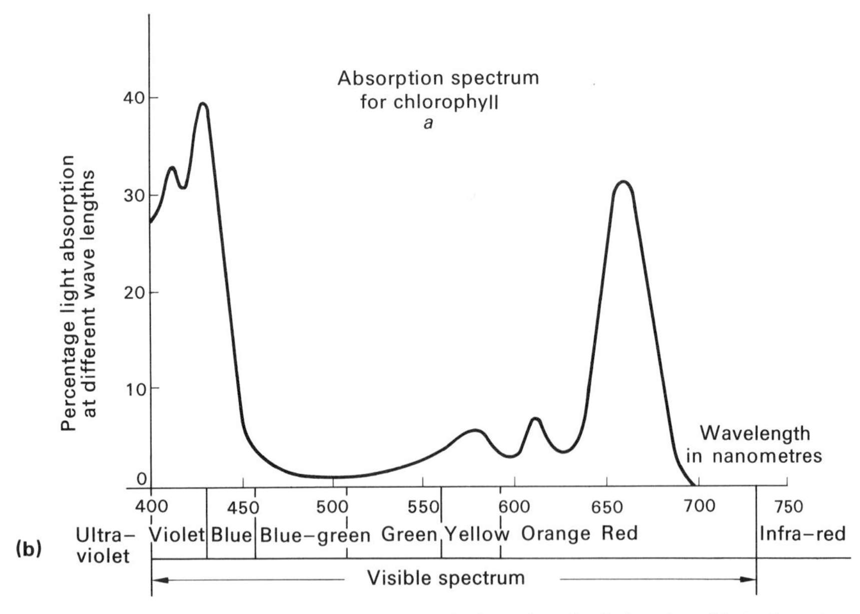

Fig. 3.22 Englemann's experiment in which a thread of the alga *Cladophora* is placed in a light spectrum. (a) Motile bacteria collect where most oxygen is being produced. (b) This corresponds closely with the light absorption spectrum for chlorophyll. (After D. A. Coult, in *Molecules and Cells*, by courtesy of Longmans.)

Chlorophyll *b* receives its ground, or low-energy, electron from the hydroxide ion of water. Water ionizes weakly to hydrogen and hydroxide ions:

$$H_2O \rightleftharpoons H^+ + OH^-$$

The loss of an electron from the hydroxide ion converts it to an unstable hydroxide radical which breaks down to oxygen and water:

$$H_2O \dashrightarrow \text{ionized} \dashrightarrow 2OH^- \xrightarrow{2e^-} 2OH \longrightarrow H_2O + \tfrac{1}{2}O_2$$

The oxygen, of course, is excreted from the plant into the atmosphere where it provides the oxygen which all living aerobic organisms need for respiration (p. 55).

Emerging from the chlorophyll *b* molecule the electrons, now in a 'high-energy' form, enter a hydrogen acceptor, probably **plastoquinone,** together with the hydrogen ions from water, and reduce it to **hydroquinone**. The hydroquinone is then reoxidized by **cytochromes** to plastoquinone.

In the chloroplast there are probably two cytochromes operating at different redox potentials. Their role is to liberate hydrogen ions and electrons (now in a low-energy state). The reduction-oxidation potentials of these changes are sufficiently exergonic to allow ATP to be created from ADP and inorganic phosphate.

Chlorophyll *a* now enters the picture. It accepts as its ground electron the low-energy electron emitted by the cytochromes. This is energized and used to reduce **ferredoxin** which in turn reduces NADP to NADPH using the hydrogen ions liberated during ATP synthesis in the cytochromes. The net result of the light reaction, therefore, is the formation of both ATP and NADPH.

Cyclic Photophosphorylation. In this form of light reaction, which has been demonstrated in isolated chromatophores from photosynthetic bacteria, ATP only is generated. Light-excited chlorophyll is the source of high-energy electrons which are passed through an electron acceptor and reduce flavine adenine dinucleotide, or FAD, to $FADH_2$. This reduced nucleotide then transfers the electrons to an electron cascade system involving cytochromes, probably very similar to the redox system described for mitochondria and chloroplasts. In this system ATP is generated, as shown in Fig. 3.24.

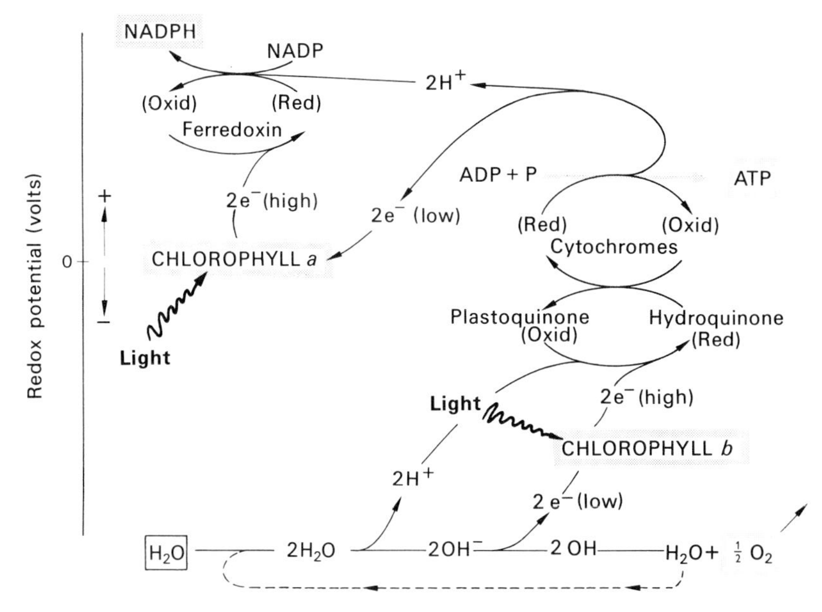

Fig. 3.23 Principal stages in non-cyclic photophosphorylation (higher plants).

However, it differs from electron transport in the mitochondria in the final stages in that the electrons, now at a low-energy level are not used to form water but are taken back into the chlorophyll as ground electrons. Thus a cyclic electron flow is established, being driven by light, in which there is no net gain or loss by any external electron donor or acceptor.

A further form of non-cyclic photophosphorylation found in the

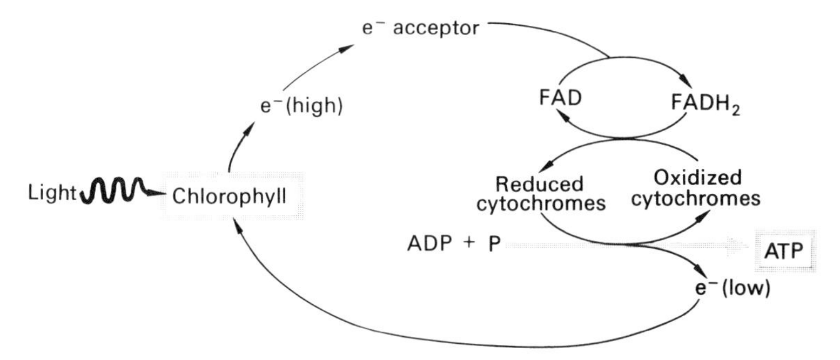

Fig. 3.24 Summary of cyclic photophosphorylation (compare with Fig. 3.25).

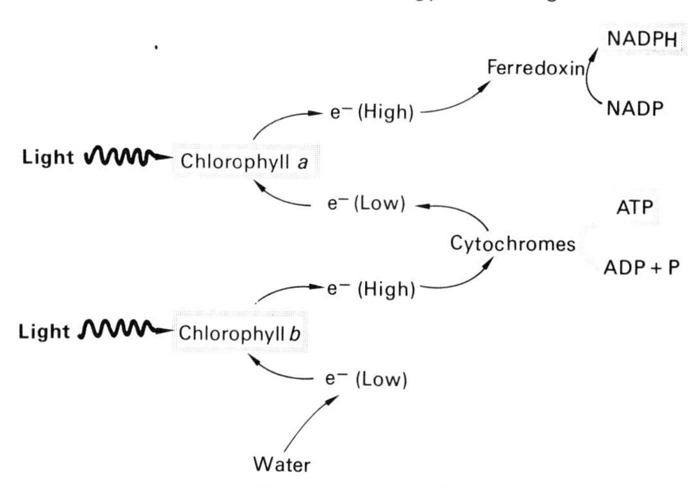

Fig. 3.25 Summary of non-cyclic photophosphorylation (compare with Fig. 3.24).

bacterium *Chromatium* deserves consideration (Fig. 3.26). This bacterium uses succinate or thiosulphate ion as an electron donor and generates ATP by passing electrons through a cytochrome system. The electron is then energized by illuminated chlorophyll, taken on to a hydrogen acceptor (a pyridine nucleotide) and is used to reduce hydrogen ions (to hydrogen gas) or nitrogen (to ammonia), (Arnon, 1960).

The significance of this sequence is that it may have been a method whereby primitive organisms, living on the earth in primeval times in the anaerobic atmosphere that prevailed before the photolysis of water provided oxygen, could have generated ATP much more efficiently than by simple glycolytic fermentation (see Chapter 10).

The Dark Reactions of Photosynthesis

The dark reactions comprise a pathway whereby carbon, in the low energy-containing compound carbon dioxide, is progressively reduced in a series of endergonic reactions to form high energy containing sugars. Hydrogen for the reductions is provided by NADPH and the energy for the endergonic steps comes from the hydrolysis of ATP. Both these substances are formed in the light reactions.

The pathway of carbon in the dark reactions of photosynthesis has been elucidated by the use of the radioactive isotope of carbon, C^{14}. This work was pioneered by Ruben and Kamen in 1941 and continued by Calvin, Bassham and co-workers (Bassham and Calvin,

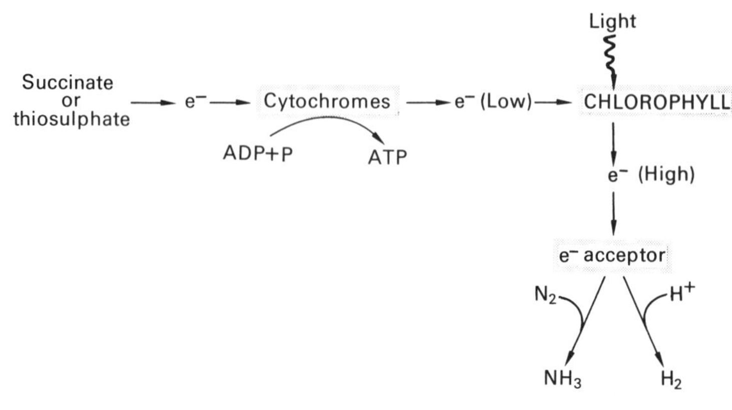

Fig. 3.26 Non-cyclic photophosphorylation in the bacterium *Chromatium*.

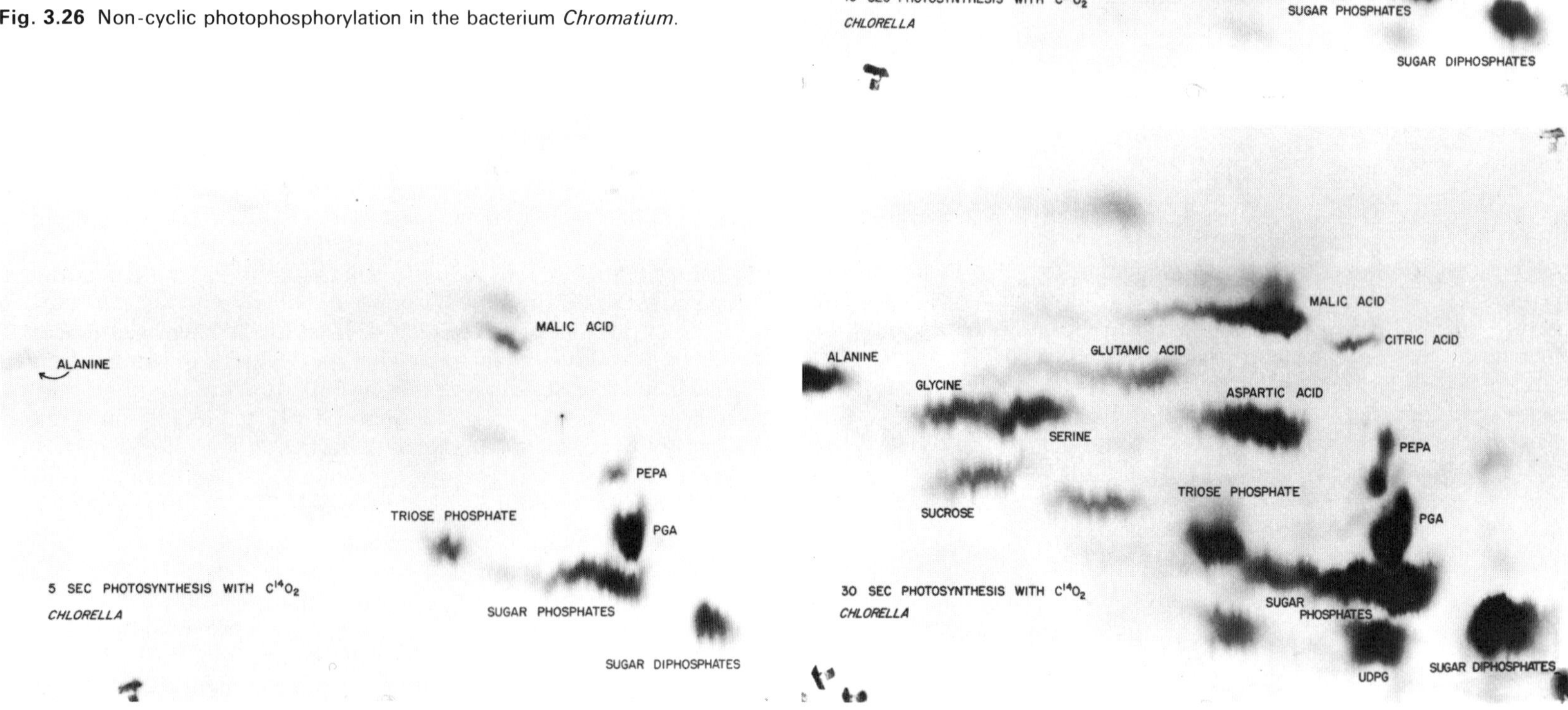

Fig. 3.27 Three autoradiographs taken from chromatograms which give good evidence of the sequence of compounds that are synthesized in the dark reactions. (*left*) The compounds formed five seconds after adding radioactively labelled bicarbonate ion ($HC^{14}O^-$). (*upper right*) The compounds formed after ten seconds and (*right*) after thirty seconds. The formation of 3-carbon sugars precedes the synthesis of the more complex sugars and amino acids. (By courtesy of J. A. Bassham and Scientific American.)

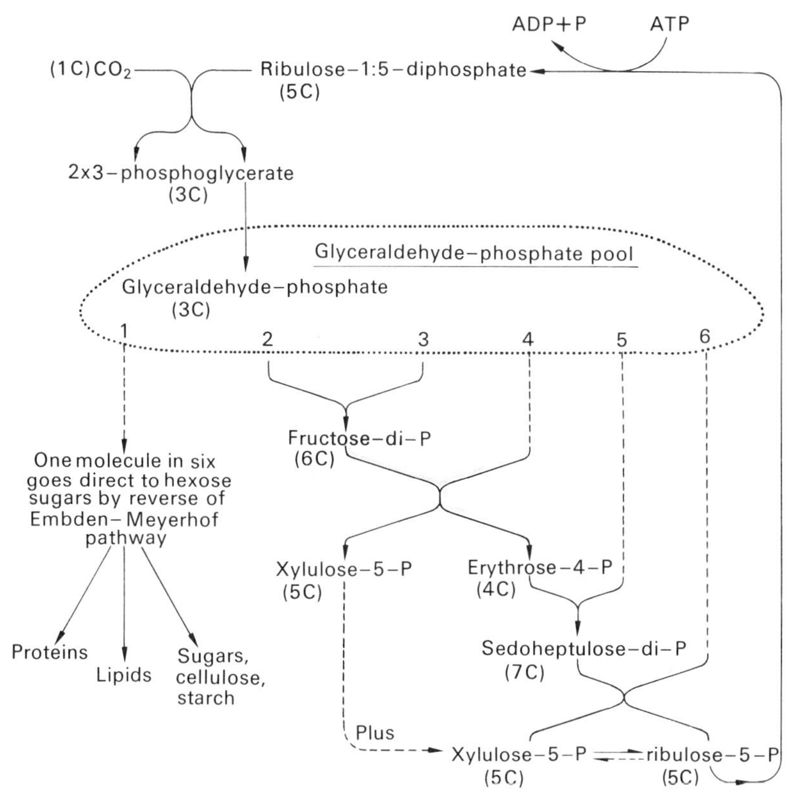

Fig. 3.28 Summary of the Dark Reactions of photosynthesis (the tinted dotted lines show those parts which coincide with the Warburg-Dickens Pentose-phosphate Pathway, compare with Fig. 3.6).

1957; Bassham, 1962). The plant mainly used for this work was the unicellular alga *Chlorella*. It was grown in continuous culture, in suspension, in an agitated culture vessel. Samples of the culture were removed at prescribed intervals, after exposure to CO_2 containing the radioactive carbon, and fixed in alcohol to stop the biochemical reactions. Sugars were then extracted from the samples and separated by paper chromatography. Those sugar spots on the chromatograms which contained the radioactive C^{14} were then identified by placing a photographic plate against the chromatogram, forming what is termed an **autoradiograph** (Fig. 3.27). The principle of autoradiography is described later in Chapter 8 (p. 153) in connection with the identification of DNA in replicating chromosomes.

Five seconds after exposure of the photosynthesizing algae to $C^{14}O_2$ the radioactive carbon was found in the 3-carbon sugar, 3-phosphoglycerate. Further work revealed, however, that the carbon dioxide was not taken up by a 2-carbon acceptor, as might be thought, but by a 5-carbon acceptor, ribulose-1,5-diphosphate. This formed an unstable 6-carbon compound which, by dismutation, formed two molecules of the 3-carbon sugar.

The 3-carbon sugar contributes to what might be termed a glyceraldehyde-phosphate 'pool', from which further reactions in the dark pathway proceed. Since glyceraldehyde-phosphate lies on the Embden–Meyerhof pathway it would seem reasonable that it should proceed in the reverse direction to form sugars. Only one molecule in six does this, however.

Of the remaining five molecules, two interact to form a 6-carbon fructose-diphosphate. This in turn accepts a third molecule of glyceraldehyde-phosphate and transmutes to form a 5-carbon sugar, xylulose-5-phosphate, and a 4-carbon sugar, erythrose-4-phosphate.

The erythrose-phosphate then combines with a fifth glyceraldehyde-phosphate molecule to form the 7-carbon sugar, sedoheptulose-diphosphate. This then collects the last glyceraldehyde-phosphate and transmutes to two 5-carbon sugars, xylulose-5-phosphate and ribulose-5-phosphate. These sugars are familiar as the 'starting-point' of the pentose-phosphate pathway and the similarities between the two pathways are clear (Compare Figs. 3.6 and 3.28).

The two molecules of xylulose-5-phosphate are converted to their isomer, ribulose-5-phosphate (making three molecules in all) which are then phosphorylated by ATP to form ribulose-1,5-diphosphate. This cycles again as the CO_2 acceptor. Thus, in net terms, three molecules of CO_2 enter the sequence and give rise to one molecule of 3-carbon sugar which leaves the system.

Section 2

The Continuity of Organisms: Patterns of Reproduction

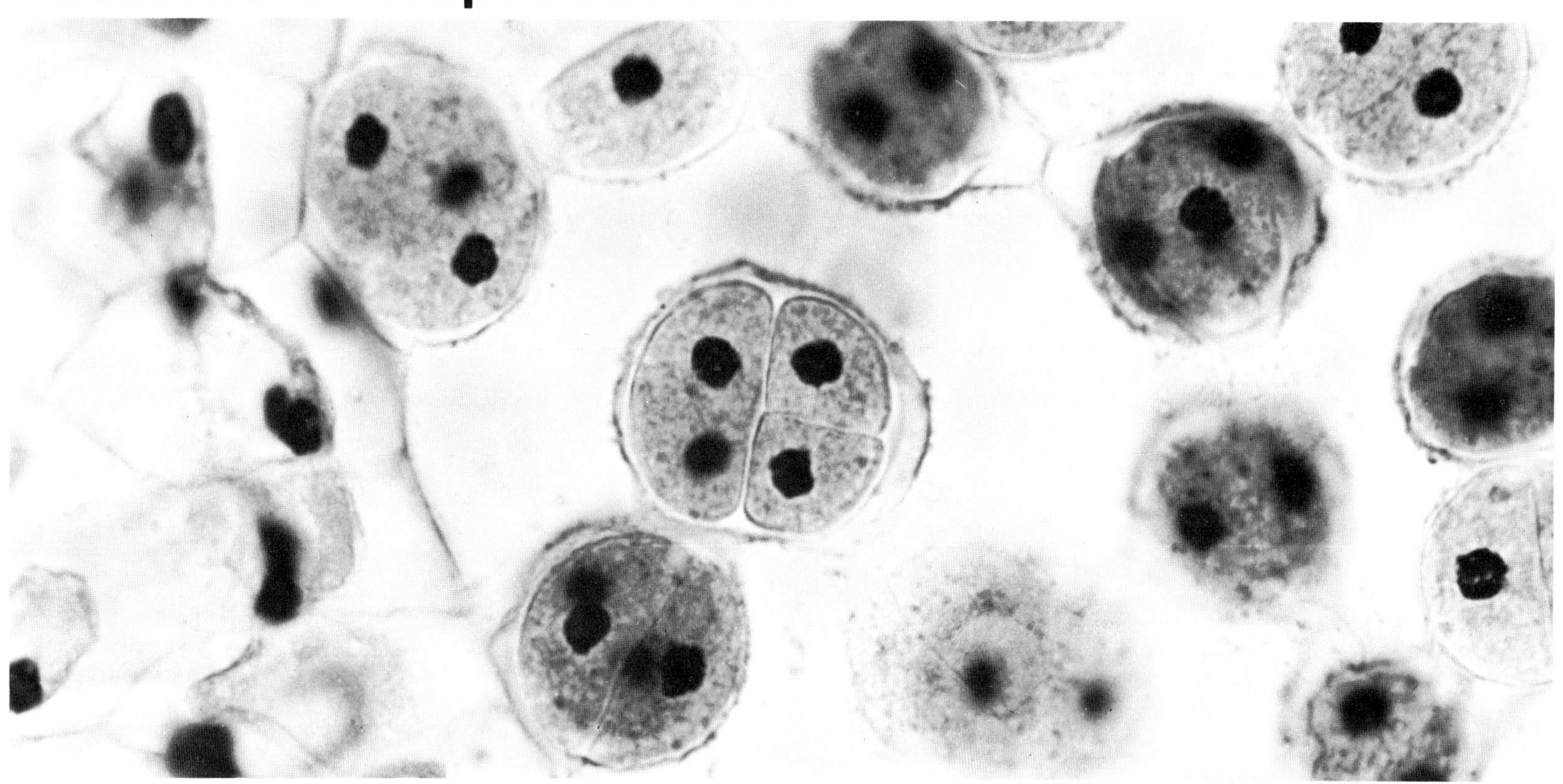

(*Photograph on preceding page*) Meiosis in the pollen mother cells in the anther of *Lilium*. The characteristic four cell stage represents Telophase II; each daughter cell differentiating to form a pollen grain. (Courtesy of Gene Cox.)

Modes of Asexual Reproduction

The second section of this book deals with the patterns shown in the biology of reproduction. Chapter 4 deals with modes of asexual reproduction, stressing the relationship that exists between this form of reproduction and growth; one is the increase in biological mass, or **biomass**, of the population and the other is the increase in biomass of the individual organism. Environmental conditions that favour one usually favour the other.

Any discussion of biology from an evolutionary point of view tends to emphasize **sexual** reproduction because this is an important means of creating genetic diversity in populations. Sexual reproduction leads to continuous mixing of genes so that original combinations occur upon which natural selection can act. But the role of asexual reproduction should not be ignored. Since asexual reproduction is associated with rapid population growth, organisms which retain the faculty for this kind of reproduction are capable of exerting strong reproductive pressures in relation to other organisms in their habitat. The capacity of micro-organisms to increase their biomass rapidly under favourable conditions is one of the prime stabilizing factors in any ecosystem, particularly where those micro-organisms which cycle matter, such as nitrogen, are concerned. Indeed, without such micro-organisms not only would evolution not be possible but life itself, as we know it, would become extinct.

4.1 Asexual Reproduction and Growth

Sexual reproduction, that is the formation of offspring by the fusion of gametes, is easily the commonest, most widespread, way in which organisms propagate their kind. But other forms of reproduction which do not involve the use of gametes and are termed **asexual**, or non-sexual, do exist and merit discussion.

Asexual modes of reproduction are employed in most groups throughout the plant kingdom. In the animal kingdom asexual reproduction occurs only in the lower phyla and appears to be restricted to relatively simple levels of body organization, such as the diplo-

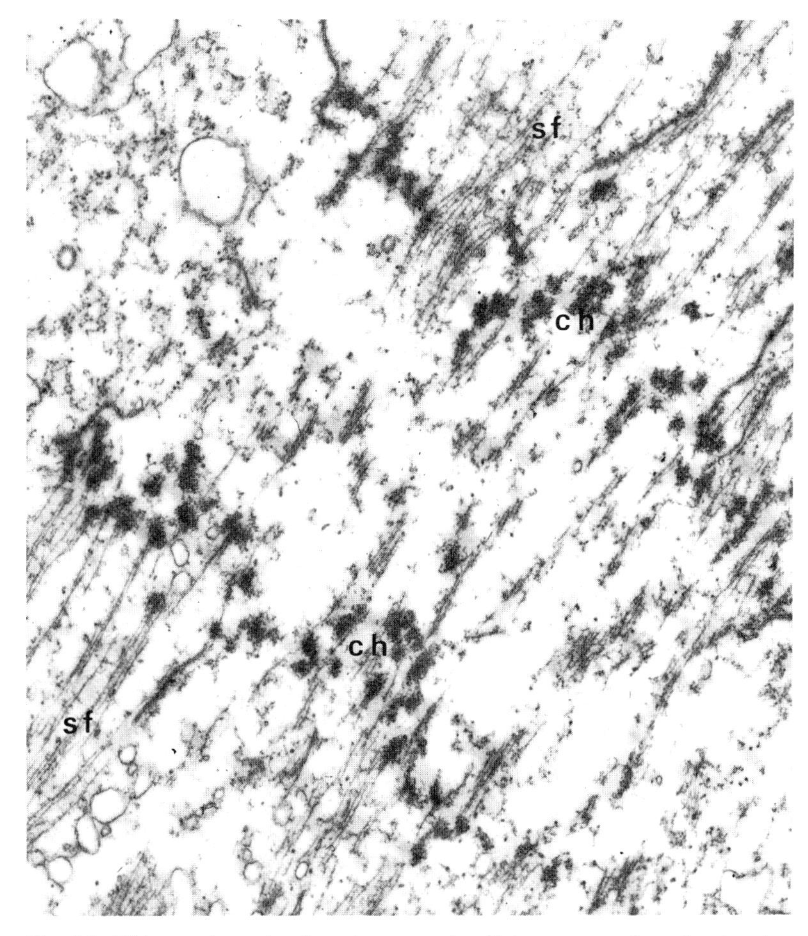

Fig. 4.1 E.M. anaphase in the giant amoeba *Pelomyxa carolinensis*, showing separation of the chromosomes (chr) and the tubular nature of the spindle fibres (sf) attached to the chromosomes. (Courtesy of L. E. Roth.)

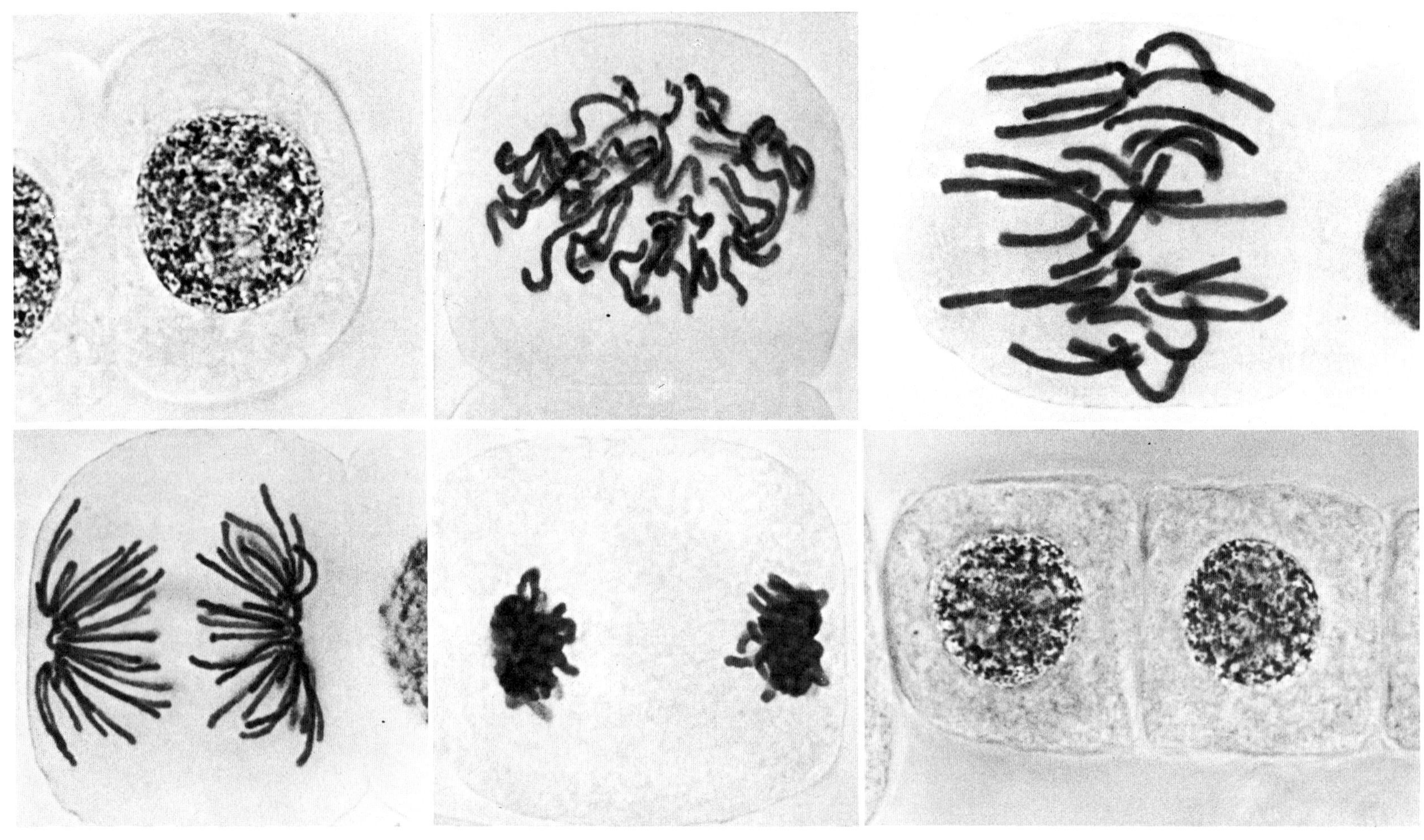

Fig. 4.2 The movement of chromosomes in mitosis. (Top left) Prophase, the chromosomal DNA undergoes replication and the chromosomes condense, becoming visible under the light microscope. In the cytoplasm a spindle forms. (Centre) In Late Prophase the nuclear membrane disintegrates and the chromosomes, seen as pairs of chromatids, migrate to the spindle. (Top right) In Metaphase the chromosomes are attached to the equator of the spindle by their centromeres. (Lower left) In Anaphase the chromatids from each chromosome separate to opposite poles of the spindle. (Centre) In Telophase the chromosomes coalesce, nuclear membranes re-form and cross wall formation begins (cleavage in animal cells). (Lower right) In Late Telophase the cross wall is completed and the chromosomes adopt the dispersed nature characteristic of interphase. These daughter nuclei are genetically identical. (Photographs by courtesy of Dr. B. Snoad.)

blastic forms and below. Asexual reproduction in animals seems to be commensurate neither with mobility nor apparently with a complex level of body organization, such as is found in the triploblastic coelomates.

The patterns of growth in higher plants and animals differ in one important respect. Animals tend to grow to what is termed an adult or mature size, at which growth ceases; growth is relegated to a juvenile stage. Perennial plants, on the other hand, tend to grow continuously, that is seasonally, throughout their lives. For this purpose meristematic tissues occur close to the surface over the whole plant, for growth in length and in girth of the shoots and roots. This abundant meristematic tissue lends itself more readily to sys-

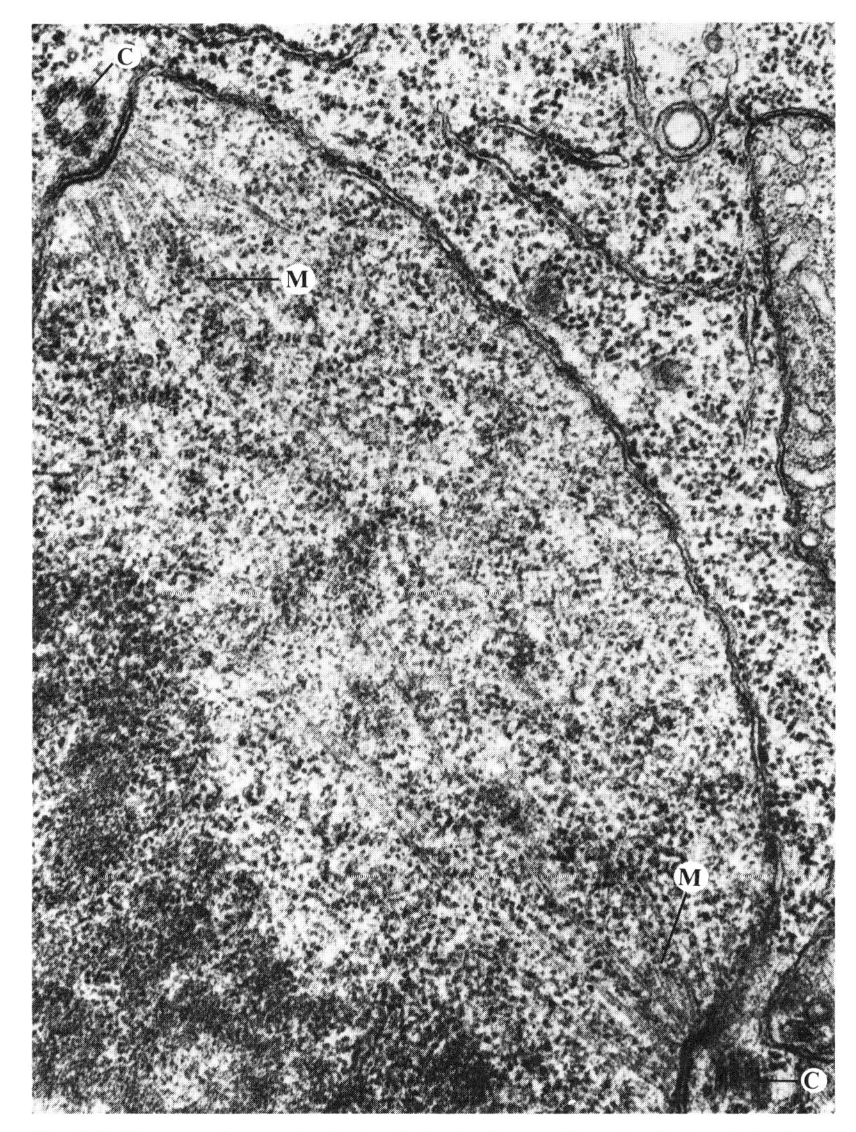

Fig. 4.3 Electron micrograph of mitosis in the fungus *Saprolegnia* in which the nuclear membrane, atypically, remains intact. The microtubules (M) appear to originate at the centrioles (C), but the positioning of centrioles at the spindle poles may be less concerned with spindle formation than with ensuring that the daughter cells each receive a centriole. Cells with no centrioles, such as those of higher plants, cannot form microtubular organelles, such as cilia and flagella, but form normal mitotic spindles. (With permission of J. B. Heath and A. D. Greenwood and Thomas Nelson Ltd.)

tems of asexual reproduction, such as the formation of runners and suckers, than do the differentiated tissues of adult animals, in which mitosis is inhibited.

All forms of asexual reproduction have a number of features in common. First, asexual reproduction is directly related to growth. The kind of cell division associated with asexual reproduction is **mitosis**, which is the manner by which organisms increase their cell number when they grow (Fig. 4.1, 4.2 and 4.3). Growth is associated with favourable conditions in the environment, such as ample food (or the facility for making food in the case of green plants) and suitable temperatures. Asexual reproduction occurs in the same conditions.

The spindle fibres are composed of aggregations of ultrastructural elements known as **microtubules**, which are known to be associated with cell movement. Microtubules are found in cilia and flagella and lie adjacent to the surface membrane in actively locomoting tissue-culture cells. In dividing cells the microtubules may possibly be organized by the centrioles from which they seem to radiate.

Asexual reproduction, therefore, can be seen as a compromise between a continuation of growth and the limitations on size (in the case of animals) or on rate of growth (in plants) imposed by the genetic, and therefore physiological, capacity of the organism. In conditions of ample food a *Hydra*, for example, does not grow beyond a certain size, in a dimensional sense, but can continue to increase its **biomass**

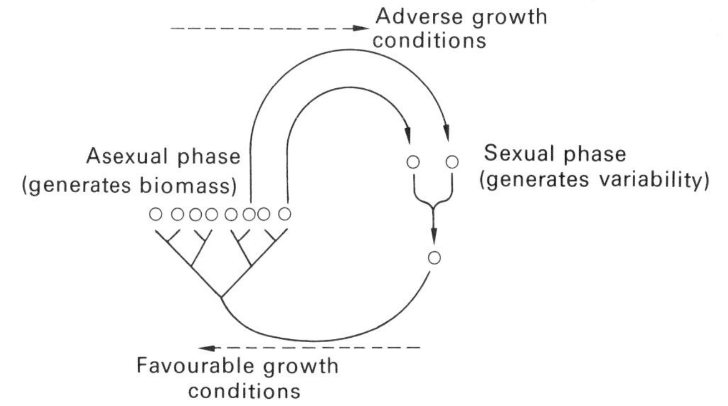

Fig. 4.4 Generalization of a life cycle involving alternation of sexual with asexual reproductive phases.

by producing one or more buds, by localized continuation of mitosis.

A second feature of asexual reproduction is that it is true reproduction in the 'copying' sense. Since offspring are produced mitotically they are genotypically and phenotypically identical to the parent.

Repeated asexual reproduction from a single parent gives rise to a genotypically identical colony, or **clone**. The disadvantages of this feature in the long-term, arising out of the reduction in variability and adaptability, are discussed in Chapter 5, but the short-term advantages are obvious. Asexual reproduction, unlike sexual reproduction, is exponential and prolific. Those organisms in an asexually reproducing population that are 'best-fitted', in the Darwinian sense, are those which most easily exploit their environment for growth. They therefore increase in number the most rapidly. Ideally, therefore, organisms should have a sexual phase, which generates phenotypic diversity, or variation, alternating with an asexual phase which most rapidly exploits the environment using the most suitable variants (Fig. 4.4).

Such a system is used widely in the plant kingdom and in many micro-organisms. In many algae and fungi, for example, the sexual phase occurs when growth conditions become less favourable. In this way the sexual process, which is essentially a **reduction** process. occurs when numbers might be expected to decline in any event and it usually results in a resistant, dormant phase, such as a **zygospore** (Fig. 4.5).

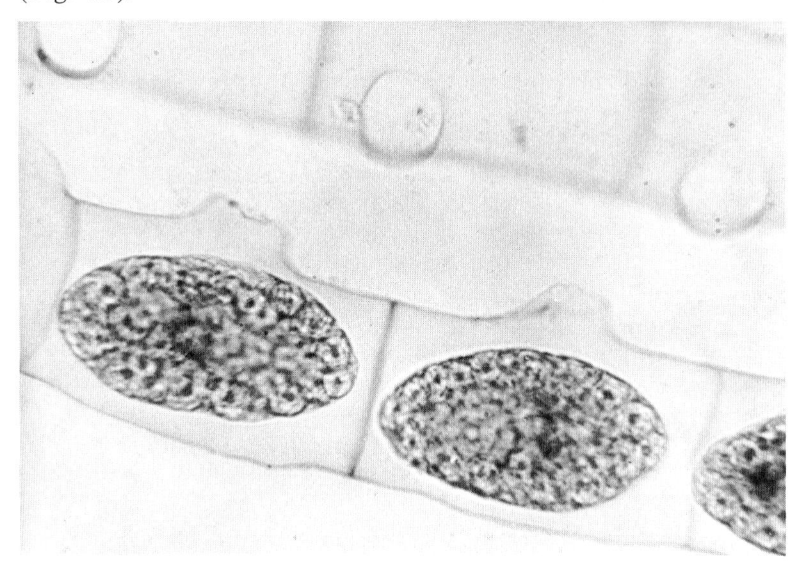

Fig. 4.5 The sexual phase in *Spirogyra* results in a zygospore. In this photomicrograph the conjugation tubes, here fragmented, can be seen together with the empty cells of the adjacent filament.

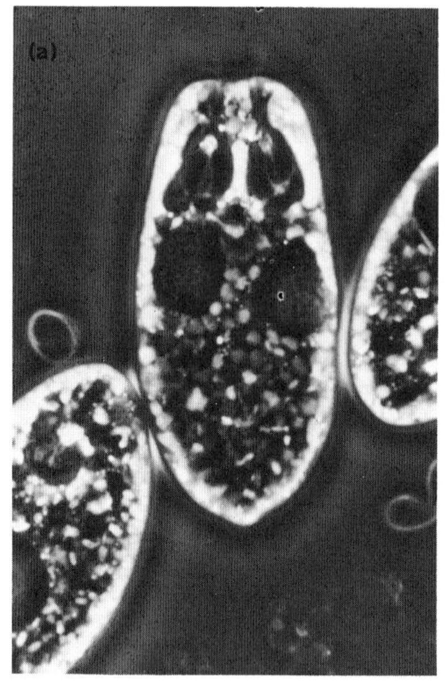

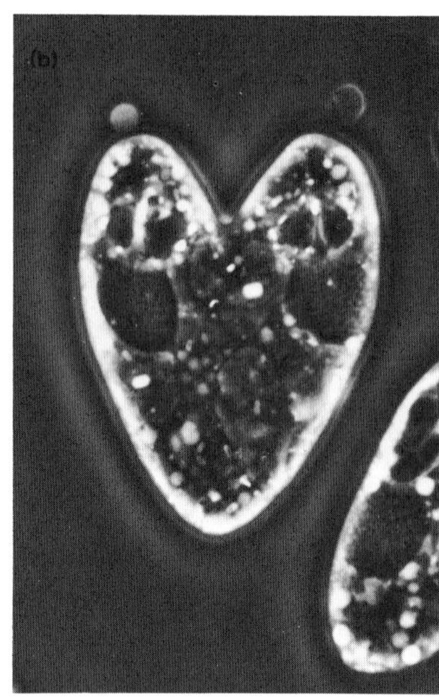

Fig. 4.6 Photomicrographs of fission in *Euglena gracilis*. (a) The cell in telophase shows the beginning of cytoplasmic cleavage with the separation of the flagellar bases and reservoirs, and (b) shows cleavage proceeding from the anterior end. Magnification × 1200. (Courtesy of Dr. G. F. Leedale.)

4.2 Reproduction by Fission

In bacteria, protozoa and protophyta asexual reproduction usually consists of division of the parent organism into two approximately equal parts (Fig. 4.6). The onset of division may be triggered in response to limits imposed either by the ratio of cytoplasmic volume to nucleus (p. 13) or by the ratio of volume of organism to surface area. The former is probably the more important.

When organisms that reproduce by fission are added to fresh culture medium in the laboratory the growth of the population (increase in number, or density, or in population biomass) follows a predictable pattern. Usually there is an initial lag phase in which little or no growth takes place. This is pronounced when organisms are

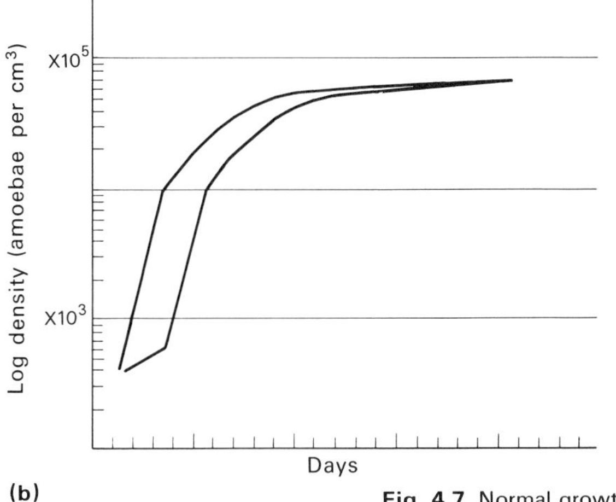

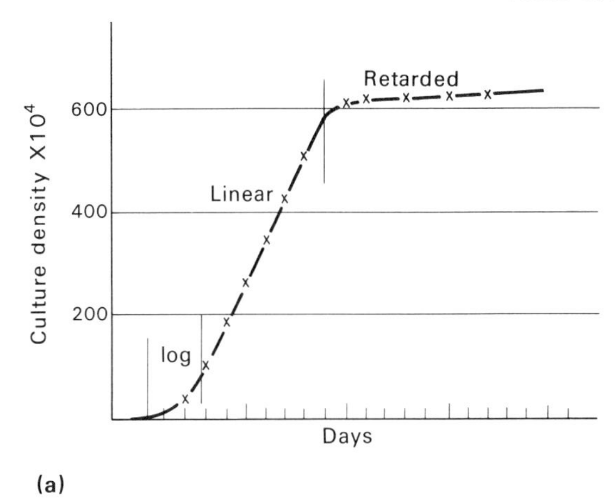

Fig. 4.7 Normal growth curves of amoebae in culture solutions, with culture density expressed on a linear scale (a), and on a log scale (b). The two curves on the log graph represent, on the left, the continued growth when the inoculum is taken from an actively growing culture, and on the right, the characteristic lag at the beginning when cells from an encysting culture are used.

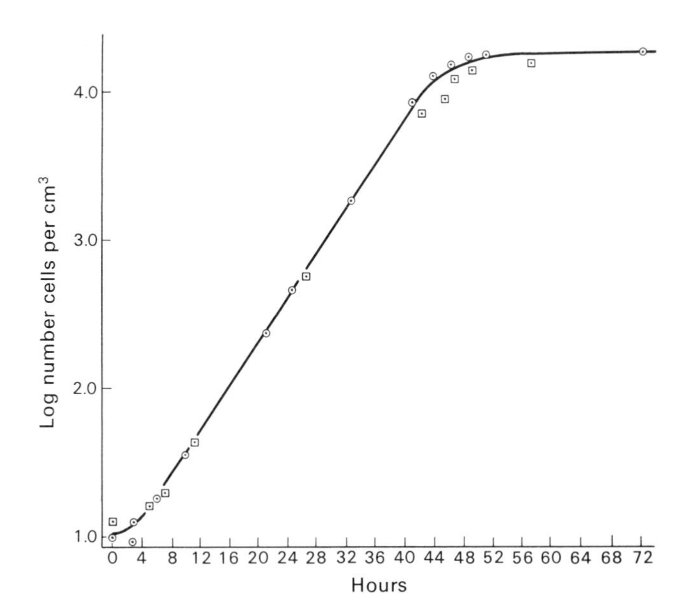

Fig. 4.8 The growth of a culture of *Tetrahymena pyriformis* in sterile peptone solution. (After Phelps, 1935.)

transferred from one type of culture medium to another and is probably associated with the adaptation of enzymes to deal with new subtrates (p. 137), although the lag phase may be absent in repeated subculturing.

There follows a logarithmic, or exponential, phase during which the population repeatedly doubles in unit time. There may follow a retarded phase in which growth is approximately linear and which may represent an interaction between the growth of the culture and an increase in some growth-inhibiting factor. The culture then enters a stationary phase in which there is little or no growth and reproduction and which is usually followed by degeneration and death of many organisms in the population (Fig. 4.7 and 4.8).

The reasons for the change from the logarithmic phase to the retarded or stationary phases may vary, but all involve the cessation of growth and reproduction. Depletion of food or of some trace ingredient and the accumulation of toxic metabolites are common factors in growth inhibition. Surface inhibition effects caused by crowding may be influential in limiting population size and many cell populations can be grown to greater densities when the cells are kept in suspension by constant agitation of the culture (Fig. 4.9).

4.3 Reproduction by Budding

In multicellular organisms asexual reproduction shows a more sophisticated growth pattern, involving localized regions, rather than fission of the whole organism. Thus the budding process of *Hydra* (Fig. 4.10a) is analogous to vegetative reproduction in, say, the strawberry plant (Fig. 4.10b). When the parent organism reaches 'full-size' growth continues in the bud. In the case of *Hydra* the bud develops from undifferentiated interstitial tissue in the ectoderm, which proliferates, then differentiates to form a miniature version of the parent. When the offspring develops a mouth and tentacles it forms a temporary 'colonial' association with the parent, since the enterons are continuous in the parent and the bud. Thus the growth resources, in terms of food animals caught, are shared on a cooperative basis until the bud leaves the parent, forms a basal disc and adopts an independent existence (see also Fig. 4.11).

In flowering plants which reproduce by suckers, runners or stolons, precisely the same principles apply. The daughter plant is first supported by photosynthetic products of the parent's leaves, transmitted in the phloem of the runner. When the daughter plant has differentiated its own photosynthetic organs a colonial status is formed, then the runner withers and decays.

Most of the flowering plants that reproduce vegetatively do so by means of modifications of the stem, such as runners (Fig. 4.10b), rhizomes, corms and tubers. Vegetative reproduction by root modifications occurs in the *Dahlia* and in *Ranunculus ficaria*, the lesser celandine. An unusual form of asexual reproduction occurs in *Bryophyllum calycinum*, a xerophytic member of the Stonecrop family, Crassulaceae. Small plantlets, complete with roots, develop from buds in notches along the leaf margins. They fall to the ground and take root to form new plants (Fig. 4.12).

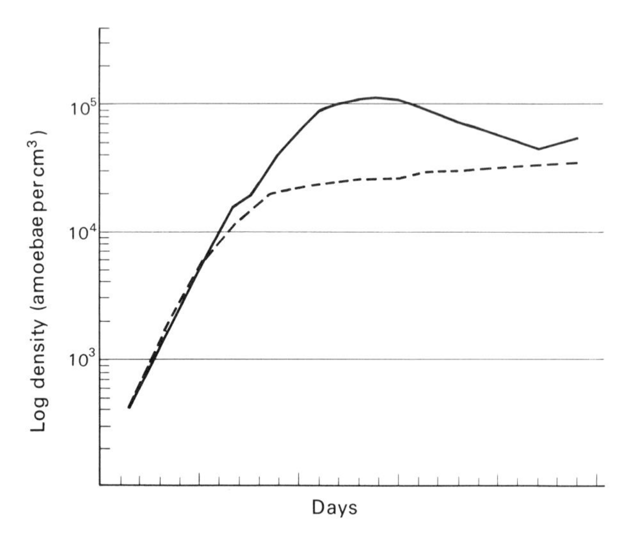

Fig. 4.9 Growth curves of amoeba cultures grown in suspension (continuous line) and in a stationary control (dotted line).

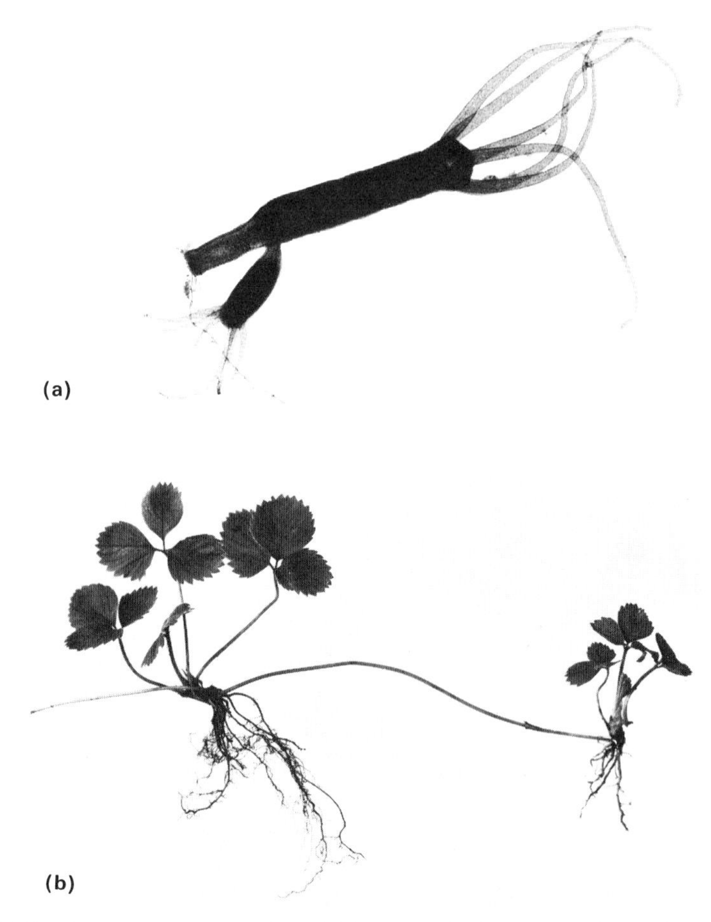

Fig. 4.10 The budding process of *Hydra* (a) is essentially the same as the production of a daughter plant by a runner in the strawberry (b), see text. (*Hydra* photograph, courtesy of Gene Cox.)

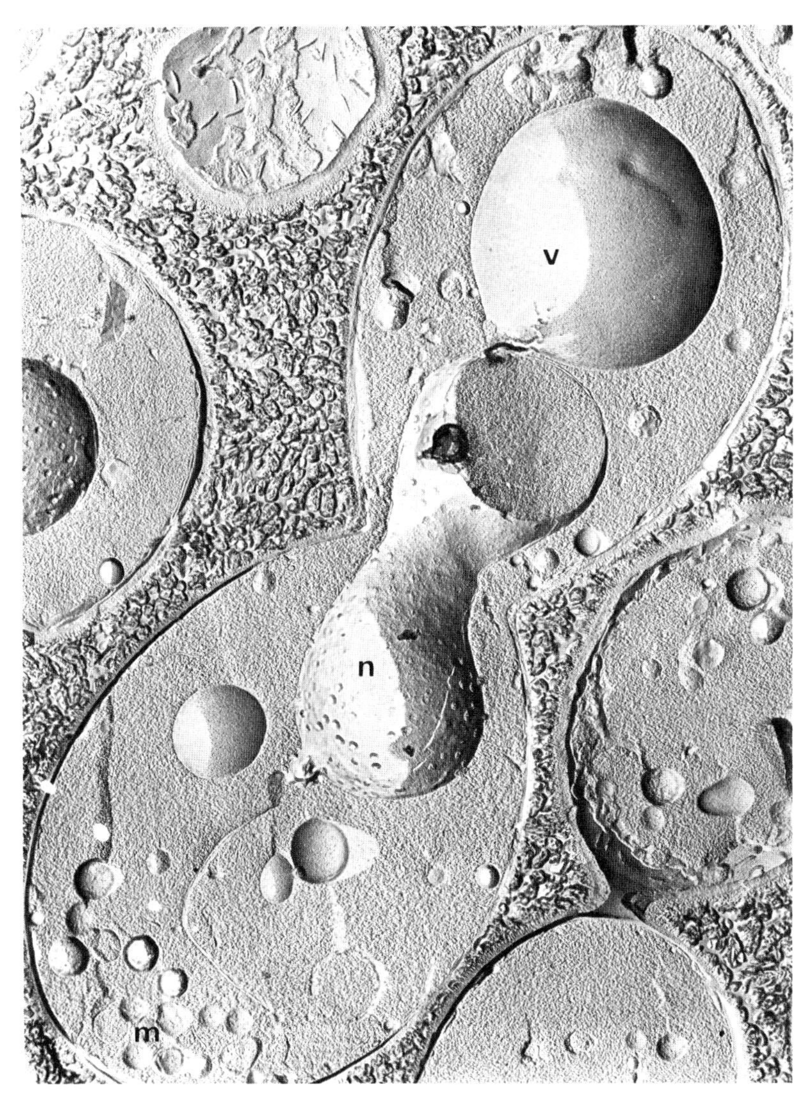

Fig. 4.11 In this freeze-etched preparation a budding yeast cell, *Saccharomyces cerevisiae*, is shown with a dividing nucleus (n) and vacuole (v). The presence of mitochondria (m) shows that the cells were grown aerobically. (E.M., courtesy of H. Moor.)

4.4 Reproduction by Spores

In the fungi there has evolved a prolific method of asexual reproduction involving the formation of spores in specialized organs called **sporangia**. The spores are microscopic and contain a nucleus and cytoplasm; they possess a relatively resistant wall able to withstand dry environmental conditions.

In the bread mould, genus *Mucor*, good growth conditions induce the formation of hyphae which grow perpendicular to the substrate (Fig. 4.13a and b). The tips of these 'upright' hyphae swell and the cytoplasm in the tip gives rise to spores, the nuclei of which are formed mitotically from the haploid nuclei of the hypha.

Fig. 4.12 Vegetative plantlets on the leaf margins of *Bryophyllum*.

The swollen hyphal tip, or sporangium, turns black and becomes brittle and turgid. When the spores are mature the central columella swells and the sporangium bursts (Fig. 4.13a and c). In dispersal the spores may be air-borne (Fig. 4.14), or, in some species such as *M. hiemalis*, the spores may remain in a gelatinous mass and are dispersed by water or insects.

An interesting mode of spore dispersal occurs in the dung fungus *Pilobolus*. The sporangium is relatively large (*c.* 2 cm) and conspicuous, having a bright yellow collar. A high turgor pressure is

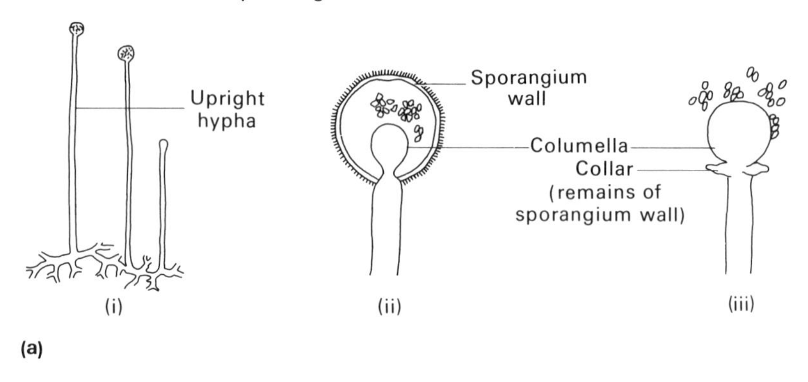

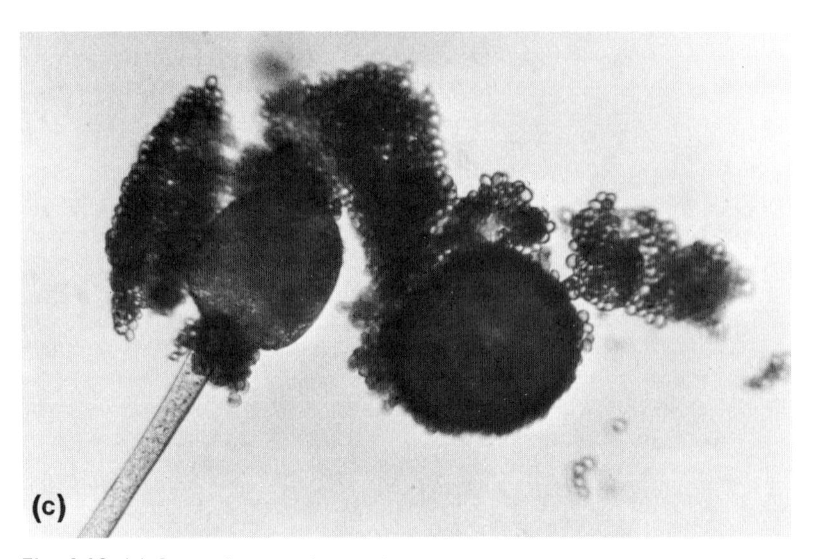

Fig. 4.13 (a) Stages in asexual spore formation in *Mucor hiemalis*, i, the formation of upright hyphae from the mass of the feeding hyphae in the mycelium; ii, the mature sporangium containing spores, in a mucilaginous slime, and the columella, and iii, after the sporangium has burst the enlarged columella remains with a collar, the attachment of the sporangium case. (b) Photomicrograph of *Mucor* sporangia growing on bread, and (c) a burst sporangium showing the sporangium case, the mucilaginous mass of spores and the columella attached to the hypha.

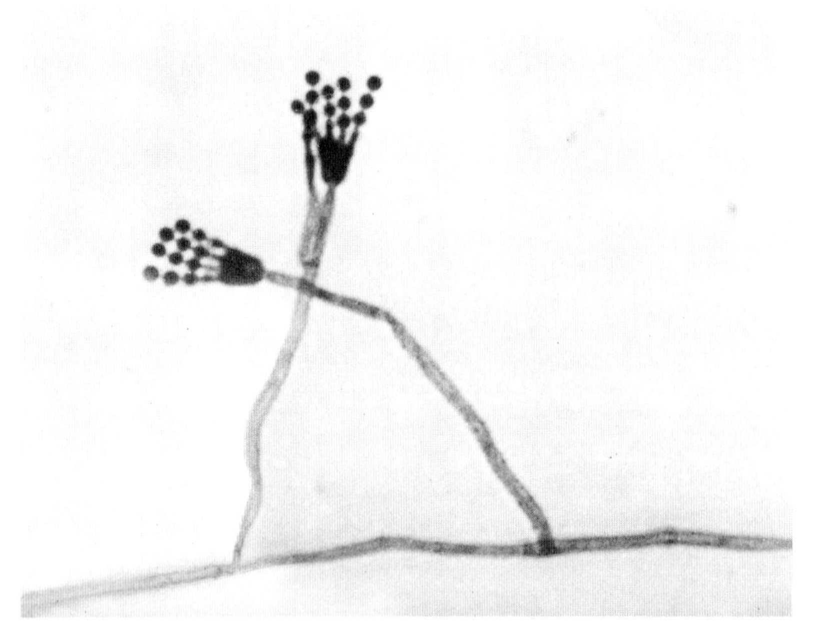

Fig. 4.14 Conidiospores of *Pencillium*. (Courtesy of Gene Cox.)

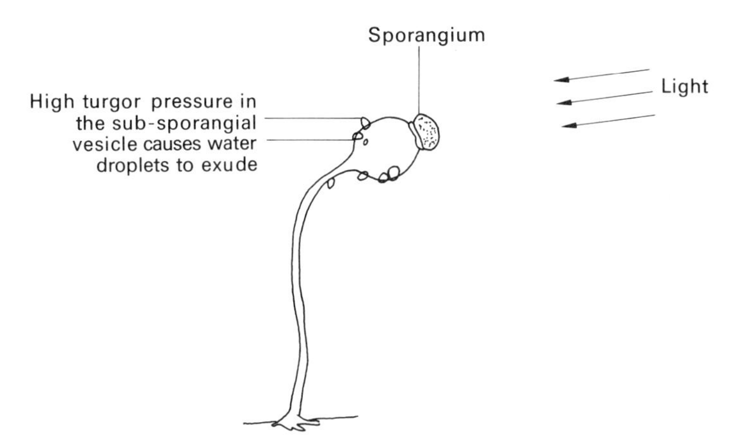

Fig. 4.15 The sporangium of *Pilobolus*, which is strongly phototropic and in which a sub-sporangial vesicle provides an explosive distribution mechanism for the spores. (After Stevenson, 1967.)

built up in a sub-sporangial vesicle which bursts explosively and ejects the sporangium and its contents for distances of up to two metres (Stevenson, 1967). The sporangiophores are strongly phototropic, if illuminated from the side they bend in that direction and shoot towards the light (Fig. 4.15).

A distinction must be drawn between **asexual** spore formation, the formation of sporangiospores and conidiospores, in which the nuclei are produced mitotically, and **sexual** spore formation. In the fungi sexual reproduction occurs by the fusion of **gametangia**, which are specialized hyphae, usually produced on different mycelia (heterothallism). In the gametangia the nuclei fuse to form diploid zygote nuclei. These divide **meiotically** to form haploid ascospores, basidiospores, etc. The spores produced in higher plants (see Chapter 6) are also produced as a result of meiotic division and although they may constitute a dispersal phase in the life cycle they cannot be described as a mode of asexual reproduction within the terms of this chapter.

5 The Evolution of Sex

Any discussion of the evolutionary origin of sex is bound to be largely speculative because of the absence of paleontological evidence about what is essentially a genetic and physiological mechanism. Nevertheless it is known that sex has played a vital part in the evolution of life on the planet and the role of sex as a generator of variability is so important in evolutionary theory that some discussion about its origins is worthwhile.

In this chapter some ideas about the origin of sex are aired and plants and animals are compared with respect to hermaphroditism. Later in the chapter hermaphroditism is discussed in relation to reproductive capacity and tissue economics. The advantages to motile animals of separation of the sexes extend beyond tissue economics to the creation of different endocrine climates in the different sexes which in turn helps to bring animals together for mating. Students interested in the origins of sex will find a number of references to general discussions, such as Maynard Smith (1958), which can be read in conjunction with this chapter.

5.1 Speculations on the Origin of Sex

It is clear that sexual modes of reproduction evolved when life on the planet was still at a relatively primitive level. Initially reproduction must have been asexual, probably by fission or budding processes similar to those seen in some extant protozoans and fungi. This is true reproduction in the 'copying' sense. The offspring are genetically identical and the characters expressed by the genes, the phenotypes, are also identical to the parent.

At some early stage a complex of mutations, that is changes in the coding sequence of the DNA, must have occurred which allowed exchange of nucleic acids between organisms, such as has been observed in the colon bacillus, or the pooling of nucleic acids by fusion of the organisms. In either case the offspring produced by subsequent divisions would possess original combinations of genes and therefore show richer variations in phenotype (Fig. 5.1).

The advantages of sexual reproduction are explained in the gene pool concept (p. 170). The pooling of genotypes in a sexually reproducing population makes it possible for mutated genes which arise in different organisms to come together in various combinations in succeeding generations. In asexually reproducing populations a mutated gene is confined to a single cell line so that combinations of advantageous mutated genes can only occur through progressive mutations in that line. Genetic analysis has shown that there is an increase in genetic variability and consequently of adaptability following conjugation (Jennings, 1929). The other advantages of sexual reproduction arising from the construction of gene complexes with selective advantage are discussed in Chapter 9.

Clearly the original mutations which resulted in the ability to combine genes would themselves be heritable. Since there is great selective advantage in possessing the capacity for variation (especially in a changing environment), such mutated genes would increase in frequency in successive generations. This is to be expected because it is now well established that selection pressures act not only on the physical and physiological features of organisms but also on the genetic mechanism itself. For example, natural selection affects the gene complex and through genic interaction can give rise to the evolution of dominance (p. 202).

Maynard Smith (1958) has argued that on one hand the advantages of sexual reproduction are both long term and concern the benefit of the population as a whole rather than the individual upon which selection acts. On the other hand the disadvantages of sexual reproduction are immediate, representing a loss in number through fusion and a handicap to the individual. He reasons, therefore, that sex is unlikely to have evolved and spread through a population if the only advantages were long-term ones. He concludes that different selective forces may have operated when sex evolved to confer short-term advantages and cites cannibalism as a possible mechanism. In this

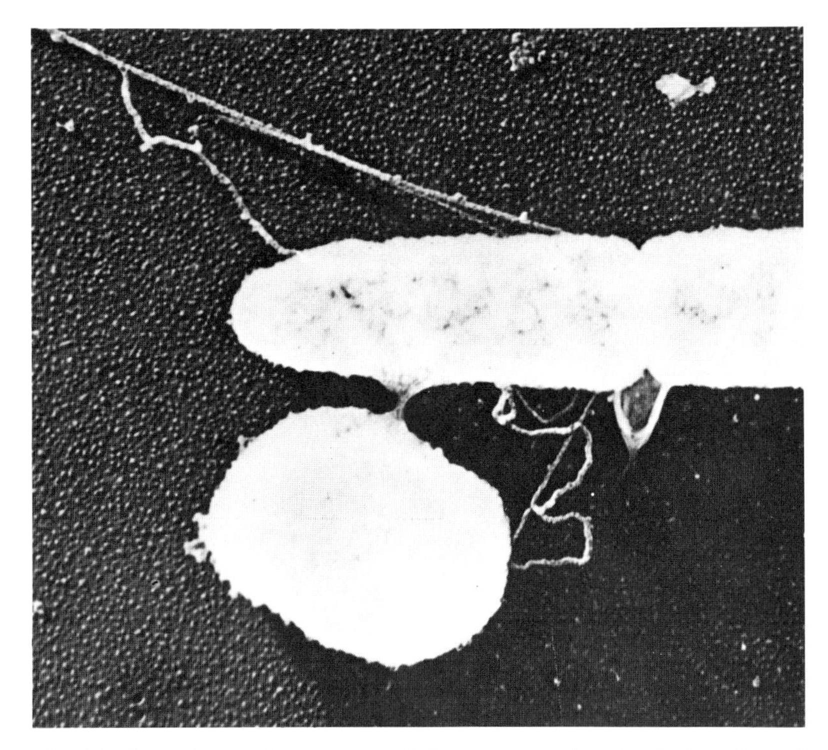

Fig. 5.1 The exchange of genetic material between organisms is a feature of sexual reproduction. Conjugation is known to take place in the colon bacillus, *Escherichia coli*. In this E.M. the Hfr 'male' cell is the long one on the right and it is connected by a temporary conjugation bridge to the shorter F^- female cell to the left. (Courtesy of Thomas F. Anderson, Elie L. Wollman and François Jacob.)

system one organism, the potential female, would ingest its mate destroying all but the genetic material which would be incorporated into the female genome. The advantage would then be immediate both in respect of nutrition and in the creation of competitive advantage through hybrid vigour.

Such a scheme is attractive and the analogy between the ovum absorbing the sperm leading to fusion of the nuclei and the cannibalistic cell is a good one. However, unless the cannibalistic cell feeds exclusively on members of its own reproductive pool (surely a selective feature considerably more disadvantageous than pairing or reduction division), it is difficult to see how the pooling of 'foreign' genetic material, which would lead to endosymbiotic associations, loss of diversity and the breakdown of speciation, could be prevented. It is certainly difficult to visualize nuclease enzymes exercising a selective role over those nucleic acids belonging to 'self' species and those belonging to other species.

In any case, the advantages of sexual reproduction need not be exclusively long-term. Jennings's experiments with *Paramoecium* showed that the rewards of conjugation were an immediate increase not only in genetic variability but also in **adaptability**. Sexual reproduction would therefore be highly advantageous in the short term when a population faced either a changing environment or a diversified one.

The pooling of genetic material from two individuals in sexual reproduction must meet two important criteria. First, it must occur between individuals of the same **species**, which form a reproductive community or pool. Members of a species can show considerable variability, but they share common features, notably from an evolutionary point of view, a definable ecological status in the biosphere (p. 229), by which they can be distinguished from other species, that

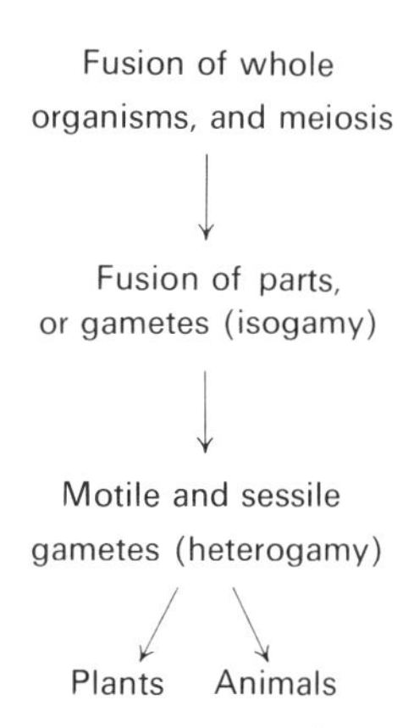

Fig. 5.2 A speculative sequence in the origin of sex.

is other organisms with which they do not reproduce. Barriers must operate to prevent the mixing of genotypes that are too unlike, since this would lead to a loss of diversity in the biosphere as a whole, at what might be termed the taxonomic level. Organisms evolving sexual reproduction, therefore, had to incorporate some mechanism of genetic recognition between 'like' and 'unlike'. Without these ırriers (discussed on p. 222) evolutionary divergence and speciation ׳ould not have been possible.

The second criterion is that the quantity of genetic material per cell nucleus must remain relatively constant from generation to generation in order that the species retain its essential and identifiable characteristics. Changes in genetic content can be a barrier to further reproduction, that is entry into the reproductive pool of the species. This necessitated the evolution of a special form of cell division in which the gene content is reduced by half either before or after nuclear fusion. That this mechanism of reduction division, or **meiosis**, is identical in plants and animals would seem to indicate that it, and the sexual process as a whole, originated before the divergence of plant and animal life forms (Fig. 5.2).

Sexual reproduction evolved not at the expense of, but in addition to, asexual reproduction. Extant primitive organisms rely mainly on asexual modes for increasing population size and use sexual modes when the environment becomes unfavourable for growth. A scheme of this nature reduces the main disadvantage of sexual reproduction, that is loss of population biomass by fusion, since it takes place when poor growth conditions lead to such a loss in any event. It also relates sexual reproduction, with its resultant increase in adaptability, directly to times when the environment changes. There is evidence, however, that in some protozoans the number of generations produced asexually is limited. There must then follow a conjugating or sexual generation, irrespective of the favourability of the environment. This can be seen to represent a clear step towards obligatory sex and alternation of generations.

Throughout the plant kingdom asexual and sexual modes of reproduction may be employed side by side. The former is associated with the spread of the plant when growth conditions are good and the latter with the formation of zygospores, seeds and so on, for overwintering. In the lower plant phyla, or divisions, in which the gametophyte generation is the more dominant, asexual or vegetative reproduction is relatively casual, depending largely on fragmentation. In higher plants vegetative reproduction is better defined, as described in the previous chapter, and is associated exclusively with the diploid, sporophyte generation.

By contrast, in the animal kingdom asexual reproduction is not normally found above the Coelenterata; its abandonment in favour of sexual reproduction therefore coincides with the origin of the triploblastic level of body organization. It seems reasonable to argue that triploblasty, particularly with the advent of the coelom, presents too high a level of complexity and cellular differentiation to allow the formation of buds like those of the coelenterates. Similarly, budding is associated with sessile or very slow-moving animals; the presence of a budding offspring might be expected to handicap seriously the competitiveness of a mobile animal, particularly in an aquatic environment where streamlining is important.

5.2 Hermaphroditism versus Separation of the Sexes

Hermaphroditism in animals and the **monoecious** state in plants are terms used to describe the occurrence of both male and female reproductive organs in a single organism. In animals hermaphroditism is largely restricted to those lower phyla in which asexual modes of reproduction, such as budding, are found. As such it is mainly confined to diploblastic animals although it does occur in some early offshoots of the coelomate stock, such as the marine arrow-worms, Phylum Chaetognatha. But it is also found in the Phylum Mollusca (notably in the land and marine snails of the Class

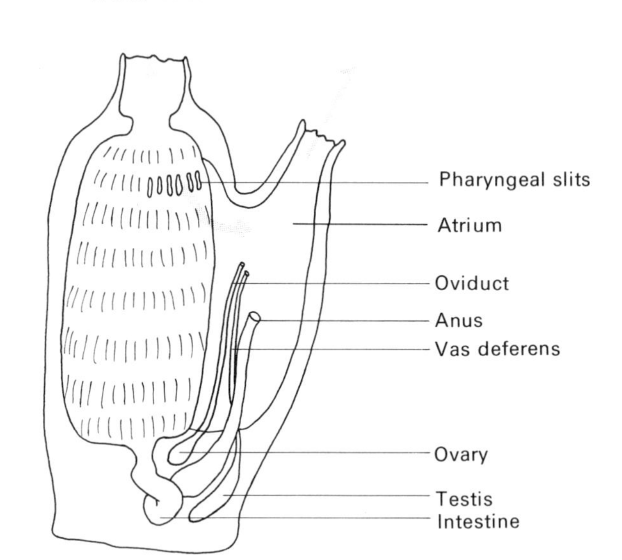

Fig. 5.3 The British sea-squirt or ascidian *Ciona intestinalis*, a hermaphroditic member of the Phylum Chordata.

Gasteropoda), the Phylum Arthropoda (in the crustacean Class Cirrepedia, which includes the barnacles) and in the Phylum Annelida. In these phyla it is conspicuously associated with sessile or slow-moving classes. Thus in the Phylum Mollusca the snails are hermaphroditic whereas in the squid the sexes are separate. Similarly in the crustaceans the sessile barnacles are hermaphroditic whereas in the motile shrimps and crabs the sexes are separate.

The obvious reason for the association between hermaphroditism and the sessile habit would seem to arise from the problems of cross-fertilization. A population consisting of all females, for example the parthenogenetic generations of some insects such as aphids and stick insects, is capable of increasing at twice the rate, or of exerting twice the reproductive pressure, of a population consisting of half males and half females. In the same way a hermaphroditic population consists wholly of egg-producing individuals and, in consequence, would be expected to exert a higher reproductive pressure. Sessile animals cannot 'find a mate' and generally shed sperm and ova (usually not concurrently) into the water in which they live, which is relatively both haphazard and wasteful. Hermaphroditism, with its attendant higher reproductive pressure, may have evolved and been retained as a counter-measure to this high loss factor.

The occurrence of what must be secondary adaptation to the hermaphroditic habit in several phyla would seem to bear this out. In the Phylum Chordata separation of the sexes is general and the behavioural processes associated with courtship, mating and care and protection of the offspring reach the greatest degree of complexity. Nevertheless, the sessile sea-squirts in the Sub-phylum Tunicata are hermaphroditic. In each individual there is a testis and an ovary, each of which opens, via a duct, into the atrium through which water is exhaled (Fig. 5.3).

In motile animals the commitment, in terms of what might be called tissue and organ economics, to reproductive tissue is surprisingly small. In the adult male rat, for example, the testes and associated reproductive structures (spermatic cord, gubernaculum, epididymis, vasa deferentia and conglobate glands) comprise less than 3% of the body weight and the testes alone form about 1% (Fig. 5.4). In the fish the comparable figure is about 2%.

		Whole body wt. (g)	Testes	Ovaries	Gonads as % of total body weight
Rat	♂	310	3.3	—	1.06%
	♀	199	—	0.04	0.02%
Pike	♂	1369	29.8	—	2.18%
	♀	1583	—	246.5	15.6%

Fig. 5.4 Table of weights of gonads in a mammal and a fish expressed as a proportion of total body weight. (Class results.)

Two other interesting points emerge from these data. First, the commitment to gonads is approximately fifty times greater in the male rat than in the female. It is well known that the male mammal produces vast quantities of sperm resulting in what has been termed a high 'failure rate'. The male rat ejaculates about two million sperm cells which at best can be expected to conceive a litter of ten, thus the sperm to zygote ratio **R** (redundancy ratio)

$$= \frac{20\ 000\ 000}{10} = 2\ 000\ 000$$

In man this ratio **R** is considerably higher. An ejaculation contains approximately 300 million spermatozoa and about one ejaculation per thousand results in conception, thus the value of **R** = 300 000 000 000. This wastage factor is not so uniformly high in the animal kingdom. In *Drosophila* a male introduces about 4000 sperm into the female which may lay up to 2000 eggs. In this case, **R** = 2.

The reasons for the large number of spermatozoa, especially in mammals, probably lie in developmental abnormalities during spermatogenesis. In man the average number of spermatozoa per cm^3 of semen is approximately 100 million. If this figure falls below 60 million it is an indication of probable sterility. Abnormal spermatozoa with mis-shapen bodies or double heads or tails are common and if these amount to about 20% of the total sterility is probably present (Hooker, 1960). Therefore, although only one spermatozoan is needed to effect fertilization, it can be argued that large numbers of normal sperm are required to assist, probably in some digestive capacity through the acrosome (p. 96).

A further source of error in spermatogenesis has been investigated by Cohen (1969). He has related the frequency of chiasmata in meiosis with the numbers of sperm produced and concludes that organisms produce the number of sperm that will about compensate for a one-third failure rate of chiasmata as well as a 40:1 failure rate because of other hazards (Fig. 5.5).

The numbers of spermatozoa appear large, but in relation to the amount of chromosomal variability (p. 181) the numbers approximate to what might be expected. In man (23 pairs of chromosomes) there are potentially 2^{23} or about 10 million combinations of chromosomes possible in sperm. The frequency of chiasmata is about 60 (Cohen, 1969) which would give a total genetic variability of about 600 000 000 which is comfortably in excess of the sperm number in any single ejaculation.

The second feature to emerge from the table of data on the weights of gonads is the great contrast between fish and mammalian ovaries. In the former the tissue commitment is approximately eight hundred times greater, which is probably chiefly a reflection of the importance of care and protection of offspring exercised by the mammal.

Since natural selection applies Occam's Razor, or the law of Parsimony, stringently to biological systems, it can be argued that

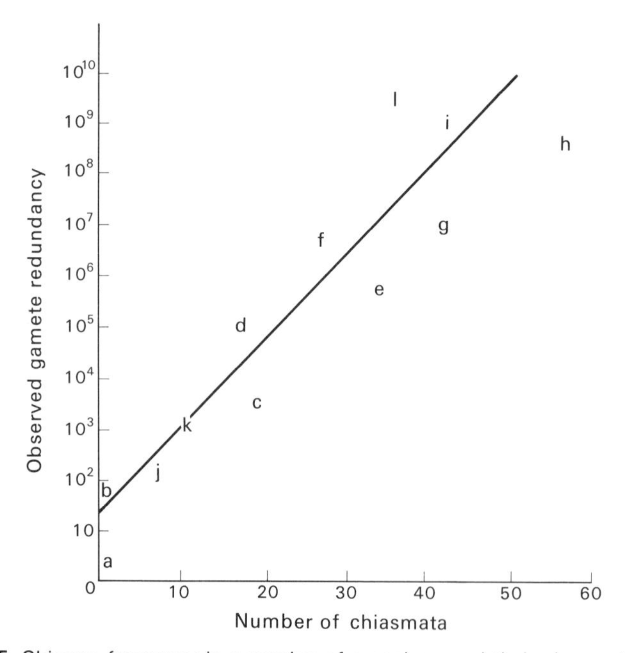

Fig. 5.5 Chiasma frequency in a number of organisms and their observed gamete redundancy. (a) *Drosophila* male, (b) Bee male, (c) Locust male, (d) Chinese hamster, (e) Mouse, (f) Golden hamster, (g) Pig-tailed macaque, (h) Man, (i) Rhesus monkey, (j) *Drosophila* female, (k) Flowering cherry pollen, and (l) Bull. A slope of about 1.5 fits the data which suggests that organisms produce a surplus of gametes that compensates for a failure rate of about one-third in the chiasmata. The slope intersects the y-axis at about 40 suggesting a 40:1 failure rate in the gametes for other reasons. (After Cohen, 1969.)

these values for the percentage of body weight given over to gonads are the minimum figures conducive to continuance of the species. Hermaphroditism, although conferring a potentially higher reproductive pressure, would therefore require approximately a doubling of the tissue commitment to reproductive organs with a complementary reduction in some or all other organs. Clearly this would be an important factor in lower chordate animals and invertebrates.

There is a more important argument against hermaphroditism in non-sessile animals. The motile habit requires that animals be brought together by shoaling, in pelagic aquatic animals, or for the exchange of sperm, in land animals, and the greater the degree of motility that is involved the greater must be the attraction between individuals. In this respect the separation of the gonads into male and female animals has created the possibility of a greater distinction between the endocrine climates of the two sexes and more powerful attractive forces as a result.

The plant kingdom would appear to fit into this pattern in that the monoecious, or hermaphroditic, state is the more common together with a non-motile habit. Also plants take measures to effect cross-fertilization as do hermaphroditic animals. However, the situation in plants is more complex and it is not possible to draw simple homologies with the animal kingdom. Plant life-cycles exhibit an alternation of gamete-producing (gametophyte) generations and spore-producing (sporophyte) generations, a feature which is associated with compromise between the advantages of sexual reproduction on one hand and diploidy on the other (Fig. 5.6 and p. 92).

5.3 Sex Ratios

In hermaphroditic animals and monoecious plants the genes determining sexual features are spread over the whole chromosomal complement. The separation of the sexes appears to have been accompanied by the linking of sex-determining genes on to sex chromosomes. This in turn has given rise to what is, in effect, a **chromosomal polymorphism**. Polymorphism has been broadly defined elsewhere (p. 211). In this case the male and female sexes represent the discontinuous or polymorphic phases in the population. The polymorphism is 'balanced' in that natural selection pressures ensure that the frequencies of the sexes and therefore of the sex chromosomes themselves, remain relatively constant. In a natural population any tendency of one sex to increase in frequency at the expense of the other is bound to be opposed by selection (Ford,

1965).

In mammals the proportion of males to females in a population tends to be approximately equal. In lower chordates this is sometimes not the case. Populations of lake trout were found to show variations in sex ratio from lake to lake, the most extreme being 700 males to 100 females (Kyle, 1926). In the tropical and sub-tropical Cyprinodont fishes the sex ratio can vary widely, being extreme in the case of *Mollienisia formosa*. Populations of this species are exclusively female and their eggs are activated (stimulated to divide rather than being fertilized in the sense of nuclear fusion) by the sperm of related species with which they shoal and upon which they are what amounts to 'reproductive parasites'.

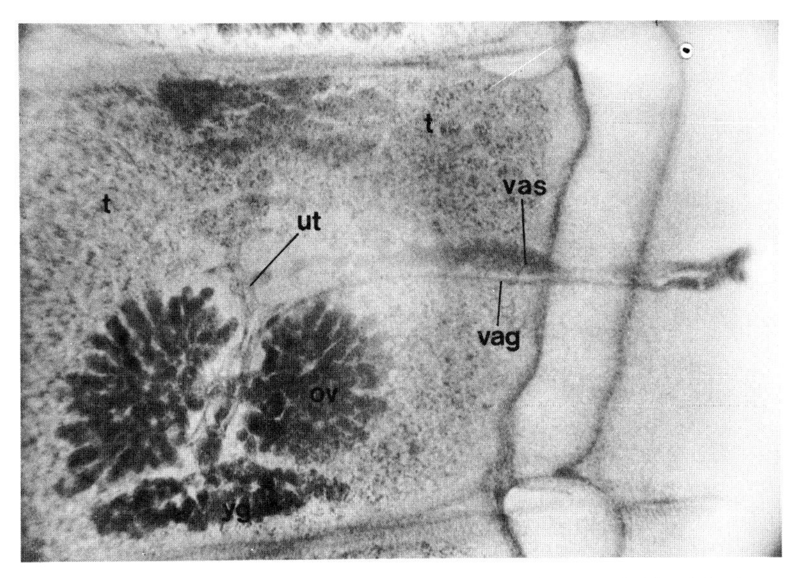

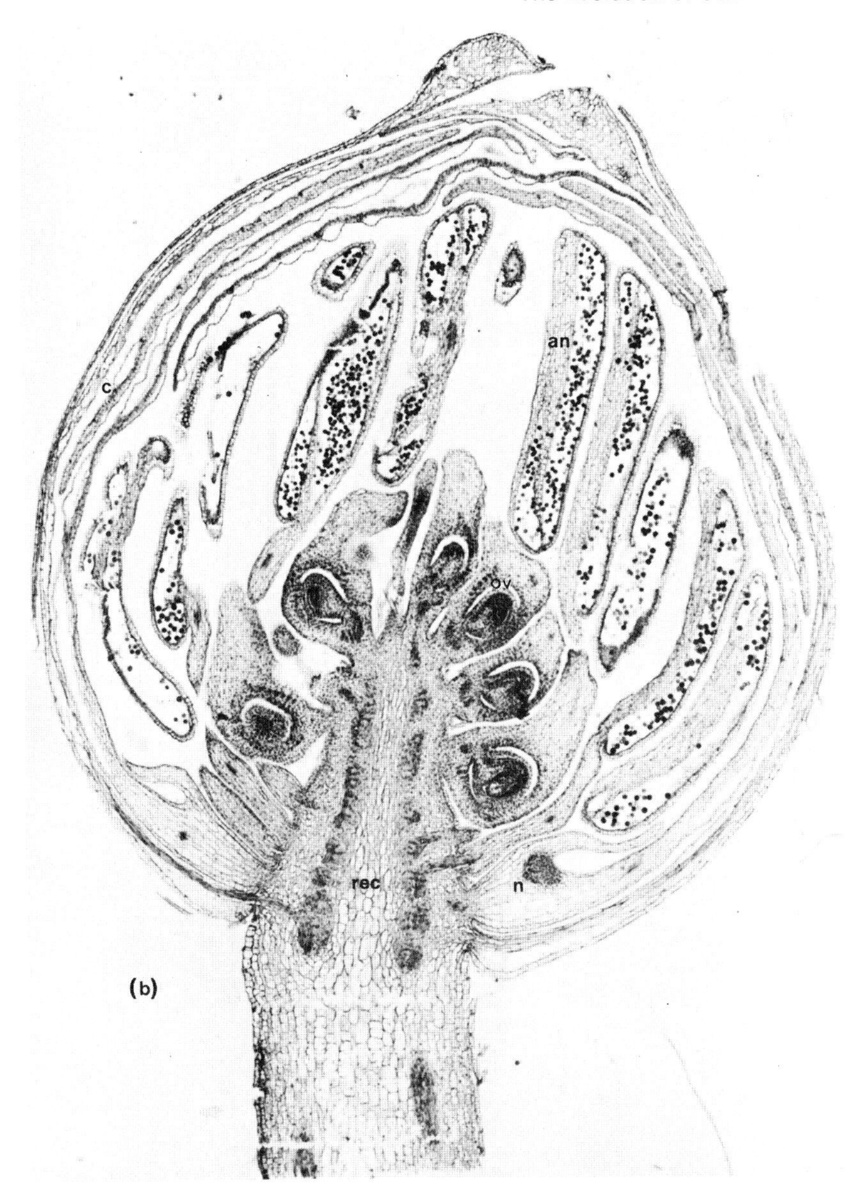

Fig. 5.6 Hermaphroditic organisms, (a) the sex organs in a proglottis of the tapeworm, *Taenia*, where self-fertilization is generally the rule because the animal is isolated in the host's intestime: (t) testis, (ut) uterus, (ov) ovary, (yg) yolk gland, (vag) vagina and (vas) vas deferens. (b) the sex organs in the flowering plant. The male gametes are contained in the pollen grains in the anthers (an) and the female gametes reside in the ovules (ov). Both sets of organs arise from the receptacle (rec). Unlike *Taenia*, the buttercup is cross-fertilized; the corolla (c) and nectary (n) are involved in making the flower conspicuous and in providing a bait for the insects which carry out pollination (By courtesy of Gene Cox.)

Modes of Sexual Reproduction

No one who takes more than the most cursory glance at nature can fail to be astonished by the extraordinary variety of life forms on the earth. A plankton net drawn for ten minutes through the surface waters of the sea, or an insect trap set on a summer's day, or observations of the bird life in a wood and the flowers of a hedgerow, yields such a great wealth and variety of organisms as to engage the attention of a naturalist for several lifetimes.

Yet at a cellular level these various organisms exhibit many common biochemical pathways and have in common many genes coding for the enzymes in those pathways (see Chapters 2 and 3). That such a uniform organic basis should produce this variety of organisms is due to the addition of what might be called supplementary genes of different qualities and in different quantities. The role of sex in evolution has been to keep mixing these supplementary genes and provide a source of variation upon which natural selection could act. It is with this in mind that sexual reproduction has been stressed as a fundamental mechanism in the evolutionary patterns discussed in this book.

This chapter deals with some of the mechanics of sexual reproduction. It begins with an analysis of **meiosis**, the form of cell division found only in association with sexual reproduction and continues with a description of the formation of male and female gametes in plants and animals. This is followed logically through fertilization and embryo formation in plants and animals. In conclusion, there is a discussion of differentiation and the concept of the **organizer** in the development of the embryo.

6.1 Sexual Reproduction and Meiosis

Sexual reproduction occurs throughout both animal and plant kingdoms, with the exception of some protozoans in which the chromosome number is so great that the pairing of chromosomes in meiosis cannot take place. It is probable that most bacteria do not reproduce sexually, although exchange of DNA has been recorded in *Escherichia coli* (p. 149). In principle sexual reproduction involves the union of genetic data, that is nucleic acids, from two organisms, or parents. The offspring of sexual reproduction possess a genetic complement which is accrued equally from each parent.

Such offspring, unlike those produced from asexual reproduction, differ genetically from their parents and from each other. Populations that reproduce sexually cannot give rise to clones and such populations exhibit wide genotypic and phenotypic variation (Fig. 6.1).

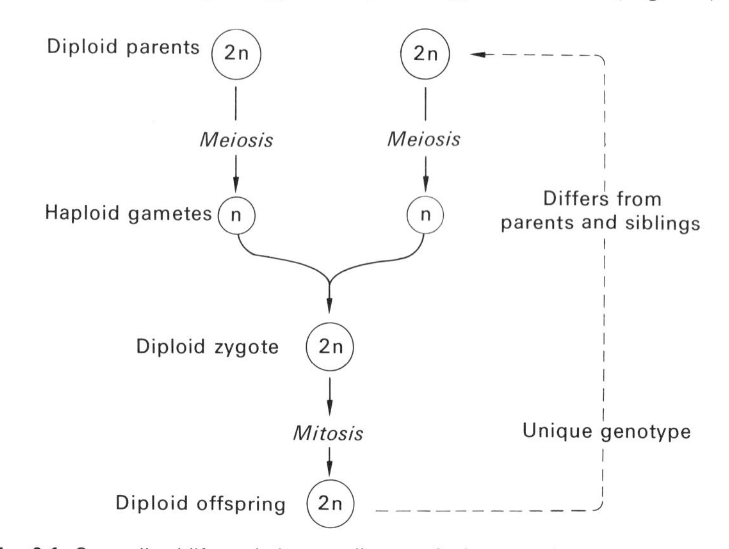

Fig. 6.1 Generalized life cycle in sexually reproducing organisms.

6.2 Meiosis

Meiosis is a relatively rare occurrence. It takes place only once in the life cycle of a sexually reproducing organism whereas mitosis occurs

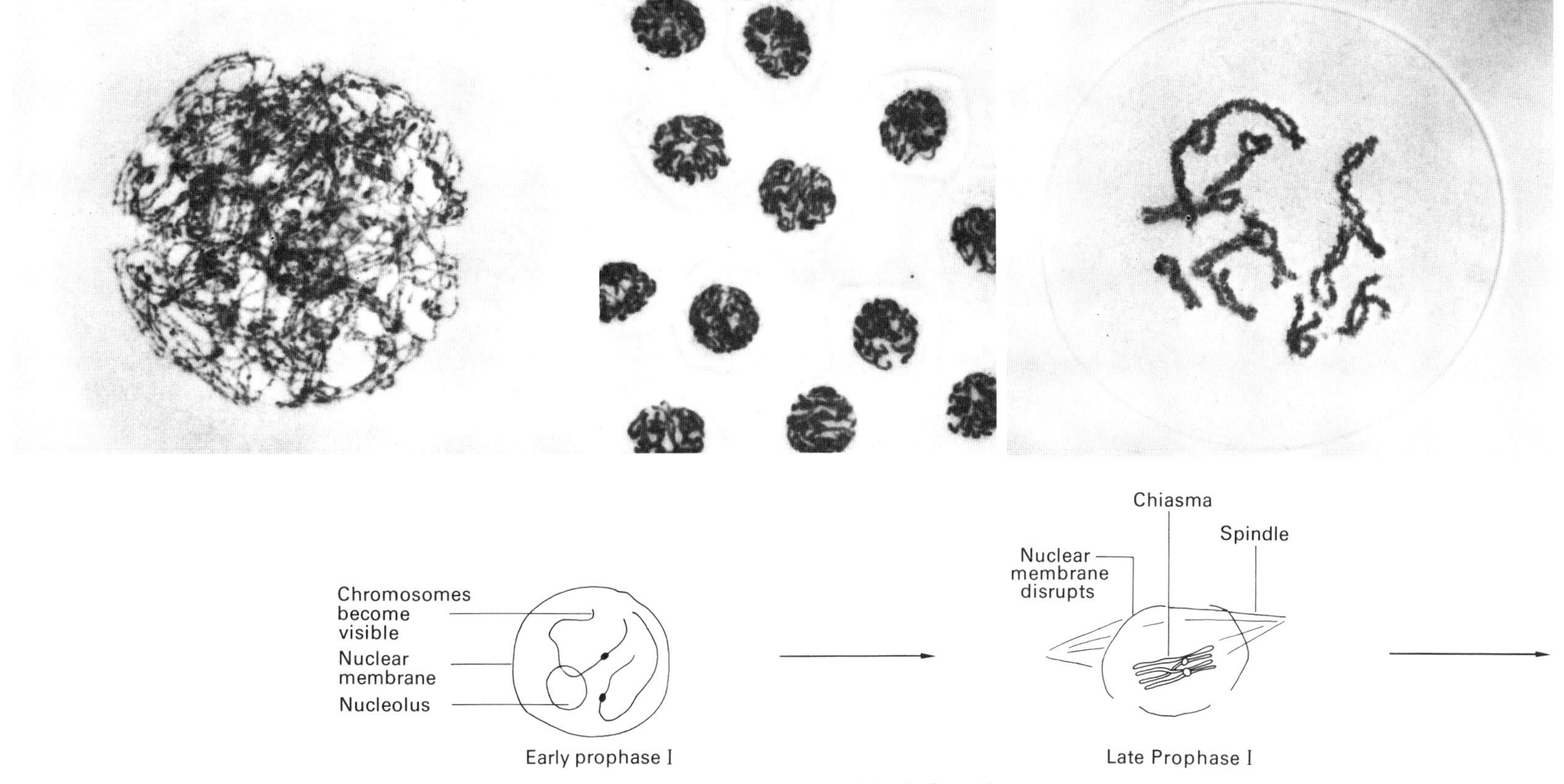

Fig. 6.2 (Pages 85 to 88) Meiosis (for explanation, see text). (Photographs by courtesy of Dr. B. Snoad.)

repeatedly in growth. Meiosis consists of two distinct nuclear divisions (I and II) resulting in four daughter nuclei. Chromosome behaviour during Meiosis I differs from mitosis in that the chromosomes form homologous pairs (called 'tetrads', since each chromosome is split lengthways into two chromatids) which separate at Anaphase I. In Meiosis II, however, chromatids separate and the process more closely resembles mitosis (Fig. 6.2).

Prophase I

Condensation of chromatin in early prophase makes the chromosomes visible under the light microscope as thin threads. **Synapsis** then occurs. Each chromosome 'zips' accurately to its homologue so that the centromeres are adjacent and allelic loci are accurately apposed. A pair of synapsed chromosomes is termed a **bivalent**.

The chromosomes of a bivalent are split lengthways, except at the centromeres, forming a **tetrad**, and chiasmata (p. 181) occur commonly. During these changes in the nucleus a meiotic spindle has been forming in the cytoplasm. At the end of Prophase I the tetrads compact and shorten, and the nucleolus and nuclear membrane disappear.

Metaphase I

The tetrads move to the spindle equator and the centromeres become attached to microtubular fibres from opposite poles.

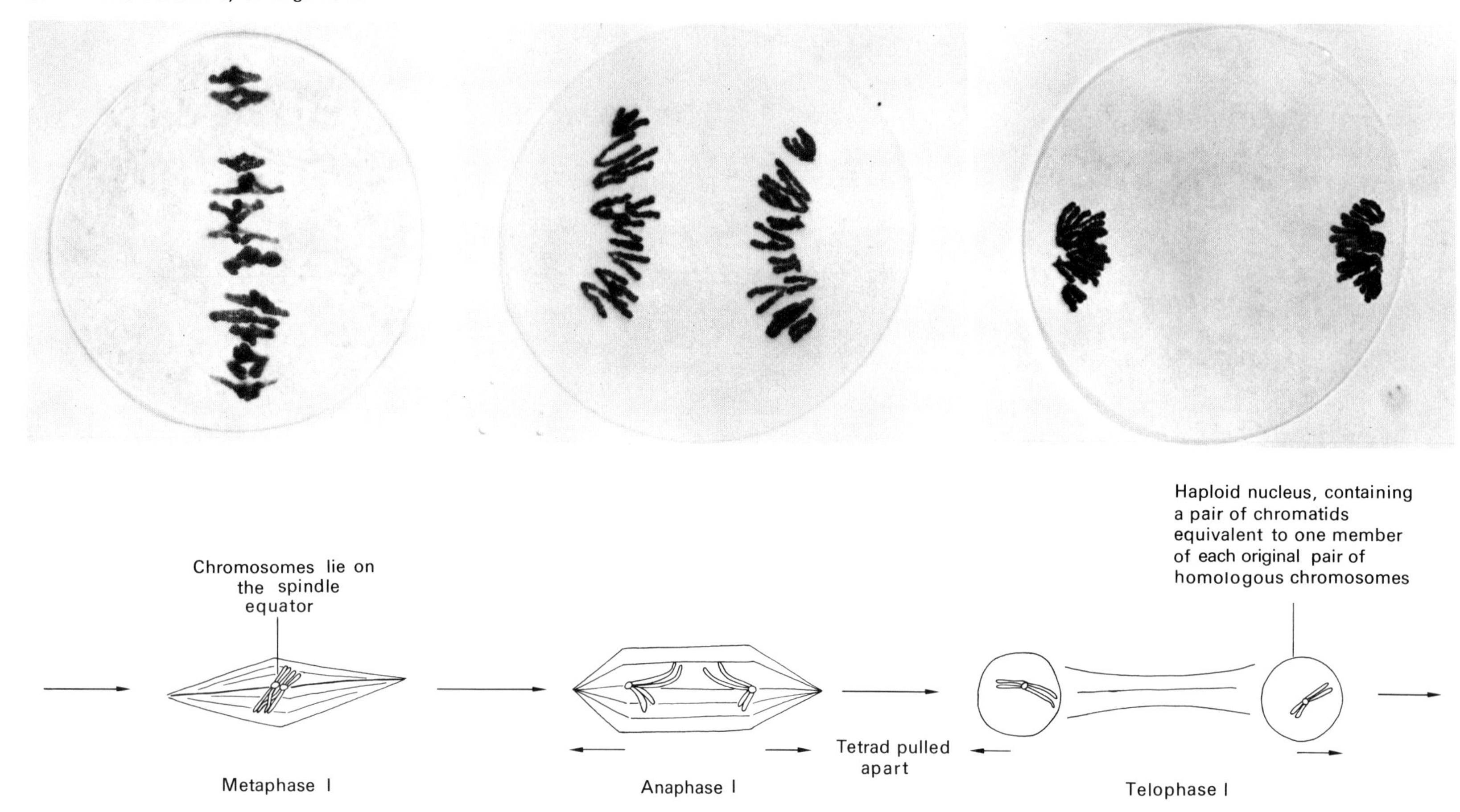

Fig. 6.2 (continued)

Anaphase I

At anaphase the centromeres are pulled apart. To each centromere is attached a pair of chromatids corresponding to the original bivalent, but of different genetic constitution than in early Prophase owing to the exchange of chromatid material in crossing-over. This stage represents the key to the meiotic process in that homologous chromosomes are separated and will reside ultimately in different daughter nuclei, or gametes.

Telophase I

As far as the nuclei are concerned this represents the reverse of the events of Prophase; the chromosomes disperse and become invisible under the light microscope and nuclear membranes re-form. Cleavage of the cytoplasm takes place. In spermatogenesis this is symmetrical but in oogenesis it is asymmetrical forming one daughter cell which has most of the cytoplasm and a polar body (p. 98).

The interphase which follows Telophase I is usually brief, making both meiotic divisions a continuous process.

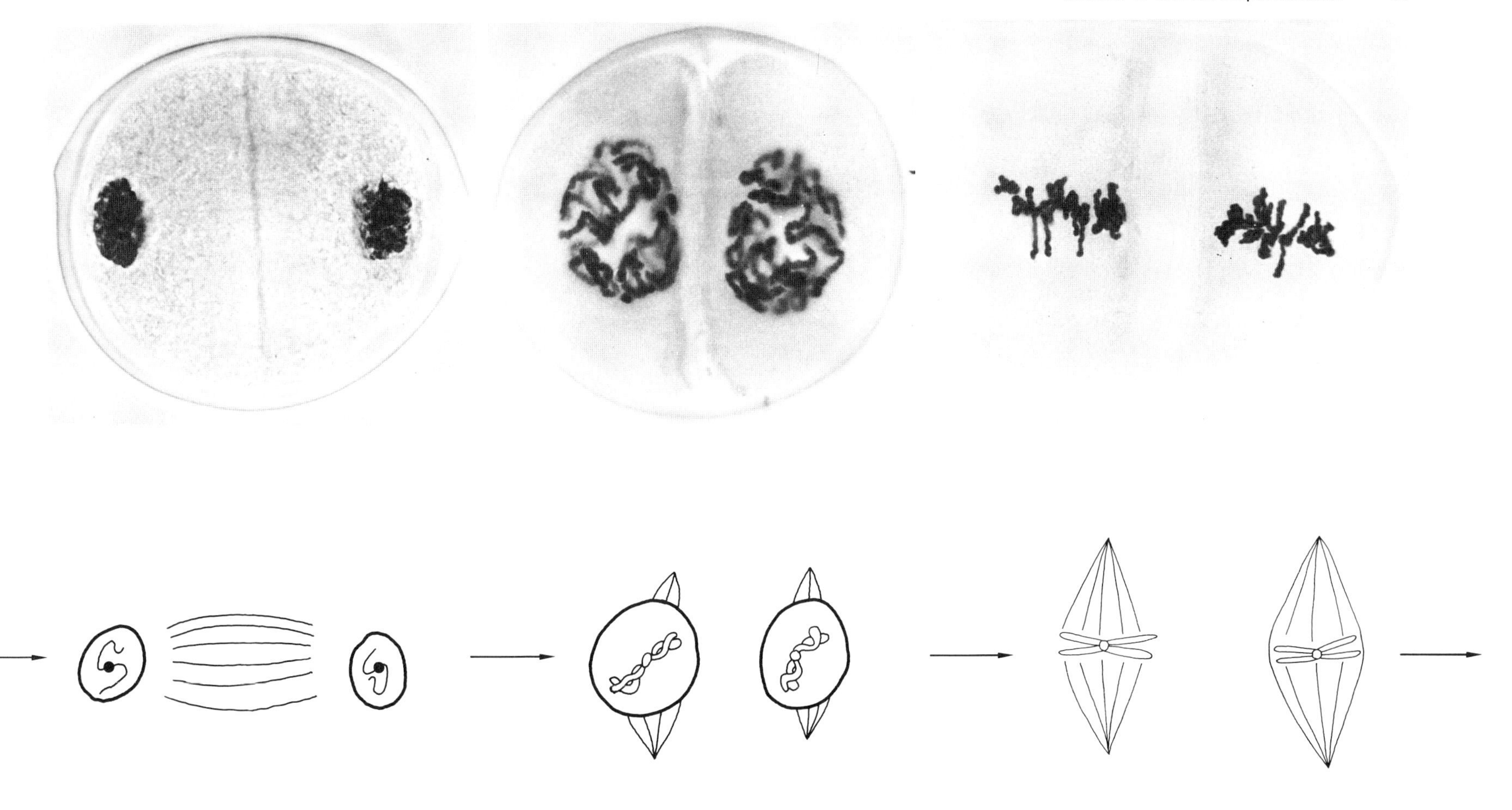

Prophase II

The chromosomes condense and become visible, the spindle re-forms and the nuclear membrane and nucleolus disappear. The behaviour of the chromosomes differs from that in Prophase I in that there is, of course, no pairing of homologous chromosomes, since only one member of each pair is present in a single nucleus.

Metaphase II

The chromosomes, as pairs of chromatids, move on to the spindle equator, forming the Metaphase II plate, and are attached to the contractile elements at the centromeres.

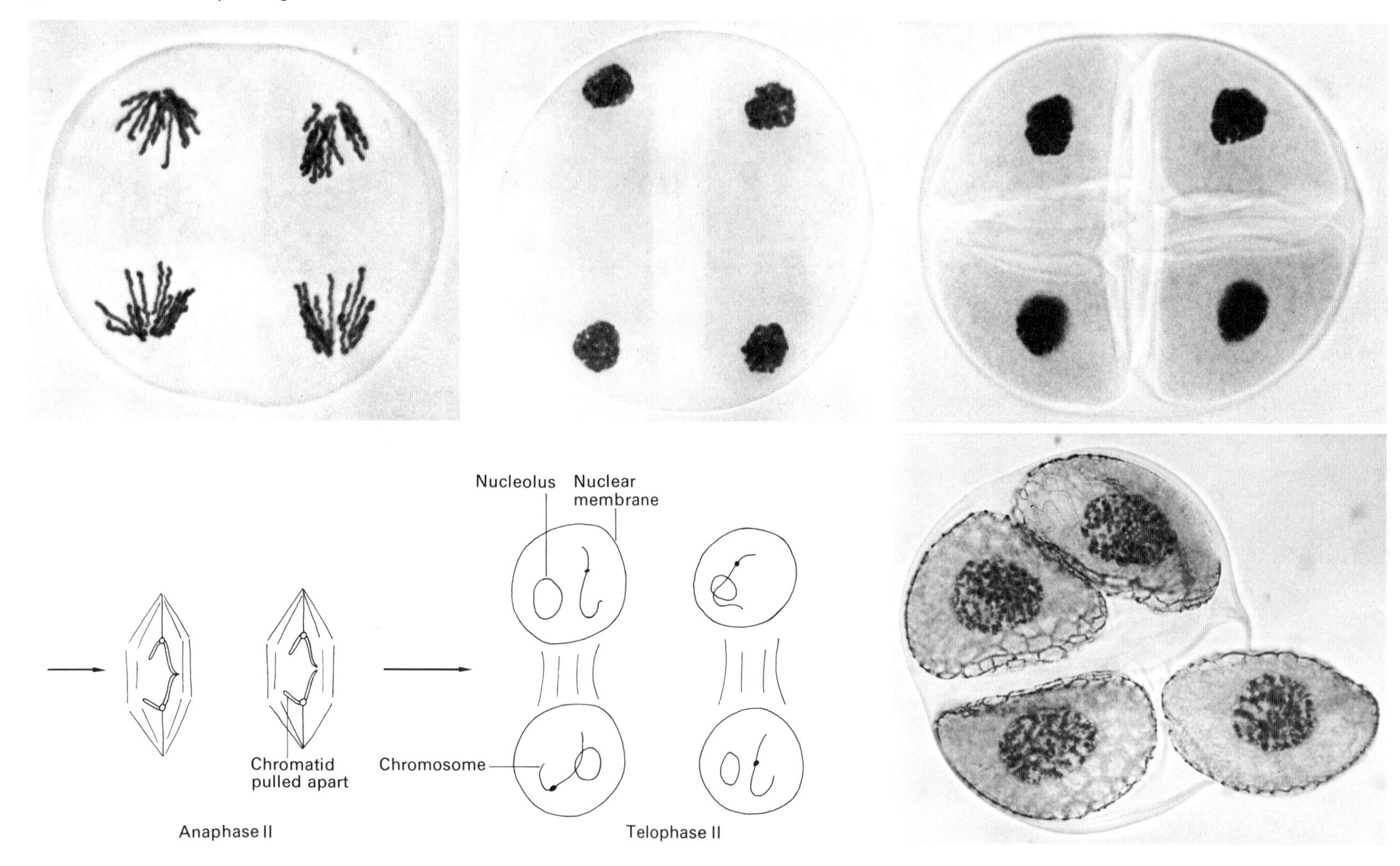

Fig. 6.2 (continued)

Anaphase II

This involves the splitting of the centromeres and the separation of the chromatids, each of which ultimately becomes a single chromosome.

Telophase II

This is essentially the same process as occurs in Telophase I. At the conclusion of Telophase II four daughter nuclei have been formed. In the subsequent formation of male gametes these usually give rise to four gametes, although in the formation of female gametes only one of the nuclei, by retaining most of the cytoplasm, is destined to be a gamete and there are three accompanying polar bodies.

6.3 Alternation of Generations

Throughout the plant kingdom, from some members of the Thallophyta to the Angiosperms, there exist two distinct stages or generations in the life cycle, one diploid and the other haploid, which alternate one with the other. In the animal kingdom, however, no such alternation of generations is found and organisms are diploid, gametes being produced by meiosis (Fig. 6.3).

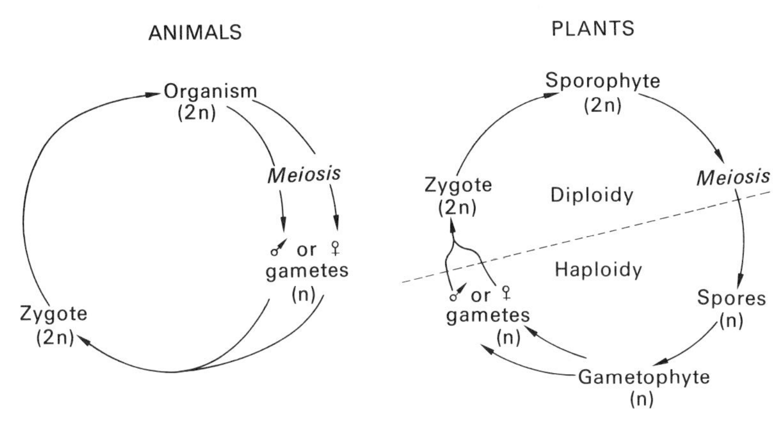

Fig. 6.3 Animal and plant life cycles showing the haploid gametophyte which exists in the plants. Some insects are said to show alternation of generations, for example the alternation of parthenogenetic and oviparous females in the aphid annual cycle, but this is not an alternation of haploid and diploid generations since both are diploid.

The advantages of diploidy in terms of the evolution of a population are obvious; it doubles the genetic complement of the gene pool (p. 170) and it permits graded phenotypes according to the penetrance of the alleles concerned. This confers a great degree of variation in a population in which mutant alleles give some measure of heterozygosity. Natural selection pressures therefore are more likely to induce effective adaptations.

Animals, from quite an early stage in evolution, took evolutionary advantage from both diploidy and sexual recombination (p. 180) as sources of variation. Plants, on the other hand, appear to have opted early in their evolution for a predominantly haploid life form in which sexual reproduction occurred. After fertilization of the gametes a diploid sporophyte generation was formed as a dispersal phase, producing haploid spores by meiosis. In primitive plants this sporophyte generation was insignificant (in a physical sense) and wholly dependent upon the gametophyte.

This pattern is reflected in the life cycles of extant liverworts and mosses in which the prominent generation is the haploid gametophyte (Fig. 6.4).

In the more advanced Pteridophyta (ferns, horsetails and club-mosses) alternation of generations is still conspicuous, but the roles are reversed. The sporophyte has become the prominent generation and the gametophyte is reduced. This change in pattern can be examined in terms of the conquest of land. Compared with an aquatic environment terrestrial environments show greater extremes in many respects, such as in temperature, concentrations of essential elements and compounds and so on, and there is a greater variety of habitats. Since the exploitation of different habitats and the survival of species in changing habitats demands the capacity for adaptability the importance of the diploid sporophyte generation as the stage in the life cycle most exposed to natural selection pressures becomes obvious.

In the shield fern *Dryopteris felix-mas* the fronds, which rise from a short, thick rhizone, represent the sporophyte generation. On the underside of the frond spore mother cells are produced in sori; these cells divide meiotically to form haploid spores. After dispersal the spores germinate and develop into a small gametophyte similar to the thallus of the liverworts.

The gametophyte, or **prothallus**, possesses chlorophyll and is independent of the sporophyte. It bears antheridia and archegonia similar in structure to those of the Bryophyta. The male gametes have numerous flagella and swim to the archegonia through the water film covering the prothallus. After fertilization of the female gamete the resultant diploid zygote develops into a diploid sporophyte, or fern plant (Fig. 6.5 and 6.6).

Origin of Heterospory

A few extant species of club-mosses, notably *Selaginella* spp. produce two kinds of spores. In **microsporangia** many spore mother cells divide meiotically to form microspores, each of which subsequently gives rise to a microthallus.

In **megasporangia**, on the other hand, only one spore mother cell divides, forming four megaspores; the other cells fulfil a nutritive role in the formation of four large spores. These spores subsequently give rise to megathalli.

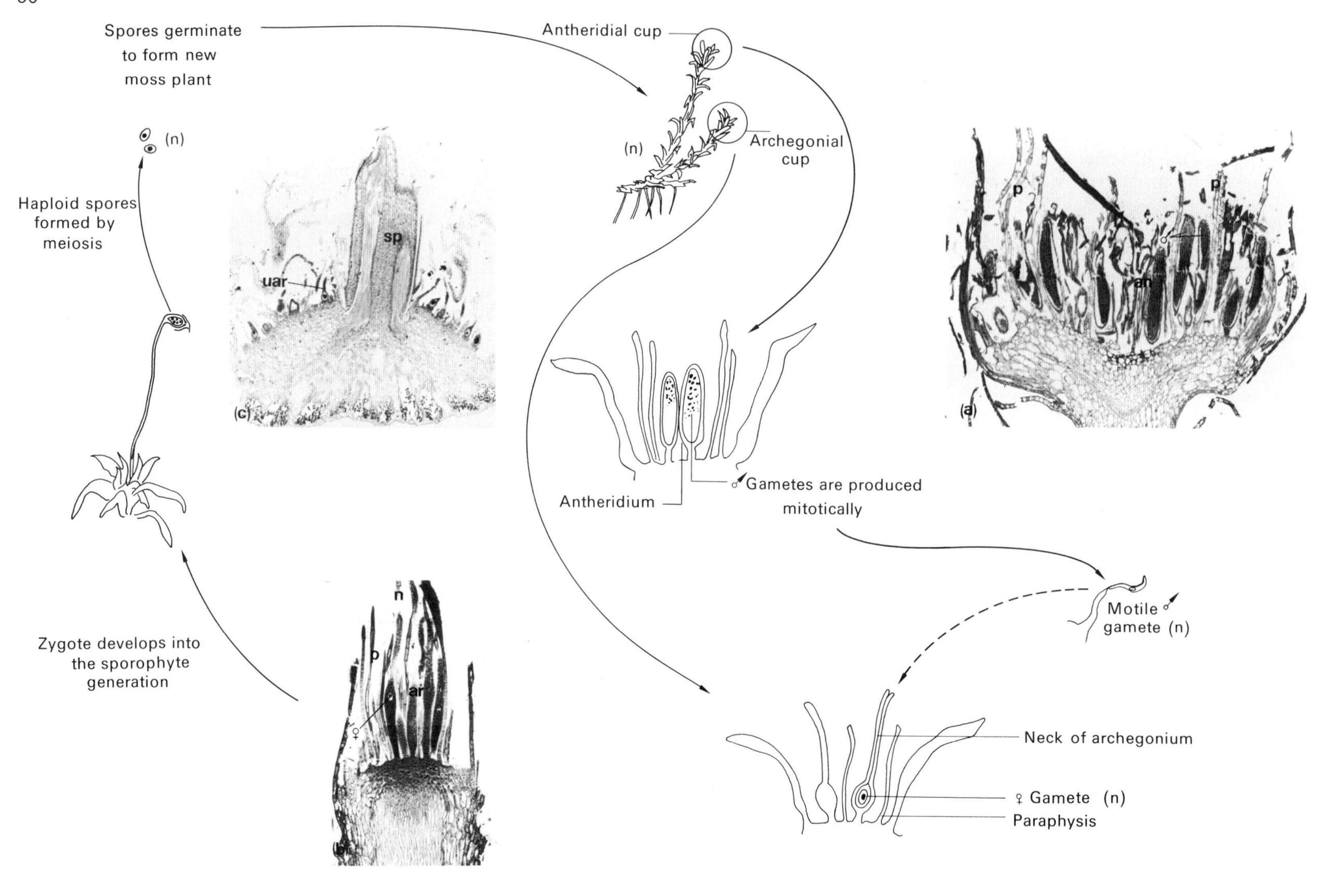

Fig. 6.4 Life cycle of the moss *Funaria hygrometrica.* (a) Section of antheridial cup of the moss gametophyte showing antheridia (an) bearing ♂ gametes and sterile hairs or paraphyses (p). (b) The archegonial cup of *Funaria* showing archegonia (ar) bearing the ♀ gametes. (c) On the archegonial cup one of the fertilized ♀ gametes develops into the sporophyte (sp) and the rest of the archegonia remain undeveloped (uar). (Photomicrography, courtesy of Gene Cox.)

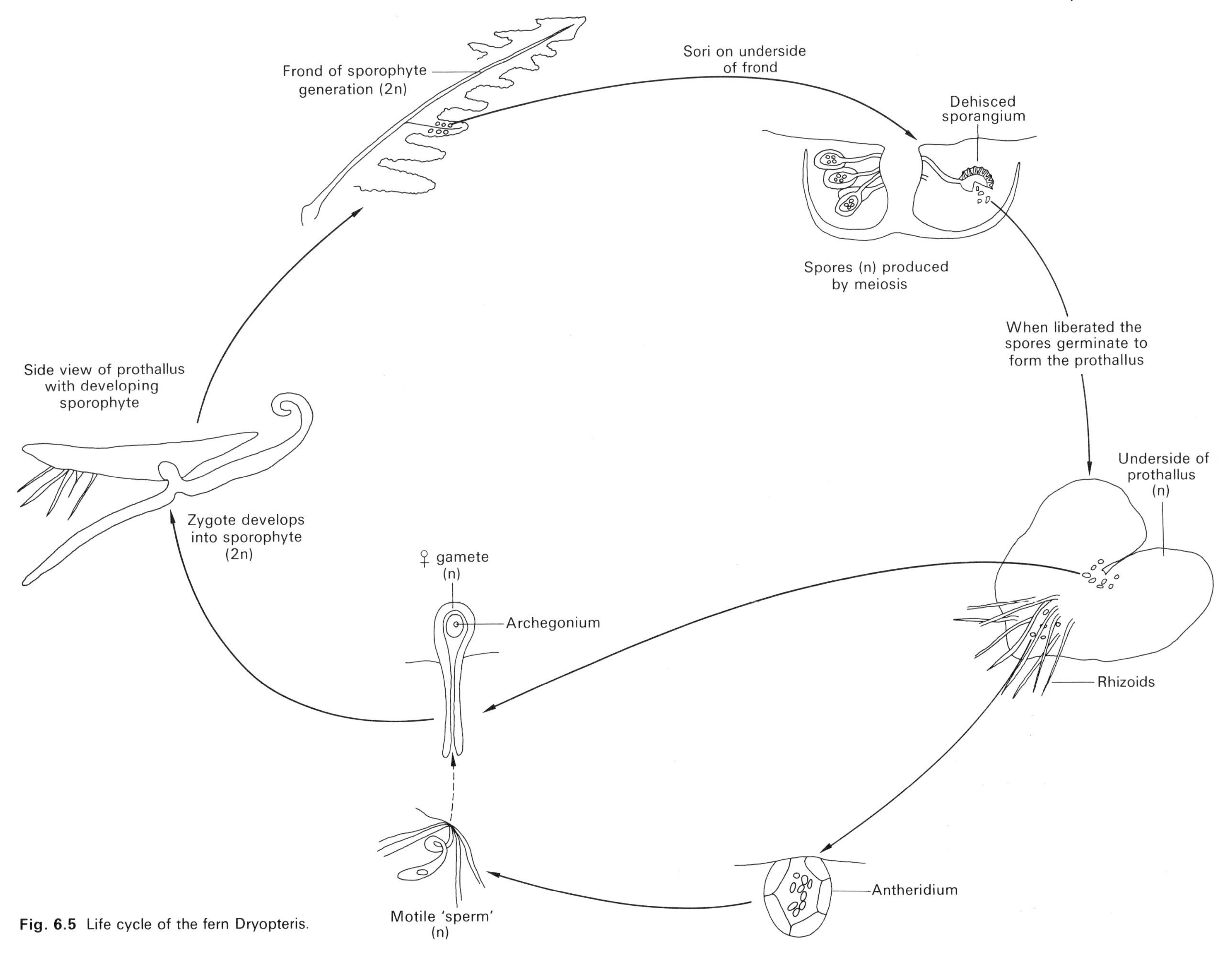

Fig. 6.5 Life cycle of the fern Dryopteris.

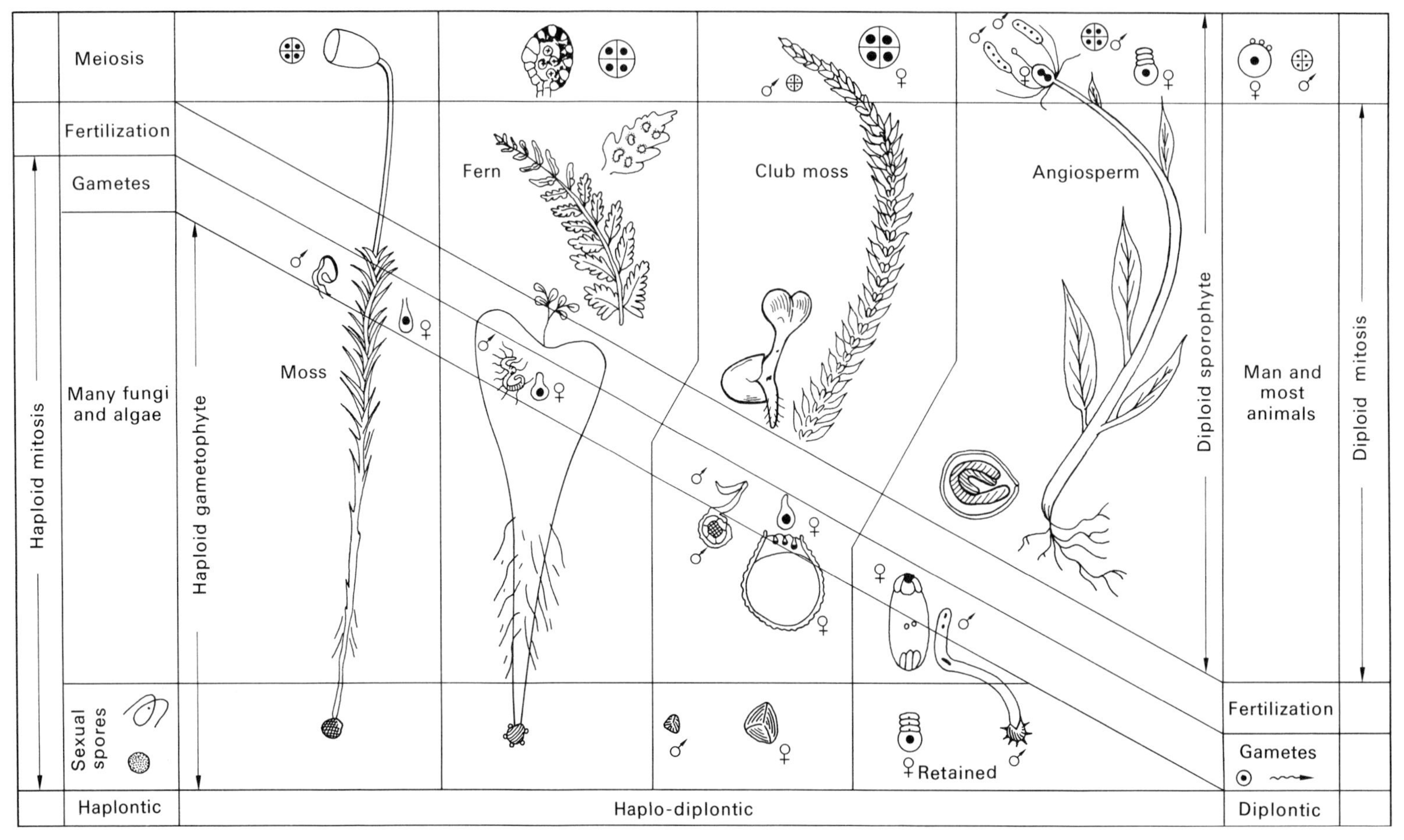

Fig. 6.6 The life cycles of the main plant groups showing the decrease in the gametophyte generation and the corresponding increase in the sporophyte as one proceeds up the evolutionary scale. (Courtesy of Dr. Roger Kemp.)

The microthallus is reduced to a few cells and carries one antheridium. The megathallus is more fern-like and carries several archegonia. Fertilization and subsequent development follows the typical Pteridophyte pattern (Fig. 6.7):

As far as *Selaginella* is concerned heterospory would seem to introduce the additional hazard of sperm transmission from one thallus to another. The advantage of heterospory lies in the increased likelihood of a zygote being formed by the fusion of gametes coming from different plants, resulting in increased genetic mixing.

Heterospory is used to advantage in the seed producing plants, the Gymnosperms and Angiosperms, which may have evolved from some Pteridophyte ancestor similar to the heterosporous club-mosses.

Alternation of Generations and Heterospory in Flowering Plants

The conquest of land and the complete exposure of the diploid generation to the pressures of natural selection were achieved in the higher plants by the further reduction of the gametophyte generation

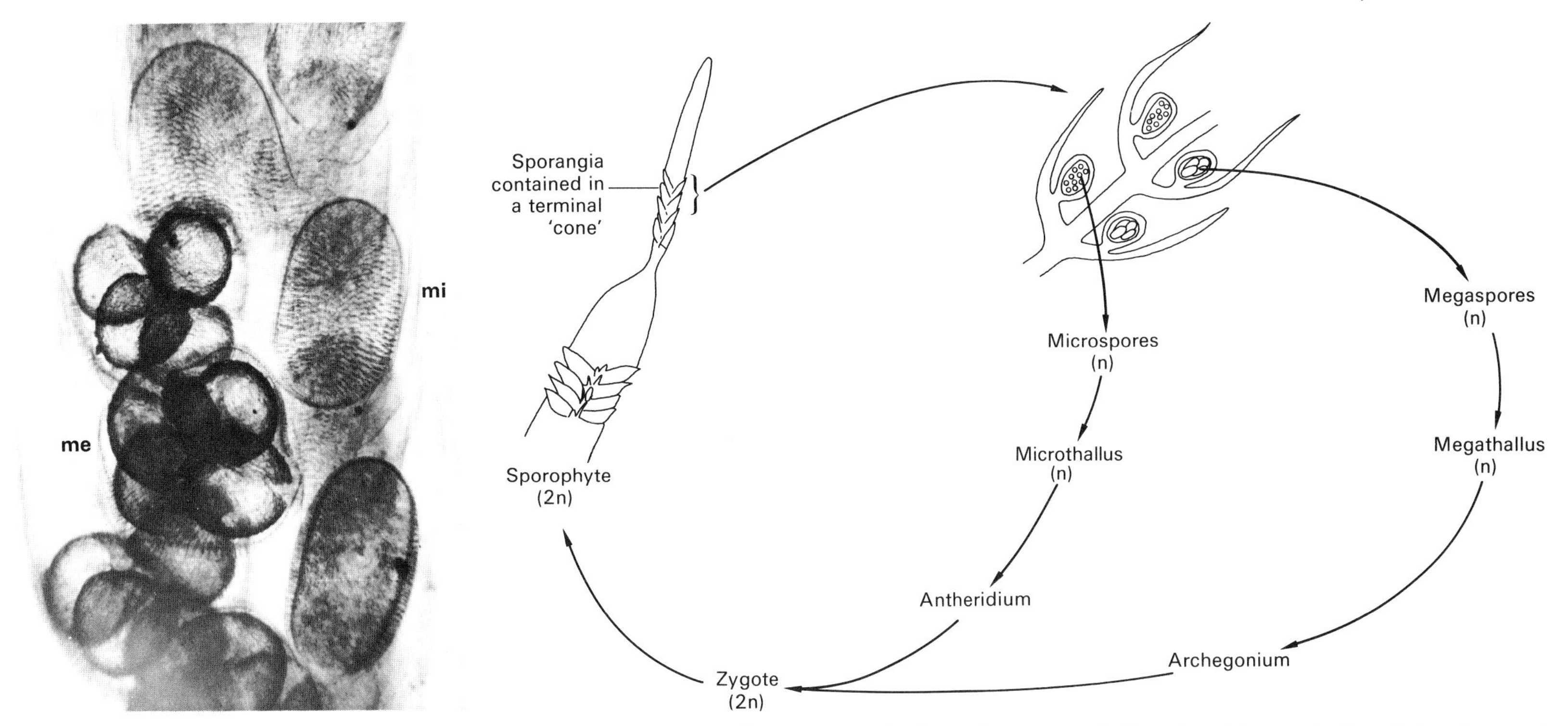

Fig. 6.7 Heterospory in *Selaginella*. (Left) Photomicrograph of part of the cone of *Selaginella selaginoides*, the heterosporous club-moss, showing megasporangia containing four megaspores (me) and microsporangia (mi). (Right) The life cycle of *Selaginella* shows the separate thalli produced by germination of the micro- and mega-spores.

and its retention within the sporophyte. In the Pteridophyta external water in the form of rain or heavy dew is essential for fertilization. In, say, the fern the independent prothallus therefore represents an obstacle in the life cycle to the exploitation of a dry terrestrial environment.

In the flowering plant microspores are produced by meiosis in the pollen sacs of the anther. The young pollen grain can be thought of as the microspore and the mature pollen grain as the micro- or male gametophyte. The male gamete is the fertilization nucleus (p. 103).

The megaspore of the flowering plant is the young embryo sac which is formed by meiotic division of a megaspore mother cell in the nucellus of the ovule. The mature embryo sac with its eight haploid nuclei (p. 99) represents the mega- or female gametophyte and the nucleus which will give rise after fertilization to the embryo is the female gamete (Fig. 6.8).

Thus the flowering plants, by a circuitous evolutionary route, come to resemble closely the typical animal life cycle. The extreme reduction of the gametophyte means, in effect, that the diploid organism produces haploid gametes by meiosis within its reproductive organs.

6.4 The Formation of Male Gametes

The process of spermatogenesis follows much the same pattern throughout the animal kingdom (Fig. 6.9). In the testes diploid spermatogonial cells divide mitotically to produce primary spermatocytes which in turn divide meiotically to form secondary spermatocytes (product of Meiosis I) and spermatids (product of Meiosis II). The spermatids differentiate to form spermatozoa or male gametes.

In flowering plants (Fig. 6.9) diploid pollen mother cells in the anther divide meiotically to produce pollen grains (Fig. 6.10). When the pollen grain subsequently germinates on a stigma, its pollen tube digests a pathway down the style. One of its nuclei, the generative nucleus, then divides to form two male gametes analogous to spermatozoa.

This mode of gamete formation more closely resembles the lower animals than the higher ones in that it is non-continuous; one batch of cells originates from a stroma and differentiates into male gametes simultaneously. When the gametes are dispersed from the organ, its life is ended. This happens, for example, in male gamete formation in *Hydra* (Fig. 6.11).

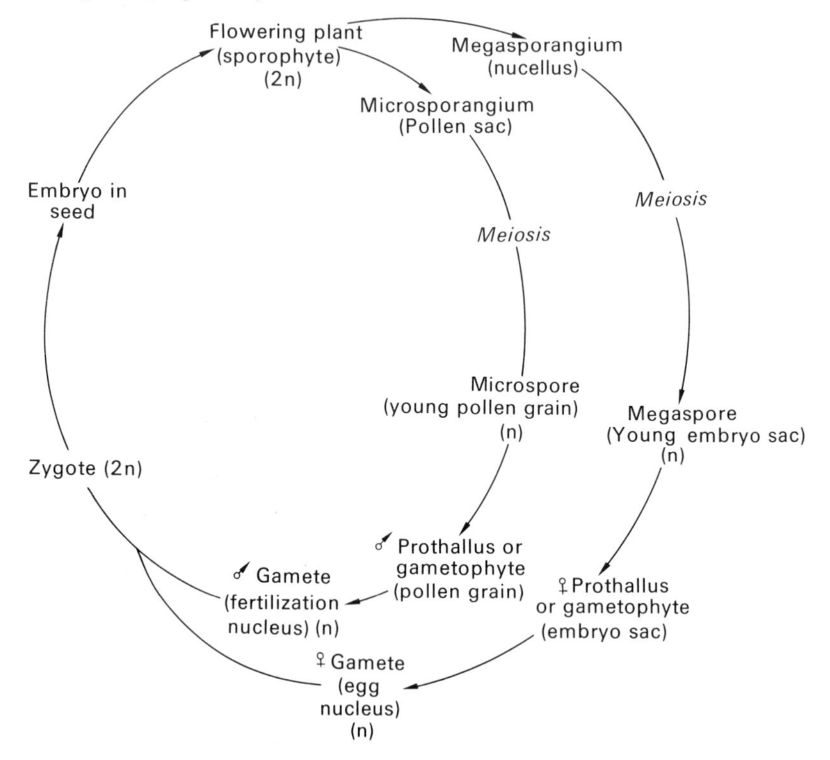

Fig. 6.8 Flowering plant life cycle.

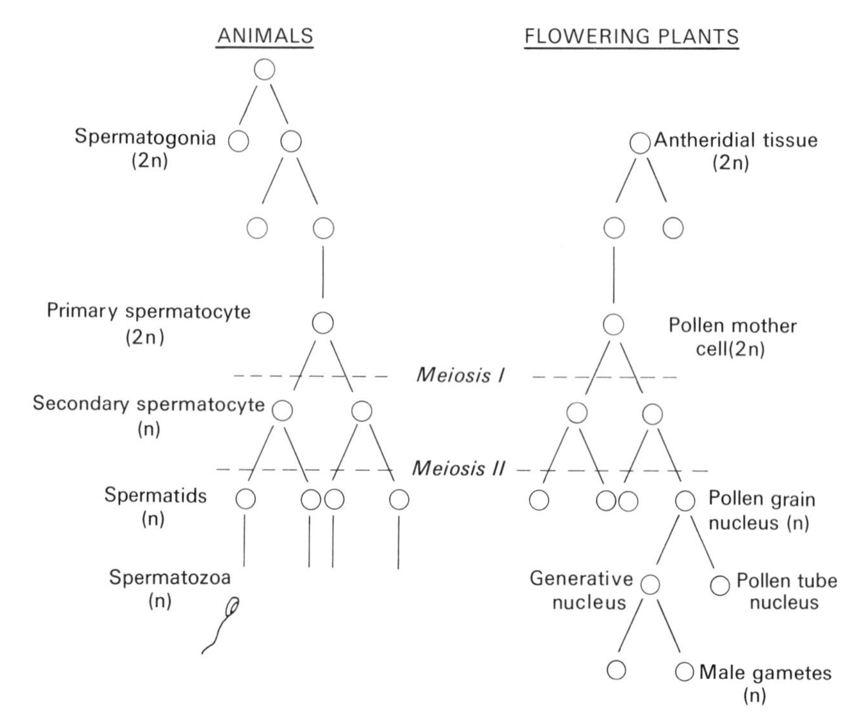

Fig. 6.9 The formation of male gametes in plants and animals showing the essentially similar patterns of events.

In higher animals, however, spermatozoa formation occurs continuously during breeding periods and the same organ persists throughout the life of the organism. In man spermatozoa are produced continuously from puberty to senility (Fig. 6.14).

In transverse section the seminiferous tubules show a peripheral layer of germinal tissue or spermatogonia in constant mitotic division, shedding cells towards the central lumen. These cells divide meiotically and subsequently differentiate into spermatozoa. In the final stages of differentiation the spermatozoa are associated with Sertoli cells which are believed to have a nutritive role and are particularly conspicuous in human seminiferous tubules (Fig. 6.13).

With a few exceptions (such as some crustaceans and nematodes which have amoeboid sperm) the spermatozoa of animals and of lower plants possess one or more flagella or tails for locomotion (Fig. 6.12).

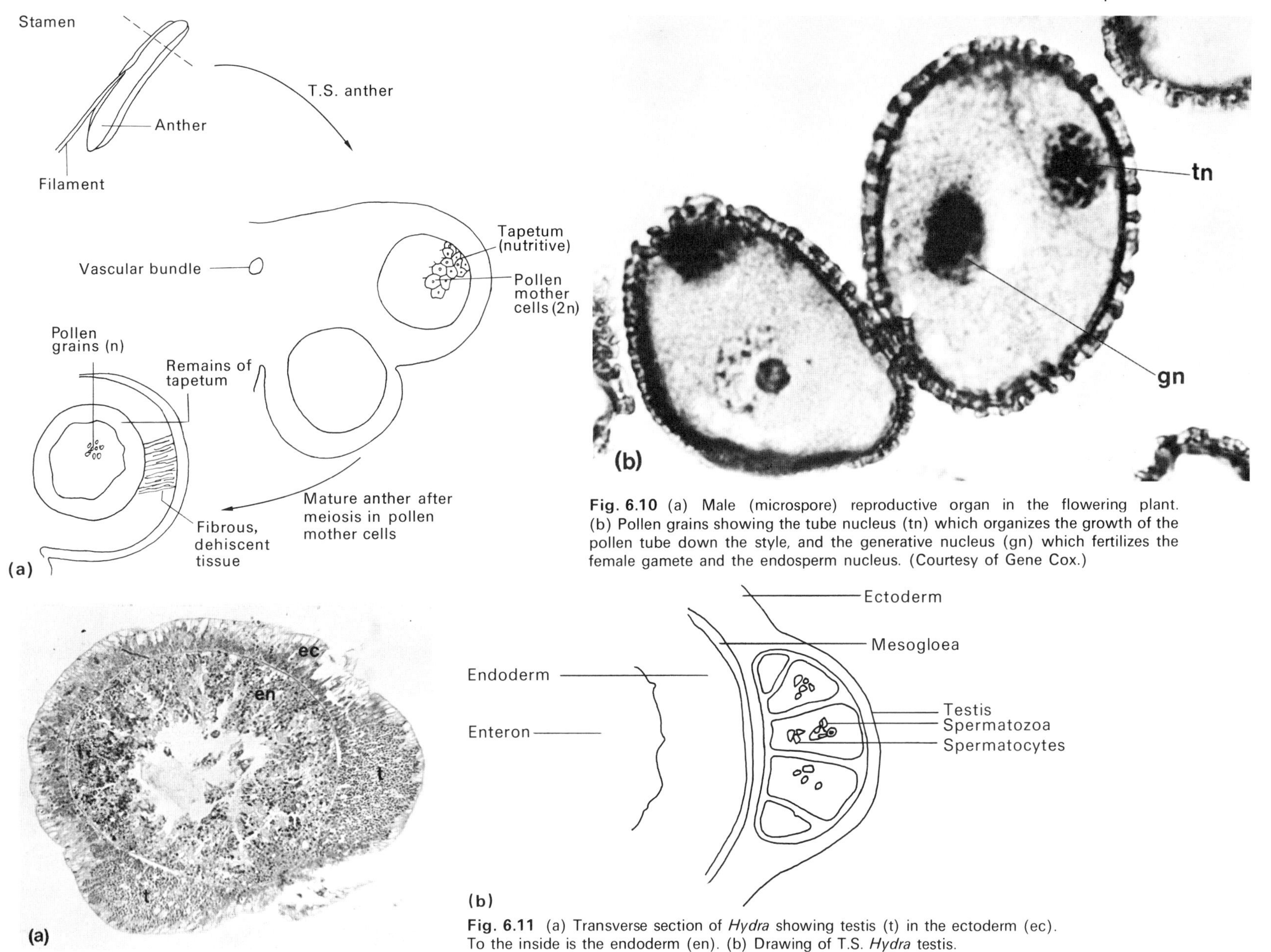

Fig. 6.10 (a) Male (microspore) reproductive organ in the flowering plant. (b) Pollen grains showing the tube nucleus (tn) which organizes the growth of the pollen tube down the style, and the generative nucleus (gn) which fertilizes the female gamete and the endosperm nucleus. (Courtesy of Gene Cox.)

Fig. 6.11 (a) Transverse section of *Hydra* showing testis (t) in the ectoderm (ec). To the inside is the endoderm (en). (b) Drawing of T.S. *Hydra* testis.

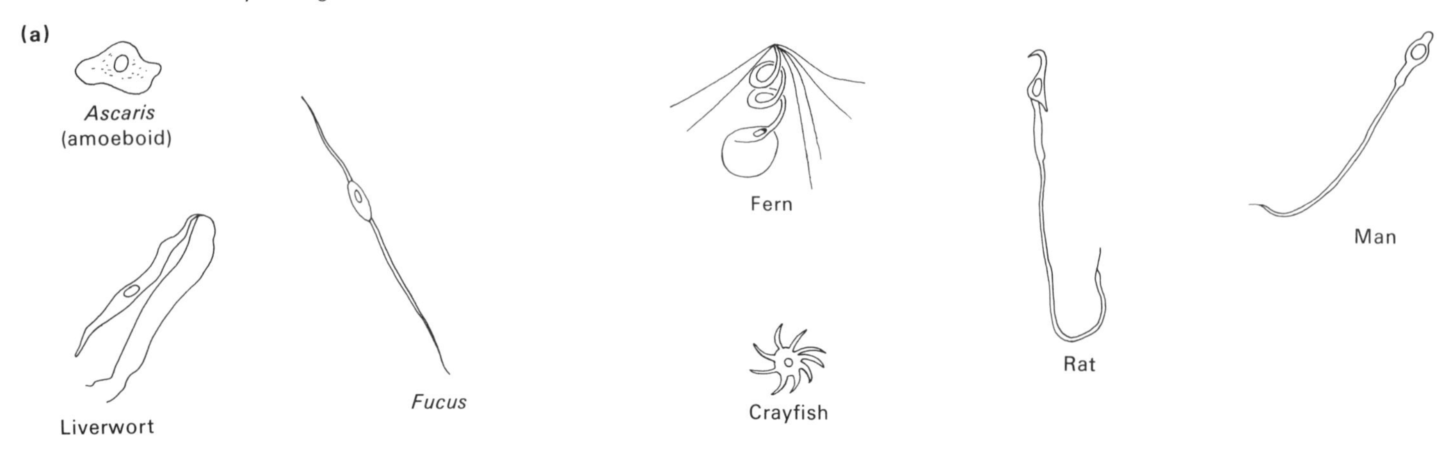

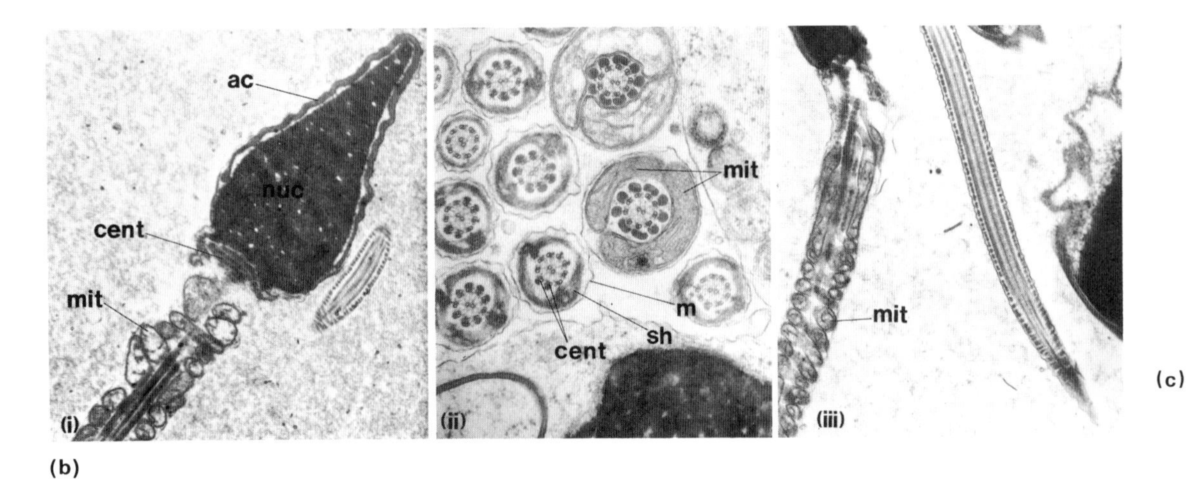

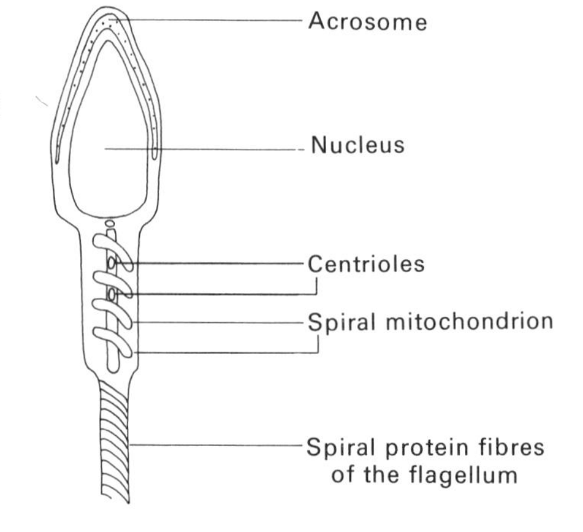

Fig. 6.12 (a) Some types of spermatozoa showing the variety of form that exists. (b) Three electron micrographs showing the structure of bull sperm: i, longitudinal section of the sperm head and body shows the nucleus (nuc), the acrosome (ac), the centriole (cent) and the mitochondria which pass in a spiral fashion round the central tubules (mit); ii, cross-section through body of sperm (centre right) showing mitochondria (mit), and tail (centre left) in which can be seen the centriolar tubules (cent), the surrounding sheath (sh) and an outer membrane (m); iii, longitudinal section of tail in which mitochondria are visible. (c) Drawing of human sperm for comparison.

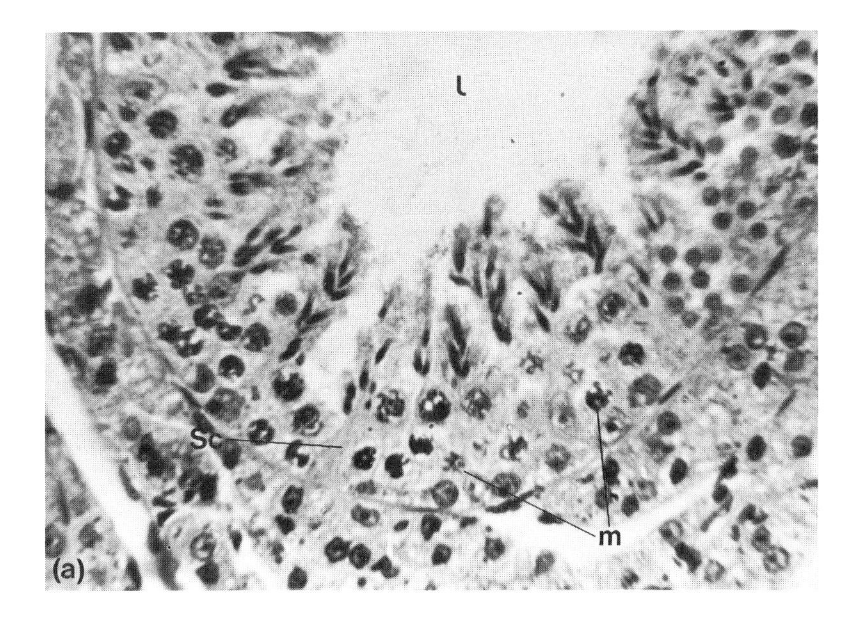

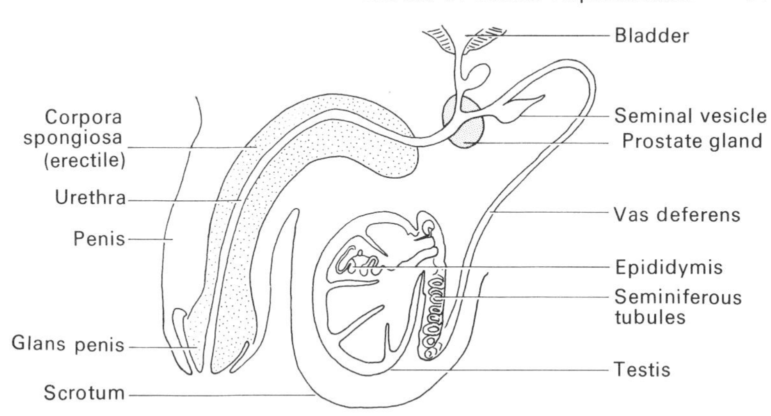

Fig. 6.14 Diagram of the testis and associated organs in man, lateral view.

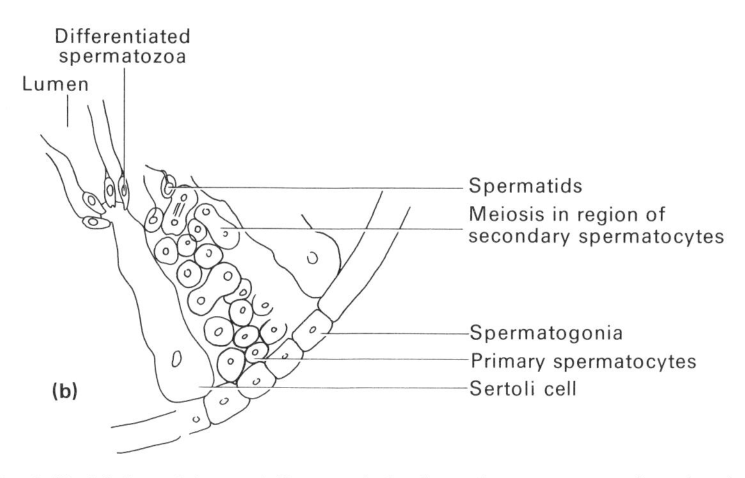

Fig. 6.13 (a) Part of the seminiferous tubule of man in transverse section, showing Sertoli cells (Sc), meiotic figures (m) and the lumen of the tubule (l). (b) Drawing of T.S. seminiferous tubule of man.

6.5 The Formation of Female Gametes

In the lower plants female gametes are produced by the mitotic division of cells of a gametophyte stage in the life cycle, or thallus (Fig. 6.15 and p. 89). The gametes, like all the cells of the thallus, have haploid nuclei.

In flowering plants the haploid, gametophyte generation is severely reduced and is retained in the diploid sporophyte. The flowering plant body, therefore, is diploid and gametes are formed after meiosis in the reproductive organ or flower. Thus, by the contraction of the gametophyte to a microscopic stage in the sequence of gamete formation, the flowering plants have come to resemble the typical animal life cycle very closely (Fig. 6.16 and p. 93). Within the ovary of the flower lie one or more ovules. In the nucellus of the ovule a megaspore mother cell (2n) divides meiotically to produce four haploid nuclei which divide again, mitotically, to form eight nuclei. These nuclei are confined in the **embryo sac** which represents the ♀ prothallus (p. 94).

One of these nuclei is the female gamete. Two others fuse together and become the endosperm nucleus. Both the ♀ gamete nucleus and the endosperm nucleus will ultimately be fertilized by ♂ gamete nuclei, giving rise to a diploid zygote (the future embryo) and a triploid endosperm nucleus (which gives rise to the future food reserve of the seed) respectively (Fig. 6.17).

In mammals the production of female gametes in the ovary is a discontinuous process, unlike the formation of male gametes in the testes. At or near birth the female mammal carries in its ovaries the full complement of primary oocytes (held in meiotic Prophase I) that the animal will require throughout its life. In man the number of primary oocytes, or 'primordial follicles', has been estimated at about 200 000, only a few hundred of which are likely to mature and be shed from the ovary during the monthly oestrus cycles between puberty and menopause (Fig. 6.18).

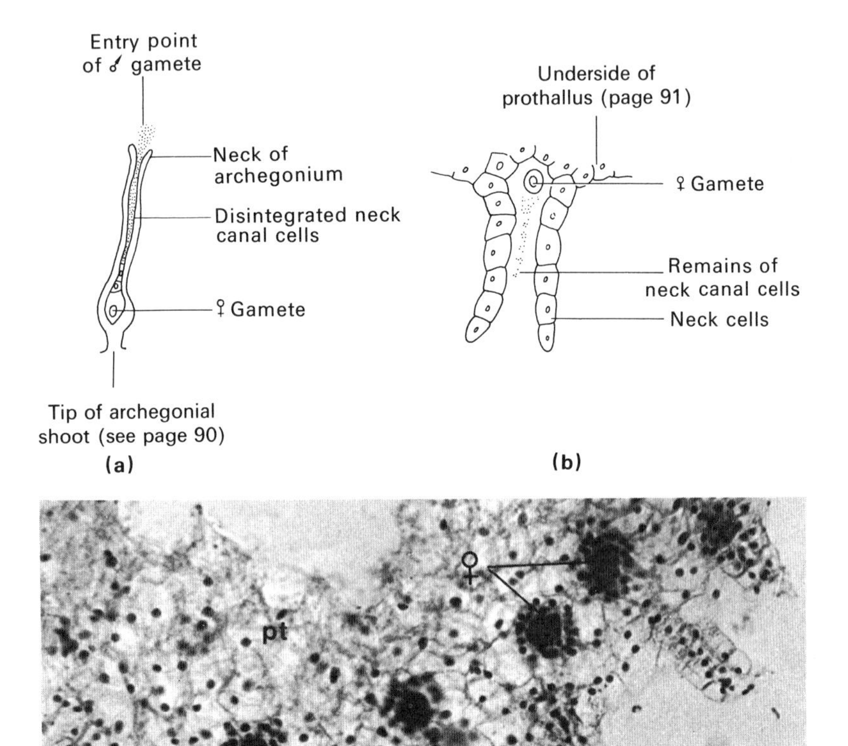

Fig. 6.15 The moss (a) and fern (b) archegonia have essentially the same structure. In the photomicrograph (c) of a section of fern prothallus (pt) the ♀ gametes can be seen in the archegonia. The neck (n) is shorter than the neck of the moss archegonium.

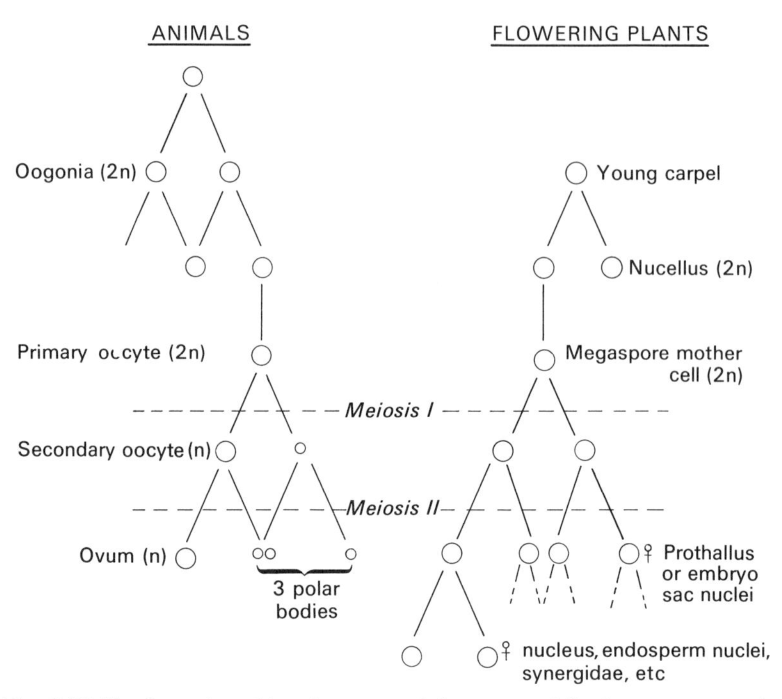

Fig. 6.16 The formation of female gametes follows essentially the same pattern in animals and plants.

6.6 The Fate of the Discharged Ovum: Evolution of the Oviduct

In the diploblastic (p. 101) Coelenterata the ovary develops from a group of interstitial cells in the ectoderm, initially in a similar manner to the formation of the testis. However, only one cell matures, apparently at the expense of the others, forming a conspicuous oocyte which divides meiotically to form a large, yolky ovum. In *Obelia*, the colonial marine coelenterate, most of the sea anemones, and the jellyfishes the ripe ova are shed into the water direct from the ovary and fertilization takes place there. In the freshwater coelenterate *Hydra* sp. the ripe ovum is retained in the ovary, exposed on the body

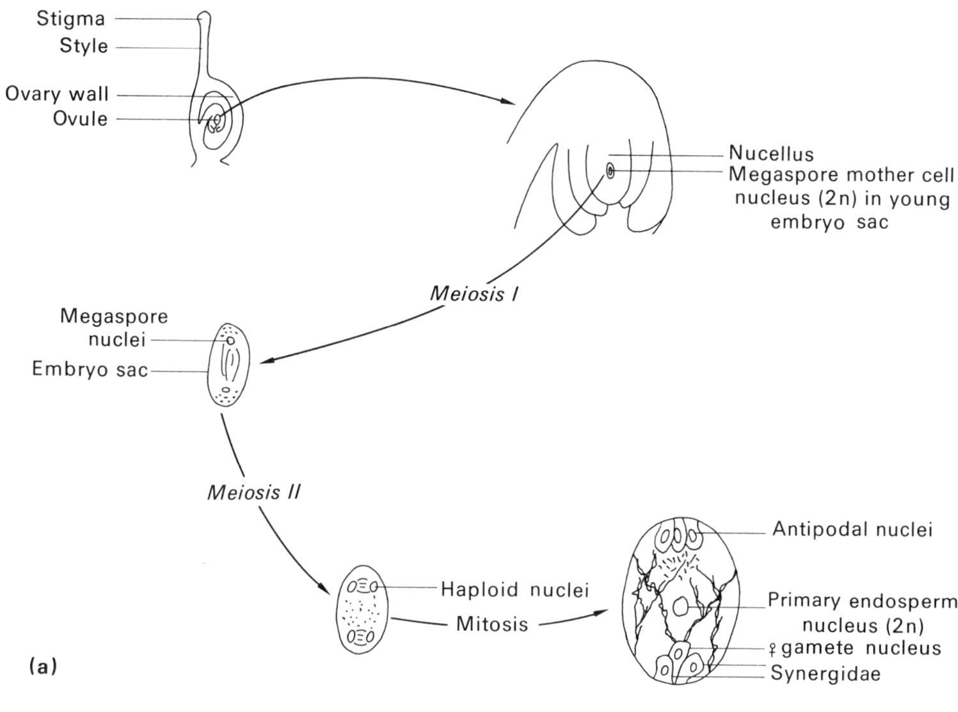

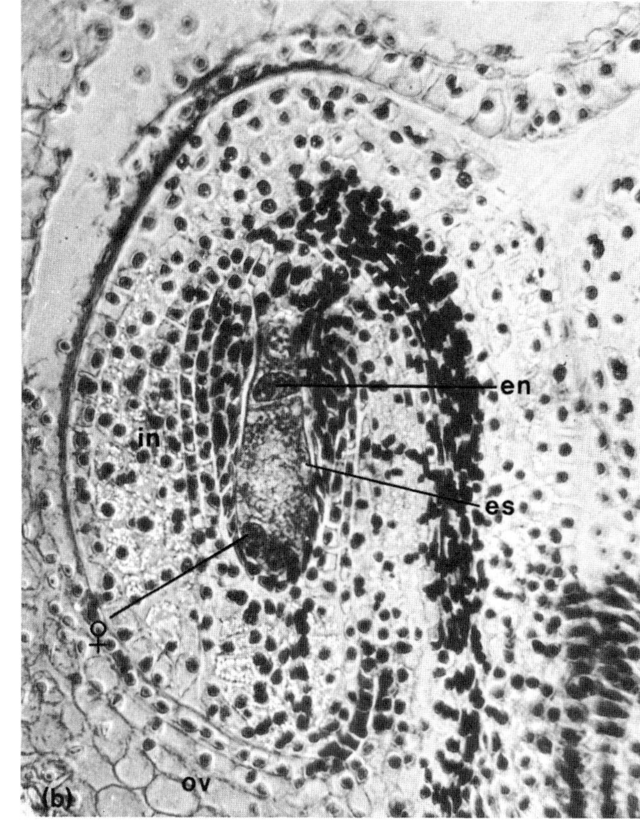

Fig. 6.17 (a) Stages in embryo sac development. (b) Photomicrograph of mature ovule showing the embryo sac (es) containing the ♀ gamete nucleus and the endosperm nucleus (en), lying within the integuments (in). Part of the ovary wall (ov) can be seen also.

wall and fertilization, together with the early stages of cleavage, takes place attached to the parent (Figs. 6.19 and 6.20).

However, in triploblastic, coelomate animals the ovary lies within the coelom, attached to the body wall. Eggs are usually shed into the coelom and pass to the outside of the animal through an oviduct. In the earthworm a pair of ovaries lies in segment 13 and a corresponding pair of oviduct convey eggs through the septum into segment 14, thence through the body wall (Fig. 6.21):

In animals which exhibit viviparity, oviviparity and internal fertilization, the distal parts of the oviduct show a variety of adaptations. In most insects the oviducts join to form a common duct, or vagina, which is the usual site of fertilization. During copulation sperm are deposited by the male in the genital pouch. The sperm pass into the spermatheca, a pouch off the vagina, and fertilization takes place some time after copulation (Fig. 6.22).

In *Glossina*, the tsetse fly, the ovaries are reduced to a single ovarian tubule. Eggs are produced singly and enter a highly glandular part of the oviduct where fertilization takes place. Here the egg is retained and subsequently develops through its larval stages nourished by secretions from the oviduct walls, or 'uterus'. The female then deposits the larva immediately before pupation.

In the mammals there is considerable variation in the structure of oviducts and associated organs. Monotremes continue the pattern shown by many fishes, amphibians and reptiles in that the oviducts,

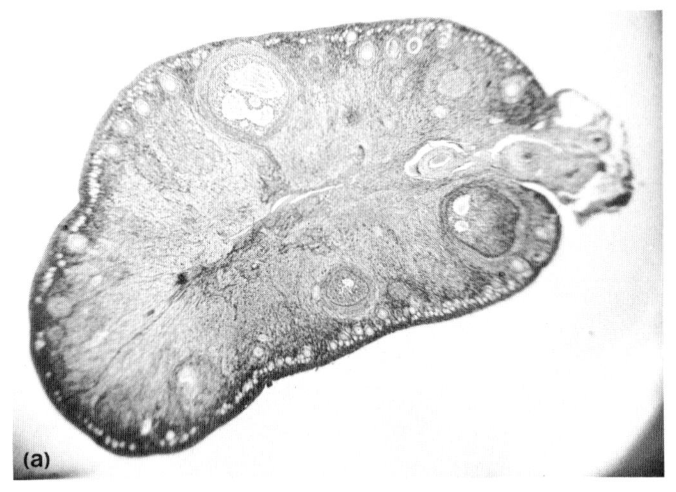

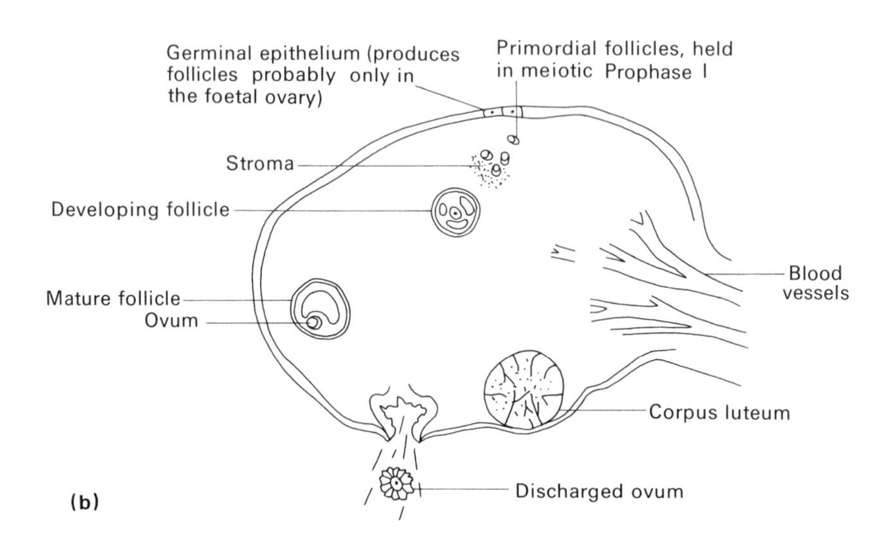

Fig. 6.18 Mammalian ovary. (a) In section. (b) Generalized drawing.

rectum and urinary ducts empty into a common cloaca. There is some thickening in the oviducts but not sufficient to merit the term uterus. In marsupials each oviduct has its uterus and vagina, the vaginae usually joining at the urinogenital sinus. In some marsupials, such as the kangaroo, a median vagina is present which forms an opening into the urinogenital sinus just before the offspring are born. Placental mammals have a single vagina but the uteri show various degrees of coalescence, being most distinct in the rodents and unified in the primates (Fig. 6.23).

6.7 Fertilization

Fertilization, in effect, is the opposite of meiosis; it is the restoration of the diploid state. In the formation of gametes homologous pairs of chromosomes are separated so that a gamete contains only one member of each pair. When the gamete nuclei fuse at fertilization homologous pairs are restored in the zygote nucleus. Clearly, the advantages of sexual reproduction would be lost if the zygote were formed by the fusion of gametes produced on the same parent body. Such self-fertilization would reduce offspring variability and would encourage both the phenotypic expression of disadvantageous recessive genes and the loss of heterozygote superiority.

Plants and animals employ various mechanisms to avoid self fertilization. Higher animals have separate sexes and, in mammals particularly, innate behavioural patterns to avoid sibling mating. In lower animals hermaphroditism is found but in many cases spermatozoa are exchanged during mating, as in earthworm, or the gonads mature at different times, as in *Hydra*. In lower plants, on the other hand, the gametophyte generation is generally monoecious, or hermaphroditic, bearing both antheridia and archegonia (see *Pellia*, *Dryopteris*, pp. 89, 91) and cross fertilization is likely only when thalli are closely packed. Most mosses are also monoecious but archegonia and antheridia are borne on different shoots (p. 90) and

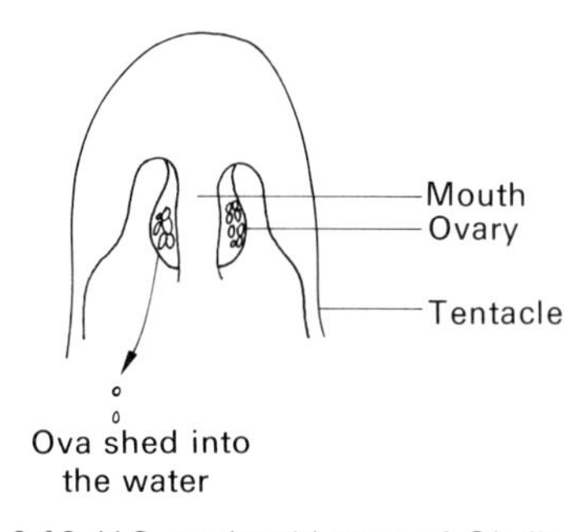

Fig. 6.19 V.S. medusoid stage of *Obelia*.

cross-fertilization could be a likely event in a closely growing community, especially as male gametes are probably distributed by rain splashes and on insect feet. In the more advanced mosses (*Mnium*, *Polytrichum*) the sexes are separate (dioecious).

Most flowering plants are monoecious. Various mechanisms exist to promote cross-pollination, although cleistogamy, or self-pollination, is found in some species (in the wood-violet, flowers produced

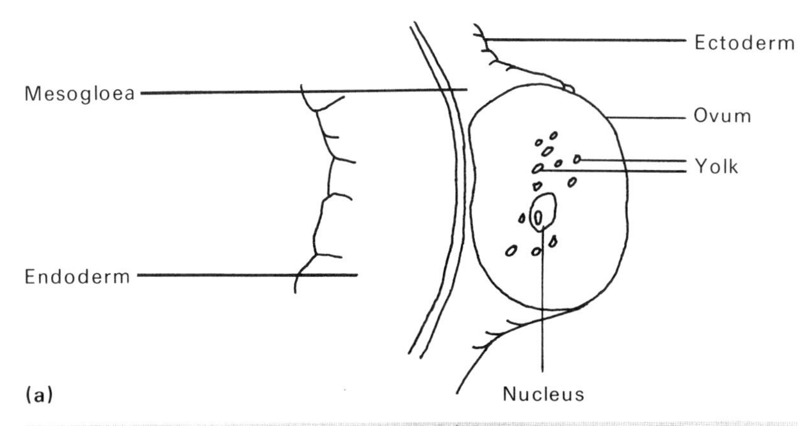

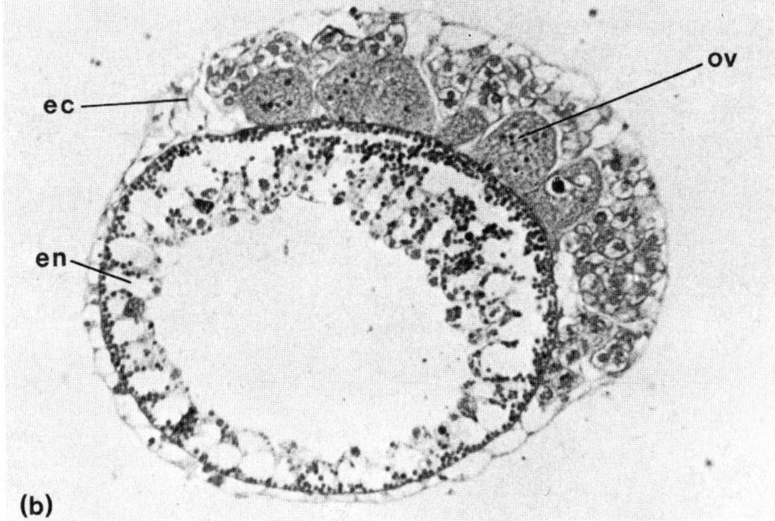

Fig. 6.20 (a) T.S. *Hydra* ovary. (b) T.S. *Hydra* showing endoderm (en) and ectoderm (ec) and the developing ovary mass (ov).

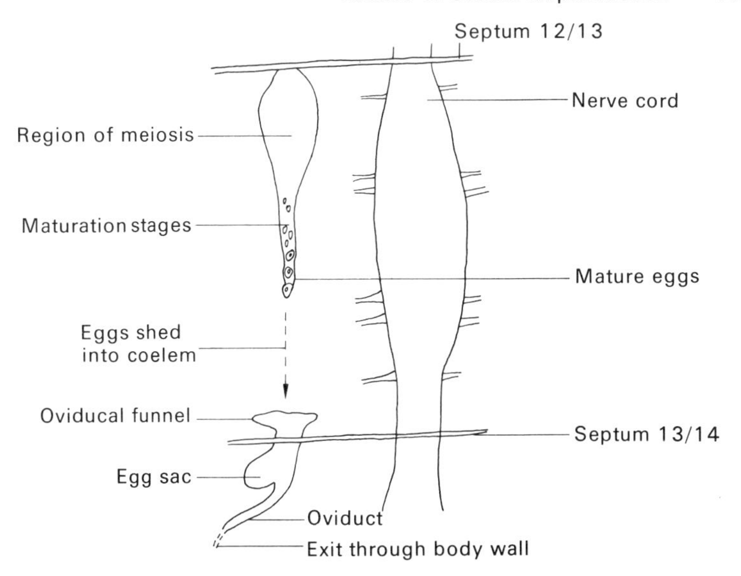

Fig. 6.21 Earthworm ovary and oviduct, dorsal view.

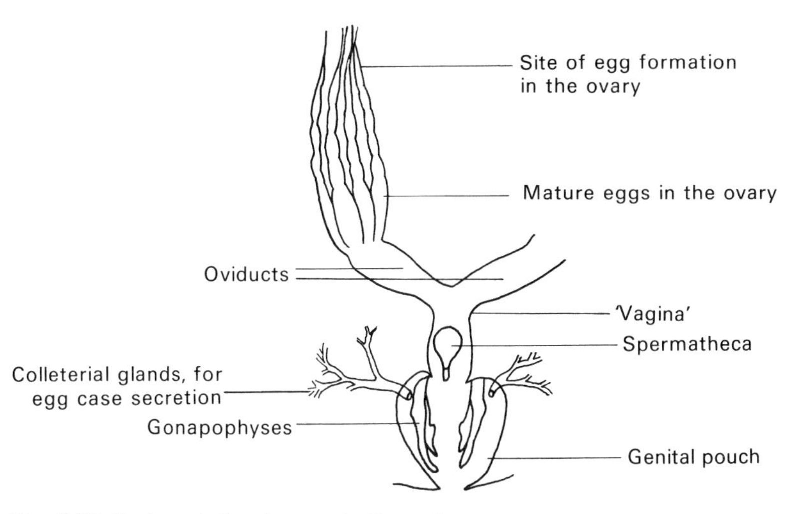

Fig. 6.22 Cockroach, female reproductive system.

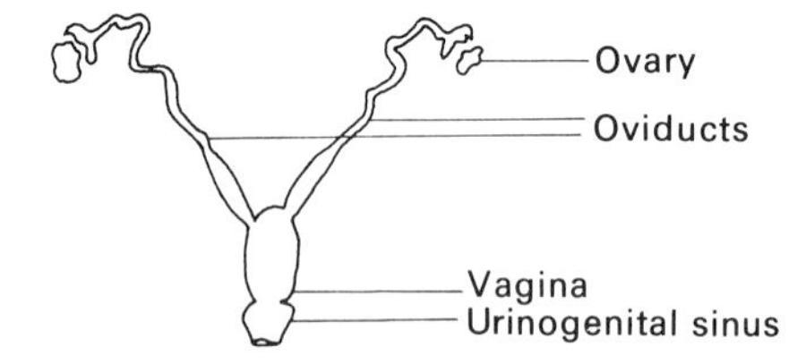

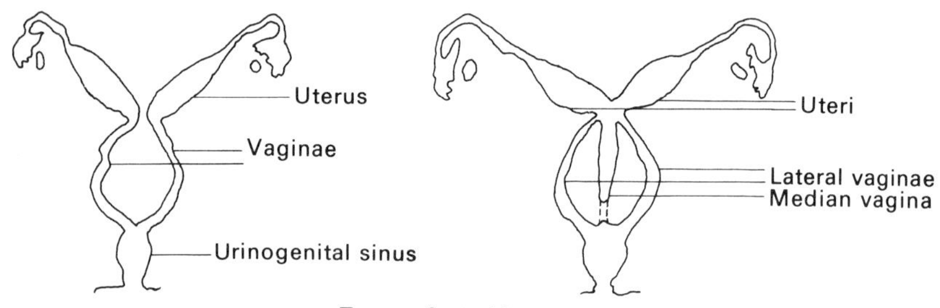

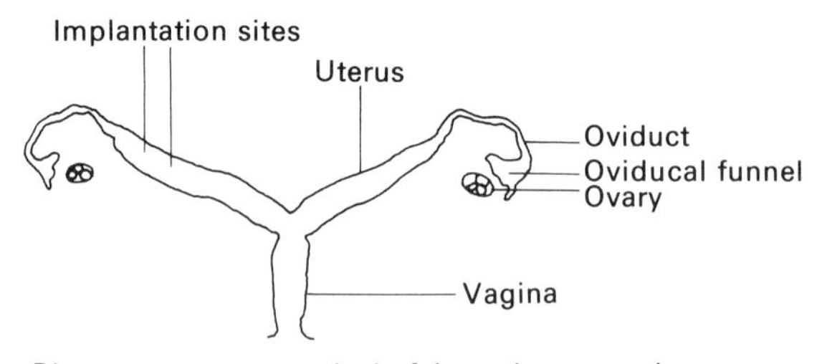

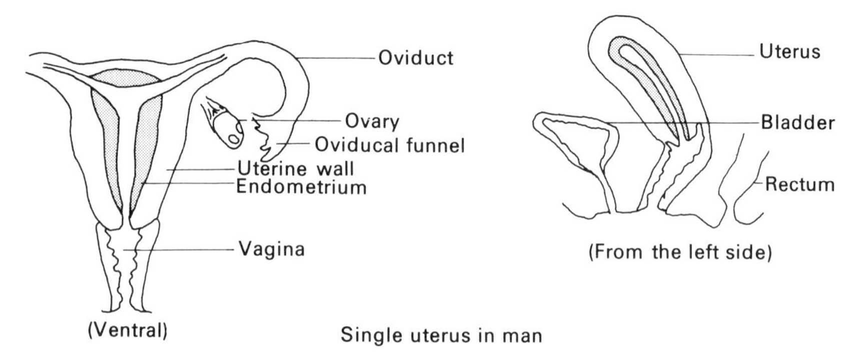

Fig. 6.23 Some female reproductive systems in mammals.

late in the summer may set seed even though the flower bud remains unopened):

(a) The sex organs may develop at different rates. The dandelion, for example, is protandrous and the flowers of one plant have shed their pollen before the style and stigma are developed. Protogyny is less common, but is shown by *Scrophularia* spp, the figworts.

(b) The stamens and stigma may be set at different levels in the flowers of different plants, so that pollen from one type of arrangement can only be transferred to stigmas of the complementary type. This pattern is found in the pin-eyed and thrum-eyed forms of the primrose flower and in the trimorphic flowers of the purple loosestrife, *Lythrum*.

(c) In many species self-sterility alleles are present (East and Mangelsdorf, 1925). If the pollen tube nucleus has such an allele in common with the style tissue the growth of the pollen tube is retarded or halted. Self-sterility is an example of multi-allelism. Taking a simple case of only three alleles (Fig. 6.24), S_1, S_2 and S_3 and a parent plant with the gentoype S_1S_2 one can see that the pollen genotypes produced from such a plant would be S_1 and S_2. Neither kind of pollen can germinate successfully on the parent style. A second plant, however, of genotype S_2S_3 would produce pollen of the genotypes S_2 and S_3. Half of these pollen, namely the S_2, would be unsuccessful in pollinating the first plant, but the S_3 pollen would effect fertilization:

Stamen genotypes	Pollen	Style genotypes	
		S_1S_2	S_2S_3
S_1S_2	→ S_1	→ no	yes
	→ S_2	→ no	no
S_2S_3	→ S_2	→ no	no
	→ S_3	→ yes	no

Fig. 6.24 Self-sterility alleles ensure that only pollen grains bearing alleles that are *different* from the style tissue genes produce an effective pollen tube.

The conquest of land presented major problems in fertilization for both plants and animals because gametes cannot tolerate dehydration. Complex behavioural patterns, under hormonal and nervous regulation, bring the sexes together for internal fertilization; such courtship is a widespread phenomenon in terrestrial invertebrates and vertebrates.

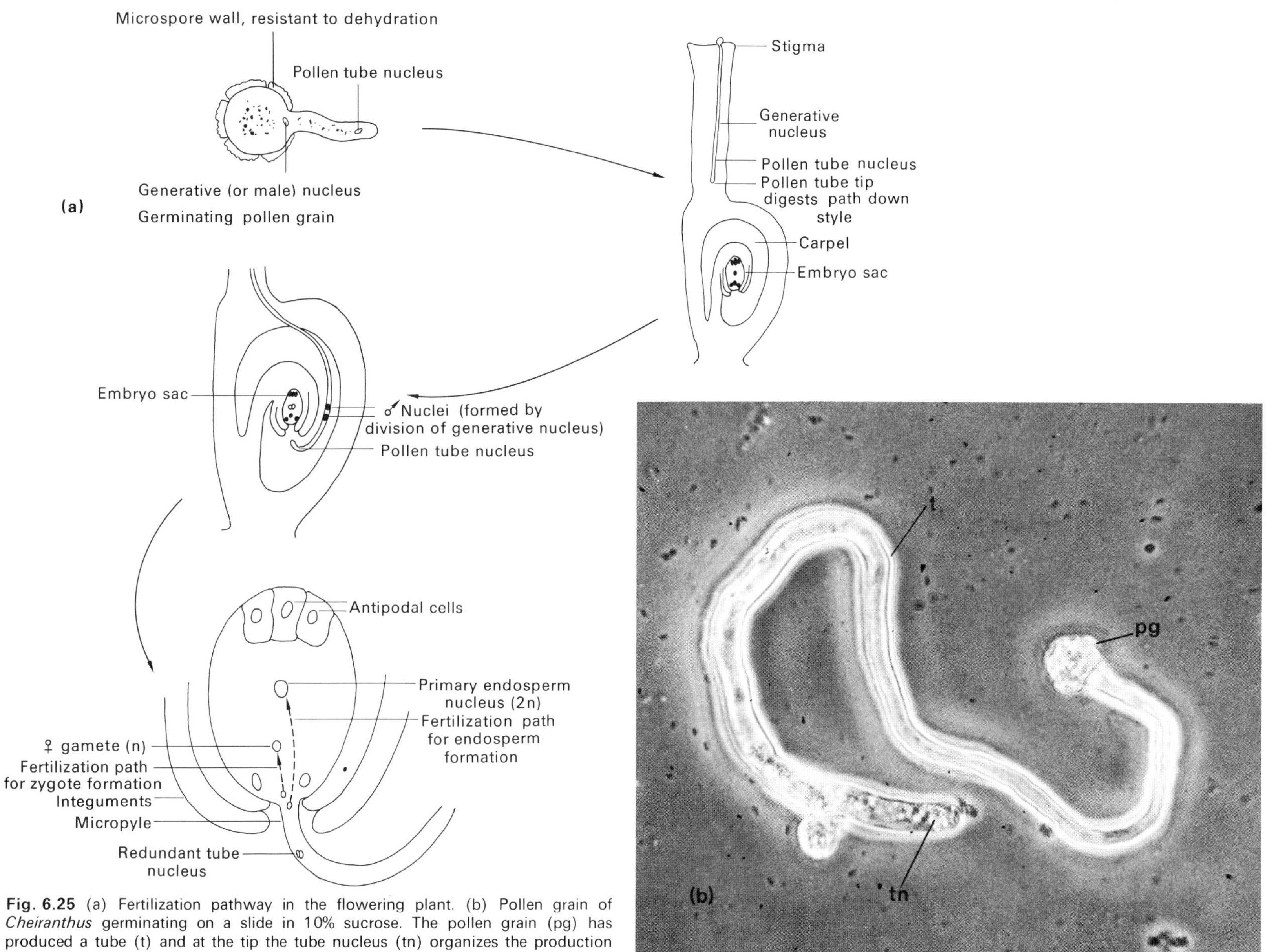

Fig. 6.25 (a) Fertilization pathway in the flowering plant. (b) Pollen grain of *Cheiranthus* germinating on a slide in 10% sucrose. The pollen grain (pg) has produced a tube (t) and at the tip the tube nucleus (tn) organizes the production of enzymes.

Flowering plants overcome their non-motility by the use of wind or animals to transfer the male gamete (protected from dehydration by the wall of the microspore, or pollen grain) from the anther of one organism to the stigma of another.

The Flowering Plant: Fertilization and Seed Formation

The male gamete nucleus is carried from the stigma to the female nucleus by the pollen tube, an outgrowth of the pollen grain (Fig. 6.25a). The pollen tube digests a path down the loosely packed core tissue of the style, using enzymes probably organized by the pollen tube nucleus. Successful pollen tube growth depends on the complementary nature of the pollen tube and style tissue nuclei, that is, they must be of the same species and must not have self-sterility alleles in common.

The tip of the pollen tube enters the carpel and meets the mature embryo sac at the micropyle (it has been suggested that the synergidae may exercise some chemotaxic role). Within the tube, the generative nucleus divides to form a male gamete, which will fuse with the female nucleus to form a zygote and later the embryo, and a second male nucleus which will fuse with the diploid primary endosperm nucleus to form a triploid endosperm mass (Fig. 6.25).

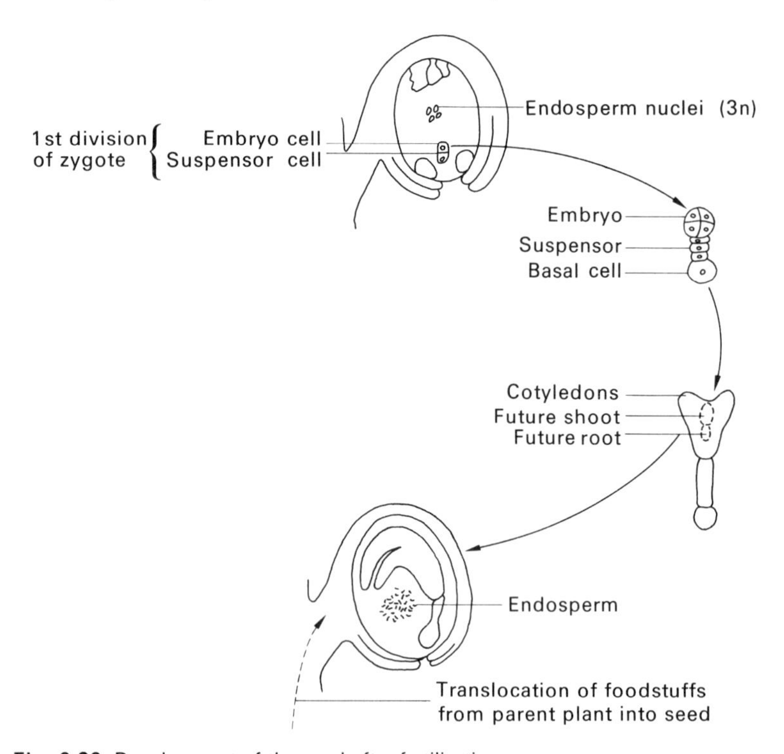

Fig. 6.26 Development of the seed after fertilization.

After fertilization the outer parts of the flower, which are concerned with attracting insects and with pollen formation, wither and the ovary increases in size. In seed formation the integuments give rise to the testa, the endosperm forms a food reserve for germination at a later time, and the zygote gives rise to the embryo. The ovary wall frequently becomes modified for seed dispersal, for example, wings for wind dispersal and edible tissue for animal dispersal (Fig. 6.26).

Plant species fall broadly into two groups with respect to the fate of the endosperm. In **exalbuminous** species (e.g. bean) the endosperm progressively diminishes and the cotyledons swell to occupy the whole seed, forming the food reserve for the embryo (Fig. 6.27b). In **albuminous** species, however, the endosperm persists as a food reserve and the cotyledons are membranous (Fig. 6.27a).

Fertilization in Animals

Studies on sperm penetration of the eggs of sea-urchins, marine annelids and fishes have revealed that the process is considerably more complicated than was once thought. Initially, a highly species-specific reaction takes place between a component of the egg coat and the head of the sperm. This appears to be akin to an antibody/antigen reaction and clearly serves to prevent egg penetration by the sperm of other animal species, an important precaution in aquatic organisms having external fertilization. The second stage appears to be an extension of the acrosome through the mucilaginous coat of the egg towards the egg membrane. The third stage is usually delayed; frequently great numbers of sperm may be associated with one egg, but in this stage (the response by the egg membrane) only one sperm is concerned. A conical blister rises from the egg membrane towards the sperm head, possibly initiated by the activity of hyaluronidase (p. 96) secreted from the acrosome. This blister engulfs the sperm head in a pinocytosis-like action. The blister then extends rapidly round the egg forming a clear region which probably serves to prevent the entry of other sperm.

The sperm tail is left outside the egg membrane. The contents of the head move towards the centre of the egg and the nucleus coalesces with the egg nucleus. It is thought that the centrioles, which are a

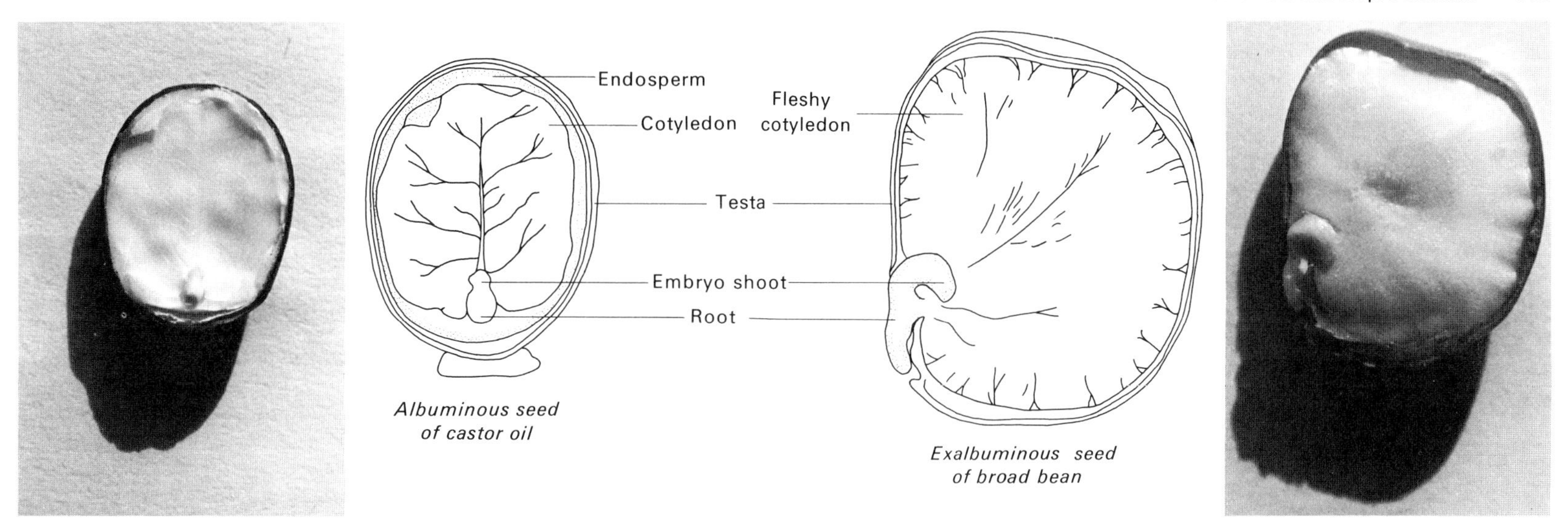

Fig. 6.27 Endosperm types. On the left the albuminous seed of castor oil in which the endosperm persists and the cotyledons are membranous structures. On the right the broad bean seed in which the endosperm reserve is taken up by the fleshy cotyledon.

conspicuous feature of the body of the sperm, may have a role in guiding the sperm nucleus towards the egg nucleus (Fig. 6.28).

Apart from its role in adding nuclear material to the egg nucleus the sperm appears to have a mechanical role. The action of penetration of the egg membrane seems to trigger a response by the egg which leads to cleavage. In the laboratory this can be done artificially by pricking eggs with a needle, in the absence of sperm. Cleavage may then commence, though in most cases it rarely procedes beyond the early stages of embryo formation, probably because the cell nuclei are haploid. Such experiments, termed **artificial parthenogenesis**, using amphibian eggs have led to the production of adult animals, but these are usually smaller and less hardy than normal adults.

Parthenogenesis is known to take place naturally in some insect species. In the honey bee the queen is capable of laying either fertilized or unfertilized eggs, which hatch into diploid or haploid larvae respectively. The former become females (workers and queens) and the latter become males (drones) (Fig. 6.29). There is some uncertainty about the mechanism whereby female larvae are converted into either sterile workers with atrophied reproductive organs or into egg-producing queens. Bee-keepers have long noticed that the workers feed queen larvae on a creamy liquid called 'royal jelly' which may contain a gonadotropic hormone. Worker larvae, on the other hand, receive little royal jelly and their diet consists mainly of 'bee-bread', a mixture of honey and masticated pollen.

This form of parthenogenesis, whereby haploid ova develop into haploid male bees, has been termed haploid parthenogenesis. In the plant lice (*Aphididae*), however, the life cycle stages taking place in the summer involve successive generations of diploid parthenogenetic females which produce diploid female offspring viviparously. This can be termed **diploid parthenogenesis**. In these summer stages, diploid ova seem to be produced by modified meiosis in which total non-disjunction takes place, thus all the chromosomes enter the egg and none enter the polar body.

In the autumn stages of the aphid a further refinement of meiotic non-disjunction takes place. The somatic chromosomes undergo non-disjunction as described above, but the sex chromosomes segregate as in normal meiosis producing an egg with the diploid somatic chromosome complement, plus a single X chromosome. This egg develops parthenogenetically into a male. Autumn stage females then produce haploid eggs by normal meiosis and these are fertilized by haploid sperm from the males. Fertilized eggs are then laid by the female and represent an overwintering stage; in the spring the eggs hatch into viviparous parthenogenetic females.

Vertebrates

In fishes and amphibians spermatozoa and eggs are usually shed independently into the water, where fertilization takes place. As a prelude, group behavioural drives or courtship activities lead to shoaling or to pairing of the sexes, which tends to increase the probability of fertilization.

In the conquest of land, the response by the reptiles to the problem of the vulnerability of gametes to dehydration was essentially the same as that of the insects and the flowering plants, namely, fertilization became internal. In most reptiles and birds the hind part of the cloaca of the male contains *corpus spongiosum* which, by vascular erection, forms a primitive penis (Fig. 6.30). During coitus, or copulation, sperm pass along grooves in the penis into the cloaca of the female. Fertilization takes place high in the oviduct.

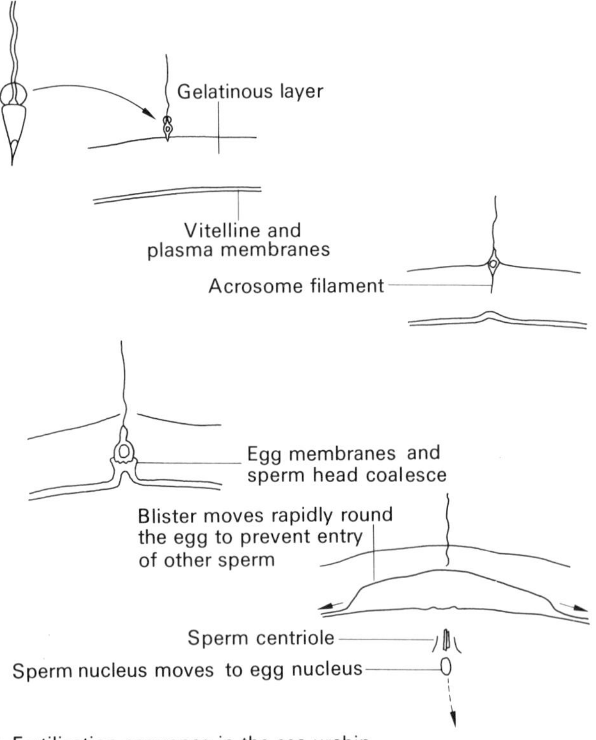

Fig. 6.28 Fertilization sequence in the sea urchin.

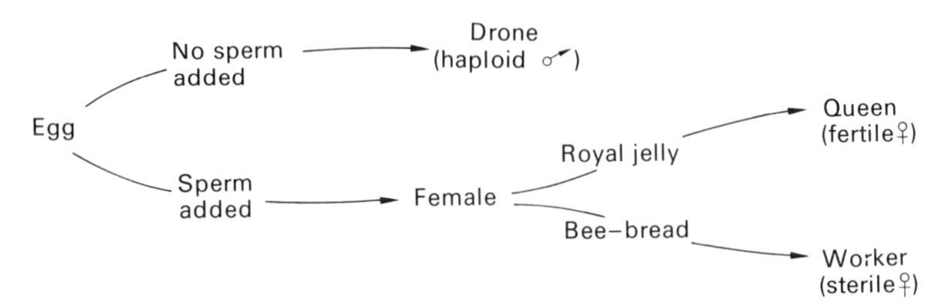

Fig. 6.29 Parthenogenesis in the honey bee, compared with sexual reproduction.

In mammals the penis is a more distinctive structure than in reptiles or birds, and the vas deferens conveys sperm to the tip instead of along grooves on its surface (p. 97). Autonomic and somatic neural pathways produce erection of the penis by dilatation of the arterioles supplying the corpora spongiosa and constriction of the venous return. In the female, autonomic nerves cause erection of the clitoris and secretion of mucus by the vulva and vaginal walls (p. 102) which facilitates passage of the penis at copulation.

Copulation culminates in **orgasm** which, in the male, involves the ejaculation of spermatozoa, mixed with seminal fluid from the prostate gland, into the vagina. If simultaneous orgasm occurs in the female it may serve to draw sperm through the cervix into the uterus (in cattle, sperm have been extracted from the oviducts approximately 0.5 metres from the vagina two minutes after copulation, compared with two hours at their unaided rate of movement).

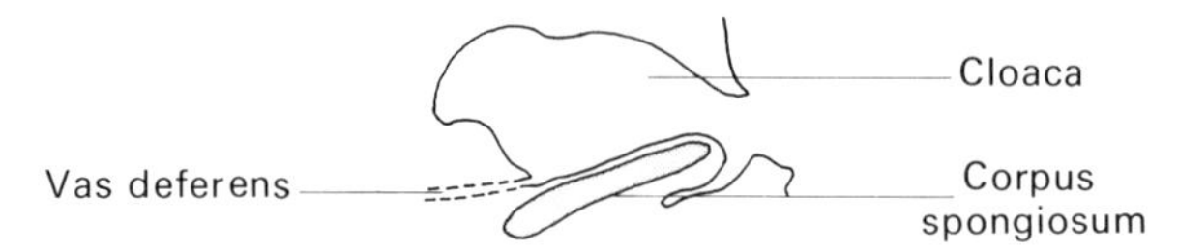

Fig. 6.30 Male reptile cloaca (lateral view).

Seminal fluid contains some proteolytic enzymes which may help to dissolve the cervical mucus plug, and 0.2% fructose which appears to act as an initiator of sperm flagellar action and as a respiratory substrate for the sperm.

As in reptiles and birds, fertilization in mammals takes place high in the oviduct. Since the ovum is surrounded by residual follicular cells (Fig. 6.31) penetration may be aided by the hyaluronidase-secreting acrosome on the spermatozoan head (pp. 96 and 104).

When the sperm head penetrates the egg membrane the tail is discarded. Changes in the egg membrane similar to those taking place in invertebrate eggs (p. 106) are thought to take place to prevent multiple penetration. The sperm nucleus then migrates to the egg nucleus and nuclear fusion takes place to form a diploid zygote nucleus. Mitosis, with complete cytoplasmic cleavage in the case of

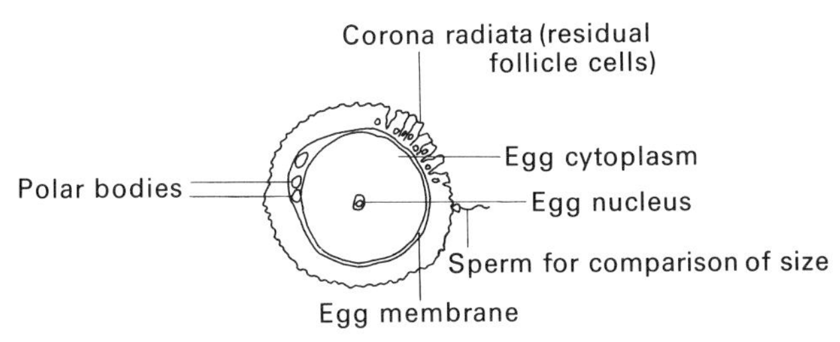

Fig. 6.31 Egg at fertilization (in man about 200 μm in diameter). The second meiotic division resulting in a female egg nucleus and three polar bodies usually coincides with fertilization.

mammals, begins immediately and is repeated without growth as the egg passes down the oviduct. Implantation in the uterine lining occurs at the blastocyst stage (p. 108); in man this is some 8–10 days after fertilization (Fig. 6.32).

Implantation in the Uterus. Implantation is the establishment of the embryo in the uterus. The form of the embryo at implantation is unique to placental mammals; it comprises an undifferentiated cell mass destined to become the future embryo and a relatively larger, fluid-filled trophoblast. The trophoblast appears to be one of the adaptations in the evolution of placentation.

The function of the trophoblast is not entirely clear, but in the passage of the young embryo down the oviduct it is in active division (the cells of the trophoblast are smaller than those of the inner cell mass) and its role is probably to form an enlarged 'membrane system' which can quickly establish metabolic exchange with the uterine endometrium as soon as implantation takes place.

Implantation shows some variation in mammals. In cattle and other ungulates the blastocyst remains in the cavity of the uterus, eventually expanding to fill it. In rats and mice the blastocyst occupies a pocket in the endometrium and there is some folding around it. In man the blastocyst passes right into the endometrium which completely surrounds it and separates it from the uterine cavity (Amoroso, 1970).

6.8 Embryo Formation and Development in Animals

The early stages of embryonic development are relatively uniform throughout the animal kingdom. These stages resemble the pattern in flowering plants (p. 104) in that mitosis of the diploid zygote nucleus occurs repeatedly and cleavage takes place without growth. Next, growth occurs at the expense of metabolites provided by the maternal organism. Thirdly, differentiation takes place, but in this respect plants and animals are relatively distinct. In animals vigorous cell migration takes place during the formation of germ layers and, later, of organs. In plants, however, there is little or no cell migration; undifferentiated meristematic regions within the developing embryo provide a source of cells for differentiation into tissues.

The end product of the first stage (repeated mitosis) in the animal egg is usually a multicellular, hollow **blastula**. The ability of the egg cytoplasm to undergo cleavage seems to depend on the amount of yolk present. In very yolky eggs, such as reptile and bird eggs, cleavage is prevented completely (p. 112). In less yolky eggs, such as those of most invertebrates the yolky pole of the egg divides more slowly, consequently the cells in that region are larger (Fig. 6.33).

When a non-spherical cell in a dividing egg cleaves, the spindle tends to lie parallel to the long axis of the cell. Thus in the second and subsequent divisions during the early segmentation of the egg, when the cells are not spherical, the orientation of the spindle is determined largely by the sequence of previous divisions.

Furthermore, it can readily be demonstrated that the plane of cleavage of the cytoplasm that follows mitotic anaphase is determined by the position of the spindle. Cleavage, however, does not simply follow the equator of the spindle; it can be shown that if the spindle is removed after the chromosomes have separated cleavage will still take place. In the newt egg it has been demonstrated that the determination of the site of cleavage by the spindle occurs some 40 minutes before cleavage begins, which is about half the duration between the first and second cleavages at room temperature.

Careful sectioning of the newt egg and examination by electron microscopy has revealed the presence of a dense band of fine filaments below the cell membrane, lying in the cleavage furrow. It has

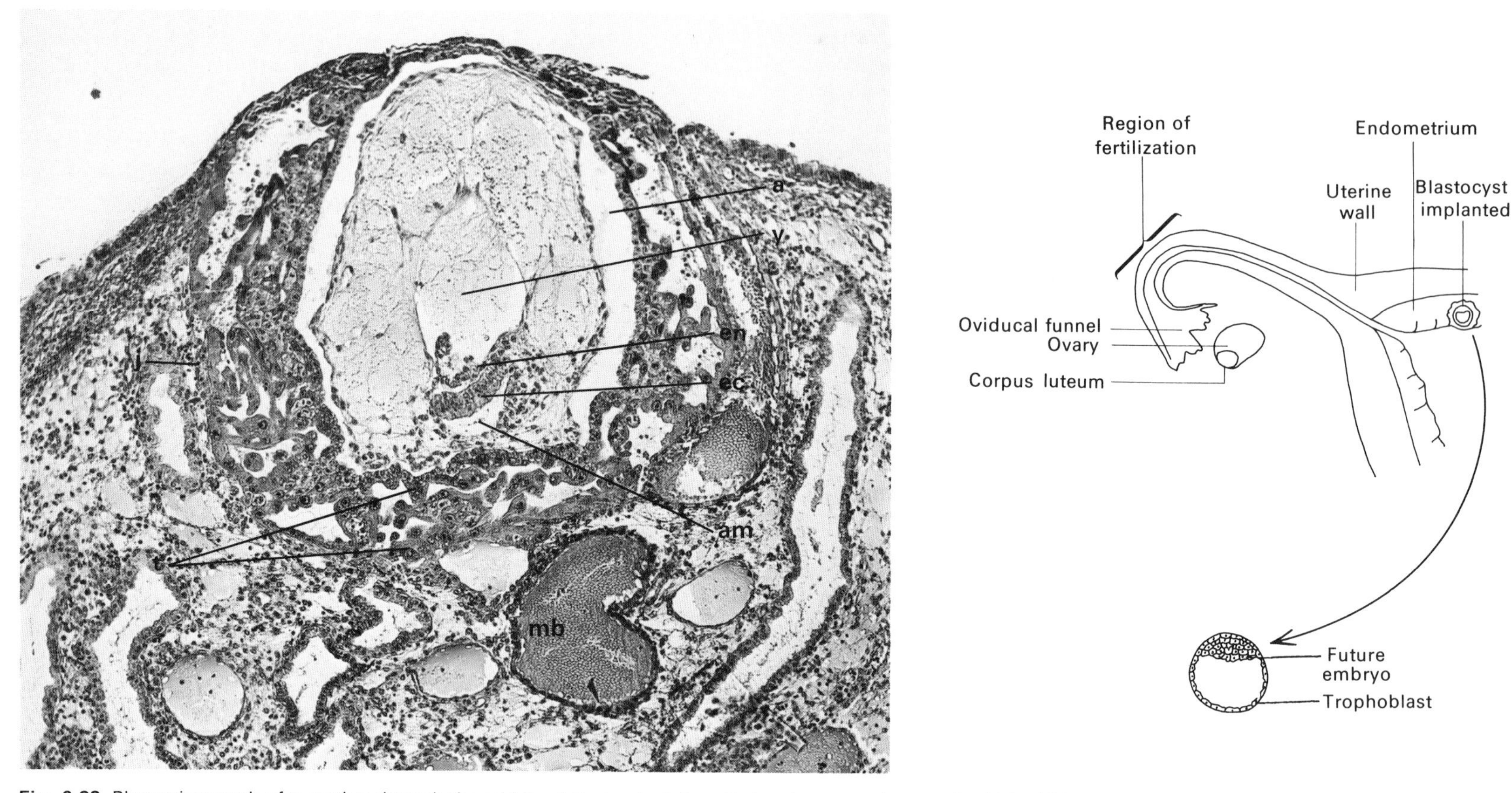

Fig. 6.32 Photomicrograph of a section through the middle of the implantation site in a human embryo at the 11th–12th day after fertilization. (a) this space between the extra-embryonic mesenchyme and the trophoblast is an artifact caused during the sectioning, (am) amnion, (ec) embryonic ectoderm, (en) endoderm, (j) junctional zone, (mb) maternal blood in an endometrial gland, (t) trophoblast, (y) yolk sac. (Reproduced by courtesy of Professor W.J. Hamilton). (*Right)* A drawing of an oviduct and part of the human uterus. Implantation occurs in the uterus, most commonly on the dorsal, or posterior, wall.

been suggested that the contraction of these filaments brings about the cleavage furrow (Selman and Perry, 1970) but a connection between the position of these filaments and the microtubules that compromise the spindle apparatus (p. 71) has not been demonstrated (Fig. 6.34).

As already described, repeated cleavage gives rise to a blastula. The blastula then invaginates to form an **archenteron** which opens at the **blastopore**. This forms the **gastrula** stage. In a number of sessile or slow moving marine invertebrates the gastrula develops into a ciliated larva which is a pelagic dispersal phase. Often the archenteron develops into a gut and the larva feeds (Fig. 6.35).

A similar dispersal phase is found in marine annelids. In *Nereis* the gastrula develops into a ciliated **trochosphere** larva. In the late larval stages the lower pole elongates and metamerism of the organs takes place (Fig. 6.36).

In Chordates, however, no such diversion occurs in the embryological sequence, although the protochordate, *Balanoglossus*, has a ciliated larval stage similar to the pluteus of the echinoderms. The

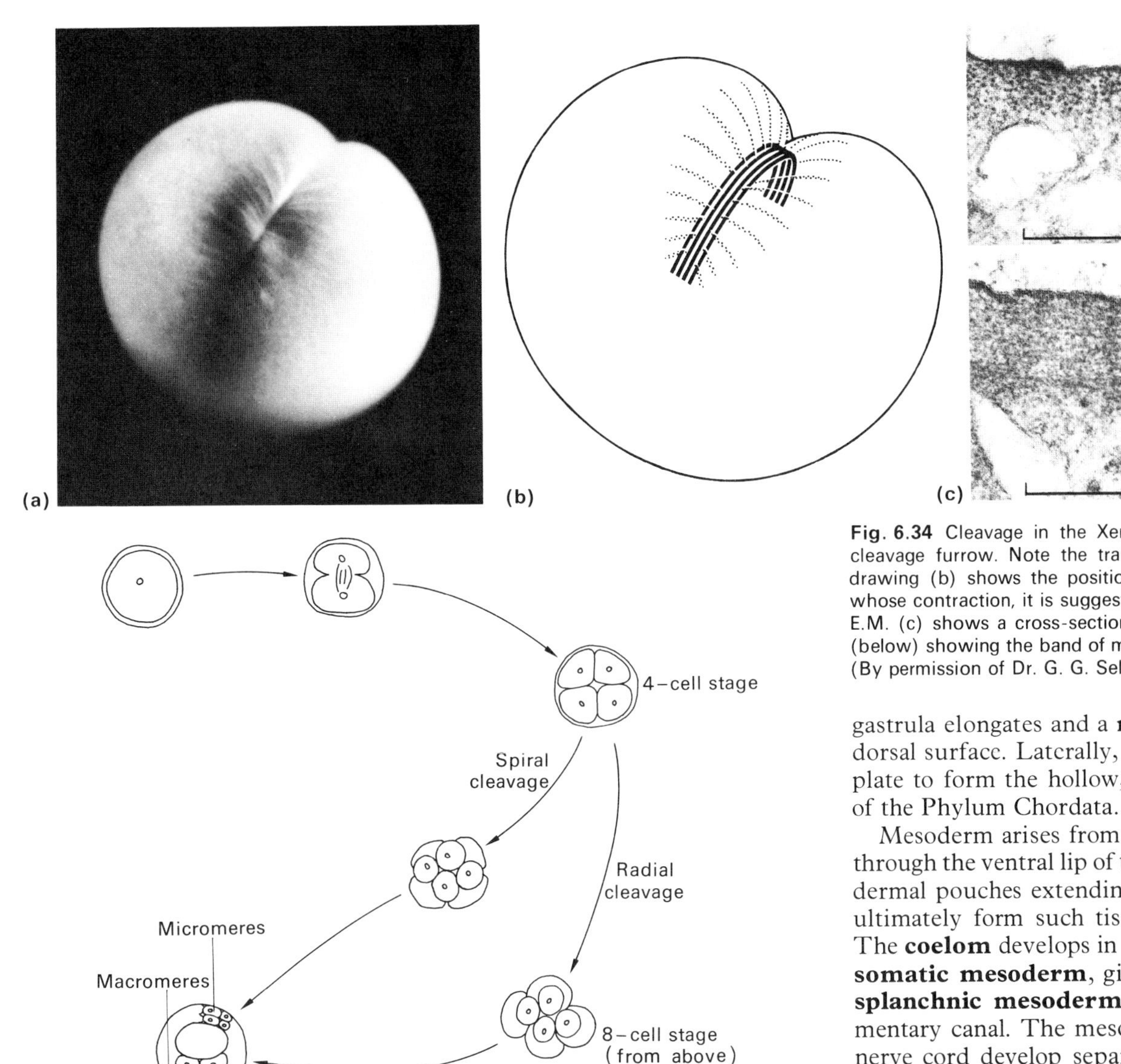

(c)

Fig. 6.34 Cleavage in the Xenopus egg showing (a) the beginning of the first cleavage furrow. Note the transverse wrinkles on each side of the furrow. The drawing (b) shows the position of a band of microfilaments beneath the furrow whose contraction, it is suggested, forms the furrow and explains the wrinkles. The E.M. (c) shows a cross-section of the furrow (above) and a longitudinal section (below) showing the band of microfilaments beneath the furrow. The scale is 1 μm. (By permission of Dr. G. G. Selman and Miss M. M. Perry.)

Fig. 6.33 Repeated mitotic division of the fertilized egg nucleus results in a hollow ball of a few thousand cells, the blastula.

gastrula elongates and a **neural plate** forms from ectoderm on the dorsal surface. Laterally, **neural folds** arise and enclose the neural plate to form the hollow, dorsal nerve cord which is characteristic of the Phylum Chordata.

Mesoderm arises from cells which invaginate late in gastrulation through the ventral lip of the blastopore. These cells later form mesodermal pouches extending between endoderm and ectoderm which ultimately form such tissues as trunk muscles and axial skeleton. The **coelom** develops in the lateral plate mesoderm, dividing it into **somatic mesoderm**, giving rise to muscles in the body wall, and **splanchnic mesoderm**, forming involuntary muscles of the alimentary canal. The mesodermal pouches lying on each side of the nerve cord develop separately to form somites, which are initially metameric. Except in *Amphioxus* and in fishes the metamerism is lost later in development and the metameres give rise to limb and trunk muscles. Since all muscles arising from myotomes are innervated from ventral nerve roots, and as the myotomes are initially segmental,

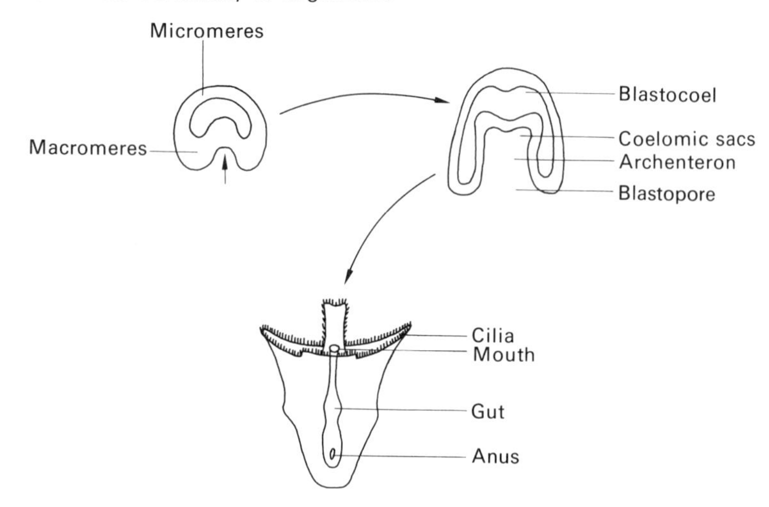

Fig. 6.35 *Pluteus* larva of the sea urchin.

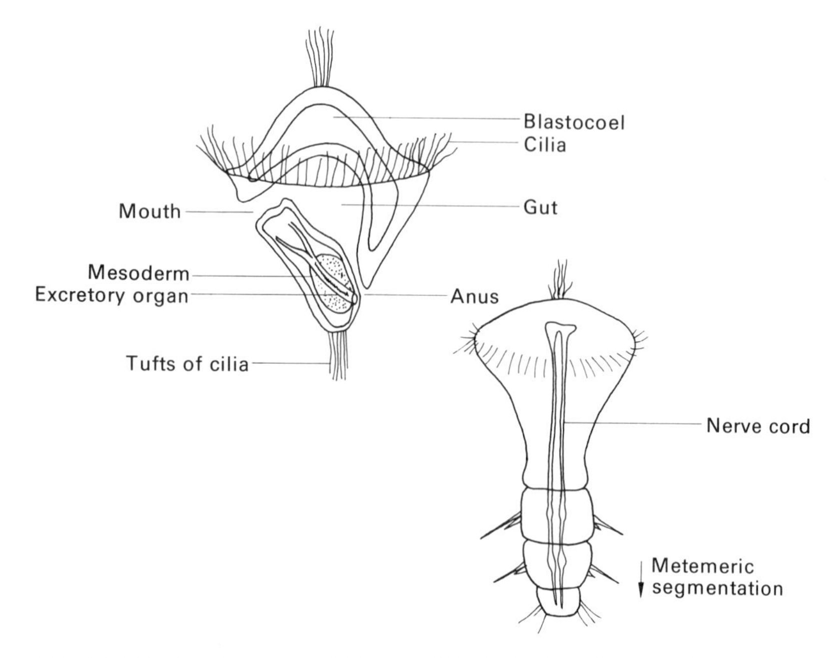

Fig. 6.36 Trochosphere larva of *Nereis* and the beginning of metamerism.

the nerve roots are also segmental, a feature which persists in the adult. The vertebral column also develops from this part of the mesoderm, enveloping the notochord and later the nerve cord.

Amphibians produce a relatively yolky egg. Although cleavage is complete, or **holoblastic**, the slow division of the yolky vegetal pole gives rise to exaggerated macromeres. Thus gastrulation differs from that occurring in invertebrates in the extent of its large-celled, yolky endoderm (Fig. 6.37).

During gastrulation changes in the centre of gravity of the egg swing it through 90° and the blastopore continues to invaginate. The last cells to enter the blastopore are the presumptive mesoderm (Fig. 6.38).

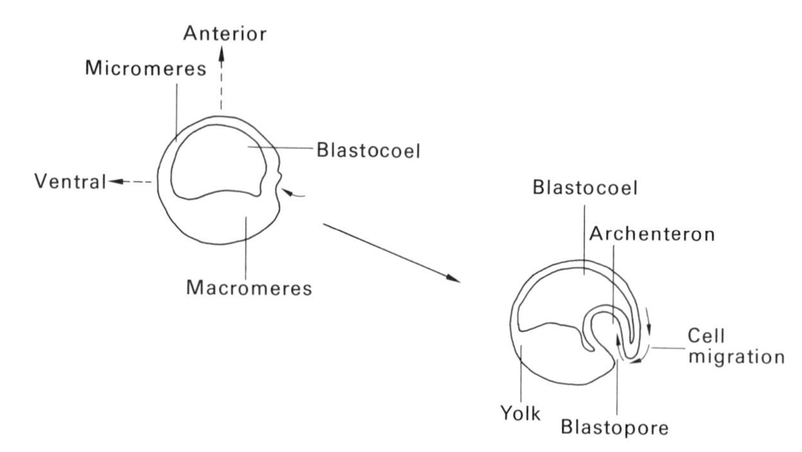

Fig. 6.37 The onset of gastrulation in the amphibian egg.

Formation of Neural Tube and Organogeny

During gastrulation, the vigorous migration of cells through the blastopore results, in effect, in a triploblastic organization. At the conclusion of gastrulation a pear-shaped region of ectoderm on the dorsal surface constitutes the **neural plate**, from which will develop the central nervous system. During the next stage, **neurulation**, the edges of the neural plate fold upwards to form the hollow brain and nerve cord. The edges of the neural folds first meet and fuse in the mid-region. They then proceed both anteriorly, to what will ultimately become the brain, and posteriorly to the blastopore. At the same time the embryo begins to elongate anterio-posteriorly.

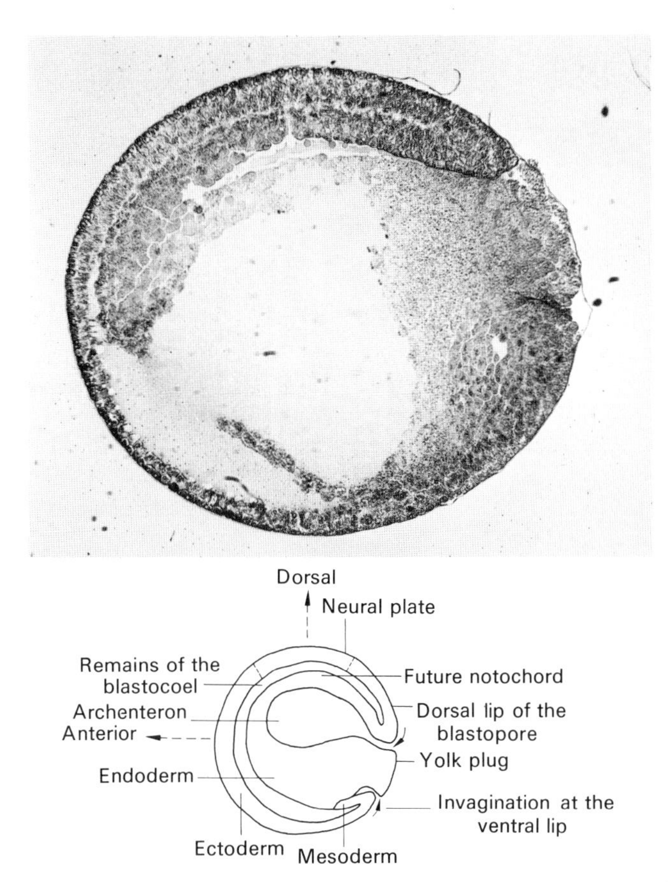

Fig. 6.38 Photomicrograph of a section of frog gastrula showing its triploblastic organisation (by courtesy of B. Bracegirdle) and, below, the disposition of tissues in the gastrula.

While neural folding is taking place some cells in the mesoderm immediately below the neural plate separate from the rest and form a rod-shaped structure, the **notochord**, which develops a characteristic vacuolated appearance. In Acranial animals, such as *Amphioxus*, the notochord forms the main skeletal structure, but in most chordates the notochord is surrounded, and to some extent obliterated, by the vertebral column (Fig. 6.39).

Before the larval amphibian hatches five visceral clefts develop on each side, in the pharyngeal region of the endoderm. From the visceral arches between the third, fourth and fifth clefts will develop the external gills. Other structures developing as endodermal outgrowths of the enteron are thyroid gland, lungs, liver and pancreas. The heart

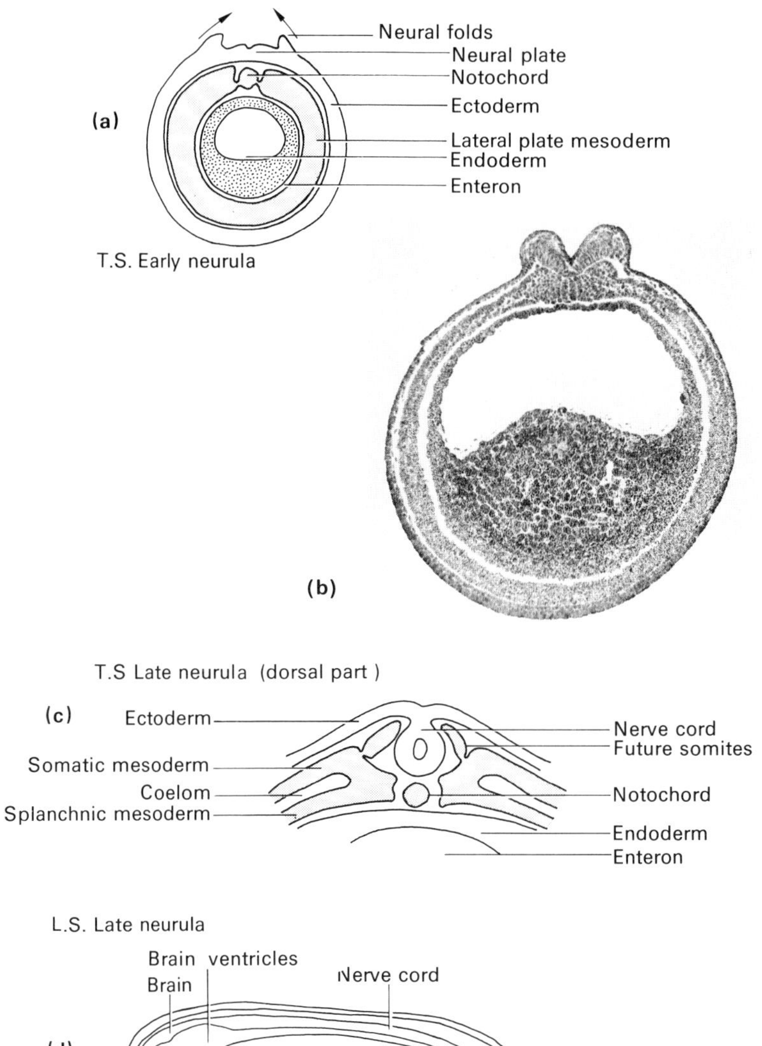

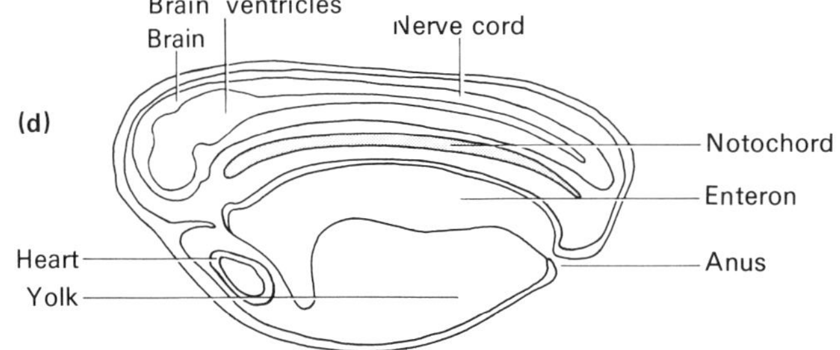

Fig. 6.39 (a) T.S. early neurula. (b) Photograph of T.S. early neurula (courtesy of B. Bracegirdle). (c) T.S. late neurula (dorsal part). (d) L.S. late neurula.

forms mainly from the splanchnic mesoderm, with an endothelial lining of scattered mesodermal cells (Fig. 6.40).

In birds considerably more embryonic differentiation takes place before hatching. This necessitates such a large yolk deposit in the egg that complete cleavage of the cytoplasm is impossible. Partial, or **meroblastic**, cleavage of the cytoplasm at the animal pole of the egg results in a disc of cells, the **blastoderm**. Further division results in an upper ectodermal layer and a lower endoderm, with a blastocoelic space between.

Gastrulation does not take place by invagination and no blastopore forms. Instead, the mesoderm is formed by the migration of the upper layer cells inwards, towards the **primitive streak**, then down into the blastocoelic space. The last cells to enter the primitive streak form the presumptive neural plate, which folds in a similar manner to amphibian neural plate to form a hollow brain and nerve cord (Fig. 6.41).

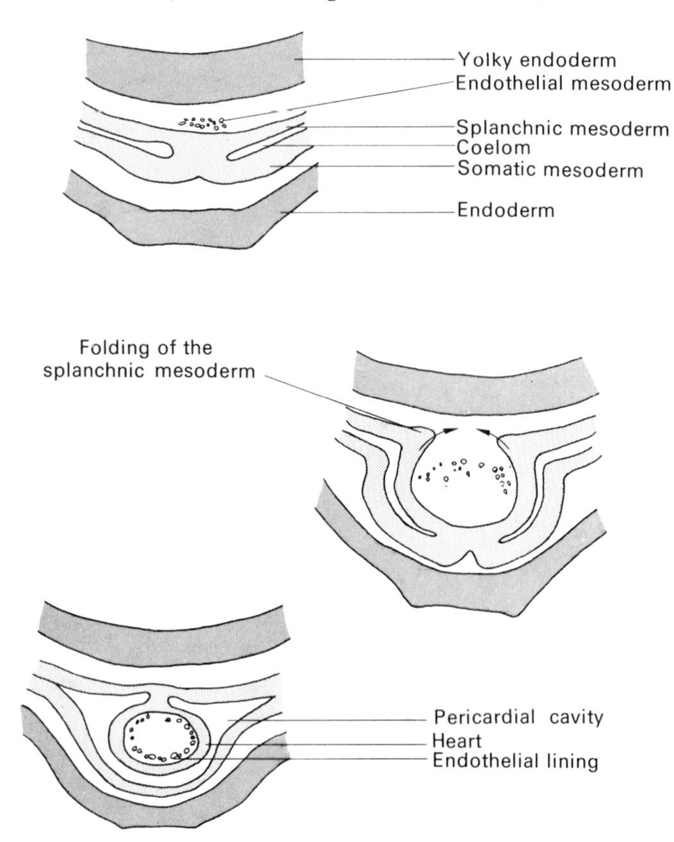

Fig. 6.40 Stages in the formation of the heart.

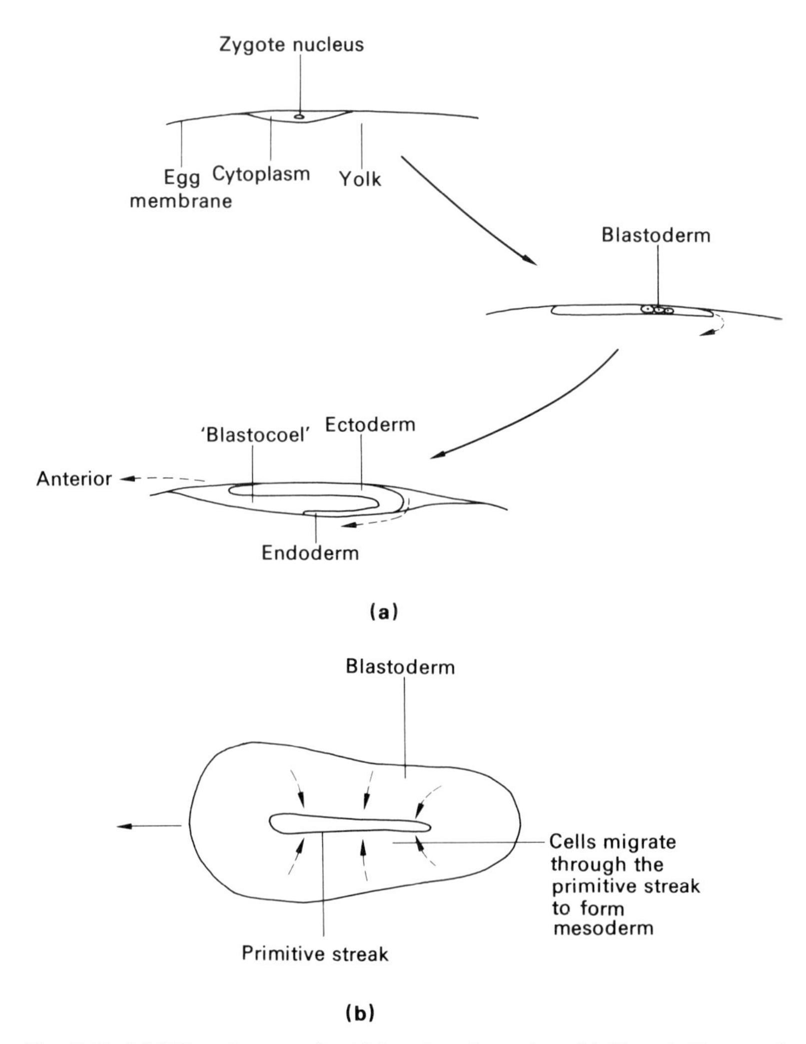

Fig. 6.41 (a) V.S. early stages in chick embryo formation. (b) The primitive streak: surface view of blastodisc. (c) see facing page.

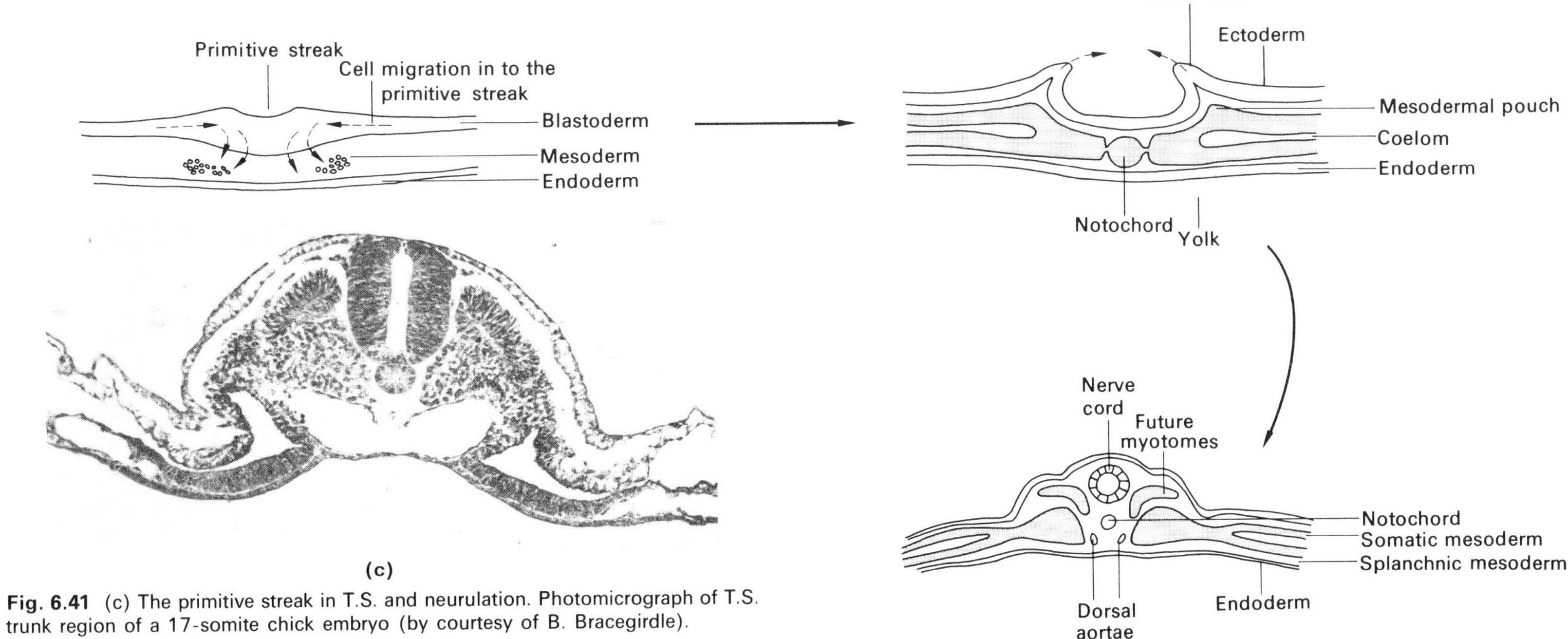

Fig. 6.41 (c) The primitive streak in T.S. and neurulation. Photomicrograph of T.S. trunk region of a 17-somite chick embryo (by courtesy of B. Bracegirdle).

The Extra-embryonic Membranes

The appearance of these membranes in evolution probably coincided with the origin of the reptiles and they persist in birds and mammals, representing an evolutionary development associated with the conquest of land. The **amnion** secretes amniotic fluid to provide a flotation chamber for the embryo, avoiding dehydration and providing a support against gravity. The **chorion**, formed at the same time as the amnion, extends round the other membranes and later fuses with the allantois to form a gas-exchange structure, the **allanto-chorion**. The **allantois**, in addition to its gas-exchange role, serves as a reservoir for the waste products of nitrogen metabolism, particularly uric acid. The **yolk sac** is an organ of yolk digestion and is highly vascular for transport of the digestion products to the embryo.

The amnion and chorion form as inner and outer layers, respectively, of folds of extra-embryonic blastoderm which arise first at the head end of the embryo, then at the tail. As the head and tail folds proceed to cover the embryo the allantois develops from the hind enteron and the yolk sac from the mid enteron (Fig. 6.42).

Embryonic Development in Mammals

Unlike reptiles and birds, the mammalian egg is not yolky and cleavage is therefore holoblastic. The development of extra-embryonic membranes, however, follows closely that in reptiles and birds (Fig. 6.43). The mother supplies nutrients throughout the development of the embryo, which can be regarded as fulfilling the role of the yolk sac in birds, and she removes metabolic wastes produced in the embryo, which in birds is the role of the allantois. In mammals the yolk sac is redundant and remains vestigial. The allantois, however, together with the chorion and associated tissues derived from the trophoblast, gives rise to the **chorio-allantoic placenta**. The amnion fulfils the same role in mammals as in reptiles and birds with regard to support.

In the maternal uterus the mammalian foetus grows in a manner similar to the growth of an isolated cell population. The human

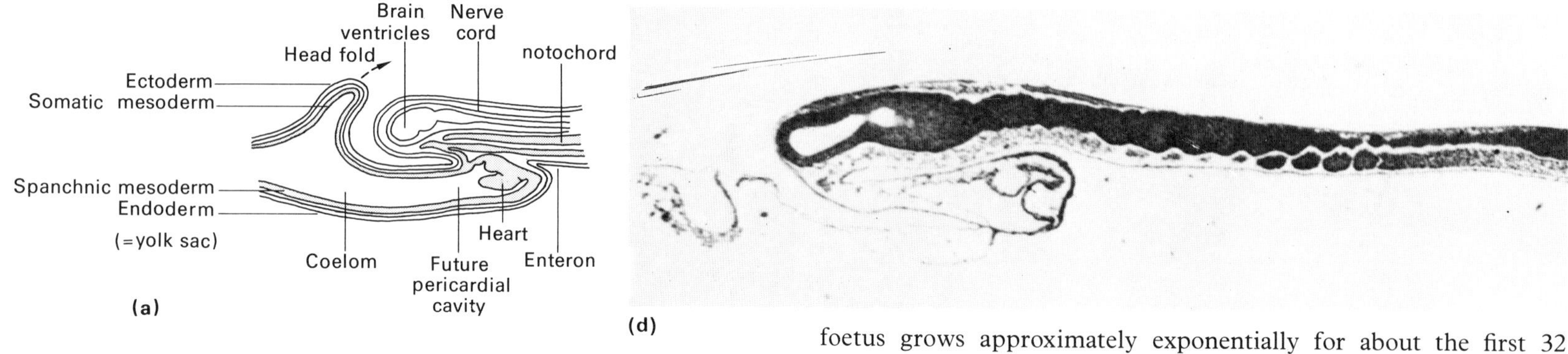

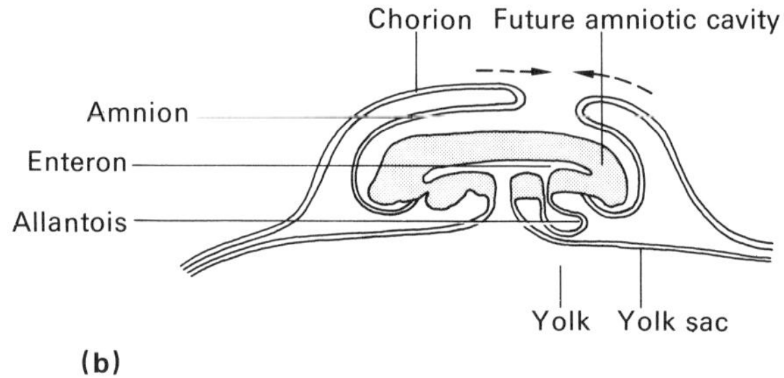

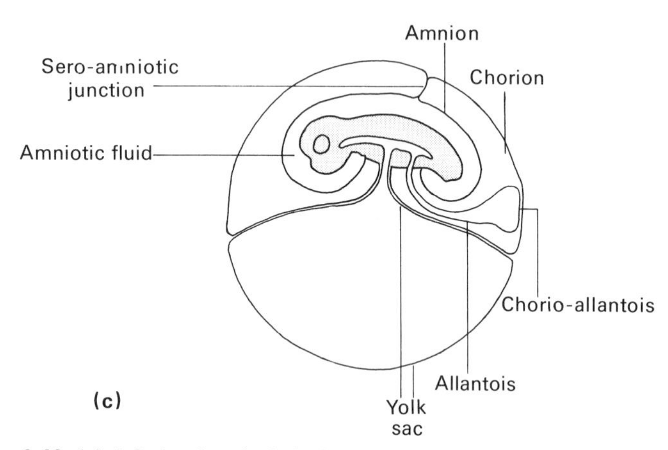

Fig. 6.42 (a) L.S. head end of chick embryo, about 36 hours. (b) Formation of extra-embryonic membranes. (c) Extra-embryonic membranes in the chick. (d) Photomicrograph of a median section of a 10-somite chick embryo (courtesy of B. Bracegirdle).

foetus grows approximately exponentially for about the first 32 weeks of pregnancy, doubling in weight about every four weeks. Growth then becomes linear, then enters a retarded phase before parturition (compare Figs. 6.44 and 6.45). The growth of the head proceeds more rapidly during the earlier stages of development. In a two month foetus the head is approximately half the length of the body and at birth it is about a quarter of the body's length. During the juvenile and adolescent growth stages the body continues to grow faster than the head, until in the adult the head forms approximately one-sixth of the body length.

The Role of the Placenta

Since the mammalian egg is not yolky the developing embryo, once growth commences, receives its nutriments from maternal tissues. When implantation has taken place the embryo exchanges solutes with the glandular endometrium, which at this stage is rich in plasma and glycogen. The formation of the chorion with its chorionic villi greatly increases the absorptive surface area. The placenta, which is the main organ of solute exchange between embryonic and maternal tissues, develops from the fused chorion and allantois at an early stage (about the fourth week in man). At birth the placenta is approximately one-eighth the mass of the foetus.

The umbilical cord, a vascular structure formed from the allantois and yolk sac, connects the embryo to the placenta and allows some mobility of the embryo. The paired umbilical arteries of the embryo supply an extensive capillary bed in the placenta. The maternal blood capillaries in the endometrium adjacent to the placenta usually degenerate to form sponge-like sinuses which are penetrated by the chorionic villi of the placenta. From the capillary bed in the chorionic villi the embryo's blood drains into a single umbilical vein (Fig. 6.46).

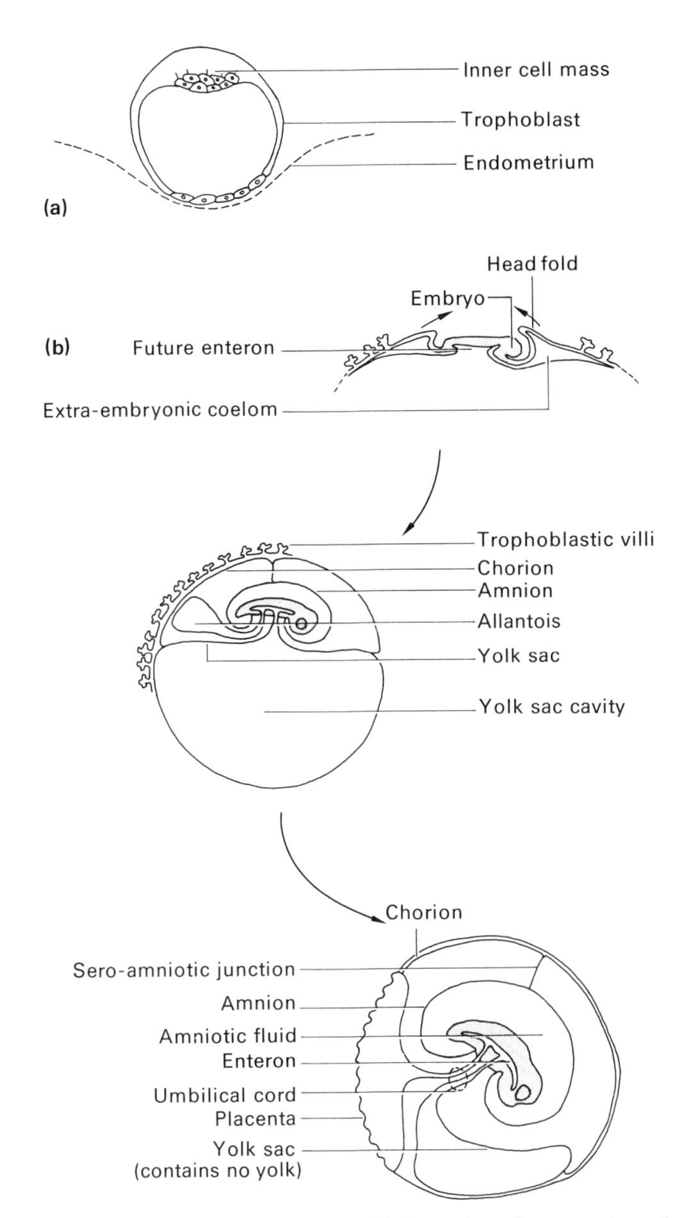

Fig. 6.43 (a) Blastocyst at implantation. (b) Formation of extra-embryonic membranes in mammals.

The rates at which solutes move across the placenta between maternal and embryonic blood depend mainly on their molecular weights, although the accumulation of some solutes against diffusion gradients (ascorbic acid, iron and calcium ions in particular) suggests active transport mechanism. Some maternal plasma proteins of relatively low molecular weights, for example γ-globulins with molecular weights up to 100 000, such as diphtheria antibodies, are known to pass from maternal to foetal blood in man. In the rabbit there is some evidence that antibodies may be transmitted to the embryo through the amniotic fluid.

White blood cells are known to have the ability to leave the blood

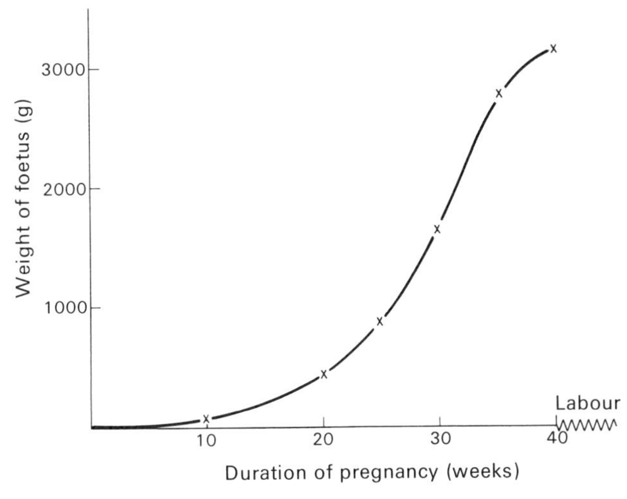

Fig. 6.44 Generalized growth curve of a human foetus. The growth rate is approximately exponential for the first 32 weeks of pregnancy, the weight being about doubled every month.

and enter tissue fluids, but the fact that offspring are not immunologically tolerant to maternal protein suggests that white cells from maternal blood do not enter the embryo in large quantities. Recently, however, white cells carrying XY sex chromosomes (p. 154) have been isolated in the blood of pregnant women who have subsequently given birth to male offspring, so the behaviour of white cells at the placenta is at present obscure.

Substances absorbed by the embryo through the placenta are broadly those absorbed by the mother from her alimentary canal. Amino acids are required for protein synthesis in the embryo;

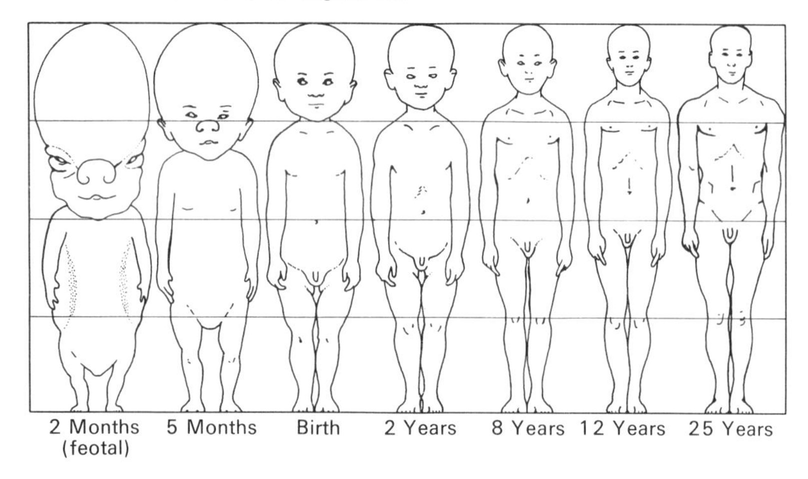

Fig. 6.45 Relative sizes of parts of the human body at different stages of growth. Growth is disproportionate, or non-allometric. (After McElroy and Swanson. Courtesy of Prentice Hall Inc.) (Right) Lateral view of the head of a human foetus (at about the 20 mm stage, about 40 days of gestation) showing the developing facial features (courtesy of Professor E. Blechschmidt and Thames and Hudson Ltd).

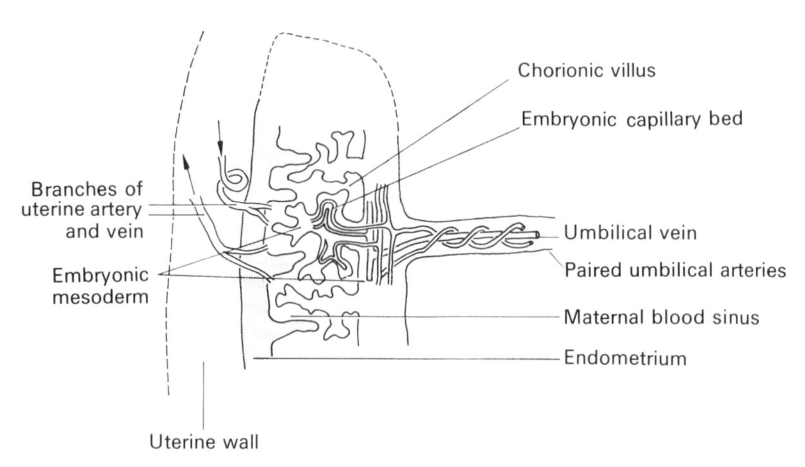

Fig. 6.46 Structure of the mammalian placenta, showing the chorionic villi which carry the embryonic blood vessels into close proximity with the blood in the maternal blood sinuses.

glycerol, fatty acids, cholesterol, and so on, are used in lipid metabolism in the embryo, for example in the building of plasma membranes. Glucose is the main respiratory substrate of the foetus. The rate of glucose passage across the placenta is greater than can be accounted for by the difference in blood sugar level between maternal and foetal blood and a glucose transport mechanism seems likely.

Gas exchange between foetal and maternal blood takes place along

	Maternal blood in the endometrium	Foetal blood arriving at the placenta
O_2 tension (mmHg)	60	15
CO_2 tension (mmHg)	43	61

Fig. 6.47 Table of the oxygen and carbon dioxide contents of maternal and foetal blood at the placenta.

diffusion gradients. Figures 6.47 and 6.48 show data obtained for sheep.

The transport of oxygen from maternal to foetal blood across the placenta is helped by the differences between the oxygen demand of adult and foetal haemoglobin. Foetal haemoglobin shows an oxygen dissociation curve shifted to the left of adult haemoglobin. This confers on the foetal blood the ability to absorb more oxygen than the adult blood, particularly at the oxygen partial pressures prevailing in the endometrial blood sinuses.

Parturition

Although the details of parturition vary widely in mammals there are certain shared underlying features. At the onset spasmodic contractions occur in the smooth muscle of the uterine wall. The contractions become coordinated in waves and cause an increase in the intrauterine pressure which dilates the cervix. The amnion breaks, releasing some amniotic fluid, and the foetus is forced from the

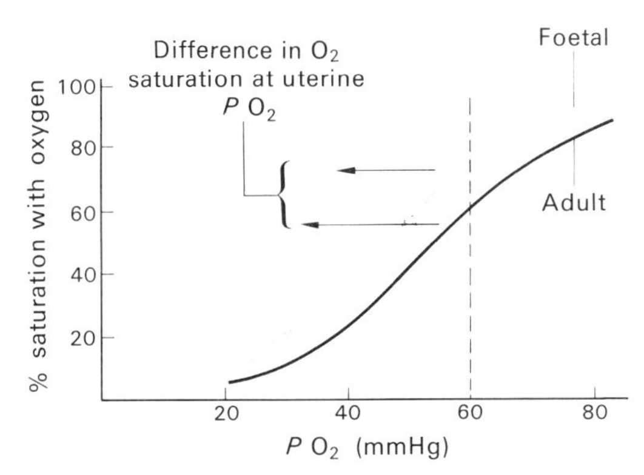

Fig. 6.48 Foetal haemoglobin, by virtue of its modified protein structure has a greater affinity for oxygen than maternal haemoglobin under the conditions prevailing at the placenta.

uterus by continued uterine contractions supplemented by contraction of the voluntary muscles in the abdominal wall. At this stage the intrauterine pressure in man has been recorded as approximately 250 mmHg.

In man the foetus is normally born head (the widest part) first, its back against the ventral part of the mother's abdomen. In ungulates the foetus is usually born fore limbs and head first, with its dorsiventral attitude the same as the mother's.

After the foetus has been expelled from the uterus the contractions of the uterine wall become quiescent for a short period (in man about 15 minutes). Contractions then recommence and the placenta, together with the remaining extra-embryonic membranes, is expelled. The placenta separates along the line of the endometrial sinuses, but usually little maternal blood is lost owing to the vigorous constriction of the associated arterioles and venules.

In the case of most mammals the umbilical cord ruptures at birth and the umbilical arteries and vein constrict sharply to prevent blood loss; the cord then withers after a few weeks. In man the cord is ligatured close to the abdominal wall, leaving a persistent scar, the navel.

Changes in the Embryo at Parturition

The traumatic change of environment at parturition induces sudden and relatively fundamental physiological and anatomical changes in the neonatal mammal. Temperature regulatory and sensory mechanisms concerned with the maintenance of homeostasis come into operation when the foetus leaves the relatively constant intrauterine environment and enters the variable external environment. Circulatory changes take place to accommodate the change from placental gas exchange and nutritional roles to lung and alimentary canal respectively.

In the foetus oxygenated blood from the placenta passes in the umbilical vein to the liver (Fig. 6.49). Some blood passes through the liver and into the hepatic vein, which carries it into the posterior vena cava. A large proportion of the blood from the umbilical vein, however, by-passes the liver in the *ductus venosus*, which also enters the posterior vena cava. In the foetal circulation, therefore, oxygenated blood enters the heart through the posterior vena cava. This is in direct contrast to the adult circulation. Most of the blood entering the heart from the posterior vena cava, however, passes through a second foetal feature, the *foramen ovale* (a hole in the interatrial septum) into the left atrium. Here the blood mixes with the small flow of de-oxygenated blood from the lungs, passes into the left ventricle and is despatched into the aorta. The heart muscle itself receives blood from the coronary arteries which arise from the base of the aorta and the oxygenated blood passes to the brain through the carotid arteries and to the tissues through the aorta.

The oxygenated blood which enters the right ventricle is also diverted round the body (only a small proportion goes to the lungs) by a third foetal feature, the *ductus arteriosus* which links the pulmonary arch with the aorta. The umbilical arteries arise from the posterior end of the aorta and thus do not carry fully deoxygenated blood to the placenta.

Thus the circulatory changes are, first, the constriction and eventual atrophy of the umbilical arteries and vein, and of the *ductus venosus*, which degenerate into fibrous strands. Second, when the flow of blood from the placenta ceases, the blood pressure in the posterior vena cava and the pressure in the right atrium falls below that in the left atrium. A valve-like fold of tissue over the foramen ovale then occludes the foramen, preventing the flow of blood from left to right. This valve adheres to the inter-atrial wall and becomes fused to it within a few days of birth. Third, the wall of the *ductus arteriosus* constricts. This constriction may not be completed for some hours, possibly days, after birth. Clearly, the change in blood pressures between the right and left sides of the heart reverses the direction of flow through the *ductus arteriosus*, so that blood now flows from the aorta to the pulmonary artery. This, together with the overall increase in foetal blood pressure caused by increased peripheral resistance resulting from the loss of the placental circuit, greatly increases the pressure in the pulmonary artery. At the same time the onset of breathing reduces the capillary resistance in the lungs by causing dilation of the alveoli. Thus the blood flow through the lungs increases massively within a few moments of birth.

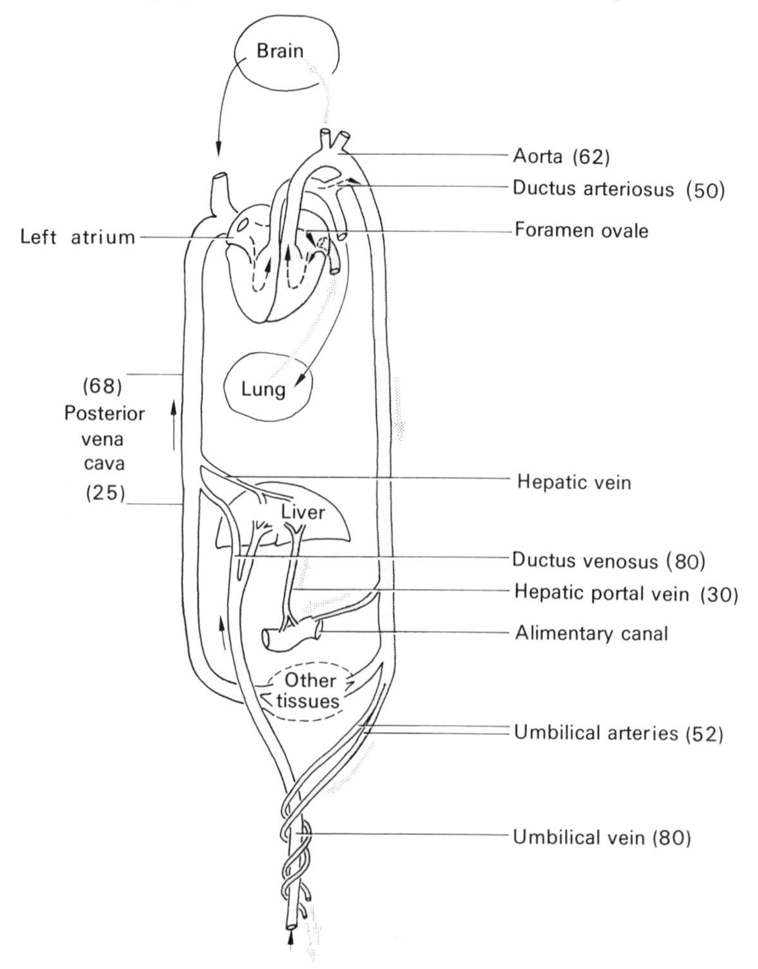

Fig. 6.49 Diagram of foetal circulation with approximate percentage saturation oxygen (brackets).

6.9 Germination: The Development of the Flowering Plant Embryo

The development of the embryo in flowering plants differs from that in mammals in the occurrence of a dormant period, expressed in the seed. The seed serves as a dispersal phase and, in temperate plants, as an overwintering mechanism. Germination involves the continued development of the embryo from the seed stage to the stage at which photosynthetic organs are produced.

Dormancy

Many plant species employ some mechanism to hold seeds in dormancy until environmental conditions most favour the subsequent survival of the seedling. Many leguminous plants have tough testas which require **scarification** before germination commences. In nature the abrasion by soil particles moving under the influence of frost, wind and rain could take several months; in the garden some seeds are usually notched with a knife before planting, for example, sweet peas. Seeds of some species of the genus *Rhus*, in the eastern United States will germinate only after a brush fire has cracked the extremely tough testa, thus giving the seedling a favourable start with little competition from other less resistant vegetation. Many species of cereals and members of the Rosaceae (rose, apples and pears) require a cold spell of four to five weeks at 1°C to 10°C before a high proportion of the seeds germinate (**vernalization**). In some varieties of lettuce germination is improved by red light irradiation in the wavelength region of 660 nm. In *Nigella* sp. and the Californian poppy, however, light appears to be an inhibitor of germination. Some desert plants surround their seeds with a chemical inhibitor of germination which is slightly soluble in water. Thus germination occurs only after prolonged washing; a mechanism which favours the survival of the seedling until its xerophytic features have differentiated.

Special dormancy mechanisms are not always present. In many species the seed contains what might be termed a premature embryo, so that germination is retarded until the onset of winter temperatures slows down the metabolic activity of the seed. In some trees, such as oak and beech, germination must proceed in the autumn in which the seeds leave the trees. It has been reported that the embryos of some willows fail to survive unless germination commences within a few days of the separation of the seed from the parent plant.

Biochemical Activity in the Seed

At the onset of germination water enters the seed, approximately doubling its weight. Enzymes in the cotyledons (or in the endosperm adjacent to the cotyledons in exalbuminous seeds) catalyse the hydrolysis of storage compounds (Fig. 6.50). Insoluble, large molecular weight lipids, starches and proteins are converted to their respective fatty acids and glycerol, sugars, and amino acids, in a manner similar to digestion in the animal gut (Fig. 6.51).

In most plant species lipids form the main store of respiratory substrate. In the castor oil seed (p. 105) lipids are mobilized at germination by esterase (or lipase) hydrolysis to fatty acids and glycerol:

$$\begin{array}{l} CH_2 \cdot O \cdot CO \cdot R_1 \\ | \\ C \cdot H \cdot O \cdot CO \cdot R_2 \\ | \\ CH_2 \cdot O \cdot CO \cdot R_3 \end{array} + 3H_2O \xrightarrow{\text{Esterase}} \begin{array}{l} CH_2OH \\ | \\ CHOH \\ | \\ CH_2OH \end{array} + R_1COOH + R_2COOH + R_3COOH$$

Lipid — Glycerol — Fatty acids

For conversion to glucose or to structural polysaccharides, or to enter the glycolytic pathway for energy formation, glycerol must be phosphorylated, and this reaction requires ATP. Since phosphorylation is an intracellular process taking place in the embryo, glycerol itself must be regarded as the translocatable product, analogous to maltose.

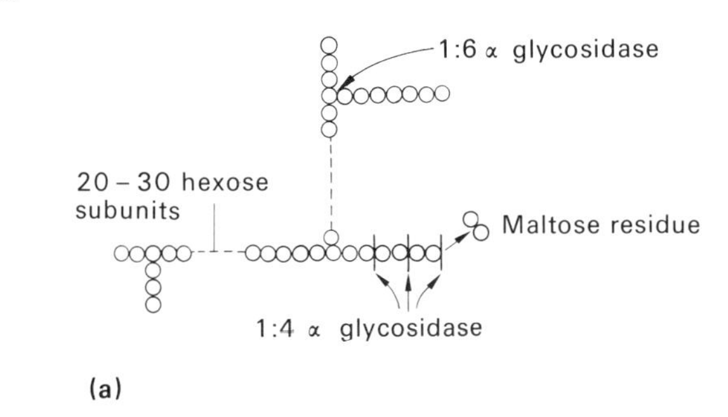

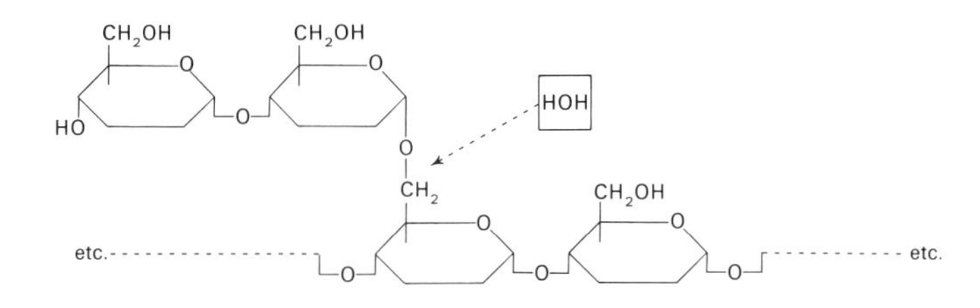

Fig. 6.50 Hydrolysis of amylopectin. (a) This branching chain starch requires two amylase enzymes: 1:4α-glycosidase, which hydrolyses the ends of chains to maltose, and 1:6α-glycosidase, which hydrolyses the branching points when the end chains have been removed. (b) Hydrolysis of the 1:6 glycosidic bond.

Mitosis and cell elongation occur first in the embryo root, which shows strong hydrotropic and geotropic responses. Shoot growth,

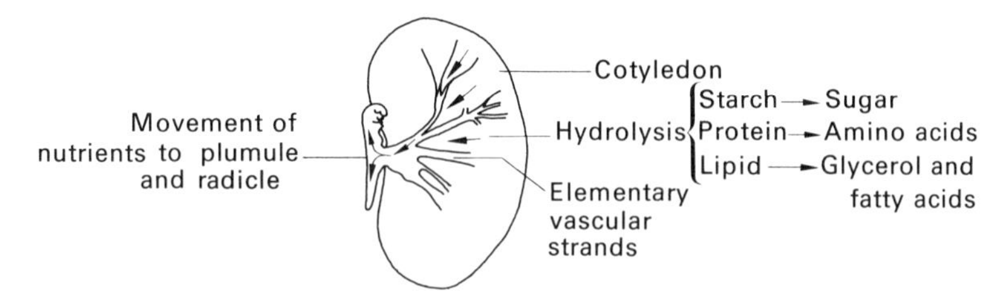

Fig. 6.51 Changes in the cotyledon at germination (compare with Fig. 6.27).

showing negative geotropic and strong phototropic responses, follows. In seeds which exhibit **hypogeal** germination, such as the broad bean, the plumule, or embryo shoot, pushes free from the cotyledons, which remain on the soil surface, or below it if the seed was buried. In seeds showing **epigeal** germination, like the French bean, the growth of the shoot lifts the cotyledons above the soil surface. In most epigeal species the cotyledons then spread horizontally, develop chloroplasts, and act as the first photosynthetic organs of the seedling. Thus the cotyledons are able to extend their supporting role of the embryo until the first true leaves form (Fig. 6.52).

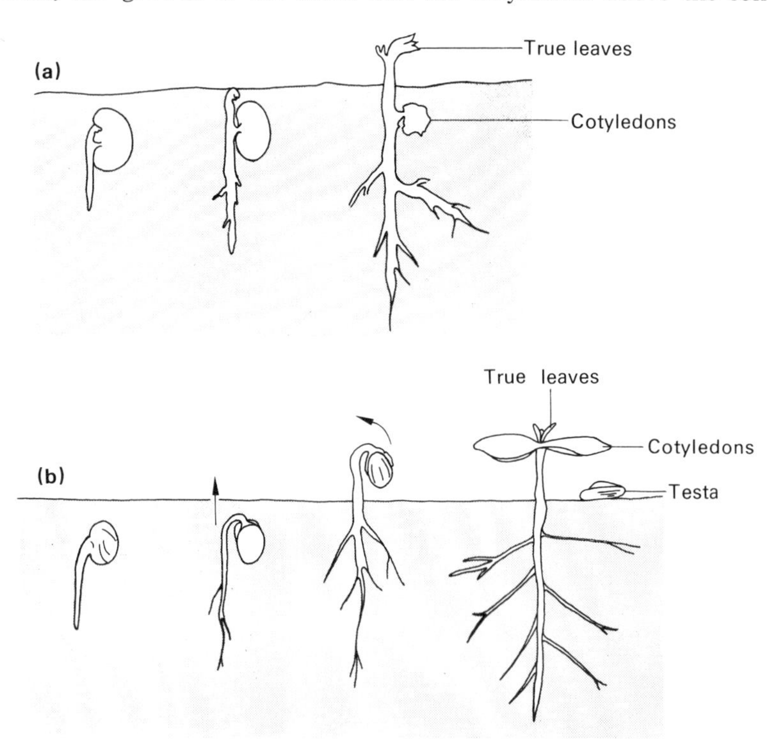

Fig. 6.52 (a) Hypogeal germination. (b) Epigeal germination.

6.10 Differentiation and the Organizer Concept in Embryology

The central dogma of mitosis is that each daughter cell of a nuclear division contains the same genetic complement as the parent cell. From this it follows that all the cells in a developing embryo, formed as they are by repeated mitosis from the original zygote nucleus, form a genetic clone and are genetically identical. Thus all cells contain the genetic information required to function as any differentiated tissue component. Differentiation into tissues, with the essential division of labour that ensues, can therefore be seen in terms of the activation of certain combinations of genes and the suppression of the remainder.

It has been known for some time that DNA in chromosomes exists in the form of chromatin, that is combined with other substances, notably protein and RNA. Work has recently been focussed on these chromatin constituents in an effort to identify the mechanism whereby some genes are 'switched off' and others allowed to operate.

Using 'molecular hybridization' techniques, by which the degree of combination, or hybridization, of DNA with labelled RNA is taken as an indication of the level of base-pair compatibility, Paul and Gilmour demonstrated that DNA was extensively masked (unable to combine with RNA and probably to code for mRNA synthesis) by basic proteins known as histones. They also showed that different DNA molecules were being masked in different tissues and organs, thus elucidating a possible mechanism for cytodifferentiation.

But although histones are known to combine readily with DNA they are not sufficiently specific to act as selective gene suppressors; only five or six kinds of histone have been discovered and they are small molecules with a molecular weight of about 20 000, which would only cover about two turns of a DNA helix. The search for specificity has led to the acidic, non-histone protein and the RNA fractions of chromatin. Paul and Gilmour suggest that histones are occluded from genes which are destined to be operational in a particular tissue by the non-histone fraction. Specificity in the non-histone fraction may lie in the acidic proteins or in a newly-discovered special class of nucleic acid, chromosomal-RNA (Fig. 6.53).

An attractive suggestion for a mechanism of embryological differentiation now emerges. If the cytoplasm of the fertilized egg were to contain a mosaic of the non-histones, the process of cleavage would subsequently equip different cells with different combinations of gene switches. Some eggs, indeed, appear to be committed to differentiation before cleavage begins. Whittaker showed that fertilized *Fucus* eggs assume an aspherical form immediately before

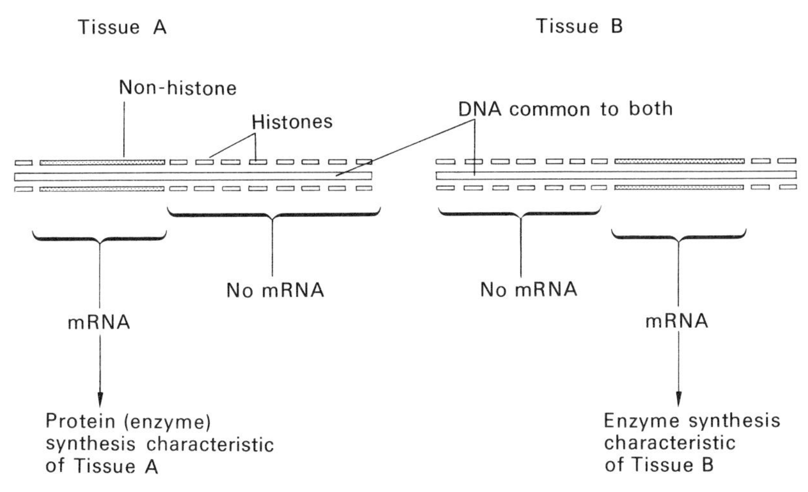

Fig. 6.53 The blocking of DNA by histones suggests a way in which genes can be switched off during differentiation.

cleavage. The bulge, or more pointed, portion develops into the rhizoid and the spherical part develops into thallus. The first cleavage plane separates the two parts (Fig. 6.54).

Whittaker's further experiments indicated that the cytoplasmic organizers present before the first cleavage were not fixed in the cytoplasm, but could be affected by environmental factors. The emergence of the rhizoidal bulge could be influenced by gradients of light intensity, pH, temperature, and the presence of other eggs at the same stage of development.

The passive role of the nucleus in the early stages of differentiation is further demonstrated by the work of Briggs and King. They removed the nucleus from a frog's egg and replaced it with a nucleus from a blastula cell. They found that the egg developed normally, indicating the absence of nuclear differentiation at the blastula stage. However, nuclei taken from later stage embryos failed to produce normal embryos when transferred to enucleated eggs. A nucleus, for example, taken from the endoderm of a late gastrula, would produce an embryo that was deficient in ectodermal tissues. This seems to indicate nuclear differentiation and raises the possibility that tissue differentiation may cause genes to be switched off in an irreversible manner.

In many organisms cytoplasmic commitments to a course of differentiation appears to occur at a later stage than in *Fucus*. Thus separation of the two cells after the first cleavage of an amphibian egg leads to the formation of two intact embryos. Monozygotic, or identical, twins in man are formed in this manner. In the armadillo

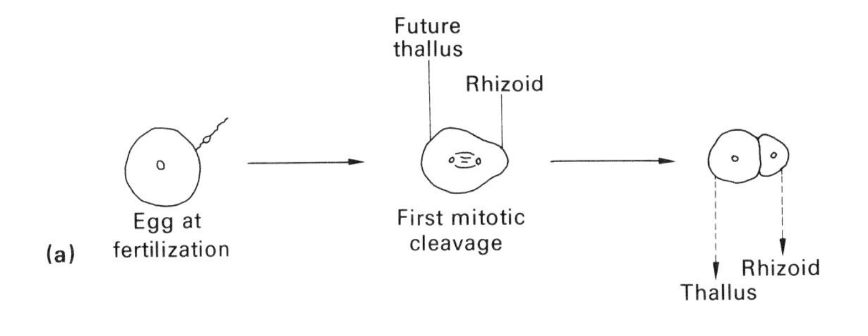

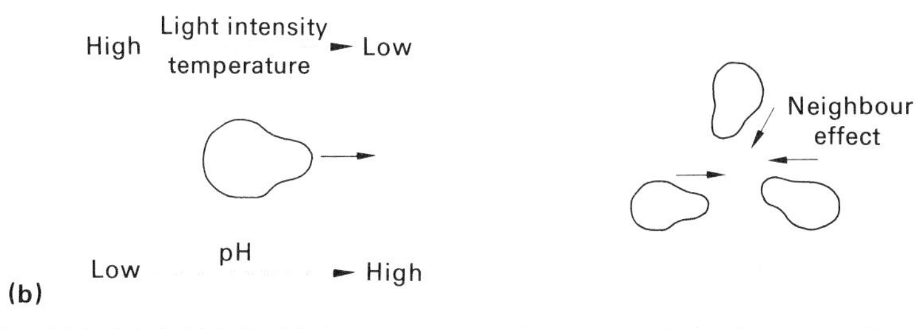

Fig. 6.54 (a) A 'rhizoidal bulge' appears on the surface of the developing *Fucus* egg just before the first division. The cell which contains the bulge subsequently differentiates into the stalk, or rhizoid, of the plant. (b) The position of the rhizoidal bulge is affected by environmental factors suggesting that the first stage of differentiation is not controlled directly by the genes.

separation after the first two cleavages occurs, to give the characteristic identical quadruplet offspring.

In all embryos, however, a stage is reached in which the removal of cells results in tissue deficiency later.

Organizers in Amphibian Embryos

Neural Tube

The classical experiments of Spemann on the induction of neural tube in amphibians led to the formulation of the organizer concept. In 1924 Spemann and Mangold reported that tissue removed from the dorsal lip of the blastopore in the later stages of gastrulation could induce neural tube formation if transplanted into other embryos. In experiments with salamander embryos they reported that the host embryo continued with normal gastrulation, but that minor invagination occurred at the site of the donor implant. At the implant site host ectoderm was found to form neural tube and associated structures (Fig. 6.55).

Organization of the Lens

In the embryological development of the eye, the optic cup forms from presumptive neural plate cells whereas the lens forms from presumptive epidermis. In the early gastrula these two cell groups are widely separated but are brought into adjacent positions by the pattern of cell migration at gastrulation.

If, in an experiment, the future optic cup is removed from one side of an embryo the lens fails to develop on that side. On the oppo-

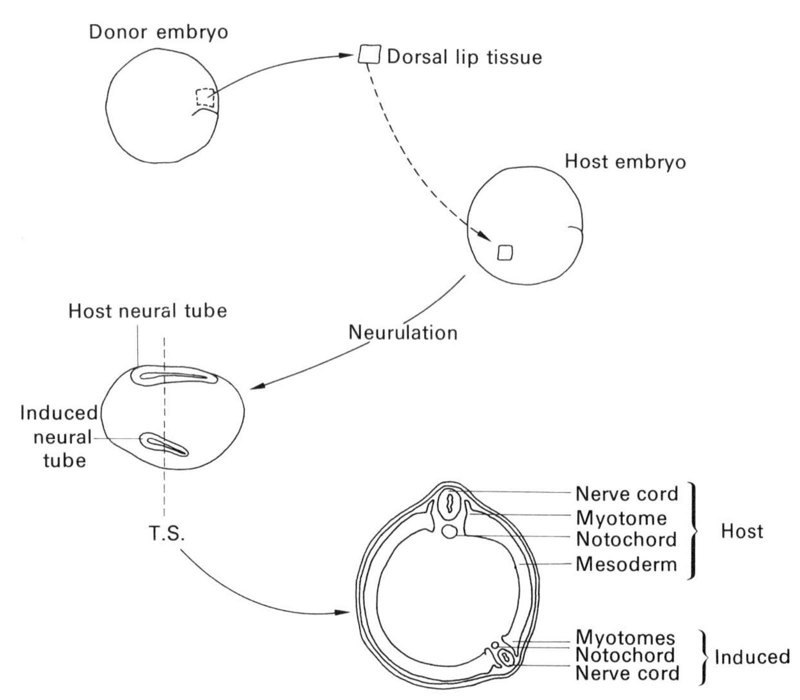

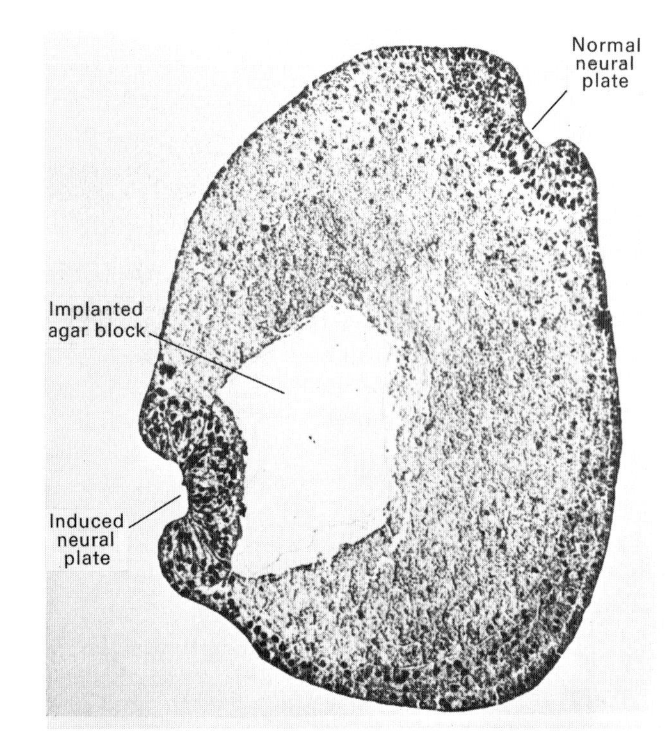

Fig. 6.55 Spemann demonstrated the organizing role of the dorsal lip of the blastopore by transferring it to the flank of a second embryo where it induced the formation of a second neural tube (left). An experiment by Wehmeier, one of Spemann's collaborators in the quest for organiser substance (right). A second neural plate has been induced in an axolotl embryo by an agar implant containing tissue extract (courtesy of B. I. Balinsky, and W. B. Saunders and Co.).

site, unaffected side eye formation proceeds normally. If the optic cup tissue which was removed is then transplanted below the epidermis on the ventral side of the embryo, lens formation is induced at that site. Such a transplant experiment produces an adult with a well-developed eye on the ventral side, but there is no neural connection to the brain (Fig. 6.56).

The Nature of the Organizer

The mode of operation of organizers is not understood. There is some evidence that the process of induction involves diffusible chemicals, but efforts to isolate such substances have given confused results. Some evidence points to nucleic acids as the chemical agent, but the possibility that it may be the 'non-histone fraction' of Paul and Gilmour is not ruled out.

Little work has been done on organizers in plant embryos, although the recent work of Bekhor, Kung and Bonner in California has revealed a histone-masking pattern in pea seeds similar to that in the calf thymus tissue used by Paul and Gilmour. Histones have not been discovered in bacteria, in which Jacob and Monod have shown a substrate/regulator-gene mechanism for the control of gene activity (p. 135), and it may be that the histone mechanism is an evolutionary feature associated with the origin of multicellularity and differentiation.

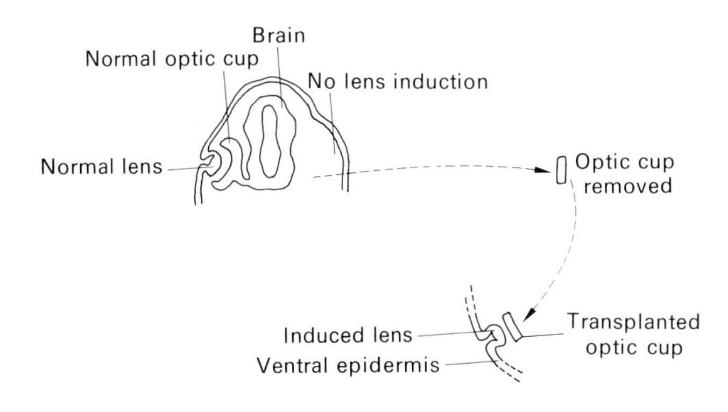

Fig. 6.56 In the formation of the eye the lens develops from a piece of ectoderm organized by the optic cup. If the optic cup is transferred to a different part of an embryo it will induce eye formation at that point.

7 Regulators of Reproduction

In the three previous chapters it has been shown that reproduction is related both to increase in numbers, or biomass, of a species under conditions which are favourable for growth, and to the necessity for generating variability in a population. In all but a few organisms, such as man, reproduction fits into a pattern of environmental factors such as daylight length, temperature and tidal effects, which act as triggers of the complex patterns of internal regulation.

In this chapter the regulators of reproduction are discussed in this context. Regulators ensure that reproduction occurs under environmental conditions which are most likely to ensure success of fertilization, development and survival of the offspring. This is particularly important in sexual reproduction, which is both numerically reducing and potentially more hazardous than asexual reproduction.

7.1 Initiation of Reproduction in Flowering Plants

A critical phase in the life cycle of a flowering plant occurs when a vegetative shoot tip transforms into a flowering shoot. In the vegetative shoot tip most mitotic activity occurs round a dome-shaped quiescent zone. When the tip transforms into a flowering shoot, however, mitosis commences in this central zone and it is from here that the flowering shoot emerges (Fig. 7.1).

Most flowering plant species fall into two broad categories with regard to the time of flowering; **long-day plants**, in which development of flower apices is promoted by the long hours of daylight in the summer months, and **short-day plants**, in which development of flower apices is promoted by the shorter daylight hours of spring or autumn. Into the first group fall the poppy, delphinium, the cereals such as wheat and barley, and the plantains on garden lawns. Into the second group fall spring flowers, such as the primrose and strawberry, and autumn flowers, such as chrysanthemums and the Euphorbias.

A number of factors are known to induce the transformation of vegetative tissue into reproductive tissue. The most important seems to be light, although temperature may be influential in some species, such as the brassicas. In sprouting broccoli, for example, flowering occurs after exposure to low temperatures; if plants are maintained artificially at 'summer temperatures' they remain in the vegetative condition.

Experiments concerning the effect of light on short-day and long-day plants by Borthwick and Hendricks at the U.S. Department of Agriculture have suggested a 'clock' mechanism involving a blue chromo-protein termed **phytochrome**. This pigment exists in two forms and exhibits photoreversibility analogous to rhodopsin in the retina of the eye (p. 122). Spectrophotometry shows that one form, designated P660, absorbs maximally in the red region at a wavelength of 660 nm, and the other form, P730, absorbs maximally in the far red region, at 730 nm (Fig. 7.2).

Flash techniques on isolated phytochrome pigment have shown the photoreversible reaction to be a rapid one, consisting of an instantaneous light reaction followed by a series of dark reactions taking up to five seconds. Persistent illumination at the wavelengths at which they absorb maximally changes each form to the other. In addition, illumination with white light converts P660 to P730 and a dark period causes the reverse reaction:

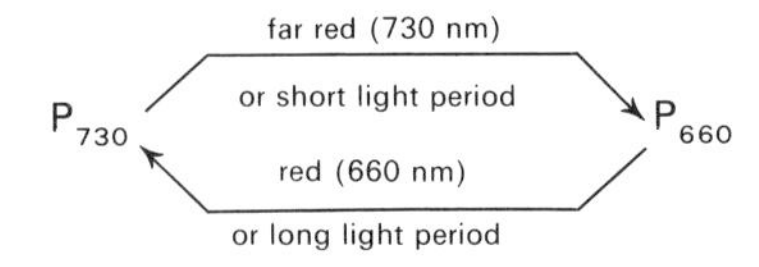

The important factor in the physiology of the plant, however, is the length of the dark period. If an experimental plant is subjected to an extended 'night' period its phytochrome is converted to the P660 form. Above a threshold concentration this form of chromatin

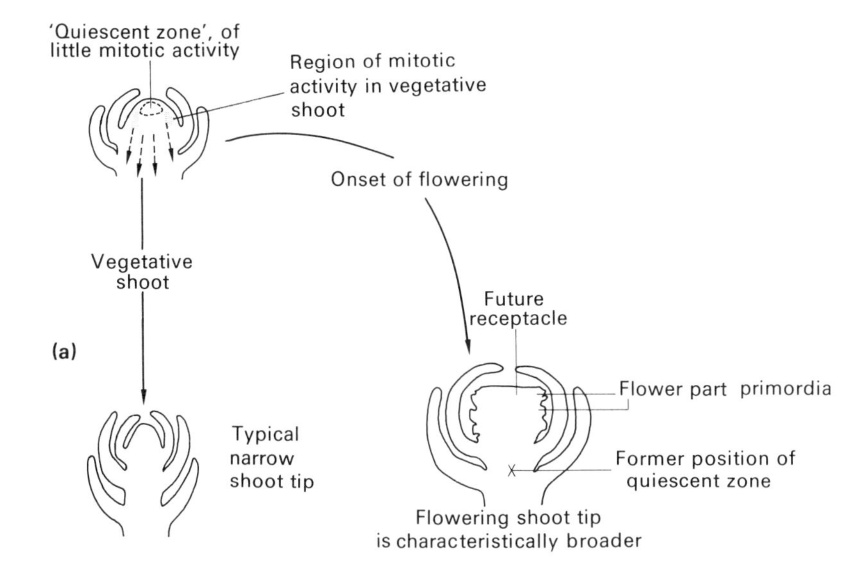

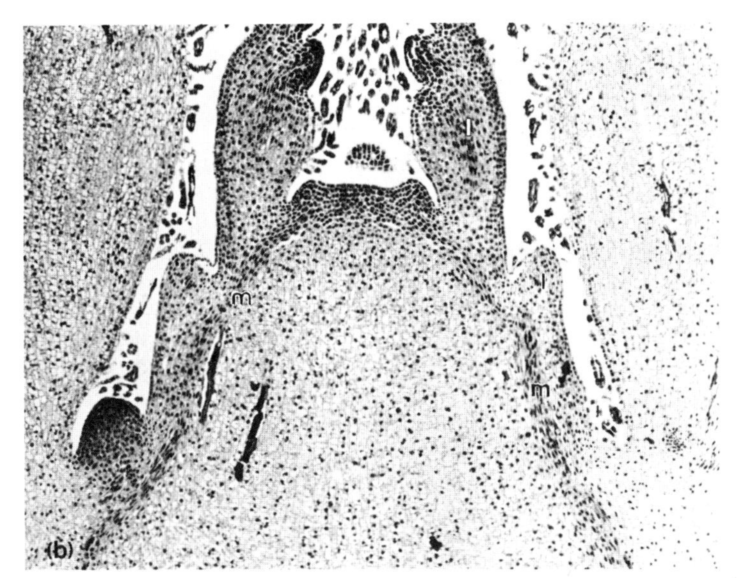

Fig. 7.1 (a) The onset of reproduction in the flowering plant is the transformation of a vegetative shoot into a flowering one. (b) V.S. of a vegetative shoot tip showing meristematic region (m) and leaf primordia (l). (Courtesy of Gene Cox.)

promotes flower production if the plant is of the short-day type, but inhibits flowering if it is a long-day plant.

Use has been made of this phenomenon by commercial growers of chrysanthemums, which are short-day, autumn flowering plants. By effectively shortening the dark period, by switching on the green-

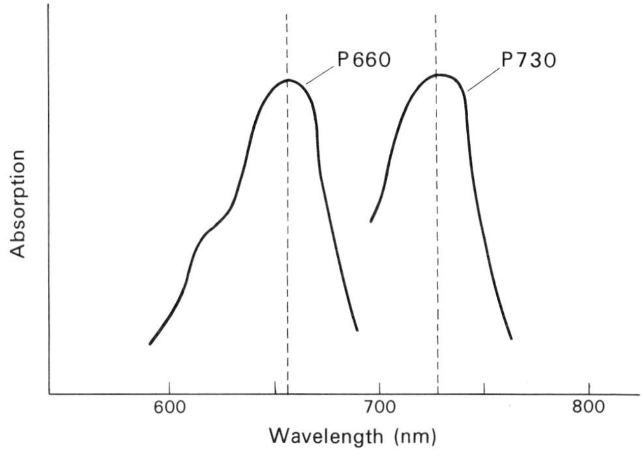

Fig. 7.2 Absorption spectra of the plant pigment, phytochrome (P). It has two, photoreversible forms whose absorption peaks are at the wavelengths 660 and 730 nm.

house lights in the late summer, flowering is postponed; this allows the growers to exploit the Christmas market (Fig. 7.3).

The mode of action of phytochromes is not known. They are believed to be instrumental in apparently widely different physiological systems, such as germination, epinasty (the spatial attitude of the leaf with respect to the stem), nyctinasty (light and dark responses, like petal and leaf movement), tuber formation, and internodal elongation. Experiments using *Mimosa pudica*, the 'sensitive plant', offer some evidence that membrane permeability is altered by phytochromes. Membranes are now known to play active roles in cellular metabolism and any modulation of membrane function, such as the initiation or suppression of selected solute transfer systems, could form an epigenetic basis for the kind of differentiation involved in flower formation.

7.2 Hormonal Regulation of Reproduction in Animals

In the flowering plants, chemical initiators (in the phytochrome

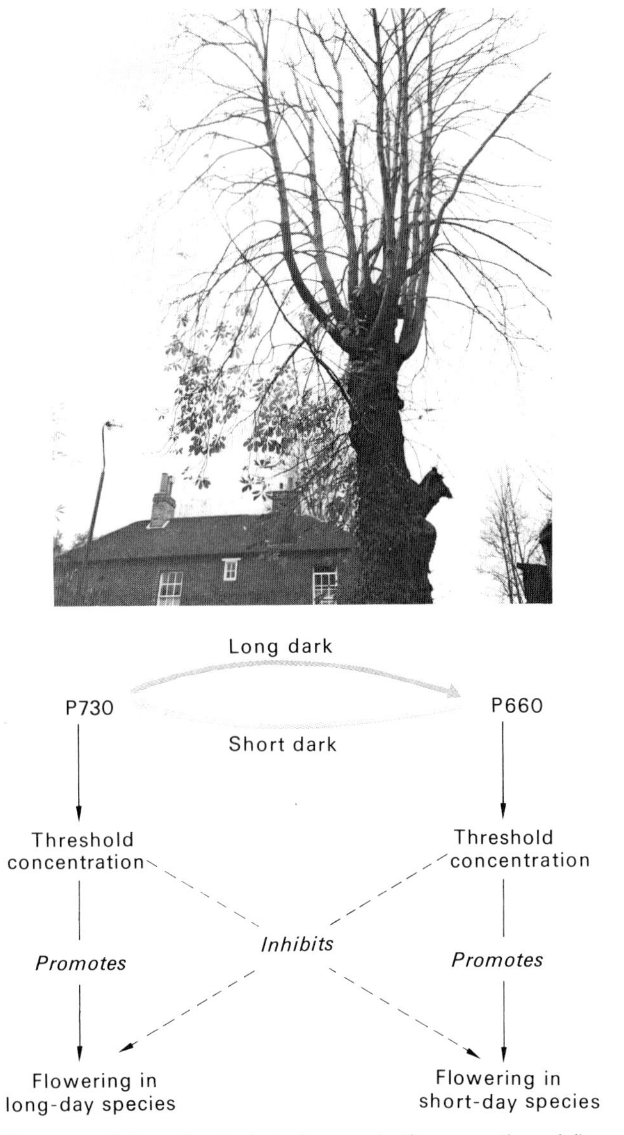

Fig. 7.3 Summary of the role of phytochromes in the promotion of flowering in long-day and short-day species. (Top) A horse chestnut tree in autumn in which leaf abscission has been delayed in branches adjacent to a street light, a localised 'long-day' phytochrome effect.

mechanism) promote flower formation at the particular season of the year at which pollination, the accumulation of seed storage substances, and seed dispersal, are most likely to be successfully accomplished. An analogous 'clock mechanism' is known to operate in animals. A number of chemical initiators, or **hormones**, is usually

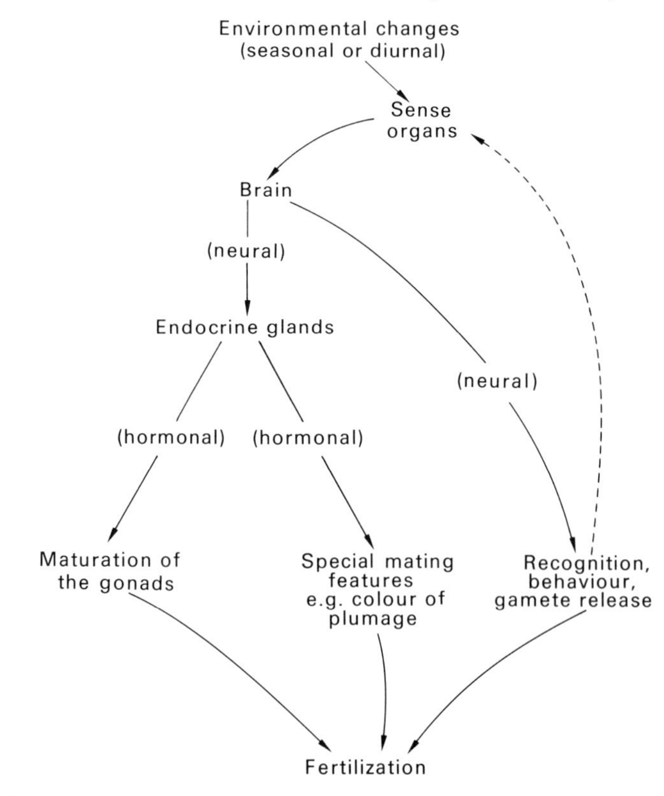

Fig. 7.4 The interaction of hormonal and neural pathways is essentially the same for a wide variety of animals.

involved but the end result (an increase in the probability of successful reproduction) is precisely the same as in plants. To achieve this ultimate aim it is necessary to activate the gonads simultaneously in both sexes; to bring the sexes together by courtship behavioural patterns; to induce the coincidental shedding of mature spermatozoa and ova (or copulation in internally fertilizing species); and to bring about all these at the time of the year at which the offspring, or the larval stages, will encounter the most favourable environmental conditions for survival.

A singularly precise clock mechanism, possibly triggered by a water pressure effect associated with height of tide, occurs in the polychaete Palolo worm, *Leodice viridis*, which inhabits burrows in the coral reefs of the south Pacific. The posterior part of the animal, in which the gametes are produced, emerges from the burrow and separates from the rest of the animal. It floats to the surface where it ruptures and releases the gametes for fertilization. The interesting feature is that most of the worms, over large areas of ocean, shed their posterior segments over a two-hour period before dawn on only one or two days of the year, at the beginning of the last lunar quarter in November. Maturation of the gametes in this worm appears to be equally closely timed; it has been reported that fertilization cannot be induced artificially two days or more before spawning.

Usually in a less exact manner, temperature changes, length of daylight, and so on act through neural and hormonal pathways to activate gonads and initiate courtship in broadly the same manner in a wide variety of animals (Fig. 7.4).

Work on the fresh water bitterling, *Rhodeus amarus*, has shown a hormonal mechanism which parallels that in mammals. In the female fish, ovarian progesterone stimulates the formation of an ovipositor, or genital papilla, an extension of the cloaca. Hypophysectomy, or removal of the pituitary gland (p. 128), however, prevents the formation of the ovipositor, suggesting a pituitary, gonadal, and secondary sexual character relationship. Fertilization is external; the female fish lays eggs into the inhalent siphon of the swan mussel, *Anodonta*,

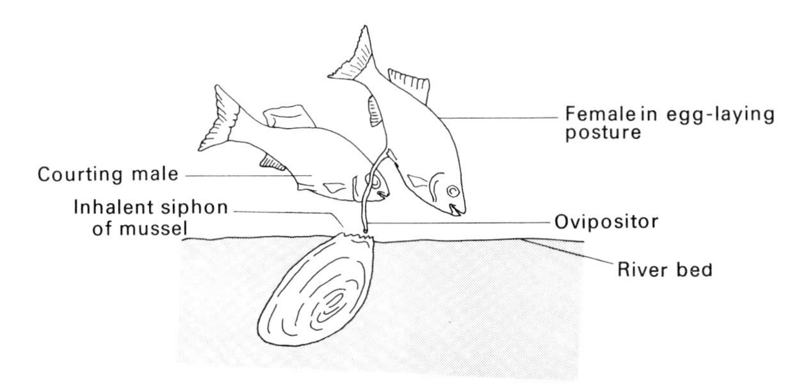

Fig. 7.5 Courtship and egg-laying in the fresh water bitterling.

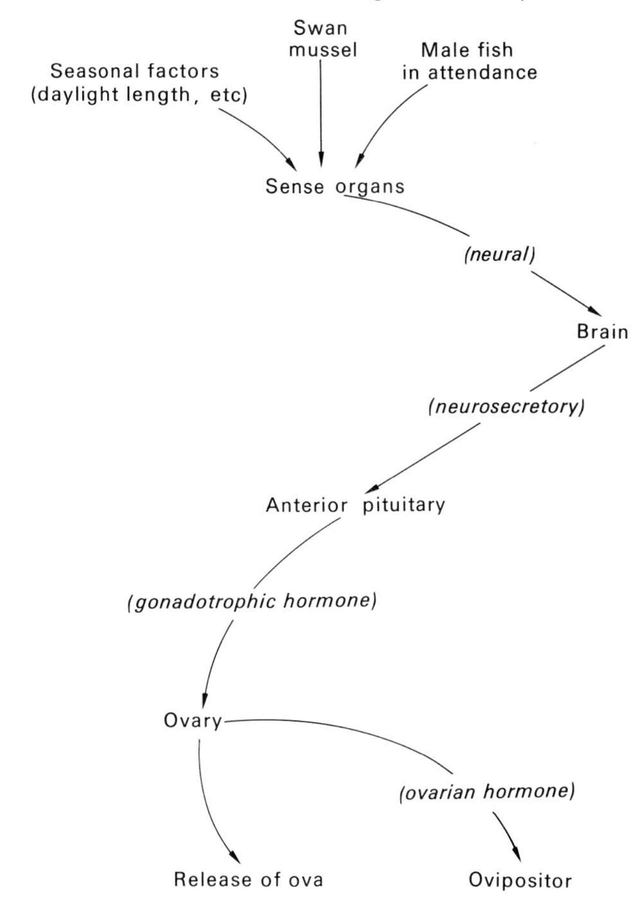

Fig. 7.6 A suggested scheme for the path leading to egg production in the female bitterling.

and the male fish sheds spermatozoa over them (Fig. 7.5). Ovipositor formation is inhibited in the absence of a male or of a swan mussel (Fig. 7.6).

The further development of the Bitterling eggs is interesting, although not strictly relevant to this discussion. The embryos develop in the mantle cavity of the mussel until the yolk sacs are absorbed, when they are ready to leave the mussel and assume an independent life. At about this time the mussel ovulates. Its eggs develop into a modified larval stage termed a *Glochidium* which becomes attached to the gills or fins of the fish. Here they lead a parasitic life, encyst and finally escape as young mussels. This has been described as a case of 'tit for tat' rather than true symbiosis or mutualism (Kyle, 1926).

Reproductive Regulation in Mammals

Internal development of the embryo necessitates preparation of the endometrium coordinated with ovulation and mating. In most mammals changes in hormonal concentrations in the blood bring about **oestrus** or 'heat', during which mating is likely to result in pregnancy. The endometrium of the uterus begins to thicken, under the influence of **oestradiol**, a steroid oestrogen secreted from the ovary, probably by the developing follicle. The endometrial cells hypertrophy, mucus and glycogen become abundant in glandular cells, and blood capillaries become distended.

The growth of the follicle in the ovary and its subsequent shedding at ovulation is promoted by two trophic hormones secreted by the anterior lobe of the pituitary, **follicle stimulating hormone** (FSH) and **luteinizing hormone** (LH). The latter hormone continues to act on the follicle after the ovum has been shed from it, giving rise to the **corpus luteum**. This body is known to secrete the 'pregnancy hormone', **progesterone**, whose function is to further the development of the endometrium, to suppress follicular development and ovulation, and generally to prepare the body for the reception and retention of the embryo. In many respects progesterone acts antagonistically to the oestrogens in the oestrus cycle. But, at the same time, experiments in animals have shown that progesterone is only effective when oestrogens are present and that the response of a tissue, for example the endometrium, depends on the relative concentrations of the two sets of hormones:

CH_3 OH, HO — Oestradiol

CH_3 $CO \cdot CH_3$, CH_3, O — Progesterone

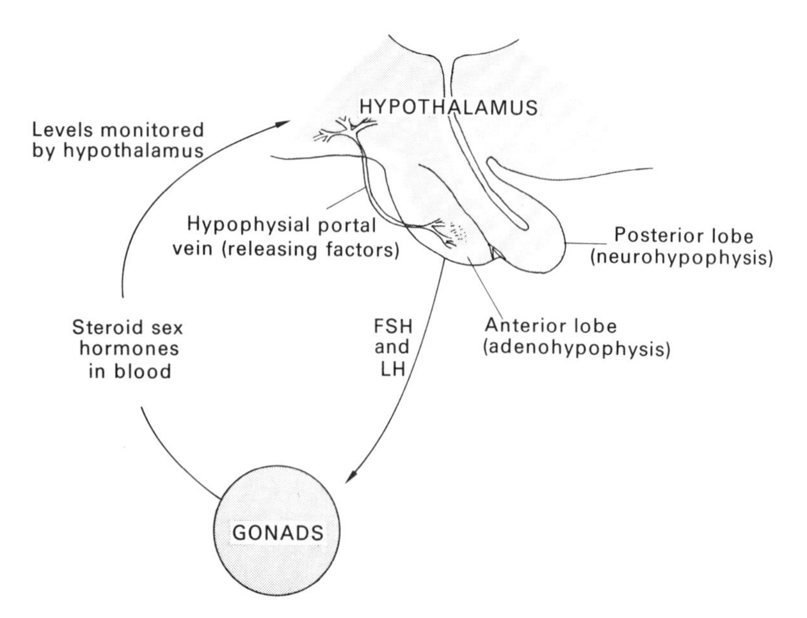

Fig. 7.7 The role of the hypothalamus in the regulation of sex hormone levels in the blood.

The production of follicle stimulating hormone and luteinizing hormone by the pituitary gland is common to both sexes. In the male FSH is believed to promote spermatogenesis, the formation of sperm in the tubules of the testes, and LH is thought to promote the production of the male hormone testosterone by the interstitial cells.

The release of FSH and LH from the anterior lobe (or adenohypophysis) of the pituitary gland is regulated by the same mechanism that applies to other anterior pituitary hormones, such as thyrotropic hormone which stimulates the production of thyroglobulin in the thyroid glands. The seat of regulation is the *hypothalamus*, that part of the brain which develops as the floor of the IVth ventricle. There is, however, no direct nervous communication between the hypothalamus and the anterior pituitary because, unlike the posterior lobe which is derived from the brain, the anterior lobe arises during embryological development from the roof of the 'mouth'.

Communication between the hypothalamus and the anterior lobe is by means of a portal vein system, the ends of which are capillary beds in the hypothalamic tissue and in the hormone secreting cells, respectively. The agent in communication is a blood soluble hormone, or *releasing factor*. The anterior lobe, then, produces FSH and LH when stimulated by FSH-releasing factor and LH-releasing factor. The levels of gonadal hormones circulating in the blood are monitored by the hypothalamus which then adjusts the output of releasing factors accordingly. Thus the level of sex hormones seems to be regulated by the sort of feedback system associated with other homeostatic systems of the body (Fig. 7.7).

Interference at the releasing factor level suggests the possibility of fertility control. The method is likely to be much finer than the rather crude administration of sex hormones to either induce or inhibit ovulation. If there exists in the male different releasing factors for FSH and LH then the selective inhibition of the former (to inhibit spermatogenesis but permit normal production of testosterone) might offer a solution to the problem of the production of a male contraceptive pill.

Two features concerning the mode of action of hormones, and sex hormones in particular, are under intensive investigation (reviewed by King, 1970). The first is that they act in very low concentrations and the second is that the structural differences between the steroid sex hormones are very small. The male-ness promoting androgen molecule can be converted to the female-promoting oestrogen by removing one carbon and six hydrogen atoms (in fact, this is what is thought to happen when oestrogens are made in the stroma of the ovary).

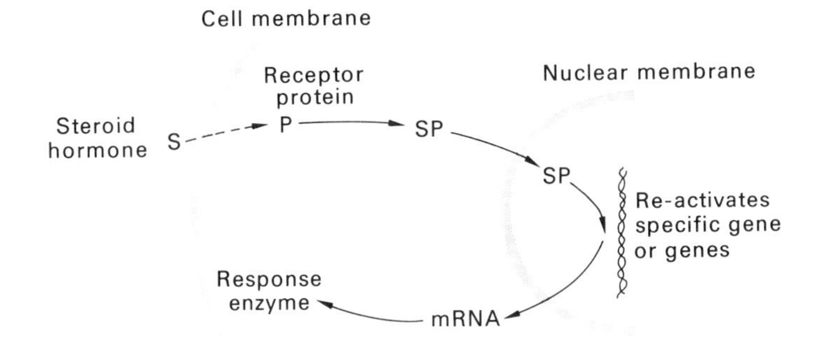

Fig. 7.8 Hormones bathe all cells in the body in the same concentration, but differential response may be achieved by the presence of receptor proteins which recognize the appropriate hormone and help it enter the cell.

There is good evidence that sex hormones work by the 'switching-on' of genes in the cells upon which they act which were previously switched off. The administration of oestrogen to uterine tissue, for example, results in the formation of new kinds of messenger-RNA and new enzymes which induce in the tissue those changes associated with an increase of oestrogen level in the body.

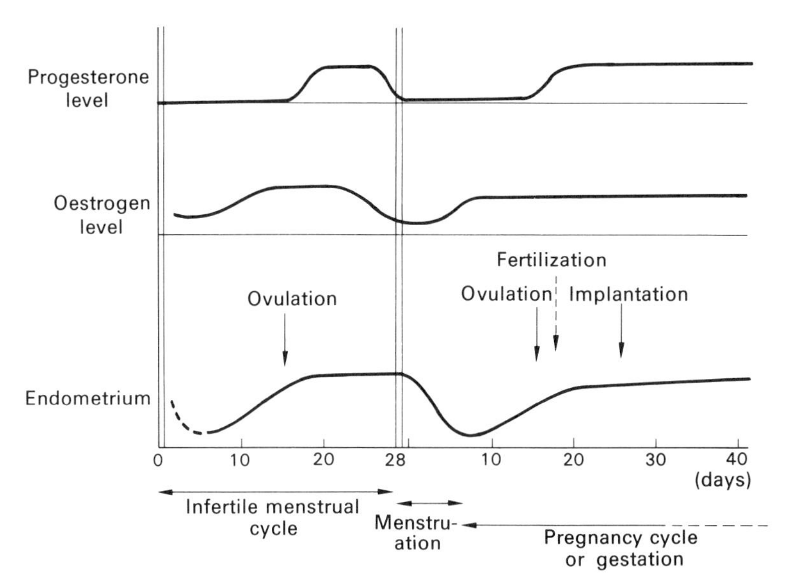

Fig. 7.9 The oestrus or menstrual cycle in man, and the onset of the pregnancy cycle.

The use of steroid hormones labelled with a radio-active tracer atom has shown that different tissues respond to hormones to differing degrees. Responsive cells appear to possess receptor molecules, which are proteins, which recognize the hormone molecule and help it to enter the cell, probably by binding on to it (Fig. 7.8).

If fertilization is successful and the blastocyst becomes implanted, the activity of the corpus luteum continues, apparently sustained by **chorionic gonadotrophin** secreted by the blastocyst. Later in pregnancy the role of the corpus luteum is taken over by the placenta, which secretes progesterone directly. Extirpation of the ovaries and their corpora lutea later in pregnancy, therefore, does not alter significantly the state of pregnancy in an animal.

If fertilization fails, however, the corpora lutea degenerate and the level of progesterone in the blood falls. The endometrium partly disintegrates and some blood, mucus and endometrial tissue is lost through the cervix and vagina; this is termed **menstruation**. In man, menstruation lasts four to six days and the whole oestrus cycle is approximately 28 days (Fig. 7.9). In the mouse the oestrus cycle lasts five days, in the dog and cat six months, and in many mammals it is an annual event.

Section 3

Continuity and Discontinuity: Inheritance, Variation and Evolution

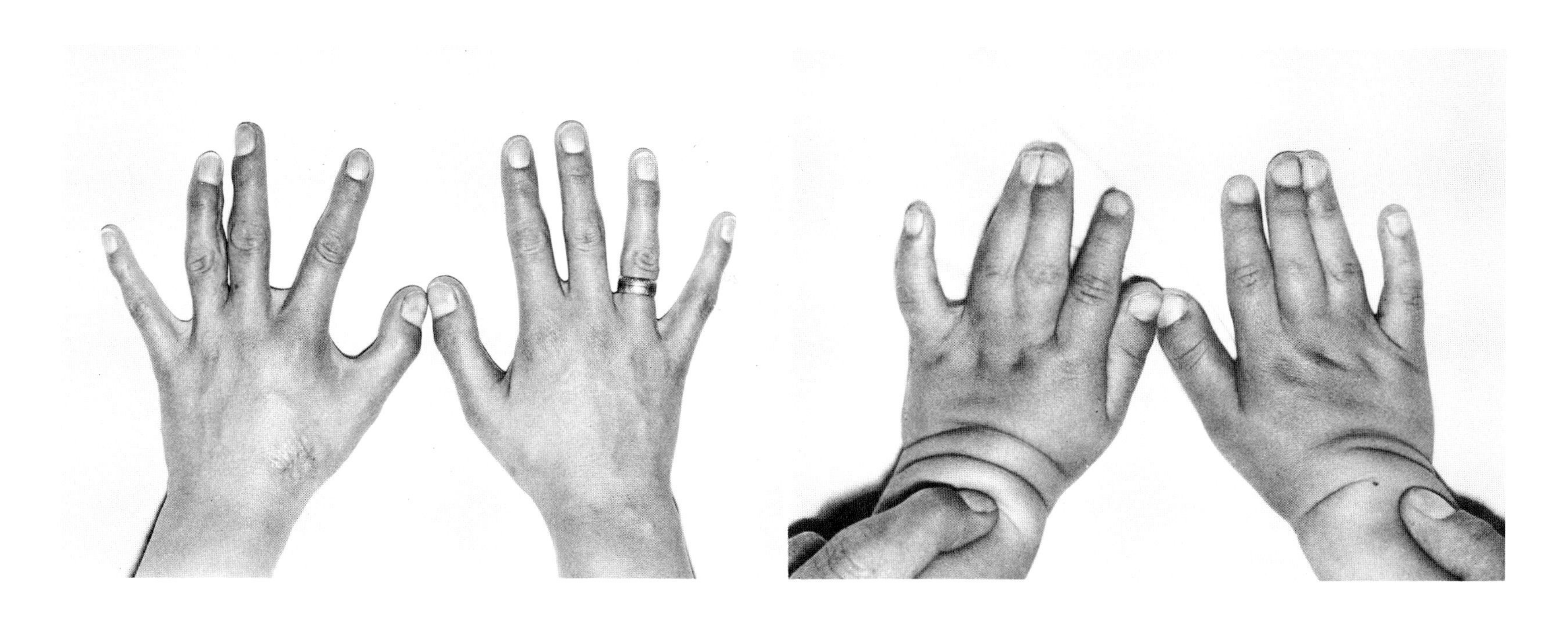

(Photograph on preceding page) Syndactyly. In this inherited disorder the digits show partial or complete joining. In the mother's hands (left) the fingers have been separated surgically but in her daughter (right) this operation has not yet been carried out. (Courtesy of Dr. R. Lax, Kennedy-Galton Institute).

The Inheritance Mechanism

The integrated activities within the organism discussed in *Patterns in Biology* depend upon and are essentially coordinated by the activities of genes. This chapter, therefore, is one of the most important ones. Furthermore, without a clear grasp of the mechanics of inheritance the process of organic evolution can make little sense. In several places in the book reference is made to the non-alignment of genetics and evolution by what amounted to historical 'mis-timing'. In Chapter 9 the events surrounding the birth of Darwinism and the concept of evolution by natural selection are described. Those events occurred before genes were discovered and before the particulate theory of inheritance, or Mendelism, was deduced.

It is logical, therefore, to begin this chapter, which is about the inheritance mechanism, with a brief account of some of the historical events leading to the concept of the gene. In the sections that follow genes are discussed at two levels. First there is a description of some of the ways in which the action of genes is controlled. These include both the temporary switching-on and switching-off of genes, such as occurs in regulation and adaptation of biochemical pathways in the cell, and the permanent sorts of gene switches associated with the programme of differentiation that occurs during embryological development.

The second level of discussion deals with the way in which genes are inherited, in what broadly is termed Mendelian genetics. In depth this discussion corresponds to that required by most examination boards at Advanced Level and embraces the familiar topics of mono- and di-hybrid crosses, linkage, and so on. In order to relate the inheritance mechanism to the evolutionary process this chapter goes on to examine the ways in which genes interact in organisms, the structure of chromosomes and some of the ways in which sex is determined genetically.

In the discussion of the origin of sex in Chapter 3 the appearance of diploidy and sexual reproduction was said to have had a major impact on the rate of evolution. But before the evolutionary process itself is analysed and the influence of these factors demonstrated we must discuss the inheritance mechanism since, in order to appreciate the differences between the characteristics of one generation and

GENESIS.

CHAP. XXX.

thee †since my coming: and now when shall I provide for mine own house also?

31 And he said, What shall I give thee? And Jacob said, Thou shalt not give me any thing: if thou wilt do this thing for me, I will again feed *and* keep thy flock:

32 I will pass through all thy flock to day, removing from thence all the speckled and spotted cattle, and all the brown cattle among the sheep, and the spotted and speckled among the goats: and *of such* shall be my hire.

33 So shall my righteousness answer for me †in time to come, when it shall come for my hire before thy face: every one that *is* not speckled and spotted among the goats, and brown among the sheep, that shall be counted stolen with me.

34 And Laban said, Behold, I would it might be according to thy word.

35 And he removed that day the he goats that were ringstraked and spotted, and all the she goats that were speckled and spotted, *and* every one that had *some* white in it, and all the brown among the sheep, and gave *them* into the hand of his sons.

36 And he set three days' journey betwixt himself and Jacob: and Jacob fed the rest of Laban's flocks.

37 ¶ And Jacob took him rods of green poplar, and of the hasel and chesnut tree; and pilled white strakes in them, and made the white appear which *was* in the rods.

38 And he set the rods which he had pilled before the flocks in the gutters in the watering troughs when the flocks came to drink, that they should conceive when they came to drink.

39 And the flocks conceived before the rods, and brought forth cattle ringstraked, speckled, and spotted.

40 And Jacob did separate the lambs, and set the faces of the flocks toward the ringstraked, and all the brown in the flock of Laban; and he put his own flocks by themselves, and put them not unto Laban's cattle.

41 And it came to pass, whensoever the stronger cattle did conceive, that Jacob laid the rods before the eyes of the cattle in the gutters, that they might conceive among the rods.

42 But when the cattle were feeble, he put *them* not in: so the feebler were Laban's, and the stronger Jacob's.

43 And the man increased exceedingly, and had much cattle, and maidservants, and menservants, and camels, and asses.

Before CHRIST 1746.

† Heb. at my foot.

† Heb. to morrow.

CHAP. XXXI.

10 And it came to pass at the time that the cattle conceived, that I lifted up mine eyes, and saw in a dream, and, behold, the ‖rams which leaped upon the cattle *were* ringstraked, speckled, and grisled.

11 And the angel of God spake unto me in a dream, *saying*, Jacob: And I said, Here *am* I.

12 And he said, Lift up now thine eyes, and see, all the rams which leap upon the cattle *are* ringstraked, speckled, and grisled: for I have seen all that Laban doeth unto thee.

‖ Or. he goats

Fig. 8.1 Facsimile of biblical passage from Genesis, Chapters 30 and 31.

successive generations, we must know how the characteristics themselves are passed on.

That certain traits can be passed on from one generation to the next has long been recognized and acted upon in plant and animal breeding. In horse breeding, for example, animals showing desired characteristics such as size, fleetness of foot, or colour, were selected as breeding stock and in-bred for the same character. Thus arose the various breeds of horse, such as the racehorse, the Percheron and the Clydesdale. The same principle has been used in cattle breeding to give high milk yields in the dairy breeds and large musculature in the beef breeds. In modern, highly competitive animal breeding industries, such as poultry and pigs, data processing and test-breeding programmes are computerized.

In Genesis, Chapters 30 and 31, there is a fascinating account of an animal breeding programme. Jacob, wandering in a distant land, met and fell in love with Rachel. But her father, Laban, required him to serve seven years for her hand. Laban then deceived Jacob and substituted the elder daughter in the marriage ceremony. Jacob was an expert in flock husbandry and consequently Laban had prospered. By this ruse Jacob was required to serve a second seven year term, ostensibly to increase Laban's wealth still further, although he was permitted to take Rachel as his second wife. But when Laban agreed to give Jacob, in lieu of wages, any non-white sheep and goats that were in the flock, the tables were turned. By agreement Jacob separated all the speckled and spotted animals forthwith and isolated them three days journey away.

In the following years the strongest animals in Laban's flocks bore more speckled animals, which Jacob duly removed, and his flocks prospered at the expense of Laban's. Jacob accounted for this by saying that he had planted stakes of green poplar, hazel and chestnut, from which he had peeled strips of bark to give a striped appearance, in front of the best animals when they were about to conceive, so they would bear striped offspring. While this explanation might have pleased Lysenko and fellow Michurinites (Huxley, 1949) there is evidence that Jacob took other steps. In Chapter 31, Verse 10 he says (Fig. 8.1), 'And it came to pass at the time that the cattle (animals) conceived, that I lifted up mine eyes, and saw in a dream and, behold, the rams which leaped upon the flock were ringstraked, speckled and spotted'. One assumes he acted on this dream and mated the best of Laban's ewes and she-goats with speckled males, presumably clandestinely since all the speckled animals had been isolated. When Laban at last realized what had happened Jacob secretly loaded his wives and children and their retinue on their camels and, with his huge flock, headed for Canaan.

This story makes fascinating reading as a piece of population genetics. In the next Chapter the behaviour of genes in populations will be discussed, in the genetic basis of evolution.

8.1 The Concept of the Gene

The concept of the embodiment of a character in an inheritable particle owes much to the work of the Augustinian monk, Johann Gregor Mendel (Fig. 8.2). In the 18th and 19th centuries it had widely been held that offspring were a blend of parental traits. Thus, for example, a cross between a red flower and a white flower produced seeds which subsequently gave rise to plants bearing pink flowers, in much the same manner as mixing red paint and white could produce pink paint. This concept of **blending inheritance**, however, failed to account for the persistence of red and white individuals in successive generations and for the reappearance of red and white offspring from crosses involving pink parental types.

Mendel's short papers, 'Experiments in Plant Hybridization' were published in 1865 and 1869. Although, by proposing a **particulate theory of inheritance** which could account for the re-appearance of original characters from a 'blended' hybrid, this work was to prove fundamental to the understanding of inheritance, it was ignored until 1900. Mendel died in obscurity in 1884.

The sequence of discoveries in this field in the last two decades of the 19th century casts an interesting light on what might be called the concept of the temporal inevitability of scientific discovery and progress. In 1875 Strasburger described the appearance of chromosomes in the nucleus of a dividing cell. Many workers followed in this field and the behaviour of chromosomes in mitosis and meiosis was described amid speculation as to their role as carriers of hereditary information, leading to the publication of Weismann's *Das Keim Plasm* (The Germ Plasm) in 1892. One feels that had Mendel's publications coincided with those of Strasburger they would have received immediate recognition. As it was, the breeding experiments which confirmed the particulate theory were performed simultaneously and independently by three workers, de Vries, Correns and Tschermak, in 1900 when the time was, if anything, over-ripe and the attention was drawn to the pioneering work of Mendel forty years earlier.

In a similar way it might be argued that the publication of the Origin of Species in 1859, when Darwin was unaware of the genetic basis of variation, led to the treatment of genetics and evolution as separate, albeit related, disciplines. Three-quarters of a century or more was to elapse after the publication of *Origin of Species* before the genetic basis of evolution and speciation began to be fully realized and expounded. In the modern synthesis of evolutionary theory, fossil evidence is no longer regarded as an indication of the way in which evolution takes place. Fossils are simply evidence that it has taken place. The way in which evolution occurs is by changes in the frequencies of the genes from which the population is constituted from generation to generation.

Fig. 8.2 Johann Gregor Mendel (1822–84). (Courtesy of the Wellcome Trustees.)

When W. S. Sutton published *The Chromosomes in Heredity* in 1903 the inheritance mechanism was given a cytological basis and the foundation of what is now termed **Mendelian genetics** was laid. At this time W. L. Johannsen, of Jena, was conducting some elegant experiments on the weights of bean seeds produced by pure lines of plants, and by crosses between pure lines. For the 'inherited particles' he coined the term **genes**, and he also invented the terms **genotype** and **phenotype**.

W. Bateson (Fig. 8.12) and R. C. Punnett in 1906 discovered that two genes could affect the same character, in this case comb shape in poultry, and that sometimes genes tended to be coupled together, or **linked**, as in pollen shape and petal colour in the sweet pea. Linkage had been postulated by Sutton in 1903 on cytological grounds, but the explanation for this coupling was offered by T. H. Morgan, who pioneered the use of the fruit fly, *Drosophila melanogaster*, as a genetical tool. He concluded that coupled, or linked, genes lay on the same chromosome and that their occasional separation was due to a cross-over, or **chiasma**, occurring between them. Chiasmata formation had been described by Janssens in 1909, so the relationships of the genes, the chromosomes and the behaviour of chromosomes at gamete formation fell neatly into place.

In 1927 H. J. Muller discovered that X-radiation could alter the nature of the gene in a permanent fashion. This was termed **induced mutation** and the effect of this discovery was to focus attention on the structure of the gene and the way in which it became manifest in the phenotype. In 1941 G. W. Beadle and E. L. Tatum, of Stanford, published what was to become a significant branching point in genetical research: *Genetics of Biochemical Characters in Neurospora*. This paper demonstrated the genetic basis of biochemical syntheses and led to the widespread use of micro-organisms as genetical tools. In the 1950's Lederberg, Hershey and their co-workers pioneered bacterial genetics and in 1961 Jacob and Monod, using the bacillus *Escherichia coli*, demonstrated enzymic induction and proposed the Regulator Gene Concept to account for the regulation of enzymic syntheses.

Hershey, Benzer and others, in the early 1960's, demonstrated genetic recombination in bacteriophage viruses. This led to a new dimension in fine genetic analysis and to the re-definition of the gene in terms of the **cistron** and its sub-units, **recons**. 'Phage genetics proved to be vital in the elucidation of the genetic code by Crick, Brenner and their co-workers in 1961.

8.2 The Control of Gene Action

Biochemical activities of the cell are determined by the concentrations of active enzymes in the cytoplasm and can be considered broadly as being controlled in several ways. The first of these is seen in the rapid control by which the biochemical pathways themselves show an autoregulatory capacity. An example of this is **feedback inhibition**, by which an accumulation of product inhibits an enzyme or enzymes involved in the earlier stages of the substrate-to-product sequence (p. 47).

The second method of control is genetic and usually involves the permanent, or irreversible, 'switching-off' of genes. This is the probable mechanism of differentiation and it affords a means whereby the cells of a particular tissue can attain the unique enzymic concentration profile by which that tissue fulfils its specialized functions (p. 120).

The third method is also genetic, but involves short-term

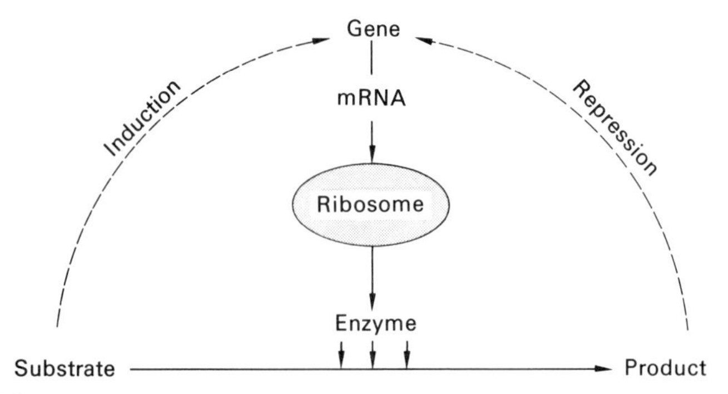

Fig. 8.3 The short-term regulation of gene action which could account for observations concerning enzyme production in bacteria.

'switching-off' or 'switching-on' of genes in a regulatory or adaptive capacity. The production of an enzyme in the cytoplasm occurs when messenger-RNA is transcribed from a gene and passes on to a ribosome (p. 37). **Repression** of enzyme production, that is, the stopping of transcription of the particular mRNA, may occur when product is added to the medium in which the cells are growing. In the bacillus *Escherichia coli*, for example, genes governing the synthesis of arginine in the cell function only when the culture medium is deficient in arginine. The addition of this amino acid to the medium causes the repression of enzyme production; the genes become 'switched-off'. The enzyme may persist in the cytoplasm for some time after transcription ceases, so this method of metabolic control is slow compared with feedback inhibition.

The opposite of repression is enzymic **induction**. In this case the transcription of mRNA and the subsequent production of an enzyme is promoted, usually by the appearance of substrate in the medium. Induction and repression almost certainly share the same method of gene control (Fig. 8.3). The classic example of induction is the production of the enzyme β-galactosidase in *E. coli* when the substrate phenyl β-galactoside is added to the culture medium (Jacob and Monod, 1961).

Arising from these studies Jacob and Monod (Fig. 8.4) proposed the **regulator gene hypothesis** to account for the extra-nuclear influence over gene action (Fig. 8.5). They postulated a complex of genes consisting of a regulator gene, an operator gene and a sequence of structural genes. The structural genes code for the enzymes in the cytoplasm. The operator gene appears to be a 'trigger-site' where the transcription of the structural gene sequence begins. The whole functional unit, of operator gene plus structural genes, was termed an **operon**. The regulator gene is believed to switch on, or off, the whole operon by freeing or blocking the trigger-site of the operon, namely the operator gene. There is evidence that the regulator gene produces a 'repressor substance' which blocks the operator gene. Inducer molecules, such as substrate metabolites, would fit into this scheme by binding on to, and therefore inactivating, the repressor substances.

This view has since been modified by the breakdown of the regulator gene to a pair of complementary genes, an **inducer** gene and a **promotor** gene (or promotor site). The inducer gene produces the repressor molecule which binds to the operator gene, occluding it and blocking the synthesis of mRNA for the whole operon. In the presence of substrate the repressor molecule is somehow altered so that it no longer binds to the operator gene (Reznikoff *et al.*, 1969). The complete role of the promotor gene is still obscure, but there is evidence that it, and not the operator gene, represents the site for the commencement of mRNA synthesis; it may be the binding site for mRNA polymerase (Roberts, 1969a).

The field of gene control has proved a fruitful one from the point of view of research. It is attracting considerable attention and the

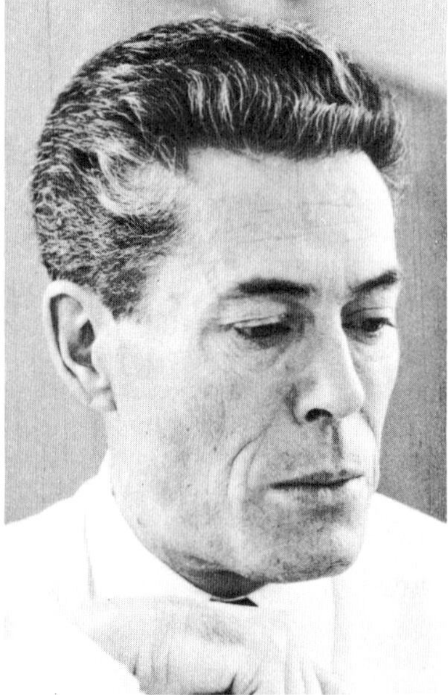

Fig. 8.4 Francois Jacob and Jacques Monod who received the Nobel Prize for their work on the elucidation of the regulator gene concept (Courtesy of Thames and Hudson Ltd. and Institut Pasteur).

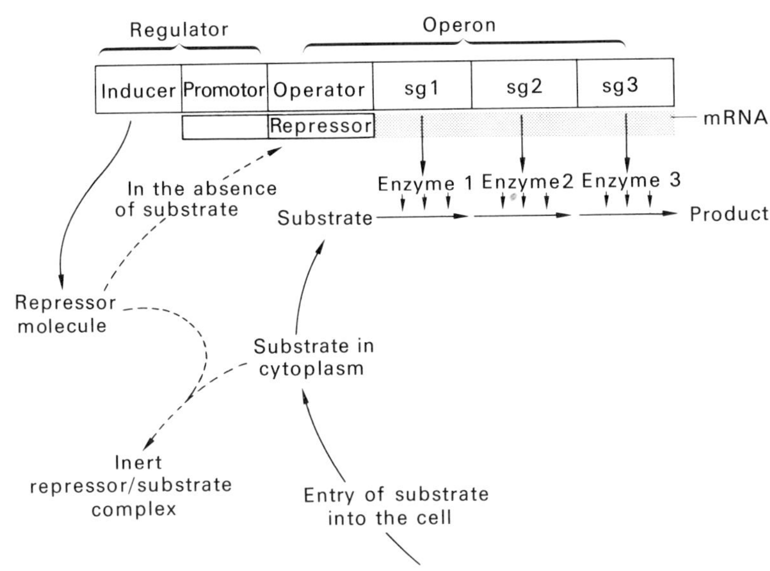

Fig. 8.5 Summary of the Regulator Gene Hypothesis (Jacob and Monod, 1961; Roberts, 1969; Resnikoff *et al.*, 1969.)

Regulator Gene Hypothesis outlined above will almost certainly be modified still further as more evidence comes to light.

The Regulator Gene Hypothesis in principle affords a useful explanation of the enzymic flexibility and speed of adaptation which bacteria in particular show to changes in their environment. The evidence reflected in this hypothesis has been gained from studies on micro-organisms, but there is evidence that a similar mechanism of gene switches may operate in higher organisms (Paul and Fottrell, 1963). The organization of the nucleus in the cells of higher plants and animals, however, is more complex. Whereas the bacterial chromosome may carry a few thousand genes, the nucleus of higher organisms may carry a thousand, or even ten thousand, times that number. In addition, such a method of gene control must be co-ordinated with, and superimposed upon, the process of differentiation which, as already described, appears to involve varying combinations of genes being permanently repressed, or 'switched-off', in different tissues.

One such mechanism operating in the nuclei of higher animals has been proposed by Tomkins *et al.* (1969). Using hormones to stimulate the production of enzymes by liver cells in tissue culture they have produced a model to account for their findings which invokes a repressor of messenger-RNA (Fig. 8.6). In this scheme a repressor gene in the nucleus would produce a repressor substance which blocks the action of mRNA produced from an enzyme gene. This repressor substance might be very short-lived unless it is bound to a hormone molecule. In this case the entry of a hormone molecule into the cell would produce a more permanent repressor complex and the mRNA of a particular enzyme gene would be switched off. A scheme such as this one, working as it does at the level of mRNA, could easily operate in conjunction with the permanent switching systems, such as the action of histones. The latter would be seen as a permanent source of differentiation in tissues and the former as a mode of genetic fine control operating within the scheme of differentiation. (See also Hormonal Regulation in Animals, p. 128).

To conclude this section on the Control of Gene Action (a field of research which is itself far from exhausted) one can return to bacteria. The Jacob/Monod regulator-gene hypothesis is essentially a system involving **repressors**, which act by turning genes off. A complementary scheme, also believed to operate in bacteria has been reported by Burgess *et al.* (1969). These workers discovered that RNA-polymerase, the enzyme that transcribes the genetic data from DNA to mRNA, can be split into two components. One part is the polymerizing enzyme and the other part they called the **sigma factor** (see also p. 35).

There is good evidence that the sigma factor, which may be only one of several kinds of factor associated with RNA-polymerase, is

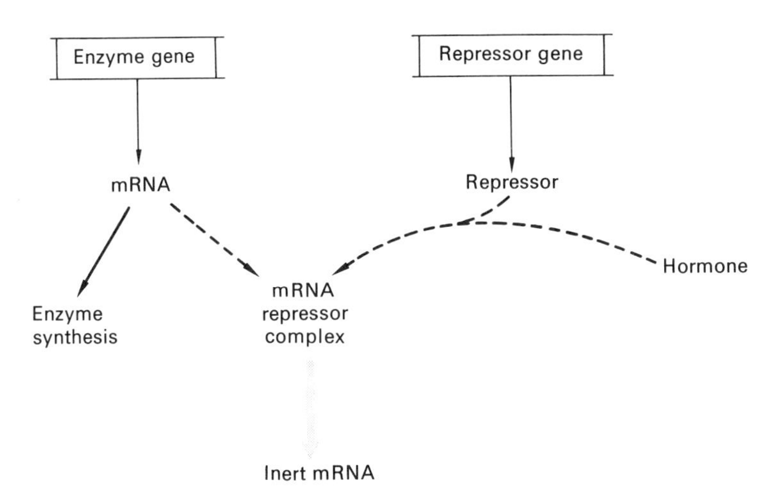

Fig. 8.6 A possible scheme for the regulation of gene action in higher organisms involving the repression of messenger-RNA.

not only essential for the polymerase to begin a transcription sequence by recognizing the 'start here' code on a portion of DNA, but also determines which parts of the DNA in the chromosome shall be transcribed. In this capacity the sigma factor could act as a regulator of gene action. It could work in conjunction with the regulator-gene system in bacteria and other unicellular organisms which, of course, have no tissue differentiation and therefore no histone mechanism for selectively switching genes off.

8.3 Mendelian Inheritance

Mendelian genetics concerns the distribution of genes from one generation to the next in diploid, sexually-reproducing organisms. Meiosis, described on p. 84, is central to this study since it concerns the separation of chromosome pairs, and therefore the genes which are carried on them, into different gametes.

In the example shown in Fig. 8.7 the parental genotype with respect to the A locus is Aa. Such a genotype, where the alleles are different (that is, they potentially code for different enzymes and

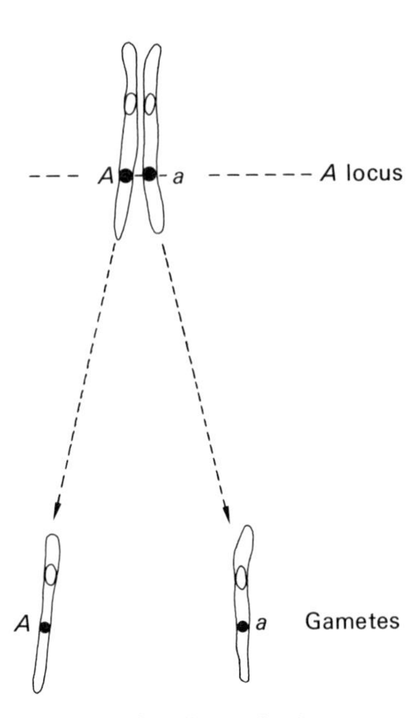

Fig. 8.7 Diagram to show how the alleles of a heterozygote *Aa* segregate during the formation of gametes. (Compare with meiosis diagrams in Chapter 6.)

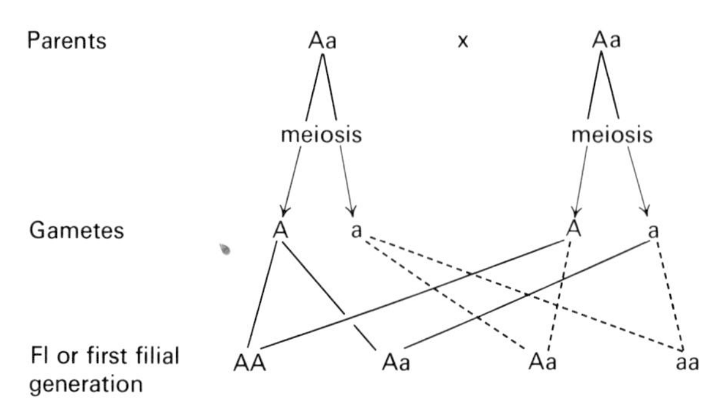

Fig. 8.8 Diagram to show that each kind of gamete of one parent in a monohybrid cross has the same probability of fusing with each kind of gamete of the other parent.

thereby produce different effects in the phenotype) is termed **heterozygous**. As the diagram demonstrates, half the gametes produced from this parental genotype will carry an A gene and the other half will carry an a gene. The gametes can therefore be described as $\frac{1}{2}A+\frac{1}{2}a$.

The Monohybrid Cross

Consider a cross between two organisms, each of which is heterozygous, as in the Aa situation described above. The gametes produced by each organism are given by $\frac{1}{2}A+\frac{1}{2}a$. Therefore, following the *Product Rule* of probability, the probable distribution of offspring genotypes with respect to this character is given by the product of the separate probabilities, or gamete distributions, namely:

$$(\tfrac{1}{2}A+\tfrac{1}{2}a)(\tfrac{1}{2}A+\tfrac{1}{2}a) = \tfrac{1}{4}AA+\tfrac{1}{2}Aa+\tfrac{1}{4}aa$$

In other words, one quarter of the offspring would be expected to be homozygous for A, half would be expected to be heterozygous, and the remaining quarter would be expected to be homozygous for a. Expressed diagrammatically (Fig. 8.8), the monohybrid cross gives the same offspring genotypic distribution.

Penetrance

In a heterozygous organism two different alleles code for two different aspects of the same character. In many cases the phenotype of such an organism is a blend of both alleles. For example, if the red and white genes determining coat colour in cattle are present in a heterozygous state, the animal has a roan coat, in which both red and white hairs are present. In other cases one gene may be expressed almost completely in the phenotype of the heterozygote, at the expense of its allele. This gene is said to be **dominant** and its allele, which is suppressed, is called **recessive**.

The degree to which an allele is expressed in the phenotype is termed its **penetrance**. A dominant allele, in theory, has 100% penetrance and its recessive allele has 0% penetrance. In practice, however, the penetrance shown by a dominant allele usually lies some way below 100%; likewise, a recessive allele usually shows a small degree of penetrance.

If, in Fig. 8.8, the A allele shows close to 100% penetrance, it may be impossible to distinguish between the phenotypes of AA and Aa. Thus the offspring phenotypic ratio of the monohybrid cross can be described as $3A:1a$. This ratio was first described by Mendel and has been termed **Mendel's First Law.** Mendel worked with the garden pea in which many characters, such as seed shape and height of plant, show high degrees of penetrance. Mendel also described the features of dominance and recessiveness.

Example 1 The Monohybrid Cross Involving High Penetrance

In the mouse the gene for the expression of coat colour shows a high degree of penetrance, or is dominant, with respect to its allele, albino. Let the dominant, coloured allele be termed C and the albino allele c.

(a) If a mouse from a pure-breeding, that is homozygous, black strain is mated with a white mouse what would be the expected genotype and phenotype of the offspring?

Parental cross: $CC \times cc$

Gametes: (all C) (all c)

Fl genotype is therefore Cc and their phenotype is black, owing to the high penetrance of the C allele.

(b) If one of these Fl offspring is then mated with a mouse of the same genotype what would be the expected genotypic and phenotypic ratios of their offspring?

Fl × Fl: $Cc \times Cc$

Gametes: $(\frac{1}{2}C+\frac{1}{2}c)\times(\frac{1}{2}C+\frac{1}{2}c)$

Expected F2 genotype: $\frac{1}{4}CC+\frac{1}{2}Cc+\frac{1}{4}cc$

Expected F2 phenotype is therefore $\frac{3}{4}$ coloured + $\frac{1}{4}$ albino

(c) What would be the expected offspring phenotypic and genotypic ratios if one of the offspring of Cross (a), above, were mated with a white mouse?

Parents: $Cc \times cc$

Gametes: $(\frac{1}{2}C+\frac{1}{2}c)\times(c)$

Expected offspring genotype: $\frac{1}{2}Cc+\frac{1}{2}cc$

Expected offspring phenotype: $\frac{1}{2}$ black + $\frac{1}{2}$ albino

Example 2 The Monohybrid Cross in the Absence of Dominance, or in which the Alleles Show about 50% Penetrance

In antirrhinums, two alleles for petal colour, p^r (red) and p^w (white) show about 50% penetrance. The heterozygous plant, p^rp^w, gives rise to pink flowers. If pollen from a pink flower is used to fertilize a second plant with pink flowers, and 100 seeds are collected and subsequently germinated, about how many seedlings would be expected to produce pink flowers in due course?

Parental cross: $p^rp^w \times p^rp^w$

Gametes: $(\frac{1}{2}p^r+\frac{1}{2}p^w)\times(\frac{1}{2}p^r+\frac{1}{2}p^w)$

Expected offspring genotypes: $\frac{1}{4}p^rp^r+\frac{1}{2}p^rp^w+\frac{1}{4}p^wp^w$

Therefore, half the seedlings, or 50, would be expected to give rise to pink flowers.

Lethal Genes

When a gene that codes for an enzyme which is vital to an organism mutates, the mutant allele may produce no enzyme or an altered enzyme. If the mutated allele has a high degree of penetrance an organism which inherits one of the genes, as a result of a mutation occurring in a gamete-producing cell of one of its parents, will die. Such dominant lethal genes are rare since they are eliminated from the population when they arise.

Lethal genes of low penetrance, however, can be carried in the heterozygous condition. One quarter of the offspring of a cross between two organisms heterozygous for a recessive lethal allele would be expected to be homozygous for that allele, and therefore to be capable only of coding for altered enzyme. If this enzyme promotes a vital metabolic or developmental pathway the organism will die. Some 'lethal' genes cause death indirectly. For example, a form of blindness in Irish Setters, retinal atrophy, is caused by a recessive lethal gene. A dog carrying this gene in the homozygous state would die in the wild, but can be fed and tended, and therefore survive, in domestication.

Example 3 A Lethal Gene of Low Penetrance

Albinism in green plants (the inability to synthesize chlorophyll) is a lethal mutant of low penetrance. What proportion of the normal offspring of a cross between two tobacco plants known to be heterozygous for albinism would be expected to be homozygous?

Let the chlorophyll-producing gene be termed C and its albino mutant allele, c.

Parental cross: $Cc \times Cc$

Gametes: $(\frac{1}{2}C + \frac{1}{2}c) \times (\frac{1}{2}C + \frac{1}{2}c)$

Expected offspring genotypes: $\underbrace{\frac{1}{4}CC + \frac{1}{2}Cc} + \frac{1}{4}cc$

Phenotypes: $\frac{3}{4}$ green + $\frac{1}{4}$ albino (seedlings die when the food reserve in the seed is used up)

Fig. 8.9 Albino and normal seedlings of the tobacco plant *Nicotiana* produced from the cross between two plants heterozygous for the albino gene.

Therefore, of the green seedlings, **one-third** would be expected to be homozygous (Fig. 8.9).

Example 4 A Lethal Gene with Intermediate Penetrance

In man the condition known as **brachyphalangy**, in which the digits are conspicuously shortened, occurs in individuals of either sex who are heterozygous for a lethal gene (Fig. 8.10). The shortening of the digits is due to reduction in length of the middle bones, so that the fingers frequently appear to have only two joints, like the thumb. Individuals showing brachyphalangy vary a good deal in the extent to which the shortening occurs. In some cases the digits are reduced only a small amount and in others the digits may be reduced to small stumps. A gene which exhibits this degree of variance in the phenotype is said to show **variable expressivity**. In the homozygous state this gene causes severe skeletal and developmental abnormalities which usually cause the death of the foetus at an early stage of preg-

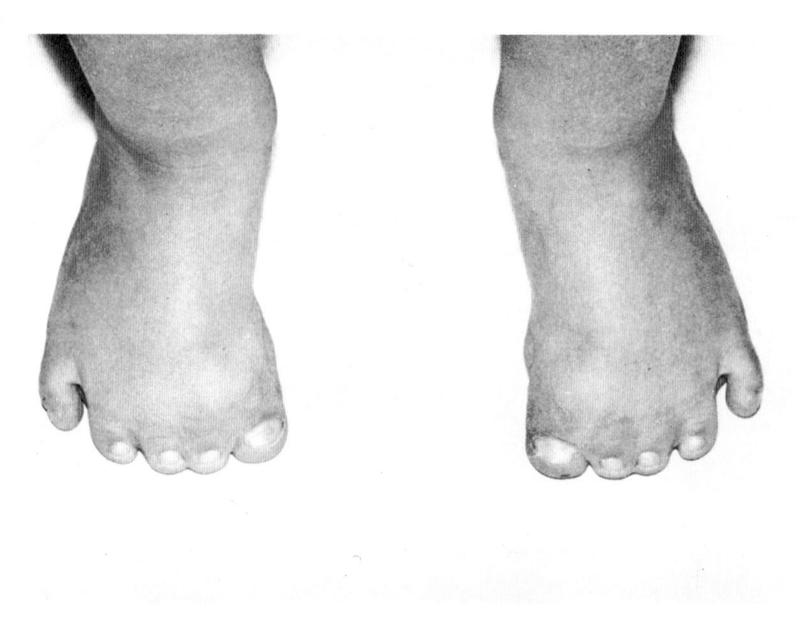

Fig. 8.10 Foot of a child showing brachyphalangy, with conspicuous shortening of the toes.

nancy.

What ratio of normal to brachyphalangic children would be expected in a family where both parents showed brachyphalangy?

Let the brachyphalangy gene be termed b and its normal allele $+$.

Parents: $+b \times +b$

Gametes: $(\frac{1}{2}+ \; + \; \frac{1}{2}b) \times (\frac{1}{2}+ \; + \; \frac{1}{2}b)$

Expected offspring genotypes: $\frac{1}{4}++ \; + \; \frac{1}{2}+b \; + \; \frac{1}{4}bb$ (dies)
phenotypes: (1 normal):(2 brachyphalangic)

Thus in the children of this family, there would be an expected ratio of one-third normal to two-thirds brachyphalangic.

Independent Assortment

The principle of Independent Assortment concerns the distribution of two or more pairs of alleles which are unlinked. Unlinked means that they are either carried on separate chromosomes, or are carried on the same chromosome sufficiently far apart for abundant chiasmata to occur between them (see Crossover Value, p. 143). Independent Assortment was first described by Mendel, whose definition of it has been called Mendel's Second Law.

In cytological terms, independent assortment means that the separation of one pair of chromosomes during meiotic Anaphase I (p. 86) is not affected by the distribution or separation of other pairs. Consider two loci, A and B, which occur on chromosome pairs. Consider two loci, A and B, which occur on different chromosomes and are therefore unlinked. If an organism is heterozygous at both loci, that is, its genotype is $AaBb$, four kinds of gamete can be formed by that organism; AB, Ab, aB, and ab, (Fig. 8.11)

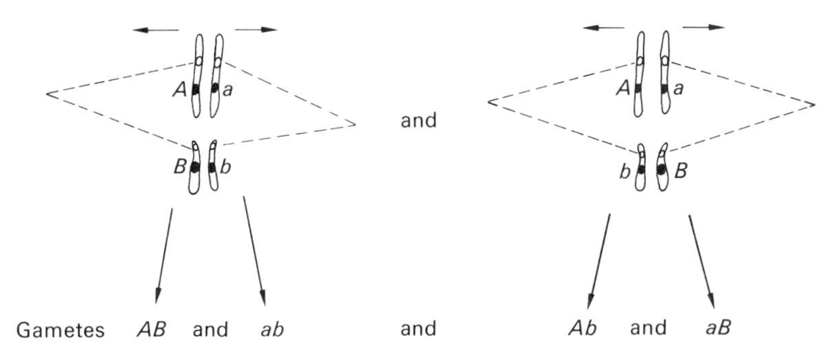

Fig. 8.11 Independent assortment of an organism with the genotype *AaBb*.

From this it can be seen that the number of kinds of gamete formed can be described as 2^n, where n = number of chromosome pairs, or allelic pairs, under consideration. It follows, therefore, that the number of kinds of offspring is $(2^n)^2$, (see Sources of Variation, p. 178).

The Dihybrid Cross

A dihybrid cross is a cross between organisms which are heterozygous for two pairs of alleles. In this case, the number of kinds of gamete produced by each parent, as described above, is 2^2, or 4. The number of kinds of offspring, therefore, is 16.

Parents: $AaBb \times AaBb$

Gametes: $(\frac{1}{4}AB + \frac{1}{4}Ab + \frac{1}{4}aB + \frac{1}{4}ab)^2$

Expected offspring genotypes:
$\frac{1}{16}AABB + \frac{1}{16}AABb + \frac{1}{16}AaBB + \frac{1}{16}AaBb + \ldots$ etc. $+ \frac{1}{16}aabb$

R. C. Punnett devised a diagrammatic method of solving the expected offspring genotypic ratios. In the figure which bears his name, the **Punnett Square**, the gametes are placed along the upper and left-hand sides, and the addition of the gametes shows the composition of the offspring in the appropriate space in the figure.

Parents: (♂) $AaBb \times AaBb$ (♀)

Gametes: ♂ AB, Ab, aB and ab. ♀, the same.

♀ \ ♂	AB	Ab	aB	ab
AB	AABB	AABb	AaBB	AaBb
Ab	AABb	etc.		
aB				
ab				aabb

The expected ratios of offspring phenotypes can be obtained by examination of the figure.

The Direct Calculation of Offspring Phenotypes

In order to determine the expected offspring phenotypes it is not necessary to calculate the genotypes first. If each allele is treated independently, in effect as a monohybrid cross, the product of the phenotypes yields the complete offspring phenotypes of the dihybrid cross.

Example 5 Phenotypic Ratios of a Dihybrid Cross

Consider a cross in cattle in which the mated animals are heterozygous for two unlinked pairs of alleles. In the first pair, **polled** (P), or hornless, is dominant to **horned** (p) and in the second pair **straight coat** (S) is dominant to **curly coat** (s). What offspring phenotypes would be expected, and in what ratios, from repeated matings?

Parental cross: $PpSs \times PpSs$

Segregation of the horn allele gives an expected offspring ratio of

$$(\tfrac{3}{4}\text{ polled} + \tfrac{1}{4}\text{ horned})$$

Segregation of the coat allele gives an expected offspring ratio of

$$(\tfrac{3}{4}\text{ straight} + \tfrac{1}{4}\text{ curly})$$

The offspring ratio is given by the product of the 'separate probabilities':

$$= (\tfrac{3}{4}\text{ polled} + \tfrac{1}{4}\text{ horned})\ (\tfrac{3}{4}\text{ straight} + \tfrac{1}{4}\text{ curly})$$
$$= \tfrac{9}{16}\text{ polled straight} + \tfrac{3}{16}\text{ polled curly} + \tfrac{3}{16}\text{ horned straight} + \tfrac{1}{16}\text{ horned curly}$$

This ratio of 9:3:3:1 of the offspring phenotypes was described by Mendel. It is characteristic of the dihybrid cross in which the dominant alleles show a high degree of penetrance.

Expected Phenotypic Ratios from a Cross involving Three Pairs of Alleles

As the number of alleles under consideration increases, the prospect of executing a Punnett Square becomes quite daunting. This is because, as has already been described, the number of gametes is 2^n, where n = number of allelic pairs, so the addition of each extra pair of alleles quadruples the size of the square. Under these conditions it becomes especially important to calculate the expected phenotypic ratios directly, especially if the alleles are heterozygous.

Example 6 Phenotypic Ratios of a Trihybrid Cross

Consider the cross described in Example 5 with the addition of a third pair of alleles determining coat colour. The genes for red and white coat colour show about 50% penetrance. The heterozygote has both red and white hairs in the coat and is termed a **roan** animal (see Example 2). If both the animals in the previous example were roan, then the parental cross is given by:

$$Pp\ Ss\ c^r c^w \times Pp\ Ss\ c^r c^w$$

The offspring phenotypic ratio to be expected is given by the product of the three alleles treated independently:

$$= (\tfrac{3}{4}\text{ polled} + \tfrac{1}{4}\text{ horned})(\tfrac{3}{4}\text{ straight} + \tfrac{1}{4}\text{ curly})(\tfrac{1}{4}\text{ red} + \tfrac{1}{2}\text{ roan} + \tfrac{1}{4}\text{ white})$$

$= \tfrac{9}{64}$ polled, straight and red,
$\tfrac{18}{64}$ polled, straight and roan,
$\tfrac{9}{64}$ polled, straight and white,
. . . etc.
$\tfrac{1}{64}$ horned, curly and white.

There are $2 \times 2 \times 3$ kinds of offspring = 12.

Linkage

When two pairs of alleles which are **unlinked** segregate during the formation of gametes they do so in a 1:1:1:1 ratio. Thus an organism with the genotype $AaBb$ will form gametes which are

25% AB, 25% Ab, 25% aB and 25% ab

However, if these two pairs of alleles are linked, that is they are carried on the same chromosome, they will tend to stay together and enter the same gamete:

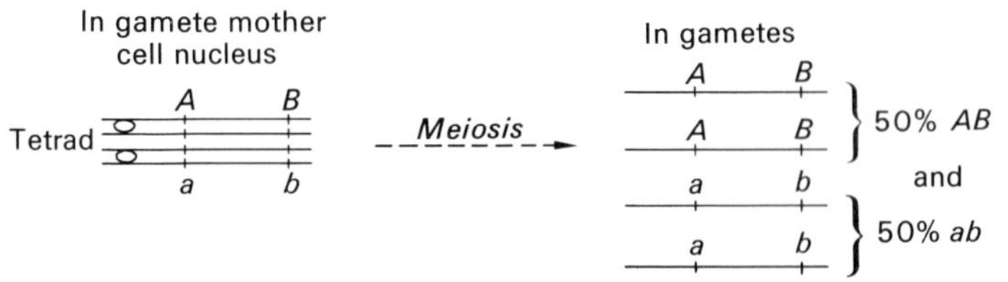

In gamete mother cell nucleus
Tetrad A B / a b
Meiosis
In gametes
A B, A B } 50% AB
and
a b, a b } 50% ab

In practice, however, chiasmata, that is the interchange of regions of adjacent chromatids, occur as a normal feature of meiotic prophase, at the tetrad stage (p. 181). In long chromosomes there may be five or six chiasmata. Clearly, then, if a chiasma occurs between linked genes the linkage group is broken and rearranged. Such rearranged groups are called **recombinants**, and the separation of linked genes is termed **crossing-over**.

Evidence that crossing-over occurs at the tetrad stage can be obtained using the fungus *Sordaria*. The nuclei in the septate hyphae of this ascomycete fungus are haploid, but at the onset of sexual reproduction hyphae from different strains, or sexes, fuse and form a heterokaryotic hypha. Two nuclei in the **heterokaryon** fuse to form a diploid nucleus which divides meiotically to form four haploid nuclei. Each of these nuclei divides once by mitosis, giving rise to eight spores arranged linearly in a fruiting body, or **ascus**.

The spores, which are ascospores, are normally black when mature. If a mutant strain, in which the ascospores are white, is 'mated' with a normal strain, the diploid nucleus in the heterokaryon will be heterozygous for spore colour. In the absence of crossing-over the spores in the ascus would be expected to be 4 black and 4 white, in that order. But if crossing-over occurs between the centromere and the spore colour locus the 4+4 pattern is destroyed. The pattern of 2+2+2+2 can only be achieved if crossing-over occurs in the tetrad stage (Fig. 8.13).

In general, the further apart that the alleles are, the greater is the likelihood of their being separated by crossing-over, thus the higher will be the proportion of recombinants. The proportion of recombinants, expressed as a percentage, is called the **Crossover Value**, or C.O.V. (sometimes called the Recombinant Value):

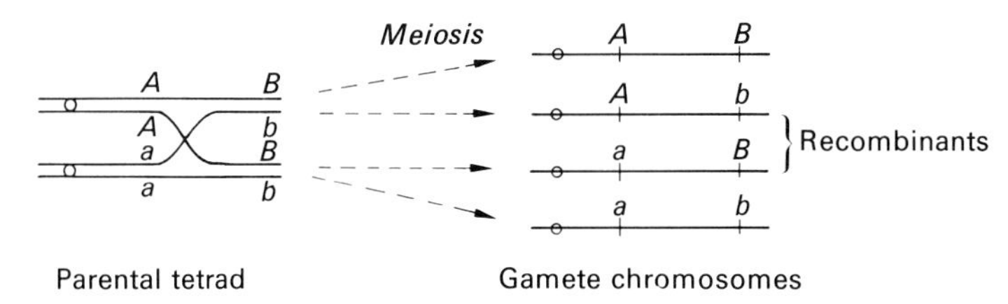

By examination, it will be seen that the maximum Crossover Value is 50%, since at this value gametes would be formed in a 1:1:1:1 ratio, in other words, they would behave as though they were unlinked.

Chromosome Mapping

In most of the organisms which have been studied crossing-over occurs during gamete formation in both sexes. In practice, therefore, it is possible to determine the C.O.V. between linked alleles by crossing an individual known to be heterozygous for the alleles concerned with a mate which is recessive for those alleles. In this way, the 'parental type' gametes and the recombinant gametes are revealed in the offspring. A notable exception, however, occurs in *Drosophila* in which, for reasons not at all well understood, crossing-over occurs only in the female fly.

By systematic, often painstaking, work the relationships of linked alleles has been determined and this has led to the construction of **chromosome maps**. Such linkage maps have been produced for *Drosophila* (Fig. 8.14), maize and the mouse in particular. Chromosome mapping has proved difficult in many higher organisms because the genic sequence in a gamete can only be observed by examination of the phenotypes of the offspring to which the gamete gives rise. Thus in mammals, for example, in which the number of offspring is relatively few, recombinants of alleles which lie close together will be rare and possibly undetectable, even in large numbers of matings.

The map distance between a pair of gene loci (the term **locus** refers to the location of a gene and its mutant alleles, if any, on the chromosome) reflects the frequency of crossing-over between them and is not necessarily a linear measure. One map unit represents 1% Crossover Value. Thus, from Chromosome I of *Drosophila*, one could demonstrate the relationship of the three linked loci **vermilion**

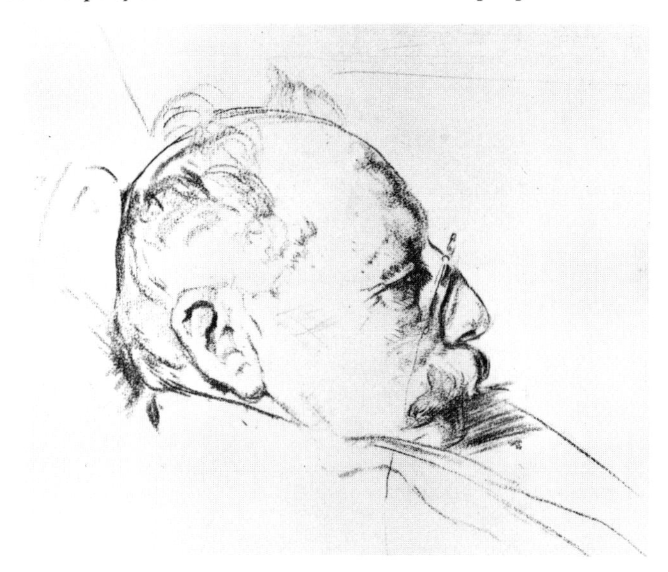

Fig. 8.12 William Bateson (1861–1926) the distinguished Cambridge biologist who investigated sources of variation in evolution and later did pioneering work in linkage genetics. (From a drawing by W. A. Forster, courtesy of the Director of the National Portrait Gallery.)

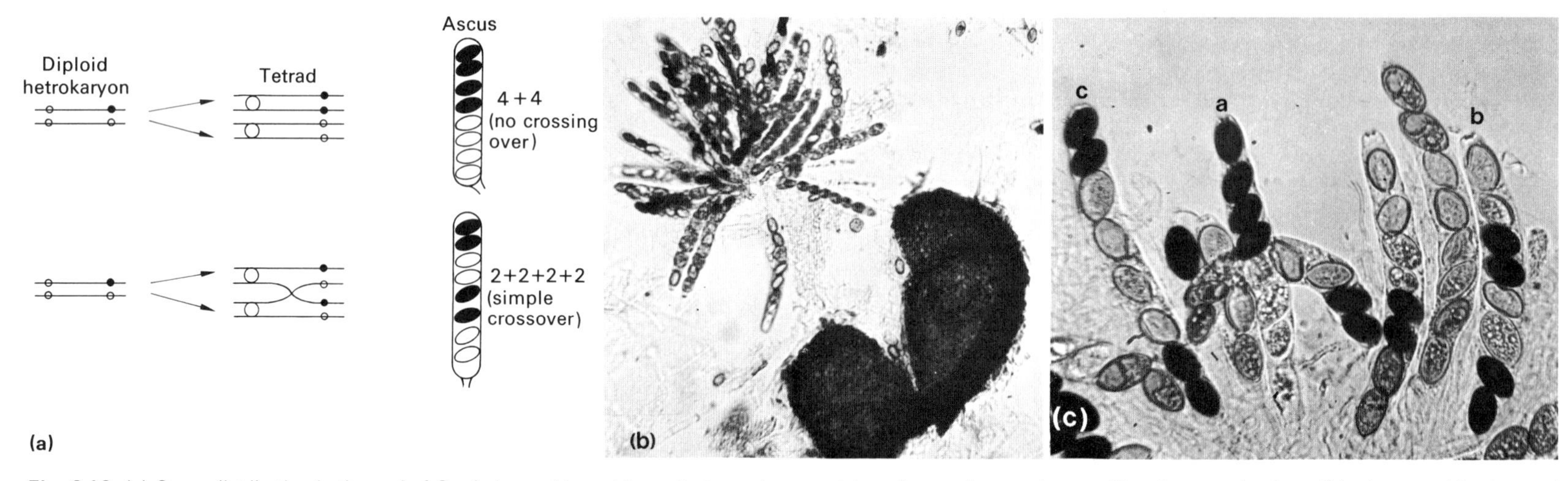

Fig. 8.13 (a) Spore distribution in the asci of *Sordaria* provides evidence that crossing-over takes place at the tetrad stage. The photographs show (b) a burst perithecium and a cluster of asci from it, (c) asci at greater magnification when the arrangements of spores can be seen; in ascus a there is no crossover, and ascus b shows the simple crossover illustrated in the diagram (a). What kind of crossover occurred in ascus c?

eyes (v), **miniature wings** (m) and **sable body** (s) by crossing a female fly heterozygous for these recessive alleles with a male fly in which they are homozygous. This cross is shown in Figure 8.15. The three mutant alleles are shown by v, m and s and their dominant, wild-type alleles (for red-eye, normal wing and grey body, respectively) are shown by the symbol +. In the diagram it has been assumed that the female fly carries the three recessive genes on the same chromosome; in other words, she represents an F1 cross between a vermilion, miniature and sable strain and a pure-breeding, wild-type strain.

From this mating one would expect six categories of phenotypes in the offspring. From the map distances shown on the chromosome map of *Drosophila* one would expect them to occur in the following proportions:

Genotypes		
+ + + / v m s	parental types	90%
v + + / + m s (crossover between v and m)	crossover between v and m	3.1%
+ + s / v m + (crossover between m and s)	crossover between m and s	6.9%

Double Crossovers and Interference

The kind of mating shown above, dealing with three linked alleles, is called a **three point cross** and has been used widely in chromosome mapping. In that mating the **vermilion-eye** locus and the **sable** locus are 10 map units apart. But when the total map distance encompassed by the three point cross exceeds about ten to fifteen map units, **double crossovers** appear. This phenomenon can be illustrated by a three point cross involving recessive mutant genes on Chromosome III of *Drosophila*:

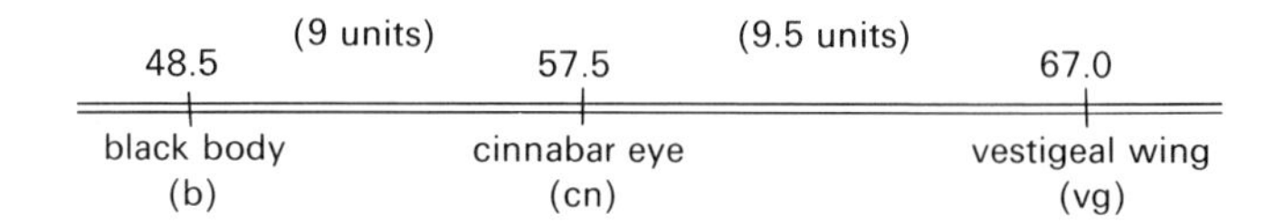

Example 7

Female flies which are heterozygous for the three alleles **black body** (b), **cinnabar eye** (cn) and **vestigial wing** (vg) are mated with male flies homozygous for these genes. From the map distances one might predict that the crossovers between *b* and *cn* would amount to 9% of the offspring and the crossovers between *cn* and *vg* would be 9.5%. However, over these distances some double crossovers would take place, that is, crossovers between *b* and *cn* **and** between *cn* and *vg*. This occurs more rarely, but gives rise to two other kinds of recombinants, namely, + *cn* + and *b* + *vg*, as follows:

$$(♀)\ \frac{+\quad +\quad +}{b\quad cn\quad vg} \times (♂)\ \frac{b\quad cn\quad vg}{b\quad cn\quad vg}$$

	offspring phenotypes	*% obtained*
+ + + / *b cn vg*	parental types	82%
+ *cn vg* / *b* + +	crossovers between *b* and *cn*	8.5%
+ + *vg* / *b cn* +	crossovers between *cn* and *vg*	9%
+ *cn* + / *b* + *vg*	double crossovers	0.5%

Therefore, to obtain the correct value for the map distances the numbers of the double crossover phenotypes must be added to **each** of the other recombinant groups. This example also illustrates the advantage of using intermediate loci in mapping. If the cinnabar locus is ignored the crossover value between *b* and *vg* is 17.5%. This is because the double crossovers increase the proportion of parental phenotypes:

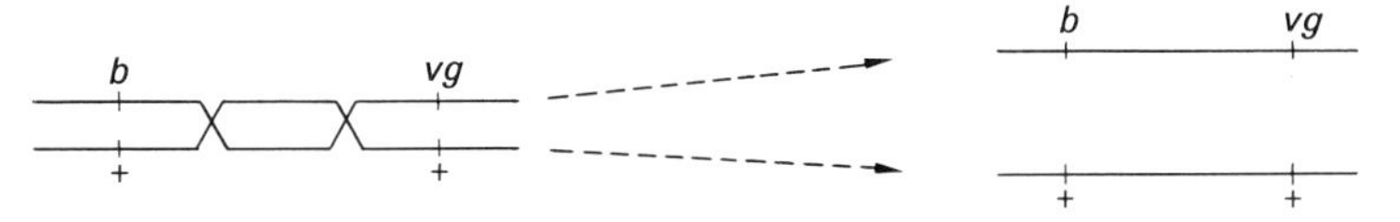

Crossing-over in the tetrad appears to be inhibited by two factors, the proximity of the centromere and the proximity of other points of crossing-over, or chiasmata. This phenomenon is termed **interference**. In *Drosophila* the presence of one crossover appears to inhibit other crossovers to a distance of about 10 map units and this effect declines thereafter with increasing distance. The reason for interference is not entirely clear but it may be simply mechanical, arising from torsion effects in the chromatids of the tetrad. It is for this reason that the observed number of crossovers is less than expected in Example 7. If the probability of a crossover occurring between the black locus and the cinnabar locus is 9% (or 0.09) and the probability of a crossover occurring between cinnabar and

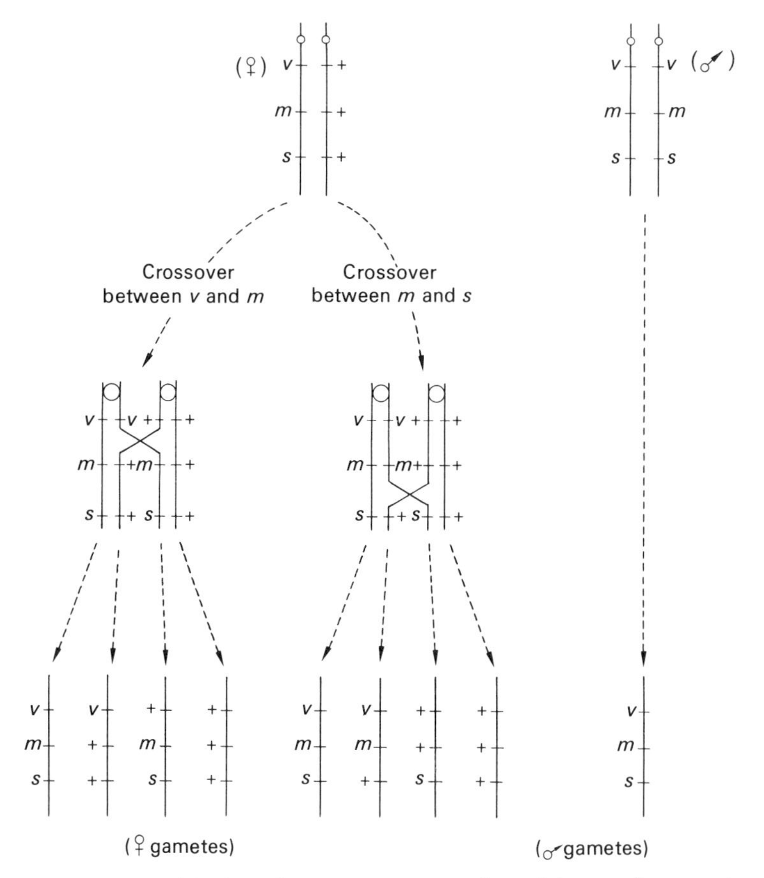

Fig. 8.15 The combination of gamete genotypes formed by crossing-over in a female fly heterozygous for the alleles vermillion eye (**v**), miniature wings (**m**) and sable body (**s**).

vestigial is 9.5% (or 0.095), then the probability of their simultaneous occurrence (being the product of their independent probabilities),

$$= (0.09)(0.095)$$
$$= 0.0086 \quad \text{or} \quad 0.86\%$$

In the example, however, the observed double crossover value was 0.5% or 0.005. In other words, out of ten thousand offspring of this cross one would have expected 86 flies to have shown the double crossover phenotype, whereas only 50 would have been obtained.

The degree of interference can be expressed in terms of the **Coefficient of Coincidence**:

$$\text{Coefficient of Coincidence} = \frac{\text{observed number of double crossovers}}{\text{expected number of double crossovers}}$$

If, in Example 7, the numbers had referred to an actual experiment, the Coefficient of Coincidence would have been given by:

$$\frac{50}{86} = 0.58$$

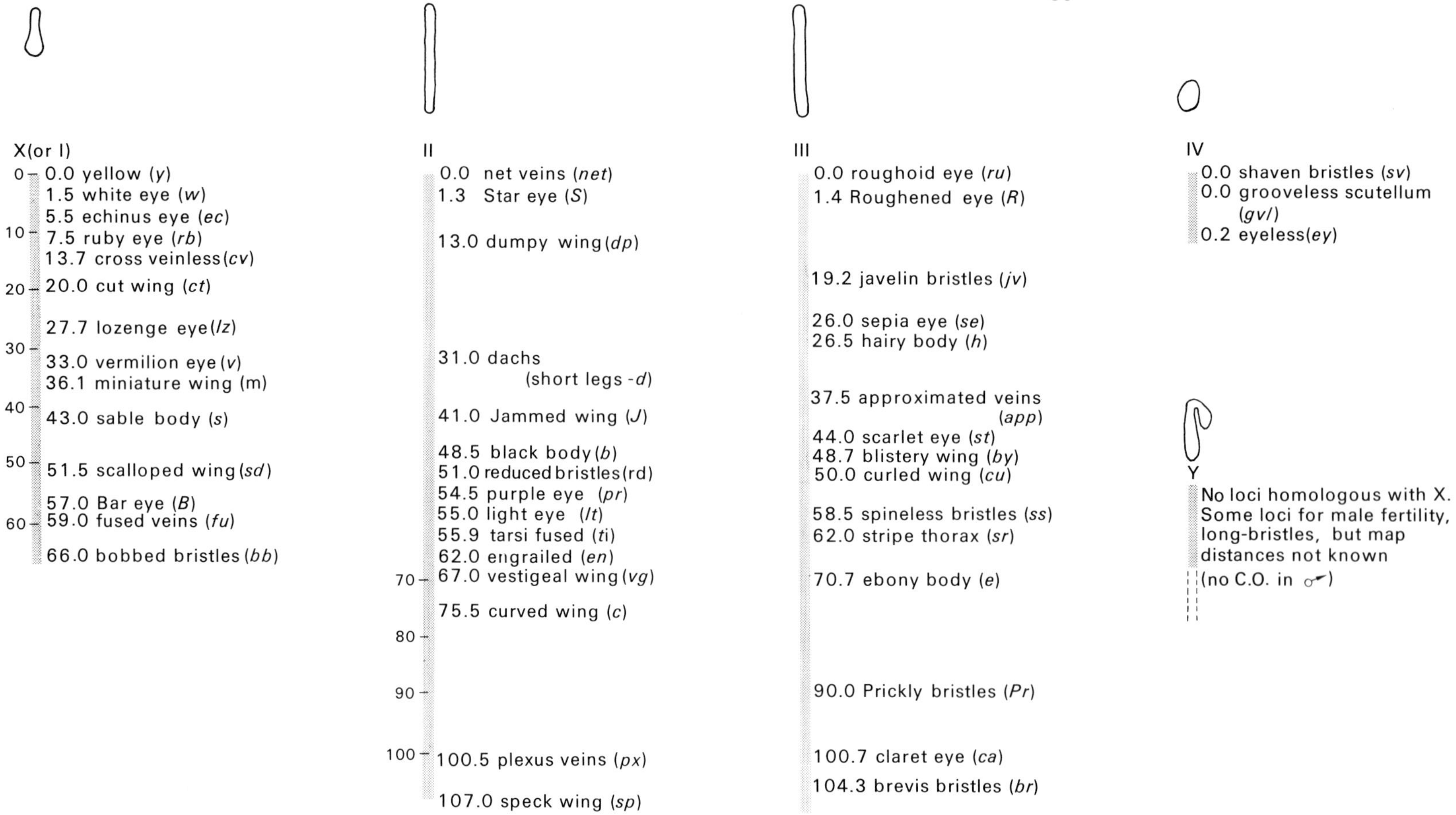

Fig. 8.14 A linkage map of *Drosophilia* derived from calculations of the frequency of recombination between loci (after Harrison, 1970).

By examination it will be seen that when interference is zero, and the observed and expected numbers of double crossovers is equal, the Coefficient of Coincidence equals 1. Similarly, when interference is complete and there are no observed double crossovers, the Coefficient of Coincidence equals 0. Thus the degree of interference is the complement of the coincidence with respect to 1, i.e.

$$\text{Interference} = 1 - \text{Coincidence}$$

In this example one could say that the degree of interference = 0.42.

In Fig. 8.14 the map of Chromosome II shows map distances up to 107, but this does not mean that one would get a Crossover Value of 107% if one were to set up a cross involving only **net veins** and **speck body**! In such a cross multiple intermediate crossovers would occur between the loci and one would probably get a C.O.V. of about 40 to 50%. As already stated, a crossover value cannot exceed 50%. Map unit distances for the long chromosomes are built up by the addition of the distances of intermediate loci.

Multiple Alleles, Pseudoalleles and the Position Effect

At this stage in the discussion it might be pertinent to point out that the mechanism of crossing-over is not at all well understood. It cannot as yet be fitted into current molecular theories nor into the complex models that have been suggested for the structure of chromosomes (Ris, 1957, and pp. 151–154). Indeed, the definition of what exactly constitutes crossing-over, in genetic and molecular terms, rather than in a cytological sense, is now rather obscure.

Until the late 1940's the concept of the allele had appeared relatively straightforward. A gene was seen as occupying a particular locus and if that gene mutated, the resultant mutant allele would also occupy that locus. In some cases a number of mutant alleles is known. The white eye locus in *Drosophila*, for example, is known to support at least 15 alleles, which give a gradation of eye colour from white to red. Such a group of alleles which could occupy one locus is termed a multiple allelic series:

Allele	Symbol
red (wild type)	w^+
blood	w^{bl}
eosin	w^e
apricot	w^a

Allele	Symbol
honey	w^h
buff	w^b
pearl	w^p
white	w

In the white-eye allelic series the wild-type shows a high degree of penetrance; the amount of eye pigment produced in an organism which is a heterozygote between the wild-type and any of the other mutant alleles is almost indistinguishable from the amount of pigment produced by the organism which is homozygous for the wild type. Heterozygotes between the mutant alleles, on the other hand, tend to show a degree of pigmentation which is intermediate between the parental homozygous types (Fig. 8.16).

In 1952, however, E. B. Lewis was able to demonstrate crossing-over between white eye and apricot eye. The frequency of crossing over was about 0.02%, that is, it would be expected to be observed only once in ten thousand offspring. This phenomenon was termed **pseudoallelism**. With the analysis of larger numbers of offspring pseudoallelism in multiple allelic series has been demonstrated in

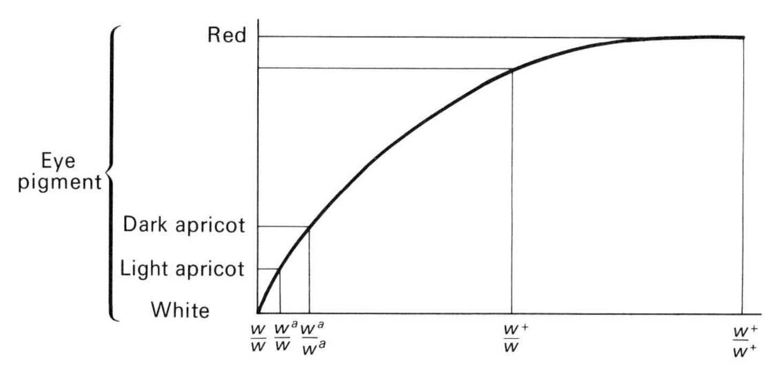

Fig. 8.16 The alleles of white eye in *Drosophilia* show a gradation of eye colour. But any heterozygote involving the wild-type allele (w^+) produces red pigment, represented by the upper part of the graph.

maize, cotton and in fungi such as *Neurospora* and *Aspergillus*. When a crossover has a value of, say, 10^{-4} it occurs once in 10 000 meioses and its chance of being detected in organisms with long generation times and few offspring, such as mammals, is remote. But in the lower organisms mentioned and particularly in bacteria and bacteriophage, which have shown themselves to be genetic tools of exceptionally high resolution, the pseudoallelic nature of multiple allelic series has proved to be commonplace rather than exceptional.

One of the notable features of the experiments concerning pseudoalleles is the **position effect**, a phenomenon first described by Sturtevant (1913). The position effect means that the phenotype of an organism is determined not only by the alleles which are present (the genotype) but by the positions of the alleles with respect to one another on the chromosomes. In other words, the same alleles can be manifest differently in the phenotype because they are arranged in different sequences.

In the case of the white-eye cross mentioned above, the normal heterozygote carrying both the white and the apricot alleles shows an intermediate eye colour known as light apricot. In the F2 generation, however, the female offspring of the light apricot female fly produced 'dull-red'-eyed offspring at a frequency of about 0.02%. Using marker genes and backcrosses, Lewis was able to demonstrate that this was due to crossing-over and not mutation. He concluded that white-eye and apricot were actually different loci and that the rare red-eyed (wild-type) phenotype occurred when the w and w^a genes were carried on the **same** chromosome. This arrangement was termed **cis** (latin: on this side). The alternative form was termed the **trans** (latin: across) arrangement:

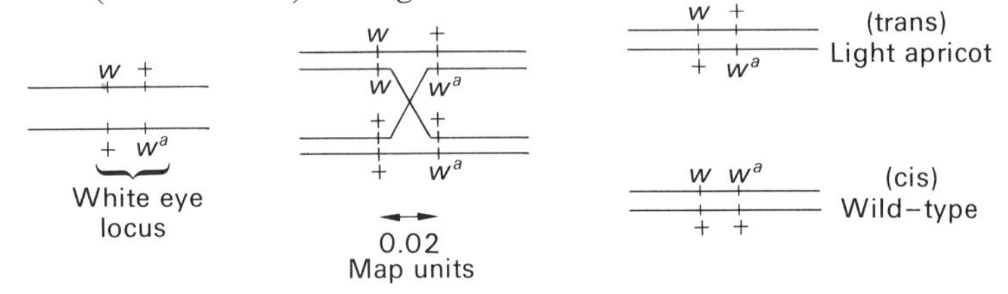

The position effect can be explained in terms of mRNA transcription (p. 35). If the eye colour locus is regarded as a complete coding unit, that is, as a single gene, then the red-eyed phenotype would occur because an intact, normal coding unit is formed. This can only take place in the **cis** state. In the **trans** state the transcription sequence is broken and consequently the production of eye pigment is impaired.

This explanation, however, suggests that crossing-over occurs **within** the gene, that is, between components, or sub-units, of the gene. Benzer (1962), using bacteriophage virus, has extended the analysis of the fine structure of the gene. Two adjacent genes on the bacteriophage chromosome have been analysed in detail. Each gene has been found to contain several hundred mutational sites, or subdivisions, between which crossing-over, or recombination, can take place. This work led Benzer to suggest new definitions of the gene and its subunits:

Cistron. This is a region of a chromosome which codes for a particular character and is synonymous with the term **gene**. Mutants occurring within the cistron show a **cis-trans** position effect.

Recon. This is the smallest unit that is interchangeable by recombination. It is indivisible; crossing-over cannot occur *within* a recon.

Muton. This is the smallest element in the chromosome which, when altered, can give rise to an altered phenotype by **mutation**.

The cistron is the largest of these units. It is a DNA sequence which codes for a complete polypeptide, so it may consist of one or two thousand nucleotides. The recon is much smaller, however, and may be only of the order of five nucleotides (Herskowitz, 1962a). The muton may be a single nucleotide.

The position effect, whereby the physical relationships of the genes on the chromosomes affect the phenotype, is relevant to discussions of the genetic basis of evolution. A number of other kinds of position effect have been studied and some of these will be discussed at a later stage.

8.4 The Structure of Chromosomes

In the discussion of linkage the linear arrangement of genes on the chromosomes has been demonstrated and some of the problems of relating crossing-over to molecular organization have been mentioned. At this juncture it is therefore appropriate to examine the fine structure of the chromosome, to attempt to relate the structure of DNA, which is known to carry genetic data by virtue of its

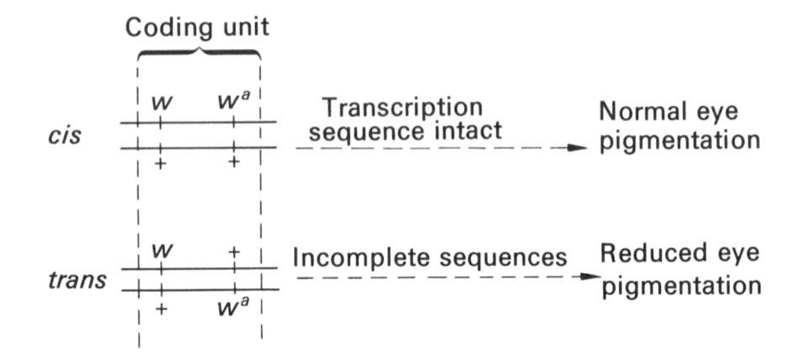

Fig. 8.17 The cis-trans position effect. In the cis position one coding sequence (+ +) is complete and can code for normal eye pigment. Crossing over within the gene, however, produces an incomplete coding sequence and faulty eye pigmentation.

structure (pp. 24–28), both to the chromosomes as they are seen through the microscope and to the chromosome as it is visualized in Mendelian and biochemical genetics.

The Bacterial Chromosome

In 1947 Lederberg and Tatum described genetic recombination in the colon bacillus *Escherichia coli*. As already stated the ease of chromosome analysis is related to the number of progeny produced from a cross. In bacterial experiments large numbers of progeny can be produced in a relatively short time and very large numbers of 'crosses' can be set up simultaneously. The gene loci investigated concerned biochemical factors related to the substrate upon which the organisms were grown, for example,

- Lac^+ the ability to use lactose in the medium,
- Met^+ the ability to synthesize its own methionine, and
- Sm^s sensitivity to streptomycin

The *E. coli* bacillus is about 2 μm in length and about 1 μm in diameter (Fig. 8.18). The cell wall consists largely of bacterial cellulose, which is less well organized than the cellulose of higher plant cells. Lining the cell wall is a lipoprotein membrane (p. 14). No nucleus is present, in the sense of a membrane-bound organelle, and there are no chromosomal configurations visible under the light microscope during cell division, such as occur in mitosis in higher organisms.

Fig. 8.18 Longitudinal and transverse sections of *E. coli* showing the amorphous, fibrous nuclear mass devoid of nuclear membrane and the densely granular cytoplasm (by courtesy of E. Kellenberger and W. H. Schreil).

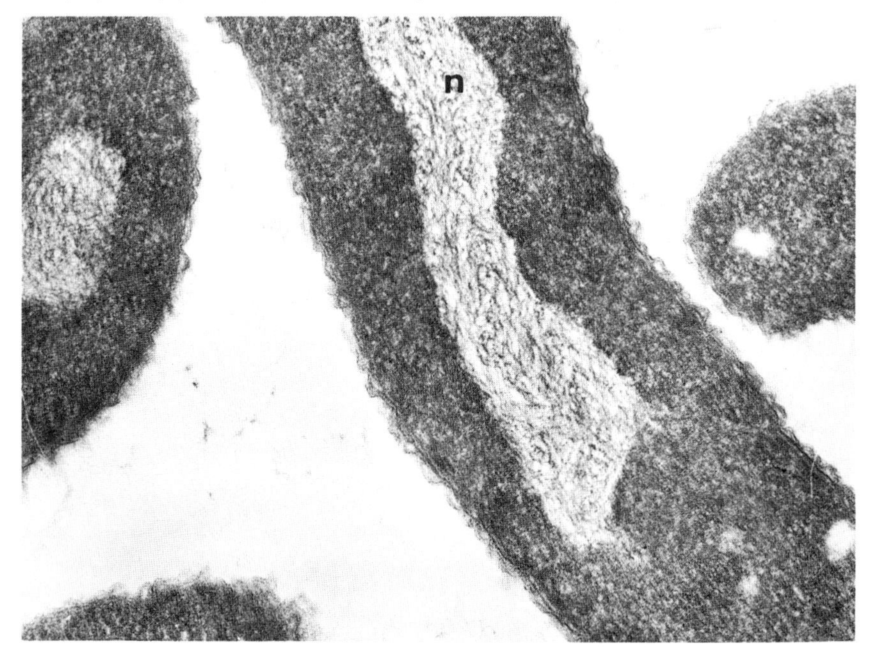

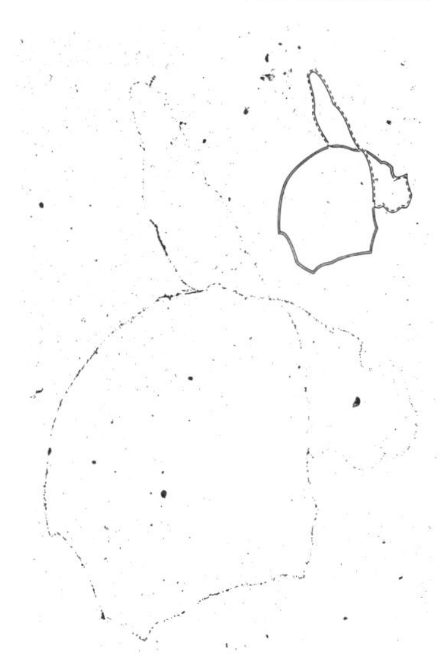

Fig. 8.19 Autoradiograph of a bacterial chromosome showing a replication loop which is clarified by the inset drawing. The main loop of the chromosome is about 250 μm in diameter (approximately a hundred times greater than the length of the cell). (Courtesy of Dr. J. Cairns.)

The hereditary information of the bacterial cell exists as DNA, as in higher organisms. Most of the DNA is believed to lie on an endless chromosome which may be of the order of a thousand times longer than the cell itself and is obviously much folded (Fig. 8.19). On this chromosome the genes are arranged linearly, as in higher organisms. Kellenberger (1960) describes the bacterial chromosome (or **genophone**) as a fibril consisting mainly of DNA, about 2.5 nm in diameter.

In some strains of bacteria a small amount of DNA (less than 4%) is known to exist as separate cytoplasmic particles, or **episomes**. These are known to be instrumental in genetic transfer. Until Lederberg and Tatum's work was published it had generally been held

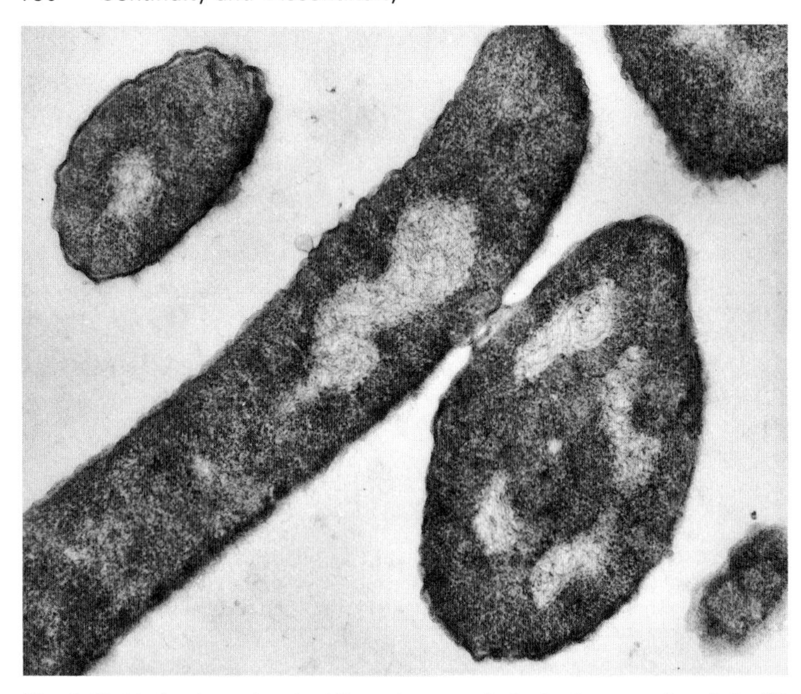

Fig. 8.20 Mating bacteria; the Hfr or donor strain is the longer cell and the F^- recipient strain is the ovoid cell. The zone of contact seems to involve only the cell walls and possibly represents the onset of conjugation. (By permission of Dr. L. G. Caro and Thames and Hudson Ltd.)

that bacteria reproduced only asexually. These workers showed, however, that very occasionally, between certain strains of *E. coli*, a form of genetic interchange could be demonstrated which was analogous to the recombination which occurs in meiotic prophase of higher organisms. Subsequently conjugation bridges between cells have been observed in electron micrographs (Fig. 8.20).

During conjugation the circular chromosome of the donor cell breaks, probably at its point of attachment to the membrane, and passes through the conjugation tube. The transfer is energy dependent and the genes are transmitted always in the same sequence. The migration of the donor chromosome through the conjugation bridge appears to proceed at a steady rate and takes about two hours. The trailing end of the chromosome probably remains attached to the donor cell membrane and there is some evidence that not all of the chromosome passes through.

Conjugation in *E. coli* takes place between a donor, or 'male', strain (the *Hfr* strain) in which the episomes are attached to the chromosome and possibly give rise to the 'break' in the chromosome, and a recipient, or 'female', strain (the F^- strain) in which episomes are absent (Fig. 8.21).

Following the migration of the donor chromosome the recipient cell divides repeatedly and the colony formed from its progeny show phenotypic features of the integrated genotypes. Thus an F^-, or recipient, cell from a Met^- strain (and therefore one which must have methionine present in the medium in order that it can grow) may give rise to a colony which is able to synthesize its own methionine if it receives a Met^+ gene through conjugation. Analysis of the bacterial chromosome makes use of the fact that the 'male' chromosome migrates at a constant rate and always from the same end. Thus if conjugating cells are submitted to strong shearing forces, such as by violent mixing of the culture in a blender, the conjugation bridge and the male chromosome are broken and the length of chromosome which has entered the recipient cell is determined by the length of time that has elapsed since conjugation began. By interrupting conjugation at different time intervals and by analysis of the substrate requirements of the progeny colonies, the sequence of the relevant genes can be determined.

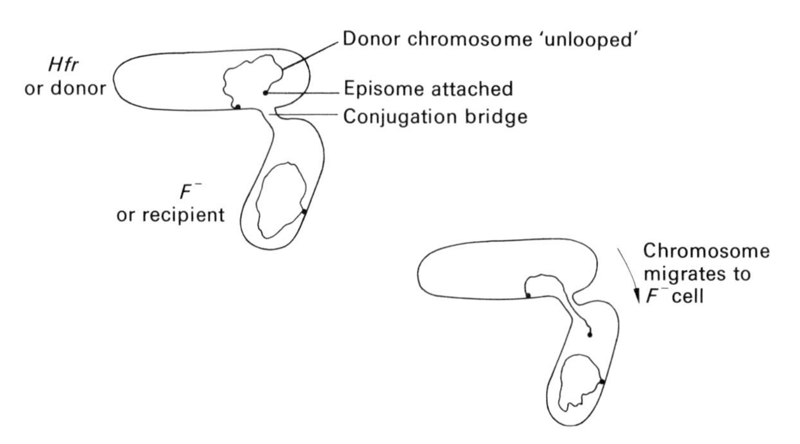

Fig. 8.21 Conjugation in *Escherichia coli* involves the migration of the chromosome from the donor through the conjugation tube to the recipient strain.

The Eukaryotic Chromosome

The linear arrangement of genes on the chromosomes of eukaryotic organisms (that is, organisms in which a nuclear membrane is present) has been demonstrated clearly. Linkage studies in a number of plants and animals have enabled chromosome maps to be constructed (Fig. 8.23) upon which the gene loci are arranged linearly. In this respect the chromosomes of higher organisms resemble those of bacteria. They differ from bacterial chromosomes, however, in one important way. When the cells of higher organisms divide the chromosomes condense into a form which is visible under the light microscope. When the cell is not in the process of division the chromosome material, or **chromatin**, is dispersed in the nucleus or possibly adheres to the inner nuclear membrane (p. 21).

The interphase, or non-dividing, nucleus has a granular appearance under the electron microscope (Fig 1.37) and chromatin fibrils cannot be detected. In this phase the nucleus used commonly to be called the 'resting nucleus'. This term has fallen into disuse owing to the fact that the non-dividing nucleus, far from resting, is actively involved in coding for enzyme synthesis.

The fine structure of the dividing cell chromosome is extremely complex. The multistranded nature of the chromosome has been demonstrated in a variety of plants and animals (Blackwood, 1956; Wilson *et al.*, 1959; Mazia, 1960). From an examination of electron micrographs of dividing cells, in squash preparations and in ultra-thin sections, Ris (1957, 1961) concluded that chromatids, which are half-chromosomes (p. 70), are composed of fibrils of 10 nm diameter which are coiled and super-coiled and twisted together like a cable. Ris and Chandler (1963) have suggested that the 10 nm fibril is the basic unit of chromosome structure and they have termed it the **elementary fibril**.

It has been stated that interphase chromosomes are in a dispersed, or hydrated, state and therefore are not visible under the light microscope. There are two notable exceptions to this, however, which have proved of great value in the elucidation of chromosome structure. The first of these is the lampbrush chromosomes found in amphibian oocytes. These structures, persisting in the interphase cells, are conspicuous and readily stained. They consist of a central strand from which arise slender side loops. The name given to these chromosomes derives from the brushes once used to clean the inside of the glass chimneys of oil lamps (Fig. 8.22).

The second exception to the rule that chromosomes disperse during interphase is the giant chromosomes which can readily be observed in the larval stages of various members of the insect order Diptera as, for example, in the fruit fly, *Drosophila*, and in the midge, *Chironomus*. These large chromosomes remain visible in non-dividing cells. They are found in actively metabolizing cells, such as salivary gland, and in the Malpighian tubules, which are excretory organs.

In these tissues growth appears to be achieved by increase in cell size rather than cell number. The balance between cytoplasmic volume and nuclear capacity appears to be maintained by repeated replication of the chromosome fibrils, forming these giant structures, which may be as much as ten times longer and a hundred times as thick as the chromosomes in ordinary dividing cells (Beerman and Clever, 1964). This process is termed polyteny. It is analogous to the way in which the cells of polyploid organisms are of greater

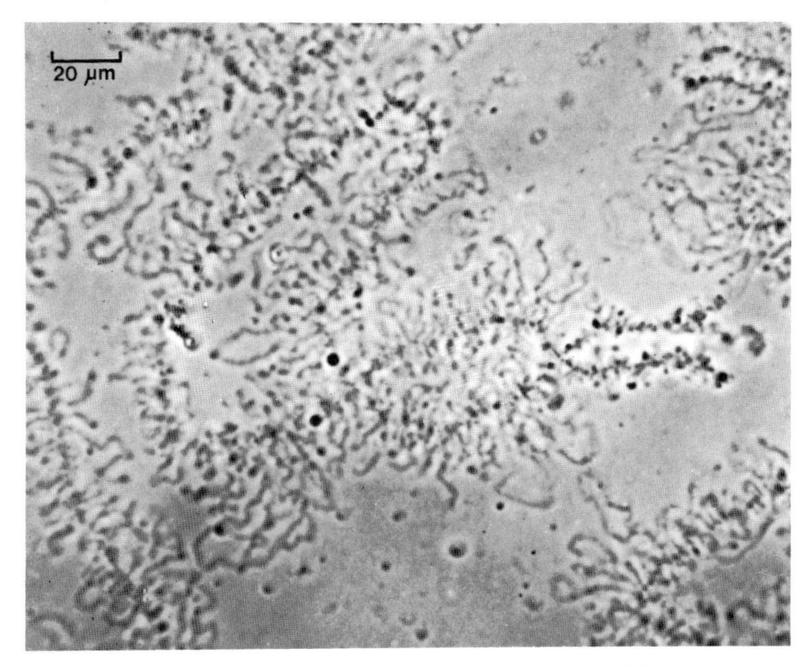

Fig. 8.22 Lampbrush chromosomes from an oocyte of the newt *Triturus*. Each loop is thought to represent duplicates of the gene lying where the loop joins the chromosome axis. These gene-slaves can amplify the action of the gene at a time when the oocyte is synthesising cytoplasmic materials very rapidly. (Courtesy of Professor H. G. Callan.)

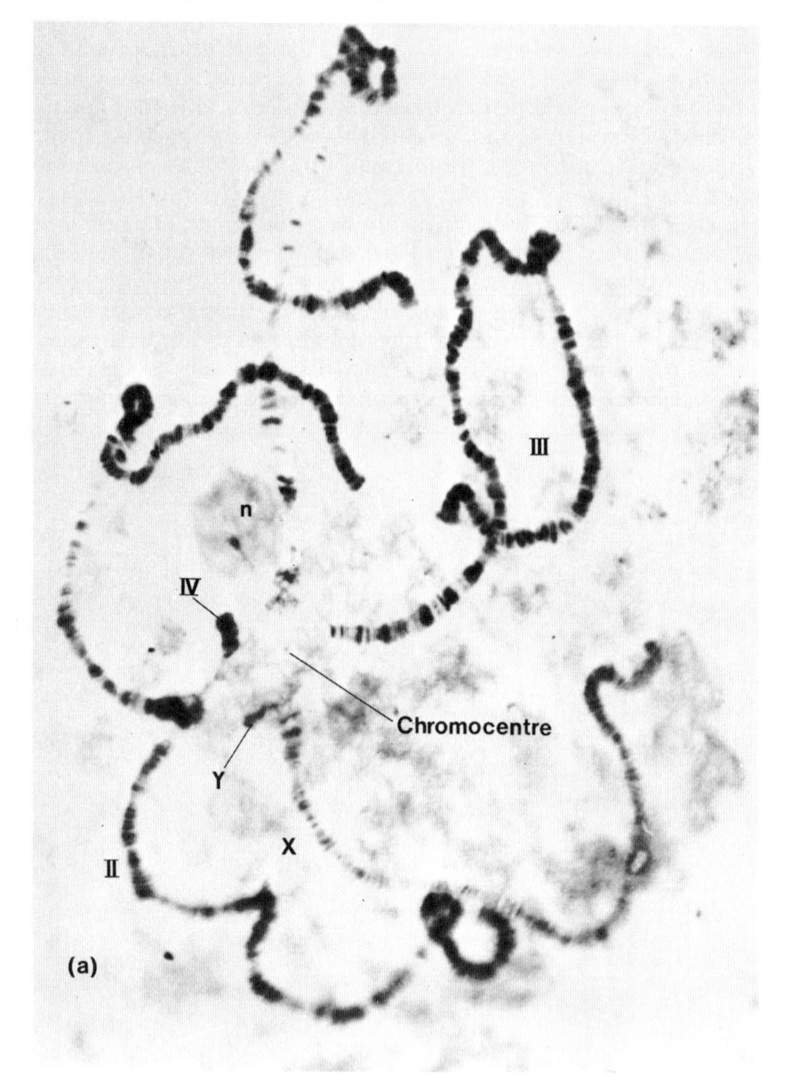

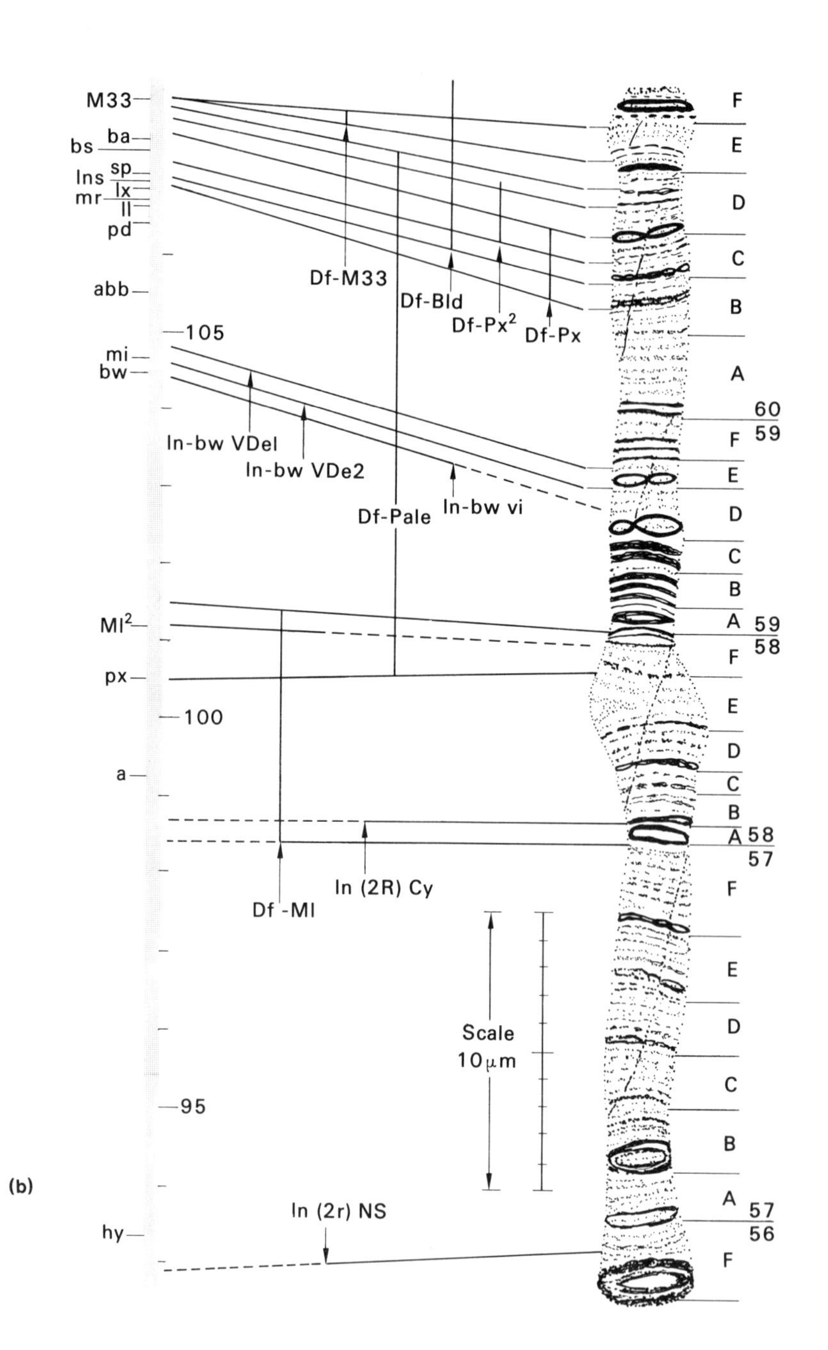

Fig. 8.23 (a) Giant chromosomes of *Drosophilia melanogaster*, the normal complement of the male. They are associated with a chromocentre and the nucleolus (n) can be seen. Note the lighter nature of the X chromosome, being only half that of the other, initially diploid, chromosomes and compare this with Fig. 8.39. (b) The linkage map and the chromosomal bands do not show an exact relationship in terms of distance.

volume than those of diploid ones (see Cell Size, p. 12). Such giant cells are, in fact, **autopolyploids**.

At intervals along the fibrils are denser regions, termed chromomeres, which are probably caused by the tight folding of the fibrils.

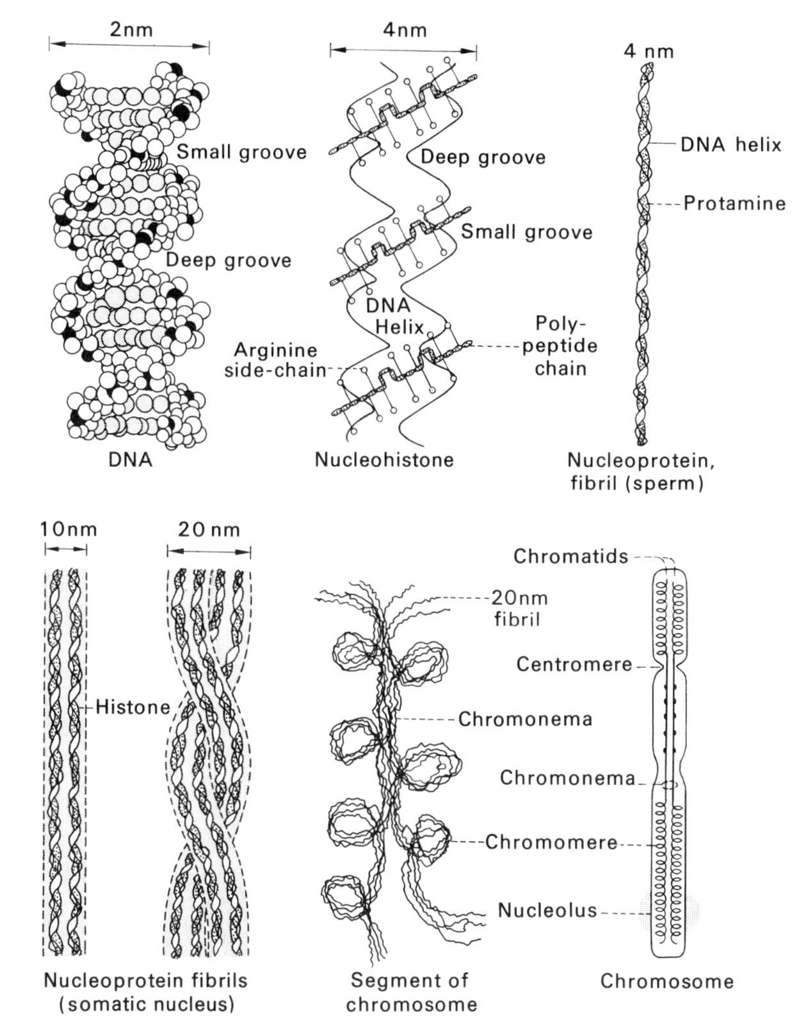

Fig. 8.24 The way in which a chromosome may be constructed from nucleoproteins, based on electron microscopic and chemical evidence. (Modified from a diagram by Dr. Hans Ris, after Paul, courtesy of Heinemann Ltd.)

In the two to six thousand fibrils that constitute the giant chromosome the chromomeres lie adjacent to one another and this gives rise to a banded appearance. Between the bands the fibrils are much less folded. Staining techniques reveal that most of the chromosome's DNA lies in the bands and the shape of the bands and their position is characteristic of the organism concerned. In the giant chromosomes of *Drosophila*, in preparations of salivary gland, Bridges (1936) was able to relate the gene for Bar eye with a particular series of bands (Fig. 8.23). At this stage, however, it should not be concluded that the bands are the genes and McClintock (1944) provided evidence which links certain genes in maize with the lightly-staining regions between the chromomeres.

DNA and the Chromosome Structure

Detailed analysis of the structure of the chromosome is difficult owing to the compact, coiled and supercoiled nature of the chromosome fibrils. From the point of view of replication a chromosome consisting of a single molecule of DNA would be the most satisfactory arrangement. DNA extracted from chromosomes, however, seldom has molecular weight in excess of 12×10^6 which, even allowing for some fragmentation during extraction, is considerably less than the total DNA in the chromosome.

Ris (1960) has suggested a model which successfully relates the dimensions of the DNA molecule to the chromosome fibril and the chromosome itself. In higher organisms DNA is closely bonded to proteins, notably to the basic proteins called **histones** (p. 120). These proteins are believed to occupy the **small** groove in the DNA helix and, in effect, serve to double the diameter of the DNA molecule. The elementary chromosome fibril is regarded by Ris as consisting of two of these nucleohistone strands held together by other protein constituents to form a nucleoprotein complex (Fig. 8.24).

If the structure of the chromosome of higher organisms really is as complex as this model suggests it would seem unlikely that the DNA molecules could replicate in the manner proposed by Meselson and Stahl for bacteria (p. 29) and achieve separation as distinct chromatids. Yet there is good evidence that this is exactly what happens. Taylor *et al.* (1957) grew bean seedlings in a solution containing the DNA base thymine which was labelled with tritium, the radioactive isotope of hydrogen, ^{3}H. The chromosomes of the root tips incorporated the tritiated thymine, in the same way as the bacterial chromosomes incorporated the ^{15}N-isotope in the Meselson–Stahl experiment. By means of autoradiographs the chromosomes

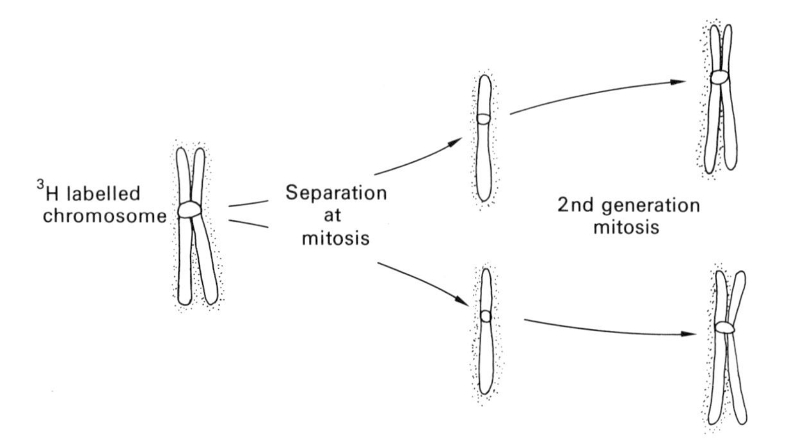

Fig. 8.25 Taylor's experiment in which autoradiographs of chromosomes were prepared after mitotic division demonstrates the semi-conservative method of replication in the chromosomes of higher plants.

taken from dividing cells in the root tip were found to be uniformly labelled with the isotope (Fig. 8.25). The seedlings were then transferred to a normal culture solution so that no further isotope could be taken up. At intervals thereafter further auto-radiographs were prepared corresponding to the generations of cell division. These revealed that, during mitosis, all the labelled bases went into one daughter chromatid, suggesting precisely the same **semi-conservative** mechanism that was demonstrated in bacteria by Meselson and Stahl.

Modifications to the Ris model have been suggested by many workers in an attempt to reconcile the evidence regarding replication and the remarkable supercoiled complexity of the whole chromosome structure (Freese, 1958; Longuet-Higgins and Zimm, 1960). An entirely satisfactory model for the structure of the chromosome, however, which accommodates all the evidence from molecular, enzymic, genetic, chemical and behavioural studies, has yet to be produced.

8.5 The Genetic Determination of Sex

In 1901 C. E. McClung reported the presence of an odd chromosome in the diploid cells of the male grasshopper, which he termed the X chromosome (X for unknown). He correctly surmised that the X chromosome was related to sexuality and later workers demonstrated that in the diploid cells of the female insect a homologous pair of X chromosomes was present. Thus, in the grasshopper, it was concluded that maleness resulted from an XO situation and femaleness resulted from XX (Fig. 8.26). During spermatogenesis the odd X chromosome would be expected to segregate into half the spermatozoa, thus the male was termed the heterogametic sex, whereas during oogenesis all the ova produced would be expected to contain an X chromosome (the homogametic sex).

Further cytological investigations led to the discovery that in many animals the X chromosome of the male was paired, at meiosis, with a smaller chromosome, which was found exclusively in the male, which was termed the Y chromosome. This situation holds in mammals; the male is the heterogametic sex (XY) and produces spermatozoa in the ratio of 50% X-carrying and 50% Y-carrying (Figs. 8.27 and 8.28).

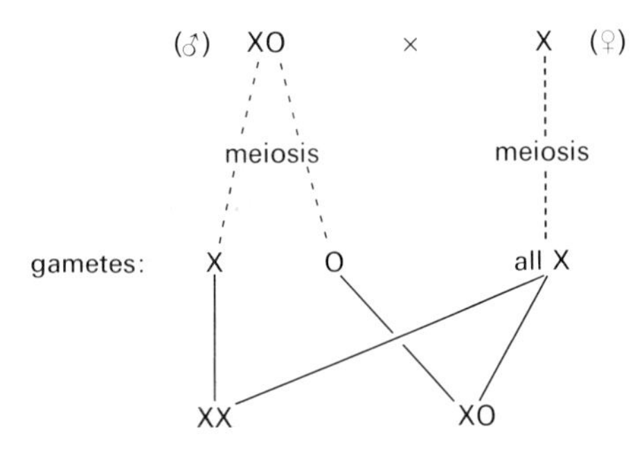

Fig. 8.26 Sex determination in the grasshopper.

The X and Y chromosomes are termed sex chromosomes. During meiosis the sex chromosomes, as might be expected, segregate randomly, that is, their orientation on the metaphase plate is not influenced by the other chromosomes (termed somatic chromosomes, or autosomes) and therefore the genes carried on sex chromosomes segregate independently with respect to genes carried on somatic chromosomes.

In the male mammal the sex chromosomes are not entirely heterologous, but a portion of both the X and Y chromosomes are homologous (Fig. 8.29). Crossing-over might occur in this region, but the evidence is inconclusive (Winge, 1923b; Tjio and Levan, 1956). In

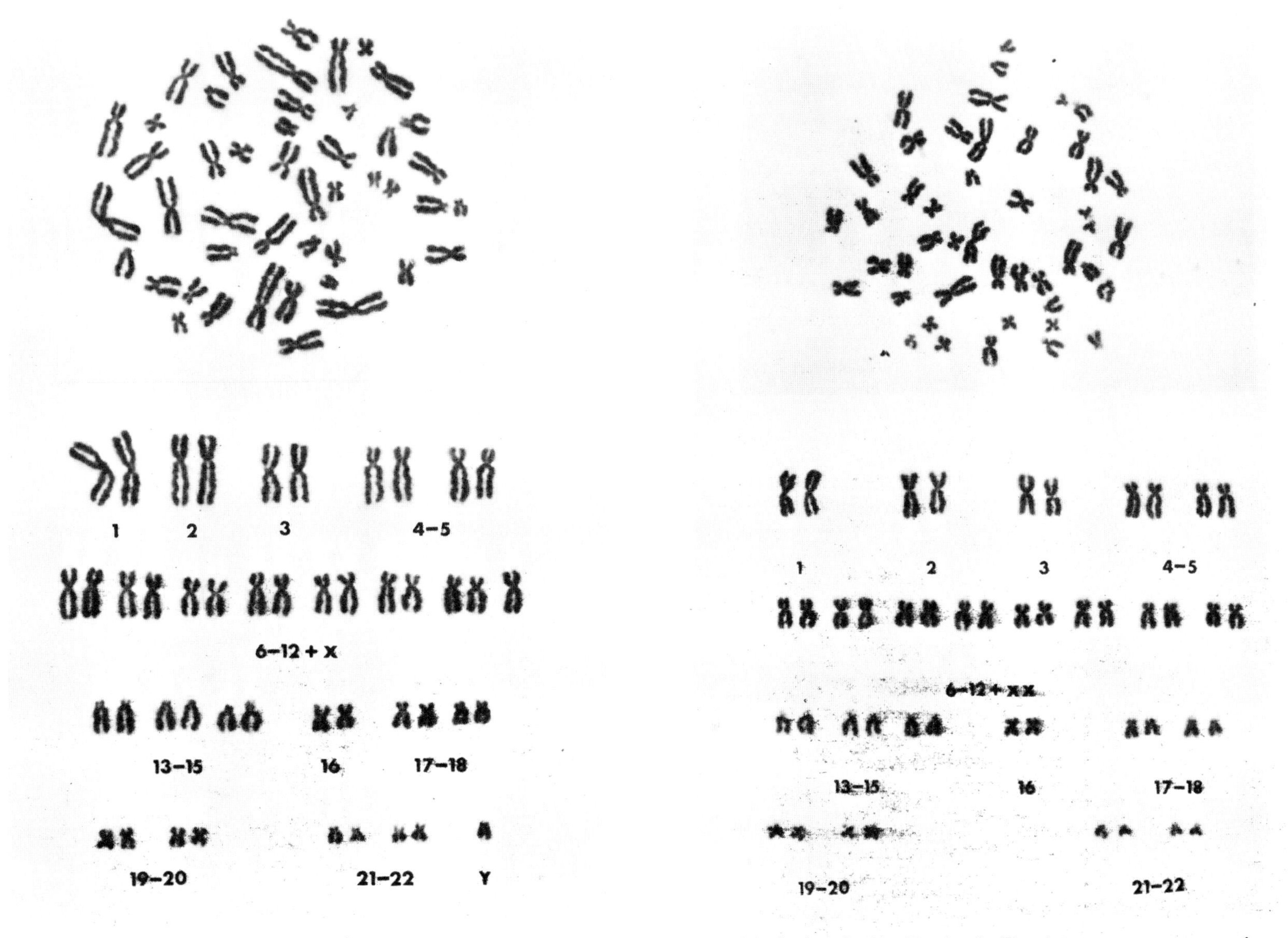

Fig. 8.28 Karyotypes of normal human chromosomes. They are prepared from the dividing nuclei of cultured white blood cells. The chromosomes are cut out from a photomicrograph (above) and pasted in groups to form the chromosomal karyotype. The male (left) has a single X chromosome in the 6–12 series and a Y chromosome in the smallest group. The female karyotype (right) shows two X chromosomes. (Courtesy of the MRC Clinical Genetics Unit, Edinburgh.)

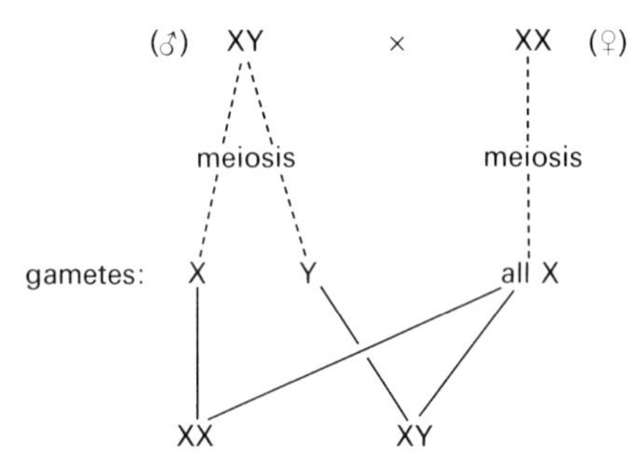

Fig. 8.27 Sex determination in mammals.

the female, the X chromosomes are homologous along their whole length.

Genes which are located in the non-homologous portion of the Y are inherited on a father-to-son basis and are termed holandric. Genes on the non-homologous portion of the X chromosome are commonly referred to as 'sex-linked' genes. In the female these genes obey the normal rules of dominance associated with the alleles on somatic chromosomes, but in males the absence of homologous alleles gives all sex-linked genes full penetrance in the phenotype.

Sex Determination in Flowering Plants

The coincidence of sexual reproduction and diploidy appears to have been achieved independently during plant evolution (see Alternation of generations, p. 89). The diploid flowering plant body is the **sporophyte**, or spore-forming, generation and strictly should not be termed male or female. However, the gametophyte generation is small and contained within the sporophyte body and, in practice, it is possible to speak of male, or staminate, flowers in which microspores give rise to male gametes, and pistillate, or female, flowers in which megaspores give rise to female gametes.

In dioecious species, those in which male and female flowers are borne on separate plants, for example the willows (*Salix* spp.) and the poplars (*Populus* spp.) sex chromosomes are present in a pattern similar to that in mammals (Fig. 8.30).

In monoecious species, on the other hand, no sex chromosomes appear to be present and the genes concerned with sex organ development appear to be distributed throughout the chromosomal complement.

An examination of the biological basis of sex determination raises a number of interesting questions. Throughout the animal and plant kingdoms the genetic mechanisms controlling developmental and metabolic processes are conspicuously uniform. The central role of nucleic acids as vital data stores for the patterns of protein synthesis are common to all organisms and there is remarkable conformity with respect to the linear arrangement of the coding units, cistrons or genes, in the form of chromosomes. It is with justification that Crick attached the label 'central dogma' to the concept of the genetic code. In the same way, meiosis and gamete formation follow the same outlines across the whole spectrum of the plant and animal kingdoms, as ubiquitous as sex itself.

It is interesting therefore to reflect upon the wide variation in modes of sex determination. *Drosophila* can have the same kind of sex chromosome arrangement as a mammal; the male can be XY and the female XX. But, unlike the mammal in which the Y chromosome seems to carry genetic data which promotes maleness, the Y chromosome in *Drosophila* appears to have no male promoting factors (with the possible exception of a fertility factor) and the genotype XO is phenotypically indistinguishable from the genotype XY. In this insect sexual phenotype appears to be a function of the ratio of X chromosomes to autosomes (see Genic Interaction, p. 198).

By contrast, in birds and lepidopterous insects the male is the homogametic sex, XX and the female is XO, or in some cases XY. In hymenopterous insects sex is a function of total chromosomal complement. In the honeybee, for example, male larvae develop from unfertilized, **haploid** eggs, although there is evidence (Mittwoch *et al.*, 1966) that chromosomal reduplication occurs widely

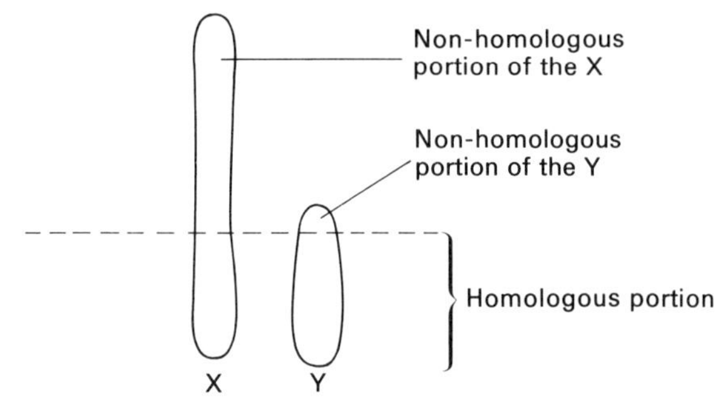

Fig. 8.29 In the sex chromosomes of the mammal the Y is thought to be only partly homologous with the X.

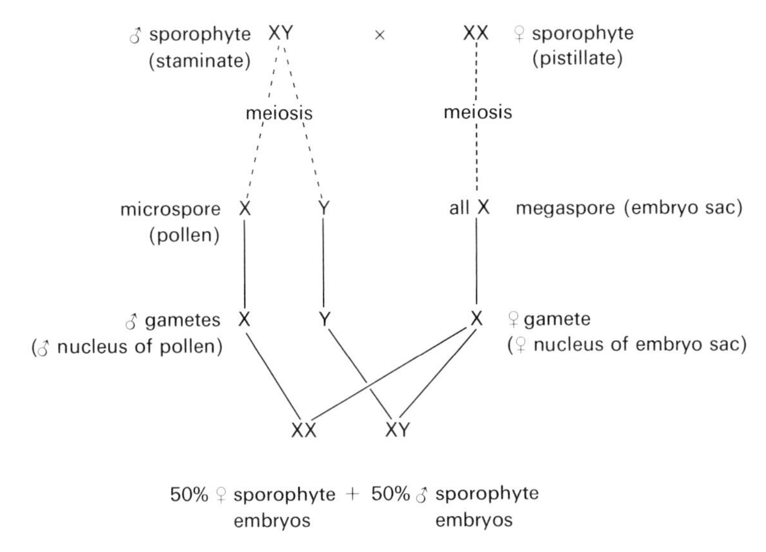

Fig. 8.30 Sex determination in dioecious flowering plants (those with separate sexes) is essentially the same as that in mammals.

during the larval stages so that many of the cells in the adult male bee may be diploid or even polyploid.

In the marine annelid *Bonellia viridis* sex appears to be determined by environmental factors. If the free-swimming larval stage commences metamorphosis in isolation it develops into a female. If the larva is in contact with a female worm, however, it develops into a male. In this case its non-reproductive tissues atrophy and the creature is reduced to little more than a male gonad, which becomes parasitic in the nephridium of the female. Environmental factors also appear to determine sex in the curious sessile marine snail, *Crepidula fornicata* (Fig. 8.31). Like *Bonellia*, the larva of this snail develops into a female if it settles on a rock, but if it settles on a female it develops into a male. Other larvae may settle and these in turn develop into males, forming an odd, twisted tower of snails. Under the pressure of 'maleness' above them, the lower males may turn into females.

In the dioecious pink species, *Silene** an XY/XX sex chromosome pattern similar to that in mammals appears to operate (Winge, 1923a). H. E. Warmke, in 1946, obtained polyploids of *Silene* and demonstrated that the Y chromosome, which was larger than the X, carried extremely strong male-promoting qualities. Unlike *Drosophila*, in which Bridges (1925) showed that the XXY complement was female, an XXY *Silene* plant produces normal staminate blossoms. Hermaphrodite flowers were only produced, however, by overloading the Y chromosome in an XXXXY complement.

* Formerly *Melandrium* then *Lychnis* (Clapham *et al.*, 1962).

Fig. 8.31 *Crepidula fornicata*, the slipper limpet. Introduced to British waters from North America about 1880, probably by ships, now prolific on the south coast and a pest of oyster beds where it smothers the oysters and competes for food. This specimen from the Solent comprises a couple of females, at the left, and a spiral of males.

Sex Determination in Vertebrates

In most species of fishes examined the sex chromosomes are indistinguishable from the autosomes under the microscope. Winge (1922) however, has demonstrated the existence of an XX female/XY male pattern of sex chromosomes in the guppy, *Lebistes reticulatus*.

In amphibians the sex chromosome pattern is less clear. Witschi (1929) produced genetical evidence that in *Rana temporaria* the female sex was homogametic and the male was heterogametic. Later, however, Weiler and Ohno (1962) showed that the female of *Xenopus laevis* was heterogametic and possessed a giant Y chromosome more than twice the length of the X chromosome. In neither fishes nor amphibians does the development of sex organs appear to be under constant regulation by the sex chromosomes. In most species studied embryologically the sex organs go through a common,

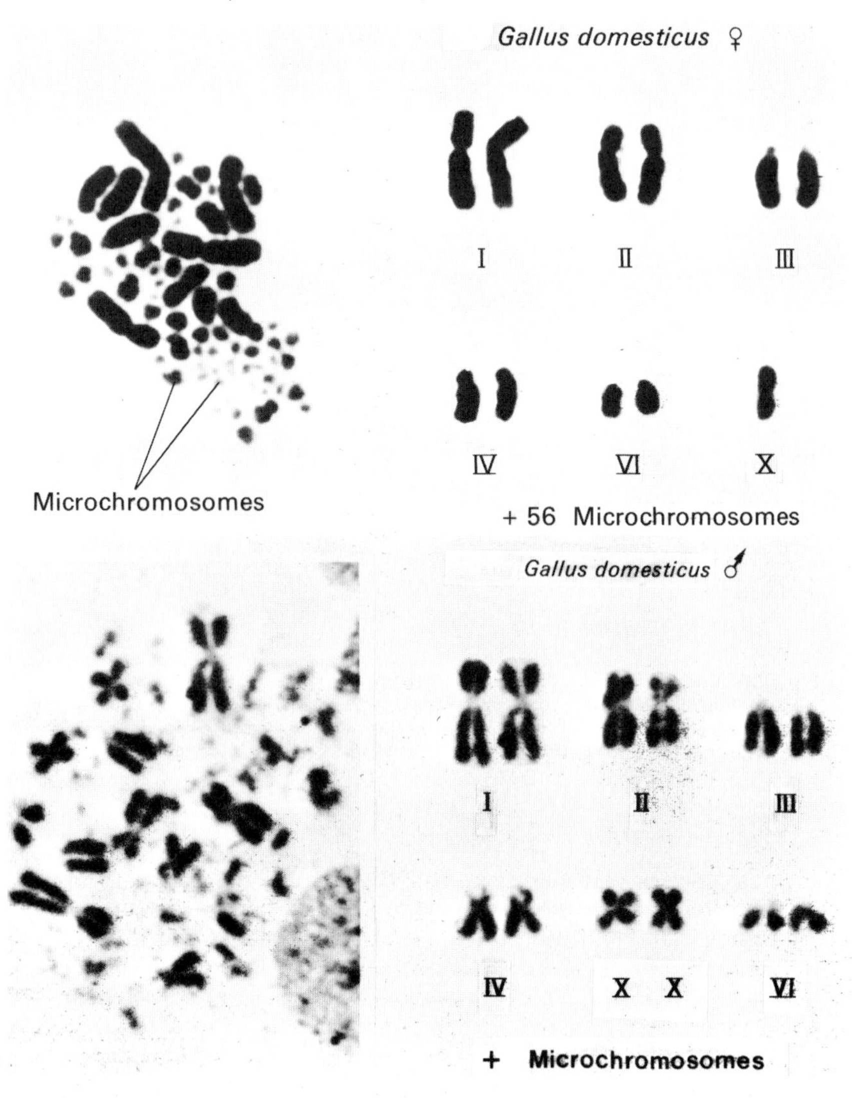

Fig. 8.32 Chromosome karyotypes of the domestic fowl. In birds the female is the heterogametic sex (unlike mammals, see Fig. 8.27) and has a single X chromosome. The male has two X chromosomes. Another difference between the bird and the mammal chromosomes is the presence of microchromosomes. These number about 56 in this species and can be seen clearly in the preparation of the female nucleus. (Courtesy of Dr. E. H. R. Ford.)

indifferent early stage, followed by a hermaphroditic or bisexual stage in which both elementary testicular and ovarian tissues are present. This is followed by the completion of one or other set of sex organs (Forbes, 1961).

The cytological examination of sex chromosomes in reptiles is made difficult by the existence of two kinds of chromosomes. In snakes, eight pairs of macrochromosomes are present with about twice that number of microchromosomes. Where sex chromosomes have been identified among the macrochromosomes, in some species of lizards and tortoises and in the adder, the female has been found to be the heterogametic sex (Kobel, 1962).

As might be expected from evolutionary evidence the situation in birds follows the reptile plan closely. Genetic evidence that the female of the fowl was heterogametic was obtained by Morgan and Goodale in 1912. More recently, karyotypes (Fig. 8.32) of fowl chromosomes have been obtained by Owen (1965) which show six pairs of macrochromosomes (of which the fifth largest are the sex chromosomes) and at least 33 pairs of microchromosomes, and there may be others present too small to be identified.

In mammals the most primitive extant order, the Monotremata, shows an affinity with the reptiles in chromosome form and number. *Ornithorhynchus*, the duck-billed platypus, has a high chromosome number consisting of macro- and microchromosomes like the reptiles and birds (Matthey, 1949), although the evidence obtained by van Brink (1959) from cytological studies on the spiny anteater, *Tachyglossus*, points to the male as the heterogametic (XY) sex.

Marsupials, on the other hand, show an affinity with the placental mammals. In the opossum, *Didelphis*, Shaver (1962) demonstrated that the chromosomes were relatively few in number ($2n = 22$) and not differentiated into macro- and microchromosomes. The female had an XX complement and the male was XY.

The best known mammal from the point of view of chromosomal studies is unquestionably man. Three almost simultaneous technical discoveries in the early 1950's had a pronounced effect on this work. Treatment of cells with the drug **colchicine**, which appears to inhibit spindle formation and therefore tends to hold the dividing cells in mitotic prophase (and also makes the chromosomes more condensed), coupled with the discovery that treatment with hypotonic solutions makes the nuclei swell and the chromosomes spread out, enabled microscope slides of unprecedented clarity to be made. These discoveries, together with improved methods of culturing cells, particularly leucocytes, *in vitro* led to intensive investigations of human chromosomes on a world scale and are leading to chromosomal studies in other organisms.

The diploid number for man was established at 46 by Ford and Hamerton (1956) (before then 48 had been widely and erroneously quoted) and the normal sex chromosome pattern was confirmed as XY (male) and XX (female) (p. 155). Chromosome studies of other placental mammals have shown variations in chromosome number, but without the macro- and microchromosomal pattern of the lower mammals, and have established that the male is the heterogametic sex (Fig. 8.33).

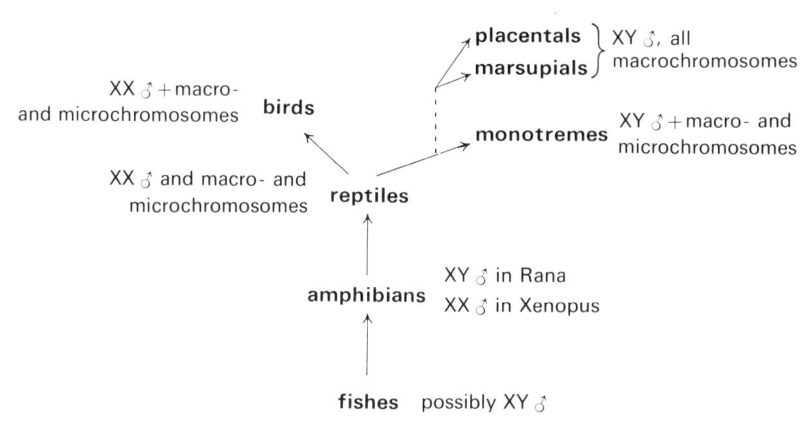

Fig. 8.33 Diagram to show the sex chromosome pattern within the accepted pattern of vertebrate evolution.

The Role of Sex Chromosomes in Development

In insects the sex chromosomes appear to exert a direct effect on the cells, and thus the tissues, which contain them. In *Drosophila*, Morgan and Bridges (1919) showed that occasionally the sex chromosomes of an XX zygote fail to segregate in one of the first few divisions, giving rise to XX and XO daughter cells. Within one fly, therefore, it is possible to have some tissues containing the XX, or female, complement, which exhibit female characteristics and other tissues containing the XO or male complement, which exhibit male characteristics. Such insects are termed **gynandromorphs** (Fig. 8.34).

In the embryological development of higher organisms, however, the sex chromosomes seem to act indirectly by promoting the development of one or other kind of sex organ from the indeterminate, or bisexual, gonadal ridge. Once formed, the gonad itself promotes the sexual characteristics of the organism by means of the hormones it secretes.

In the mammal the genital ridge develops through an indeterminate, or indifferent, gonad stage. At this stage the gonadal tissue comprises a medulla and a cortex. In the female the medulla then regresses and the cortex persists, giving rise to an ovary. In the male the opposite happens; the cortex regresses and the medulla persists, forming the testis (Fig. 8.35).

The question of how the X or Y chromosomes effect this developmental switch is as yet unanswered. There may be specific male-promoting genes on the Y chromosome which direct the formation of some testis-induction substance that acts on the indeterminate gonad. Alternatively, the 'maleness' genes may lie on the X chromosome and be activated by regulator genes on the Y chromosome (Fig. 8.36).

Sex Chromatin

In man and in other mammals that have been studied the interphase,

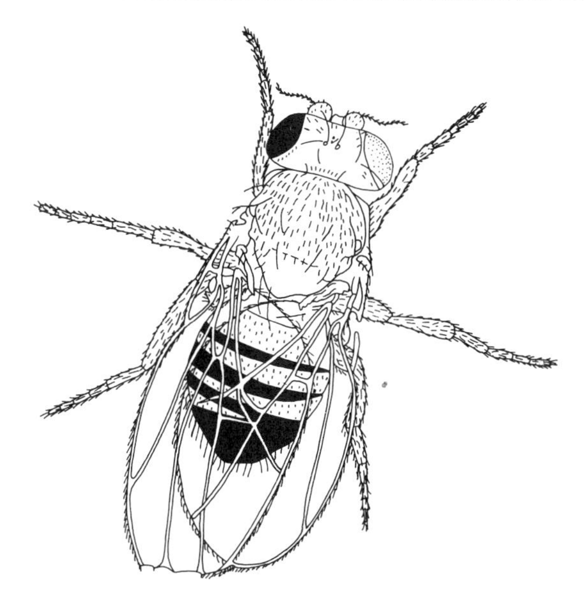

Fig. 8.34 A gynandromorph of *Drosophila melanogaster.* The left side of the body is predominantly female and the right side predominantly male. (After Bridges, from Sinnott, Dunn and Dobzhansky, Principles of Genetics, courtesy of McGraw-Hill.)

or non-dividing, nucleus may carry a sex chromatin body. This structure stains in microscopical preparations with those dyes which stain chromosomes in dividing cells, for example Feulgen and cresyl violet. It was first described in 1949 by the Canadian M. L. Barr in the nerve cell bodies of the cat, later in other tissues of this animal and, in 1950, in man. It is commonly referred to as the 'Barr body'.

Barr bodies are usually present only in female cells and they disappear during mitosis when the chromosomes are visible under

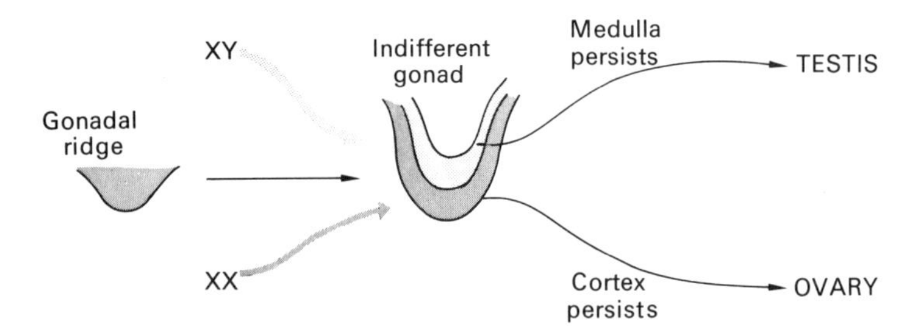

Fig. 8.35 The development of gonads in the mammal involves a common, or indeterminate stage from which the testes or ovaries develop under the promotion of the sex chromosomes.

the microscope (Fig. 8.37). The striking correlation between the number of X chromosomes in a cell and the number of Barr bodies, there being one fewer Barr bodies than X chromosomes) has led to the conclusion that the Barr body is a highly condensed and probably inert X chromosome (Stewart, 1960).

Dr. Mary Lyon has suggested that the X chromosome which is to be rendered inert is selected randomly at some relatively early stage in embryogenesis, but that once a chromosome has been selected the same X chromosome becomes inert in all the daughter cells formed by mitosis. Thus a female organism would be made up of a patchwork, or **mosaic**, of groups of cells in some of which the maternal X is active and in others the paternal X is active. The implication of this hypothesis is that, in a female, only one member of a sex-linked pair of alleles is active. This would account for the fact that an enzyme coded for by a gene on an X chromosome is produced in about the same concentration in a male as in a female.

Much of this work has been done on the mouse, but confirmation of the Lyon hypothesis comes from the work of Beutler *et al.* (1962) on man. A sex-linked mutant gene causes deficiency in the level of the enzyme glucose-6-phosphate dehydrogenase (see metabolic

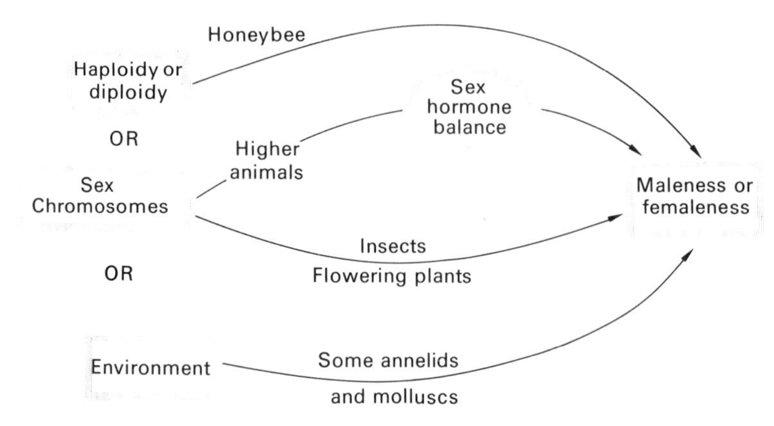

Fig. 8.36 Some patterns for the determination of sex.

pathways, p. 51) and these workers were able to demonstrate that in women who were heterozygous the enzyme level was intermediate between the levels corresponding to both kinds of homozygote (Fig. 8.38).

Heterochromatin and Euchromatin

It has been known almost since the discovery of sex chromosomes at the beginning of the century that some chromosomes, wholly or in part, have different staining properties from others and often behave differently during cell division. These atypical chromosomes were called **heterochromosomes** by Montgomery (1904), thus the substance of which they were constituted later came to be called **heterochromatin** (the normal chromosomes were termed **euchromosomes**, made up of **euchromatin**). Sex chromosomes of many organisms have been shown to consist wholly or largely of heterochromatin although many autosomes, or euchromosomes, have been found to contain regions of heterochromatin. It is thought that the chromocenter, the body to which the salivary gland chromosomes of *Drosophila* are frequently attached is made up of heterochromatin and probably embodies about one-third of the X chromosome and all the Y chromosome (Fig. 8.39).

The banded regions, euchromatin, were thought to be genetically active and the heterochromatin to be genetically inert. Recently, however, Commoner (1964) has suggested that both forms of chromatin play a dual and interdependent role in the genetic regulation of the cell's activities. He postulates that euchromatic DNA contains the 'ordinary' genes concerned in enzyme synthesis, which conform to Mendelian inheritance, whereas heterochromatic DNA acts as a genetic regulator.

Heterochromatin accounts for about a quarter of the total DNA. It behaves differently at meiosis, synapsing with other heterochromatic regions with relatively little accuracy, in contrast to the euchromatic regions which synapse with a high degree of precision. Heterochromatin does not appear to contain the major Mendelian genes and Commoner contends that the heterochromatic regions of the chromosomes do not act through the genetic code/enzyme synthesis pathway, but act directly on neighbouring euchromatic regions. He suggests that heterochromatin exercises a **generalized effect** on embryogenesis, cell size and division and a **quantitative effect** on such features as the degree of male fertility in *Drosophila* (see The Control of Gene Action, p. 135, and The Position Effect, p. 147).

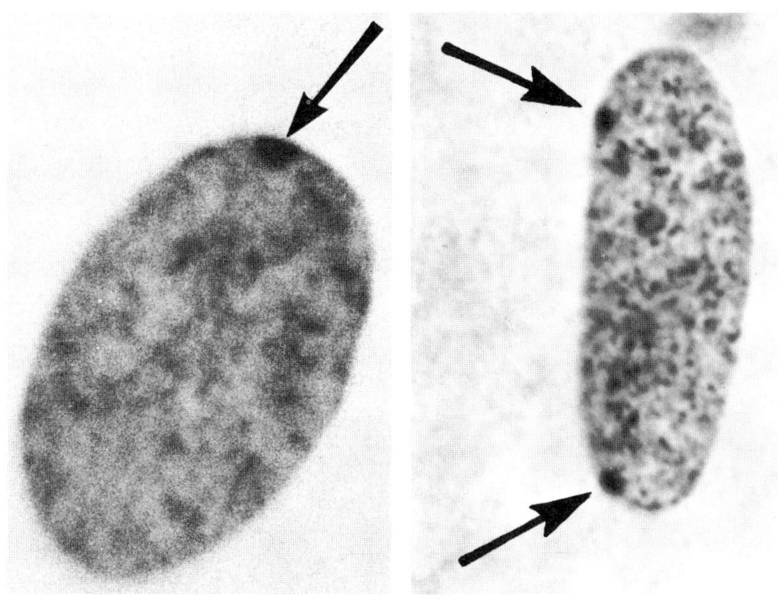

Fig. 8.37 The Barr body is a particle of densely staining chromatin closely associated with the nuclear membrane. It is thought to represent a redundant X chromosome, there being one less Barr body than the X chromosomal complement. These photomicrographs of fibroblast nuclei show (left) a single Barr body (arrowed) in a normal or XX female and (right) two Barr bodies in a woman patient known to have a triple-X genotype. (Courtesy of Dr. Ursula Mittwoch.)

Abnormal Sex Chromosome Ratios in Man

In 1942 H. F. Klinefelter and his co-workers described the clinical symptoms of 'gonadal dysgenesis' in a male patient who showed increased secretion of Follicle Stimulating Hormone, reduced spermatogenesis and some breast development. The improved methods of cytological chromosomal analysis later enabled Jacobs and Strong (1959) to determine that a patient exhibiting 'Klinefelter's Syndrome', as the symptoms came to be known, had an XXY sex chromosome complement (Fig. 8.40). This complement appears to be the commonest cause of Klinefelter's Syndrome, although other sex chromosome ratios, such as XXYY and XXXXY, have been established as leading to Klinefelter's Syndrome by Barr *et al.* (1962; 1964).

In spermatogenesis non-disjunction, that is the failure of homologous chromosomes to separate in Anaphase I of meiosis, could give

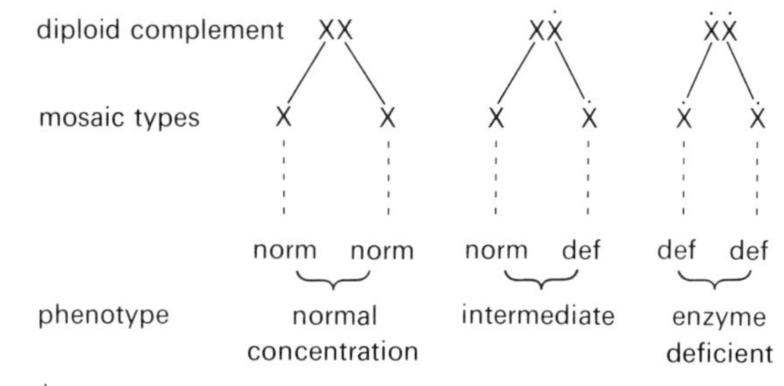

Ẋ = reduced ability to synthesize enzyme

X = normal allele

Fig. 8.38 A sex-linked gene involved in the synthesis of the enzyme glucose-6-phosphate dehydrogenase has a mutant allele which causes a defect in enzyme production. The fact that a woman who is heterozygous at this locus produces a level of enzyme intermediate between the deficient homozygote and the normal level lends weight to the Lyon hypothesis because it suggests that about half of the cells of the body produce enzyme normally and the remainder (in which the chromosome bearing the mutant allele is the functioning one) are deficient in enzyme.

rise to XY-carrying sperm, which in turn could give rise to the XXY Klinefelter condition. In the same manner non-disjunction during meiosis I of oogenesis would give rise to an XX ovum which, as a result of fertilization by a Y-carrying sperm, could give rise to an XXY zygote.

An allied condition of abnormal sex chromosome ratio in man is known as Turner's syndrome (described by H. H. Turner in 1938). Patients with this condition are females, showing gonadal dysgenesis. The ovaries are rudimentary and inert, and there is a characteristic shortness of stature and webbed neck. In such patients Barr bodies (p. 161) are absent, which would indicate the absence of an X chromosome. This was confirmed by C. E. Ford *et al.* (1959) when patients with Turner's syndrome were found to have an XO sex chromosome complement. Figure 8.41 shows that the types of non-disjunction described by Klinefelter's syndrome could also give rise to Turner's syndrome.

Since the types of non-disjunction which give rise to Klinefelter's could also give rise to Turner's syndrome one might expect them to occur in the population at about the same frequency. In fact, Turner's syndrome is significantly rarer (4 per ten thousand live births compared with 20 per ten thousand for Klinefelter's). D. H. Carr (1965) examined the sex chromosome ratios of foetuses which had been spontaneously aborted and found among them about 6% of the XO complement which were presumptive Turner's. It might be concluded, therefore, that the XO complement is normally lethal and only a small proportion survive to exhibit Turner's syndrome (Fig. 8.42).

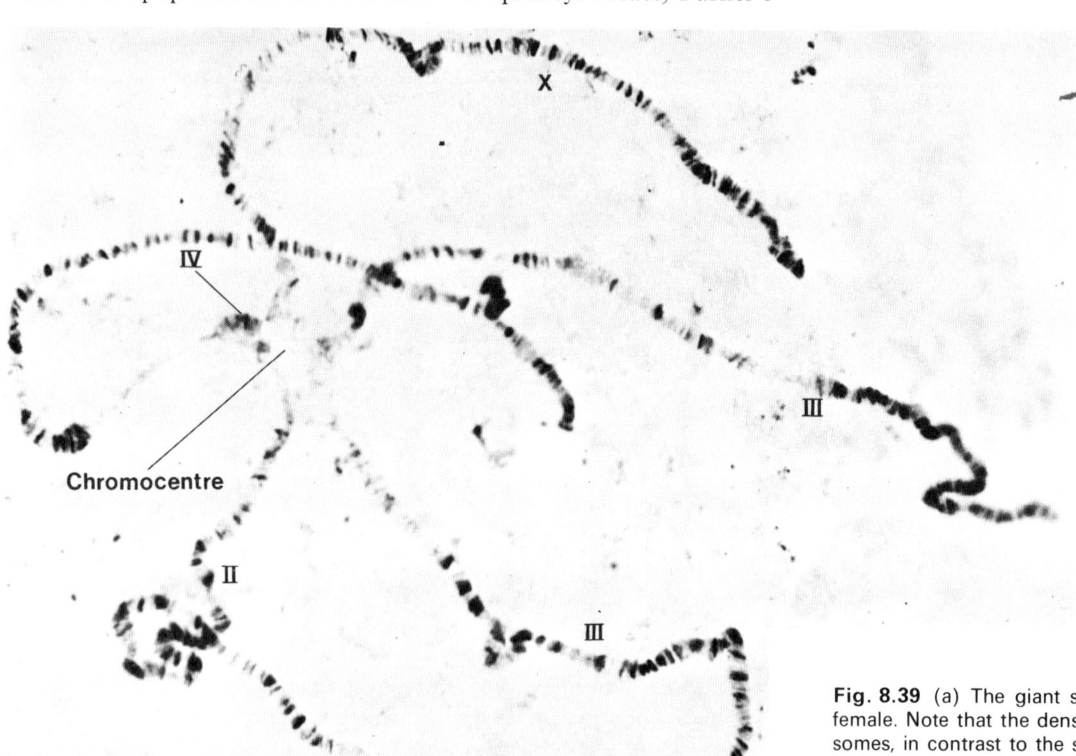

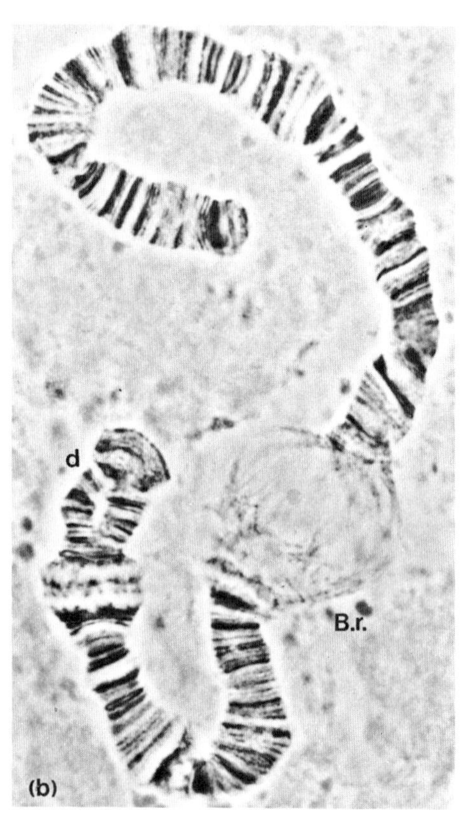

Fig. 8.39 (a) The giant salivary gland chromosomes of *Drosophila melanogaster* female. Note that the density of the X chromosome is about the same as the autosomes, in contrast to the situation in the male (Fig. 8.23a). (Courtesy of Dr. O. G. Fahmy.) (b) Chromosome III of *Smittia parthenogenetica* showing the clear banding of euchromatin and heterochromatin. Also seen is a large Balbiani ring (B.r.) and a split part of the chromosome which probably represents a deletion or duplication (d), see p. 184. (Courtesy of Professor H. Bauer and Thames and Hudson Ltd.)

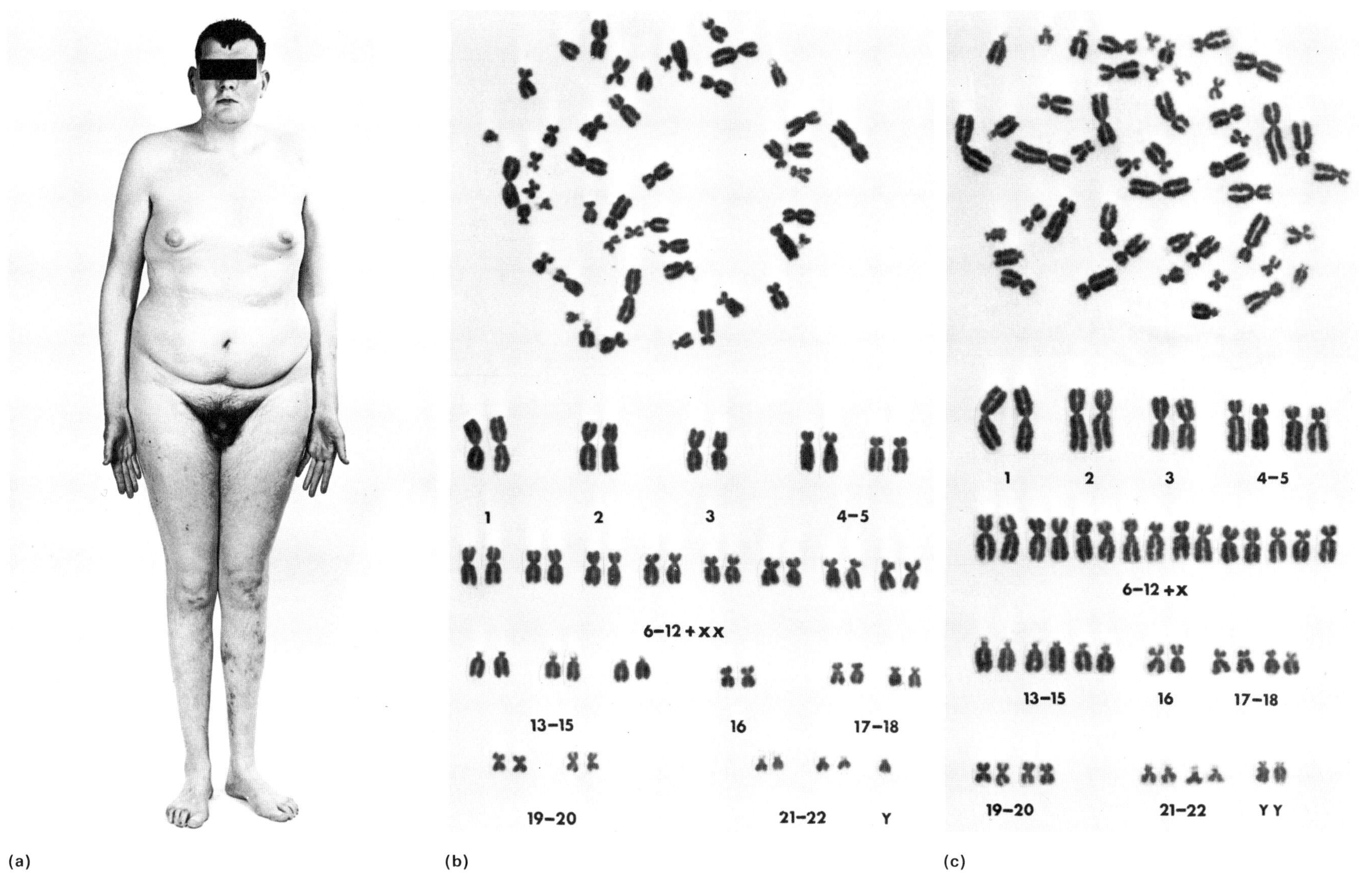

Fig. 8.40 (a) A Klinefelter patient showing the characteristic gynandromorphic features (Courtesy Dr. R. Lax, Kennedy-Galton Institute.) (b) The chromosome karyotype of a klinefelter patient showing a total complement of 47 including an XXY sex chromosome situation. (c) A chromosomal complement of 47 arising from an additional Y chromosome (sex chromosomes = XYY) producing a normal male phenotype. (Courtesy of the MRC Clinical Genetics Unit, Edinburgh.)

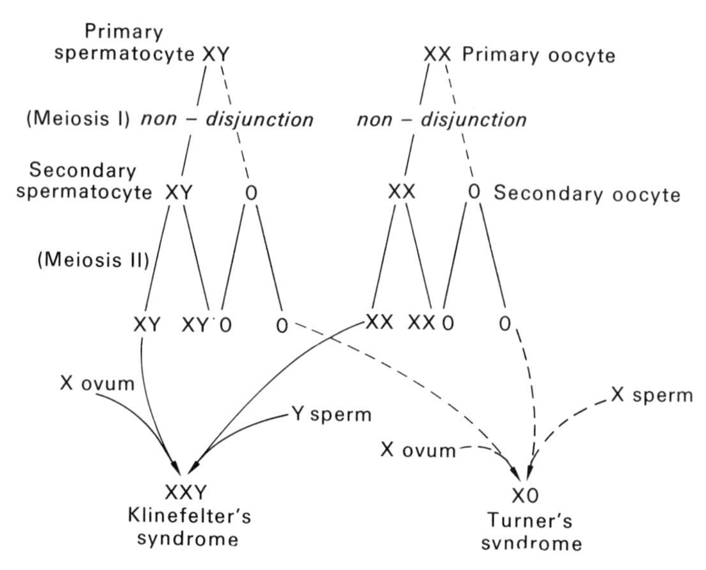

Fig. 8.41 Diagram of the way in which Turner's and Klinefelter's syndromes could arise by non-disjunction during either spermatogenesis or oogenesis.

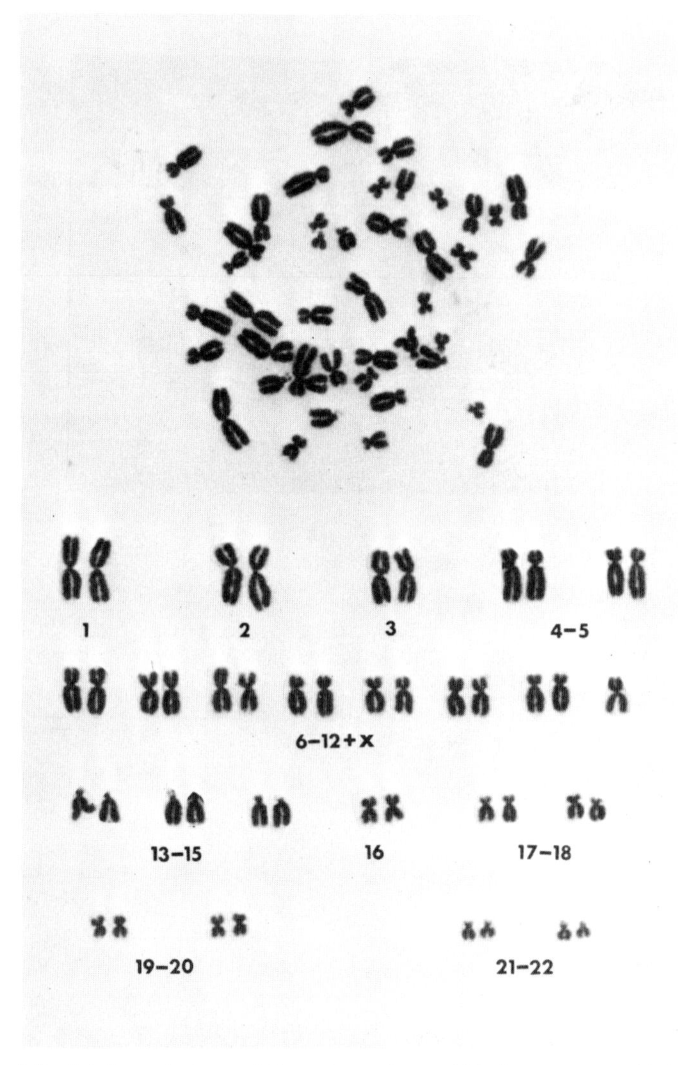

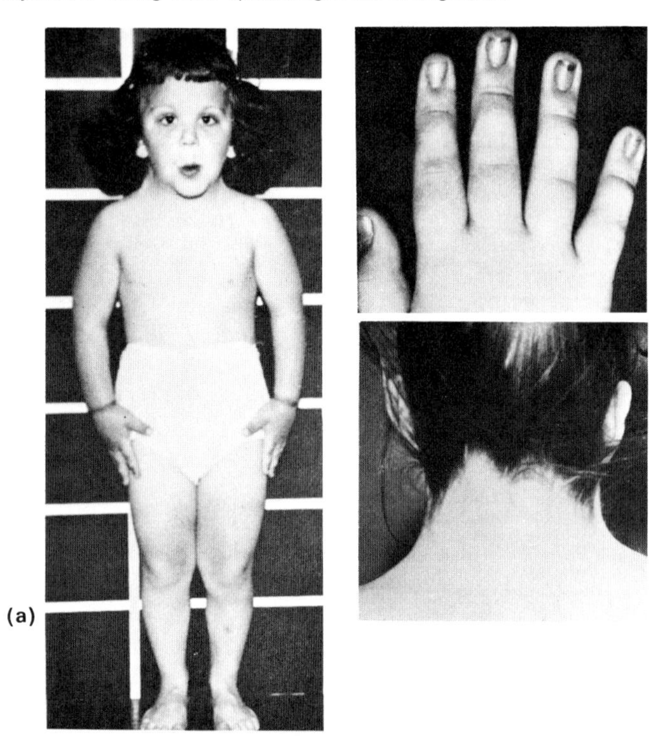

Fig. 8.42 (a) Some features of Turner's syndrome. This four year old girl shows a reduced stature and the characteristic residual lateral web neck. The low posterior hair line and puffy fingers with deep set, hyperconvex finger nails are also common features associated with the XO chromosomal complement. (Courtesy of Professor D. W. Smith, and W. B. Saunders and Co.) (b) the karyotype of a Turner syndrome patient showing a single X chromosome in the total complement of 45. (Courtesy of the MRC Clinical Genetics Unit, Edinburgh.)

The Genetic Basis of Evolution

In this chapter the work of the previous three chapters is exploited in order to develop the concept of evolution. As elsewhere in the book an historical introduction is made in order to account for the dichotomy that has existed between genetics and evolution until relatively recently. The chapter then broaches one of the areas that is too often ignored in school biology programmes, namely the behaviour of genes in populations. The mathematics involved here is simple, well within the scope of anyone who has done the subject at O Level, and centres around binomial expansion. It is important, however, and is the key to understanding how populations evolve.

The rest of the chapter deals with some of the ways in which variation can arise in the genetic makeup of populations and discusses how such variations can be continuous, that is show a spectrum effect like weight in man, or discontinuous, that is showing clear categories, like blood groups. The chapter concludes with a discussion of the origins of races and the step from races to species. In this way the origin of species is described from the springboard of genetics.

9.1 The Historical Development of Evolutionary Theory

Of all the topics broadly encompassed by 'The History of Science' there is one which most biologists find utterly fascinating. It concerns the development of evolutionary theory in Victorian England. The interest we have in this chapter in the development of biological thought lies partly in the debates that centred on the concept of evolution and which drew together, as colleagues or combatants, many of the most vigorous intellectuals of the day.

Such was the widespread interest aroused by the publication, in November 1859, of *The Origin of Species* by Charles Darwin (Fig. 9.1), that the chroniclers of the day have recorded faithfully many of the attitudes taken by both scientists and non-scientists and the debates that followed and often raged. Much has been written about Charles Darwin and his work. The attractive account by J. Huxley and Kettlewell (1965) and Darwin's autobiography (written as *Recollections for his family* and published posthumously by his son, Sir Francis Darwin in 1887) paint a somewhat confused, almost contradictory, picture of a man whose life was dominated by a single idea yet whose vacillation and procrastination almost lost the reward and acclaim that were his due.

After an undistinguished school career at Shrewsbury and as a student at the universities of Edinburgh and Cambridge he was

Fig. 9.1 Charles Darwin (1805–82) from a lithograph by T. H. Maguire done when Darwin was about 44 years of age. (Courtesy of the Wellcome Trustees.)

offered a post as unpaid naturalist on the survey ship *HMS Beagle* (Fig. 9.2). This vessel left Plymouth in December 1831 for the South Atlantic where extensive charting of the East Coast of South America was carried out. It then sailed through the Straits of Magellan and northwards along the coast of Chile to the Galapagos Islands, on the equator. From there the ship sailed westwards to Tahiti, New Zealand, Tasmania and Cape Town, returning to Falmouth almost five years later, in October, 1836.

During the first four years of the voyage Darwin was principally concerned to add to his knowledge of geology. His guideline was Lyall's classic text *Principles of Geology* and he made extensive rock collections upon which he later wrote three works. But in the last year of the voyage a five week survey was made of the Galapagos Islands and here Darwin's attention turned to biology. He was struck both by the similar features shared by the island and mainland flora and fauna and by the characteristic island patterns shown by some animals, particularly the tortoises and finches (Fig. 9.3).

It was in the Galapagos Islands that Darwin, possibly primed by the notion expressed by Lyall in his book that the geological landscape had changed over millenia, and was still changing, became convinced that the concept of the fixity, or immutability, of species was wrong. Within a few months of the completion of the voyage Darwin began to compile his biological data in a notebook which he termed *The Transmutation of Species*, and by 1839 his theory of evolution by natural selection was clearly conceived.

To a modern scientist it seems astonishing that Darwin should have delayed publication of his views about the mutability of species for a further twenty years and it is the reasons for this remarkable delay that hold particular interest for the modern biologist equipped with an understanding of the genetic basis of evolution.

Biographers have attributed the delay to Darwin's ill-health. In South America he almost certainly contracted Chagas' Disease, a

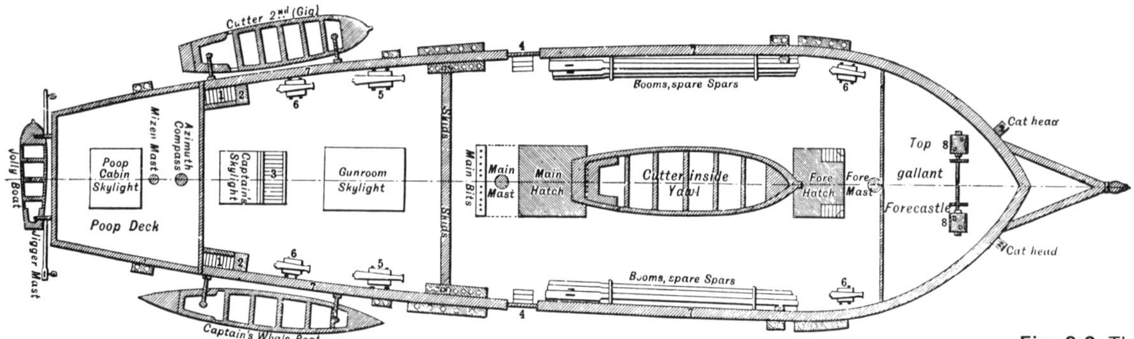

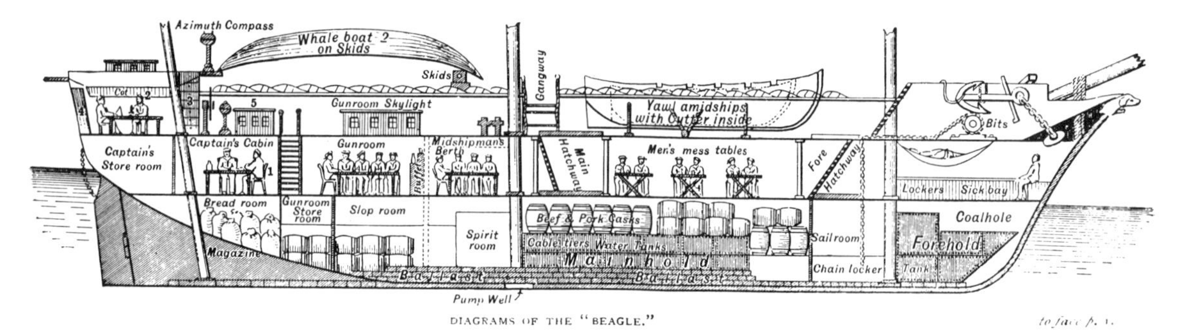

Fig. 9.2 The plan of the Admiralty survey ship H.M.S. Beagle.

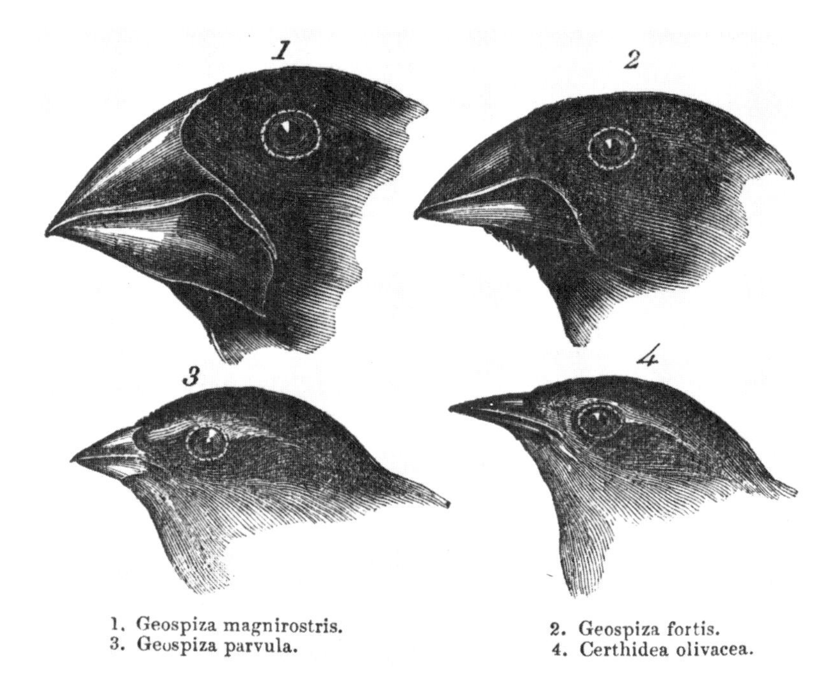

Fig. 9.3 The Galapagos finches provided Darwin with an example of divergence from a single species and of adaptation to different habitats. (From Darwin's *Journal of Researches*, 1845, by courtesy of the Wellcome Trustees.)

trypanosomal parasite of the blood transmitted by the 'Great Black Bug of the Pampas' *Triatoma infestans*, specimens of which he had allowed to feed on his arm in order to study its feeding habits. From 1837, for the remaining forty-five years of his life, he was a chronic invalid; as he wrote, 'I was not able to do anything one day in three'. It has been said that his invalidism was largely psychoneurotic. Huxley and Kettlewell suggest that '. . . once tiredness and unpleasant symptoms had developed, from whatever cause, it would be very easy for his subconscious to take refuge in invalidism in order to escape from the pursuit of social or professional duties . . .'.

Nevertheless, he was able to work during this period. His geological collections provided materials for *The Structure and Distribution of Coral Reefs* (1842), *Volcanic Islands* (1844) of which a short extract is given in Fig. 9.4 and *Geological Observations on South America* (1846) and he published two important volumes on barnacles. In

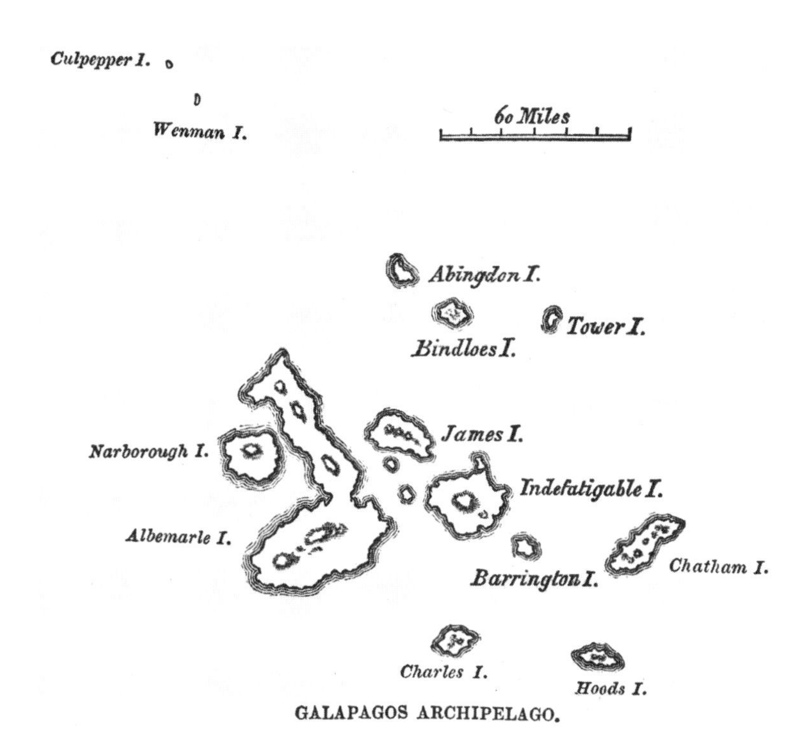

CHATHAM ISLAND. *Craters composed of a singular kind of tuff.*—Towards the eastern end of this island, there occur two craters, composed of two kinds of tuff; one kind being friable, like slightly consolidated ashes; and the other compact, and of a different nature from any thing, which I have met with described. This latter substance, where it is best characterized, is of a yellowish-brown colour, translucent, and with a lustre somewhat resembling resin; it is brittle, with an angular, rough, and very irregular fracture, sometimes, however, being slightly granular, and even obscurely crystalline: it can

Fig. 9.4 An extract from Darwin's *Geological Observations on the Volcanic Islands*, 1844. (Courtesy of the Wellcome Trustees.)

view of the fact that Darwin considered the Origin of Species to be '. . . the chief work of my life' it seems unlikely that ill-health alone could have accounted for the continued postponement of publication.

Another reason for the delay may have been his aversion for the probable controversy that would follow publication. Certainly he had never shown great resolution. At the universities he had failed to qualify either in medical studies or later in Holy Orders. He had meekly submitted to his father's objections against his joining the *Beagle* and had not his uncle Josiah Wedgwood pleaded his case for

Fig. 9.5 Thomas Henry Huxley (1825–95). (From a photograph in the Wellcome Institute, by courtesy of the Trustees.)

him (in much the same way as T. H. Huxley (Fig. 9.5) was later to argue Darwin's case against the Bishop of Oxford, 'Soapy Sam' Wilberforce, in the notorious Monkey Debate at the British Association meeting of 1860) he would never have embarked on that historic voyage.

In his *Recollections* Darwin says that 'much was gained and nothing was lost by the delay'. He felt that any opinion was unlikely to carry weight with his scientific contemporaries unless supported by as great a body of data as could possibly be amassed. This view may in part be attributed to his earlier mistake in declaring that the Parallel Roads of Glen Roy were raised marine beaches. His self-confidence and his scientific credibility in the eyes of his colleagues must have been damaged by the subsequent demonstration that the beaches marked the margins of a glacial lake.

There were, however, further solid grounds for the delay in publishing. The evidence he had amassed on the *Beagle* voyage satisfied him that evolution had taken place but was not in itself evidence for the manner in which it occurred. The transmutation of species was by that time fairly widely accepted in European scientific circles, but in Darwin's mind the evolutionary mechanism, namely **natural selection**, was largely conjectural. The relationship between sources of variation in populations of species and the inheritance mechanism must also have affected Darwin's thinking about the way in which evolution took place and may have been a further factor contributing to his hesitation about publishing *The Origin of Species*. Darwin, in company with many of his fellows, almost certainly believed that inheritance followed what has been termed the 'blending' pattern, that is the offspring of a cross represented an amalgam of the characters of both parents. This opinion was superseded by the 'particulate' theory of inheritance upon which Mendelian genetics is based.

Blending inheritance can be visualized in terms of a number of pots of red and white paint, representing two variant forms of an organism. The mixing of two red pots would give red (offspring) and likewise with white. But the mixing of red and white gives pink. Since the mixing of pink with pink gives pink also, it can be seen that the random mixing of the variant stock would give rise to a uniformly pink state in relatively few 'generations'. Given such conditions, Darwin may have argued, natural selection must act with great speed. But the geological evidence available to Darwin showed the opposite to be generally the case; that evolutionary change occurred over time spans many orders of magnitude greater than the generation times of the species concerned. This was Darwin's dilemma and surely lay at the root of his unwillingness to publish his concept of the evolutionary process.

J. S. Huxley (1942) has described the Darwinian concept in terms of three facts and two deductions:

Fact 1. A species tends to increase its numbers exponentially.

Fact 2. The actual size of a population of a species remains fairly constant over long periods of time.

Deduction 1. Therefore there must be a struggle for existence since not all the individuals produced can survive.

Fact 3. There is variation within every species.

Deduction 2. In the struggle for existence those variants better adapted to their environment leave behind them proportionately more offspring than their less well adapted contemporaries.

The major scientific loophole in the hypothesis lay in the absence of evidence bearing on the origin of variation in populations. Only the arrival of a letter from Alfred Russell Wallace (Fig. 9.6) outlining a hypothesis of transmutation of species by natural selection, identical in principle to Darwin's own, stimulated Darwin into publication. Darwin and Wallace presented joint papers at a meeting of the Linnean Society in July 1858. Within thirteen months Darwin had finished *Origin of Species* and the initial printing of 1250 copies was sold out on the day of publication. Wallace conceded the prior claim of Darwin and his major work containing, among other things, data from his collections in the Malayan archipelago, he generously entitled *Darwinism* (1889).

When the *Origin of Species* was published the popular imagination was caught by the idea of man's evolution from a primate ancestry rather than from a specially created breeding pair, Adam and Eve. But in *Origin of Species* Darwin studiously avoided any mention of man, with the exception of a short, prophetic sentence in the penultimate paragraph of the book, '*Much light will be thrown on the origin of man and his history*'. When Darwin published the *Descent of Man* in 1871 the public fury was largely spent and the book was received quietly.

The major loophole, **sources of variation in populations**, was always open to scientific criticism and provided a platform for the sceptics, like Louis Agassiz and Sir Richard Owen. Variation became an important issue in biological thinking during the last three decades of the century. Bateson (1894), of Cambridge (Fig. 8.12) published a 600-page treatise entitled *Materials for the Study of Variation* and in the introduction he posed the question of what laws govern the form of living things. He added, '*It is more than thirty years since the* Origin of Species *was written, but for many these questions are in no sense answered yet . . . In the present work an attempt is made to find a way of attacking the problems afresh . . .*' But Mendel's discovery of the particulate nature of inheritance had yet to be re-discovered. Since the genetic basis of variation in populations was quite unknown it is not surprising that Bateson's 'attack afresh' should founder and his book amounted merely to a catalogue of examples of variation ranging from segmentation in annelids and arthropods to teeth, colour markings and digits in vertebrates.

With the acknowledgment of Mendel's work, Bateson switched to genetics and made remarkable pioneering contributions, notably in the study of linkage. It is interesting to note that in 1905 Bateson defined the science of genetics as '. . . *the elucidation of the phenomena of heredity and variation*'.

In its modern synthesis evolutionary theory is regarded principally in terms of population genetics. A population is said to be evolving if the genes which are contained in the population show consistent changes in frequency in successive generations. When the organisms which constitute a particular generation of a population reproduce

Fig. 9.6 Alfred Russel Wallace (1823–1913). (Courtesy of the Wellcome Trustees.)

sexually the genotypes of their gametes can be thought of collectively as a **gene pool**. From this pool of genes the gentotypes of the next generation are derived randomly. The individual organisms of one generation, therefore, can be regarded as samples of the gene pool. It is upon the phenotypes, that is the physical and physiological characters, produced in these samples that natural selection acts. Those phenotypes which compete most successfully in the ecological framework of the population (in the Darwinian sense are **best fitted** to their environment) are more likely to contribute their genotypes to the gene pool from which the next generation is drawn. It is the natural selection of phenotypes, in Darwin's own concept, which provides the direction in which the gene frequencies move (Fig. 9.7).

9.2 Genes in Populations

Success in an evolutionary sense depends upon survival to, and through, reproduction. If the size of an animal population remains relatively constant against its tendency to increase exponentially, its members are in constant competition for food and territory and those organisms which are most successful in respect of structural, functional and behavioural factors are most likely to find mates and reproduce successfully. Post-reproductive factors have little evolutionary significance; there can be no natural selection for longevity, for example. Competition is equally demonstrable in plant populations, particularly with respect to light, and the same principle of post-reproductive redundancy holds.

Consider a gene A and its mutant allele a which shows some penetrance in the heterozygote. Under certain environmental conditions the phenotype exhibited by the AA organisms may be more successful than the Aa organisms, which in their turn are more successful than the aa organisms, or

$$AA > Aa > aa$$

The absence of the detrimental a gene means that the AA organisms, being more successful in competition with the Aa and aa members of the population, feature more prominently at reproduction. Therefore gametes carrying the A gene will form a larger proportion of the total gametes produced by the population. It follows that the gene pool from which the following generation is drawn will contain a higher frequency of A and a correspondingly lower frequency of a. If the environmental conditions remain un-

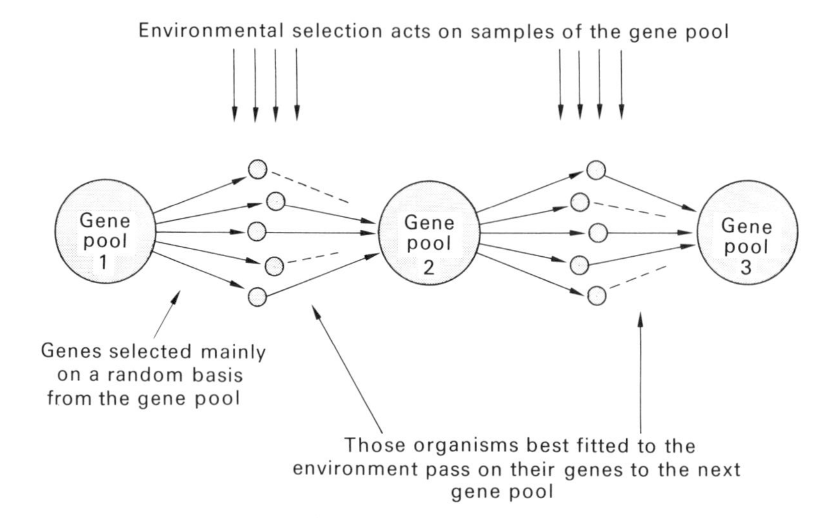

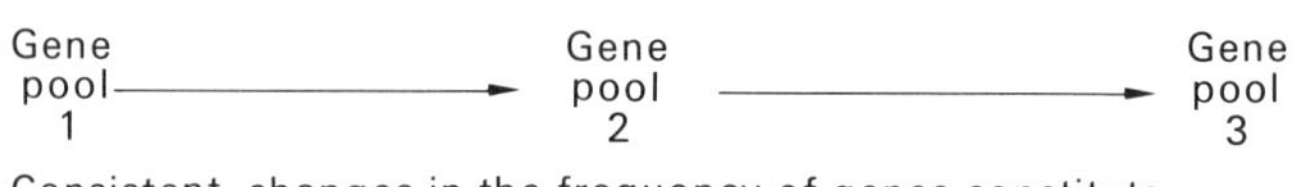

Fig. 9.7 The gene pool concept. A population of a species is said to be evolving when there are consistent changes in the constituents of the gene pool from generation to generation. The direction of the changes is dictated by environmental selection pressures.

changed selection against the a gene will continue and its frequency will decline in successive generations. In practice, many such deleterious genes are not, as might be expected, removed from the population. They tend to remain at a relatively constant frequency corresponding to a balance between elimination by competition and appearance by mutation.

Where a disadvantageous allele is of such low penetrance that the heterozygote is indistinguishable from the dominant homozygote it is likely to remain at a higher frequency in the population. In such a situation, which can be described as:

$$AA = Aa > aa$$

the frequency of the a allele will decline until the occurrence of the aa genotype is rare. At this order of frequency the proportion of heterozygotes in the population is also low, so that the probability that mating will occur between two heterozygotes (and thus give rise to homozygous recessives in an expected one-quarter of their offspring) is extremely low. In this manner a deleterious gene of low penetrance can be maintained in a population by means of **carriers**, or heterozygotes, at a higher frequency than a deleterious gene of higher penetrance.

In some cases a heterozygote may be more successful, having a greater adaptive value, than either homozygote. This condition is known variously as **heterosis**, **heterozygote superiority**, or **hybrid vigour**, and can be described as:

$$AA < Aa > aa$$

Natural selection under such circumstances results in equilibrium whereby both alleles A and a are established permanently in the population. If the three phenotypes produced by AA, Aa and aa are distinct a situation known as **balanced polymorphism** can arise whereby the three phenotypes exist in equilibrium in successive

Fig. 9.8 The effects of inter-strain hybridization in maize. (above) The two plants at the left are the parental strains and the first tall plant is the hybrid produced as a result of a cross between them. Successive inbreedings from the hybrid, however, show that the stature of the parental strains is reached after about four generations. (left) The effect of hybridization on yield. At the top of the photograph the four cobs are from inbred, parental strains. The two cobs in the middle are the F1 hybrids, showing an improved yield. The bottom row of cobs shows the still better yield produced from the inter-hybrid hybrids. (Courtesy of the Connecticut Agricultural Experiment Station.)

generations (p. 211).

It has been recognized since ancient times that in-breeding of domestic animals leads to weakness in the stock. Plant and animal breeders, by deliberate and planned out-breeding, increase levels of heterozygosity of a wide range of alleles in their stocks and consequently increase productivity and yields.

Some of the most remarkable improvements in plant breeding have been achieved in the United States in the development of hybrid maize. In 1909, Shull produced in-bred strains of maize by repeated self-pollination. By crossing these strains an **inter-strain hybrid** was formed in which the yield exceeded that of either parental strains before in-breeding began. This technique was modified in 1917 by the **double-cross** method (Jones, 1925). In this case four

	Sample of 130 offspring	
	Parental properties	Hybrid backcross
Size of litter	8	10.6
Number reared	7.4	9.8
Number of litters per year	1.8	2.15
Days to bacon, from birth	184–193	174
Food conversion ratio	4.7	3.1
Live weight at bacon age (lb.)	195	195
Carcase quality, Grade No. 1	49%	82%
No. 2	12%	12%
Grade C and overweight	39%	6%

Fig. 9.9 The advantages of controlled heterosis to the pig-breeder are obvious when a comparison is made between hybrids and parental types. (Data courtesy of Jackson's Marketing Services.)

parental in-bred strains were used. They were cross-pollinated in pairs to produce two inter-strain hybrids which were themselves cross-pollinated to produce an **inter-hybrid hybrid**. The size of the double-cross hybrid plants was increased as were the yields of corn they produced (Fig. 9.8).

Inter-strain, or inter-breed, hybridization programmes have also produced remarkable results in pig breeding. Crosses have been used involving the Landrace breed of pig, a Danish strain having a long back and therefore of value in bacon production, and the English Large White breed, which is more robust and has a deeper rib cage. Both breeds have similar breeding and growth properties, listed as **parental properties** in Fig. 9.9.

Perhaps the most promising results to date have been obtained from backcrosses between inter-breed hybrid females and males of either parental type. The economic advantages of such crosses to the pig grower arise from larger litters and more litters per year, together with a more rapid growth rate in terms of 'days to bacon'. There is also a better **conversion ratio** shown by the offspring, that is the weight of feed meal per unit of live weight gained. The final carcase of the offspring has a much improved average quality; over 80% grade as No. 1 carcase compared with about 50% in the case of the parental breeds.

In this example of pig breeding heterosis may be acting at two levels. At the first the inter-breed hybrid females, by virtue of their increased vigour, may be providing a better intra-uterine environment, which leads to larger litters. At the second level the hybridization is reflected in the greater growth rate, better conversion ratio and improved carcase quality of the offspring.

Heterosis can occasionally result in the maintenance of deleterious alleles at unexpectedly high frequencies in a population. **Sickle cell anaemia** in man is caused by mutation of a gene which codes for haemoglobin synthesis. The mutant allele codes for a slightly altered amino acid sequence (p. 204) in one of the polypeptide chains of the haemoglobin molecule. A person homozygous for the mutant allele shows symptoms of acute anaemia and his red cells exhibit a characteristic collapse, or sickling, at reduced oxygen tensions (Fig. 9.10). Other features of the phenotype, which are probably pleiotropic effects, are unusual lengthening of the limbs and physical and probably mental retardation. Death from severe anaemia and side effects usually occurs before reproductive age.

This lethal mutant allele has a low degree of penetrance. The heterozygous individual ($+h^s$) shows slight symptoms of anaemia and none of the pleiotropic effects described above. His red cells show sickling only at significantly lower oxygen tensions. Heterosis, or heterozygote superiority, occurs in areas where the malaria parasite, *Plasmodium falciparum*, is present. The mechanism whereby the red cell of the heterozygous individual is made less vulnerable to attack from the parasite is not fully understood, but has been demonstrated by Allison (1954, 1956). This situation can be described by:

$$++ < +h^s \gg h^s h^s$$

As a result of the superiority of the heterozygote the frequency of the

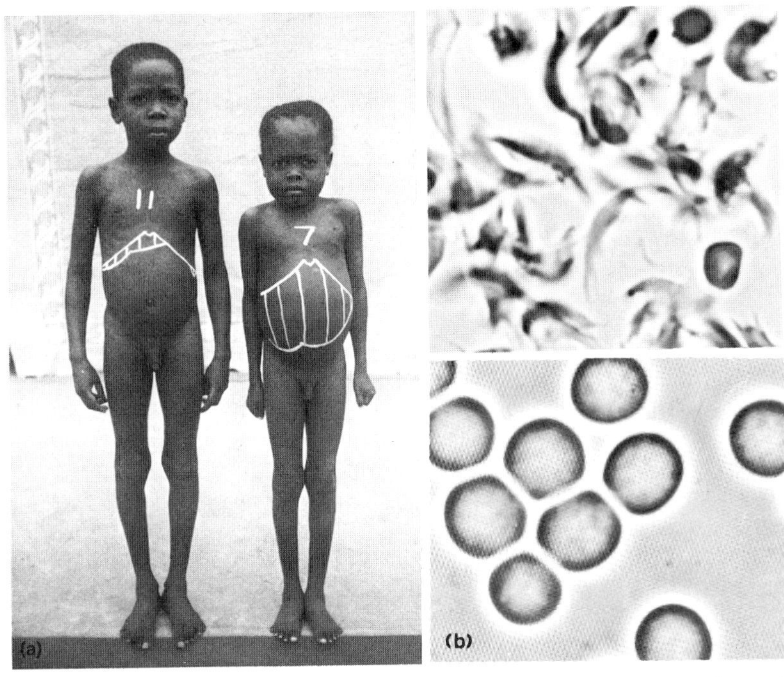

Fig. 9.10 (a) Some symptoms of sickle-cell anaemia. In these Bugandan boys of 9 years of age the one on the right (No. 7) is homozygous for the sickle-cell gene. He shows stunted growth, an enlarged liver and spleen and the characteristic bossing of the skull. (b) Red blood cells from a patient homozygous for sickle-cell anaemia showing the characteristic collapse, or sickling, under reduced oxygen tension (*above*), compared with normal red cells (*below*). (Courtesy of Dr. A. C. Allison.)

lethal gene h^s is higher in malarial regions than in the surrounding regions where the disease is less prevalent (p. 212).

An unusual situation occurs in the evening primrose *Oenothera lamarckiana* which, although largely self-pollinating, is believed to be maintained in permanent heterosis by a system of **balanced lethals**. Homozygous individuals fail to survive, so hybrid vigour is maintained in what amounts to an in-breeding population. The diploid number of chromosomes in this species is fourteen. But instead of existing as seven homologous pairs the chromosomes have undergone **heterozygous translocation**, that is one member of each pair has exchanged a portion of itself with a chromosome of a different pair (p. 189), (Cleland, 1936).

The consequence of this segmental interchange is that the chromosomes can no longer 'pair off' in the zygotene stage of meiotic Prophase I. Gametes are formed, however, by a modified meiosis in which the fourteen chromosomes are synapsed end to end to form a closed circle (Fig. 9.11). At Anaphase I all the paternal chromosomes go to one pole of the spindle and all the maternal chromosomes go to the other pole. There is no 'independent assortment' of genes since there is, in effect, only one linkage group. Only two kinds of gametes are formed, one kind bearing maternal and the other paternal chromosomes. Different sets of lethal genes carried on the chromosomes ensure that all pollen grains carrying one set of chromosomes are non-viable and all ovules carrying the other set are non-viable. Thus the only viable seed results from the union of pollen and ovules bearing different sets of chromosomes, namely the total heterozygotes; the gametes forming the seed are identical with the gametes which united to form the parent plant.

Such a system is both highly evolved and evolutionarily degenerate. The complex chromosomal behaviour achieves permanent hybrid vigour but adopts many of the evolutionary disadvantages associated with asexual reproduction. The random assortment of the gene pool which is associated with normal sexual reproduction and which

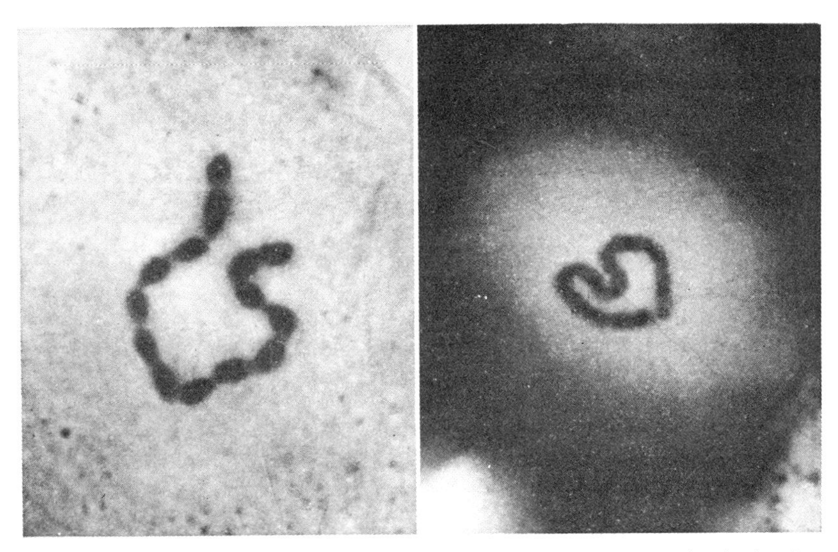

Fig. 9.11 Circle of 14 chromosomes in *Oenothera*. Chromosome number is clear in the left hand cell where the circle has broken open. (Courtesy of R. E. Cleland.)

results in genotypic variation has largely been lost and populations of *Oenothera* would be expected to show a reduced evolutionary capacity for adaptation, either to changing ecosystems or to new ones. Genotypic variation in *Oenothera* is limited to mutation, to other kinds of translocation occurring on the chromosomes (thus achieving phenotypic changes by means of the position effect, p. 147), to the slight possibility of cross-pollination with other sub-species and to polyploidy.

A similar kind of enforced heterosis has been reported in a central American population of *Drosophila tropicalis* (Dobzhansky and Pavlovsky, 1955). In this species **chromosomal polymorphism** is present. Inversions, that is changes in the gene sequence in a portion of a chromosome (p. 184) give rise to two main orders of chromosomes in which only heterozygous inversions are viable. Zygotes which are homozygous are non-viable and although this leads to a loss of half the potential members of the next generation, as in *Oenothera*, the population is nevertheless large and thriving under enforced heterosis at a large number of alleles.

The evidence, therefore, both from observations on wild populations and from breeding programmes on domestic animals and plants, points to the advantages of heterozygosity. The most vigorous populations are generally thought to be heterozygous at a large number of loci. Dobzhansky (1959) suggests that 'good' genes are those which contribute best to the adaptiveness of an organism when they are in a **heterozygous** state, almost irrespective of their effects when homozygous. In most of the organisms which have been studied genetically and most notably the higher organisms, a particular feature of the phenotype is only rarely determined by a single gene. Most of an organism's characters are determined by the interplay of a number of genes, termed **genic interaction**, and the evidence suggests that populations in which the individuals are heterozygous at many gene loci are rendered, through genic interaction, more adaptable to environmental changes. The ability of a population to adapt readily to environmental changes by virtue of the flexibility of its genetic makeup has been termed **genetic homeostasis**, by analogy with the self-regulatory ability of the mammalian body in adaptation to environmental change (Waddington, 1953a).

An important aspect of the discussion on adaptation is **genetic assimilation**, which means the adoption in the population phenotypes of features originally induced by the environment. In an important experiment Waddington (1953b) subjected *Drosophila* pupae to heat shock and noted that some of the flies that emerged were deficient in cross-veins in the wings. These cross-veinless individuals were selected as parents of the next generation. Again the pupae of the next generation were heat-treated and the proportion of cross-veinless flies was found to have increased. The selection procedure was continued over several generations until a cross-veinless strain was produced in which the cross-veins in the wings were absent even when the pupae were not given the heat shock. Waddington discounted the possibility that this result was simply selection of a mutant allele on the grounds that the cross-veinless strain differed from the original stock with respect to several genes and concluded that '. . . it is therefore actually possible to select for the capacity to respond to the environment.' (Fig. 9.12).

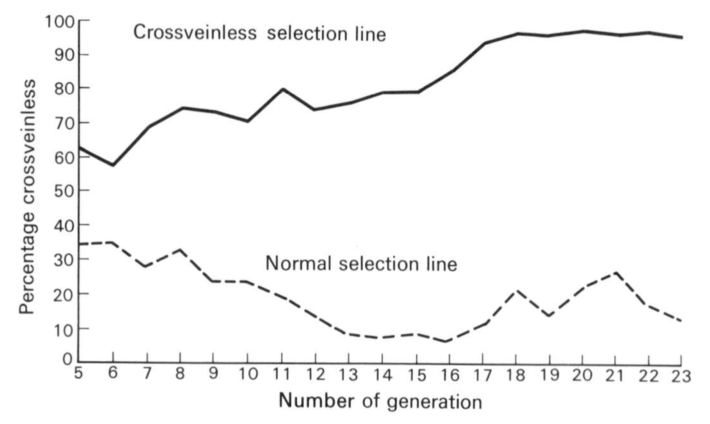

Fig. 9.12 The progress of selection for and against the ability of *Drosophila melanogaster* to produce a crossveinless phenocopy as a response to a temperature shock (from the fifth generation onwards). (After Waddington, 1953.)

A distinction must be drawn between the principle of genetic assimilation by natural selection suggested by this experiment and the inheritance of acquired characters, or **Lamarckism**. The latter principle holds that characters which an individual develops in response to the environment may become incorporated in the genetic data, for example an athlete who trains by weight-lifting and thereby increases his musculature will have children with larger muscles than they would have had if he had not trained. No experimental evidence or observations on natural or domesticated populations have verified the Lamarckian hypothesis. In Waddington's experiment, however, heat shock provided a novel environmental factor to which some individuals, by virtue of their genotypes, were able to respond by producing cross-veinless wings. It was the genetic capacity to respond in this manner that was selected over the succeeding generations and was assimilated into the gene pool. In natural populations genetic assimilation might provide a means whereby a population can quickly colonize a new habitat, adapting to that habitat initially by means of phenotypic plasticity and later by genetic assimilation (see p. 221).

Behaviour of Genes in Non-evolving Populations

The distribution of genes in successive generations of non-evolving populations was described independently by G. H. Hardy (1908) in England and W. Weinberg (1908) in Germany. These observations have been termed the **Hardy-Weinberg Principle**. This principle applies to randomly mating, or randomly pollinating, populations, that is populations in which the possession of certain alleles does not influence the choice of mating. An example in human populations is the blood group alleles, which are non-selective in mating. By contrast, skin colour alleles clearly do affect mating. The Hardy-Weinberg principle also applies only to populations which are not subject to natural selection pressures, to mutation, or to the immigration or emigration of genes from surrounding populations.

The Hardy-Weinberg principle was derived directly from the random nature of the Mendelian cross (p. 138). Consider a pair of alleles, *Aa*. Let the frequency, or proportion, of the *A* gene in the population be termed p and the frequency of *a* be termed q. Since the members of an allelic pair segregate during gamete formation their distribution in the gametes of one sex can be described as $p+q$. Thus the distribution of the alleles in the next generation, being the product of the gametes of both sexes, is $(p+q)^2$.

This binomial can be expanded to

$$p^2+2pq+q^2$$

which represents the genotypes . . .

$$AA+Aa+aa$$

And, since $p+q = 1$,

then $p^2+2pq+q^2 = 1$ (See Fig. 9.13)

Using this equation the frequency of genes in populations can be determined. Consider a human population which has been sampled for the ability to taste the chemical substance, **phenyl thiourea**. The ability to taste this substance appears to be single-gene controlled and tasting (*T*) is dominant to non-tasting (*t*). If, in the sample, non-tasters made up 36% this would represent the proportion of the population whose genotype was *tt*. So, if the frequency of the *T* gene is termed p and the frequency of the *t* allele is termed q, then q^2 in the equation is 36%.

$$q^2 = 0.36$$

$$q = 0.6$$

and, since $p = 1-q$

$$p = 0.4$$

The proportion of the population homozygous for the dominant taster allele (*TT*) is p^2, or 0.16 (16%) and the proportion of the population which is heterozygous (*Tt*) is 2pq, or $2\,(0.4)(0.6) = 0.48$ or 48%.

In summary:

p^2	$+2pq+$	q^2	$= 1$
0.16	0.48	0.36	
(*TT*)	(*Tt*)	(*tt*)	

As far as is known a person does not choose a mate on the basis of

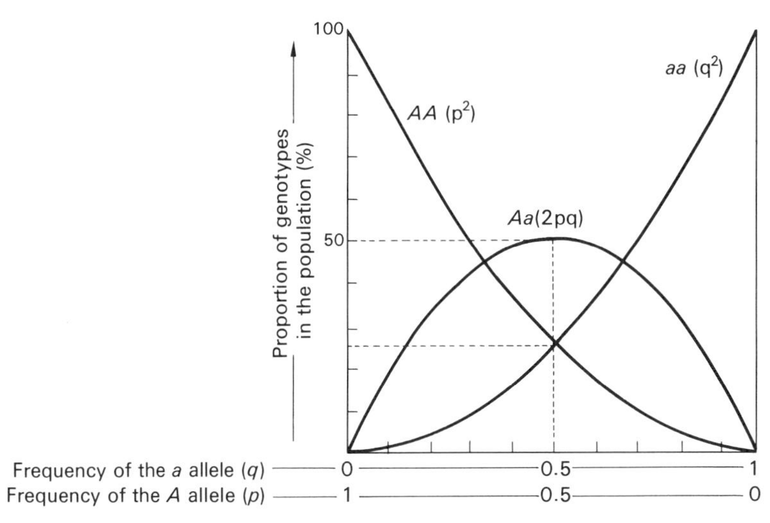

Fig. 9.13 The proportion of genotypes in a population for different frequencies of the alleles according to the Hardy-Weinberg Principle. At all frequencies of the alleles $p^2+2pq+q^2 = 1$. (After Kemp, 1970.)

his or her ability to taste phenyl thiourea, nor does the tasting genotype appear to affect an individual's reproductive capacity. In a human population, therefore, mating is random with respect to these alleles, which would be expected to remain in relatively constant frequencies from generation to generation, a state of balanced polymorphism (p. 211). This can be demonstrated from the hypothetical figures given above.

In any given generation, the proportion of sperm carrying the *T* allele is

$$\begin{array}{ll} 0.16 & \text{(all the gametes of the } TT \text{ males)} \\ +\frac{1}{2}(0.48) & \text{(half the gametes of the heterozygous males)} \\ \hline = 0.4 & \end{array}$$

and the proportion of sperm carrying the *t* allele is therefore 0.6.

The same results are obtained for the *T* and *t* carrying ova produced by the females in that generation. If mating is random the distribution of the *T* and *t* alleles in the next generation is given by the product of the proportions of the gametes:

Ova \ Sperm	0.4 *T*	0.6 *t*
0.4 *T* 0.6 *t*	0.16 *TT* 0.24 *Tt*	0.24 *Tt* 0.36 *tt*

which yields the same frequency of genotypes as in the parental generation, namely 16% *TT*, 48% *Tt* and 36% *tt*.

Changes in Gene Frequency in an Evolving Population: Selection against a Recessive Allele

Consider a case in which a mutant allele of low penetrance becomes disadvantageous because of some environmental change, so that the homozygous recessive organism has little or no chance of survival to a reproductive age. A hypothetical example might be **silver-fur** in the seal. Coat colour in a seal population may exercise no effect on survival or choice of mates until the advent of a new predator, man, who regards silver seal pup fur as highly valuable.

Let us assume that the frequency of the silver-fur gene (s) in the seal population, before predation began, was 0.4 and the frequency of its normal, or wild-type, allele ($+^s$) was 0.6. From the Hardy-Weinberg principle each generation would be expected to produce 16% of silver pups and the frequency of the genotypes in each generation would be:

$$36\%\ +^s+^s + 48\%\ +^s s + 16\%\ ss$$

If all the *ss* pups are culled from the population, however, the s genes entering the next generation do so only from the heterozygous parents. The silver pups in the next generation could only arise from matings between heterozygous animals and would be expected to form one-quarter of their offspring (p. 138).

After the culling of the silver pups the proportion of heterozygotes in the breeding population can be expressed as:

$$\frac{+^s s}{+^s+^s + +^s s} \quad \text{or} \quad \frac{2pq}{p^2+2pq}$$

The probability of a mating taking place between heterozygotes is therefore:

$$\left(\frac{2pq}{p^2+2pq}\right)^2$$

so the proportion of silver pups (ss) appearing in the next generation would be expected to be:

$$\frac{1}{4}\left(\frac{2pq}{p^2+2pq}\right)^2 = \frac{1}{4}\left(\frac{2q}{p+2q}\right)^2$$

Since $p = 1-q$ this can be simplified to

$$\left(\frac{q}{1+q}\right)^2$$

Thus, since $q = 0.4$, the proportion of pups born with silver fur in the next generation would be

$$\left(\frac{0.4}{1.4}\right)^2 = 0.082 \quad \text{or} \quad 8.2\%$$

If the culling were to continue and the homozygous silver pups removed from the population each new generation the frequency of the s gene would continue to decrease. But the rate of decrease would diminish. Selection against a recessive allele by removal only of recessive homozygotes would, in practice, be unlikely to remove the gene altogether from the population since the likelihood of heterozygotes mating becomes progressively less. After n generations in which the recessive homozygous animals are continually removed from the population before they reproduce, the proportion of recessive homozygotes appearing is given by:

$$\left(\frac{q}{1+nq}\right)^2$$

and is shown in Fig. 9.14.

No. of generations (n)	0	1	2	4	8	20
Proportion of silver pups born	.16	.08	.05	.02	.016	.003

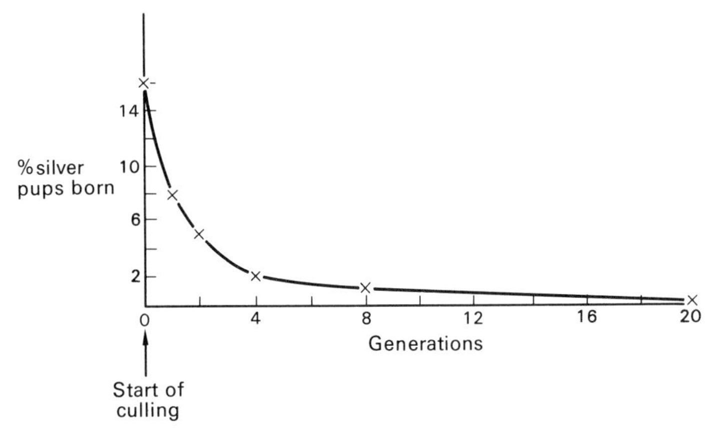

Fig. 9.14 The proportion of silver seal pups born in a population over 20 generations when they are deleted from the population before reaching reproducing age. In practice it is impossible to remove a recessive gene from a population by elimination of the homozygotes.

Random Changes in Gene Frequencies in Populations

As described in the previous sections the tendency of gene frequencies in successive generations to remain constant, as elaborated in the Hardy-Weinberg principle, applies to randomly mating populations which are not subject to natural selection pressures. Such populations must also be large. Wright (1931, 1948) has argued that in small populations gene frequencies would fluctuate to a greater degree than in large populations. Under such conditions one member of an allelic pair could be eliminated from the population and the other member **fixed**, that is assume a frequency of 1.

Kerr and Wright (1954) set up 96 small, isolated populations of *Drosophila*, each consisting of four males and four females. In each population there was, initially, an equal proportion of the sex-linked recessive gene **forked**, which produces forked bristles, and its wild-type allele which produces normal bristles. From the progeny of each generation a further four males and four females were selected at random to form the next generation. In this way the populations were maintained for sixteen generations in a constant laboratory environment.

After sixteen generations only twenty-six out of the ninety-six populations were found to contain both alleles. Of the remainder forty-one contained only the wild-type allele (that is the forked allele had been eliminated) and twenty-nine contained only the forked allele.

This chance change in gene frequency has been termed **genetic drift** by Wright. Clearly the effect of genetic drift is greater in smaller populations. In the populations of *Drosophila* used by Kerr and Wright the size of each population, eight flies, was considerably smaller than any stable natural population is likely to be. Eight flies would arise from sixteen gametes which can be regarded as having been selected randomly by the experimenters from the 'gene pool' of the previous generation. By calculating the standard error on the expectation of achieving eight gametes containing one gene and eight containing its allele an indication of the likelihood of a random change in gene frequency can be obtained.

$$\text{standard error (S.E.)} = \sqrt{\frac{(a)(b)}{n-1}}$$

where a and b represent the number of gametes and $n = a+b$. Thus

$$\text{S.E.} = \sqrt{\frac{8 \times 8}{15}} = \text{approximately } 2$$

Any deviation from the expected selection of eight gametes containing, say, the wild-type allele in excess of $2 \times$ S.E. would be regarded as **significant**. Such a selection would be unlikely to occur in more than about 5% of the times this selection procedure were carried out. In this case it would mean 8 ± 4, or twelve of one gene and four of its allele. On the other hand, a deviation of one S.E. (that is 8 ± 2) might be expected to occur in about two-thirds of the times this selection procedure were carried out. For such a small population of flies one standard error constitutes 25% of the number of gametes. Genetic drift in such a population, therefore, can be regarded in statistical terms as the result of an acceptable, indeed a usual, 'error in sampling'. In a larger population of organisms, say 50 000 individuals, one standard error is 50 000 ± 50 which is only 0.1% of the gamete number.

Since Wright first suggested genetic drift as an evolutionary mechanism, more than thirty years ago, many attempts have been made to demonstrate it in natural populations. To date, however, no differences between natural populations have been ascribed specifically and solely to genetic drift (Stebbins, 1966). The main reason for this is that the smallness of small populations is most likely to be due to their having been in a hostile environment or to their being on the fringes of a population migrating into a disadvantageous and possibly worsening environment. In such conditions the strong selective pressures which are exerted are unlikely to leave any gene unaffected for a sufficient number of generations such that changes in its frequency in the population can be ascribed to chance alone.

It has been suggested that the differences between blood group frequencies in various human populations and races may have arisen as a result of genetic drift when, at an early stage in human evolution, populations were small and isolated (Boyd, 1940). For example, the significant differences between the frequencies of blood groups in two isolated populations of Eskimos have been attributed to the action of genetic drift (Maynard Smith, 1958). Evidence is becoming increasingly available, however, which links blood groups with specific, selective advantages and disadvantages (p. 217) and it is probably no longer necessary to postulate genetic drift to account for differences in blood group frequencies. (For a discussion of blood groups see p. 216.)

The isolation of one or a few organisms which, by chance, differ widely from the norm of the population can give rise to a race in which the variational norm is significantly different from the norm of the ancestral population. This effect, which is akin to genetic drift in that the essential ingredient in the change is **chance**, has been termed the Founder Principle (Mayr, 1954) or the Adam and Eve Principle (Spurway, 1953).

Spurway has examined genetic variation in several species, subspecies and varieties of the European newt, genus *Triturus*. She suggests that gene flow throughout the species is extremely slow because of land barriers between ponds. If migration overland is slow 'new' ponds at the fringes of the population are likely to be colonized by only a few animals, the Adams and Eves of the new population. In the event of a breeding pair, or a gravid female, arriving at a hitherto unpopulated pond they would be the founders of a new population. If, by chance, these founders happened to possess gene frequencies uncommon in the main population their descendants could constitute a new race.

Chance, or random, changes in gene frequencies which become established in localized populations could account for the origin of races, or even sub-species, in cases where a population is expanding into, or colonizing, a new habitat. Under such conditions, however, natural selection pressures are likely to be particularly strong and evidence for the contribution made by random change is hard to obtain. Nevertheless, the possibility that genetic drift, or similar random events, have made significant contributions to the origin of species should not be ruled out entirely.

9.3 Sources of Variation in Populations

1. Phenotypic Plasticity

In the historical discussion which introduces this chapter mention is made of the difficulties which beset Darwin and his contemporaries with regard to the mechanisms whereby variation arose in populations. Continuity and discontinuity are integral parts of Darwinism. But the paradox of **discontinuity**, whereby the offspring of sexually reproducing organisms showed variability, and **continuity**, in which favourable characters were passed on to offspring, was unresolved until the advent of the study of genetics.

Variation is recognizable in all natural populations. It may be physical, when individuals in a population can show variation in size, form and colour, or physiological. Physical and physiological features of organisms form the substance upon which environmental selective pressures act. The gene pool, genetic recombination, mutation, pleiotropy, position effects and so on are theoretical considerations and although organic evolution is interpreted in terms of changes in gene frequency in a population, it is achieved at a practical level by interplay between environmental factors and the **phenotype**.

The phenotype of an organism is seldom determined exclusively by its genotype. Environmental factors can modify the phenotype, a phenomenon which is termed **phenotypic plasticity**:

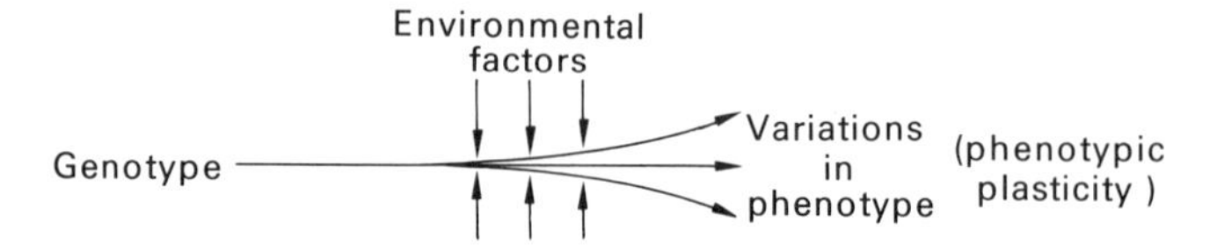

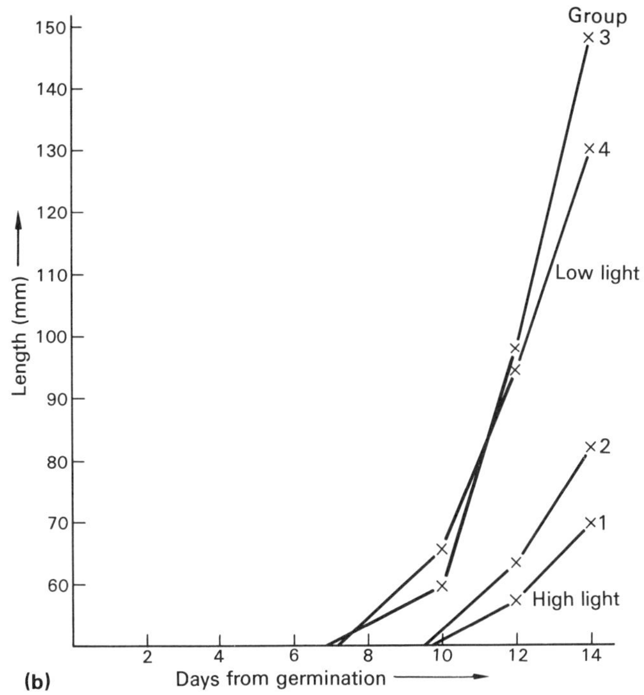

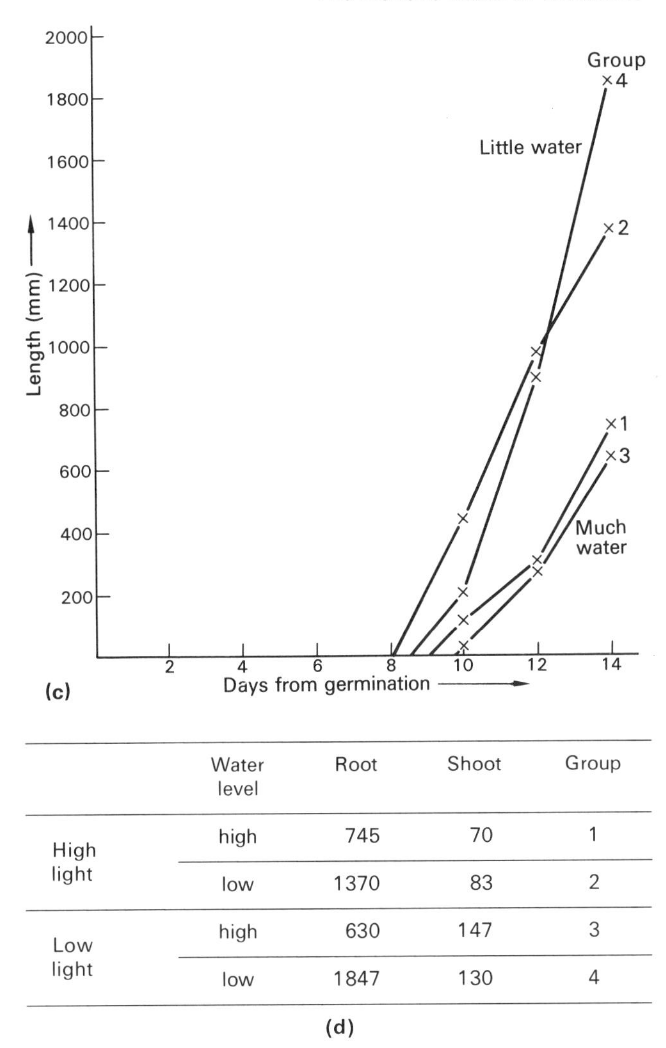

	Water level	Root	Shoot	Group
High light	high	745	70	1
	low	1370	83	2
Low light	high	630	147	3
	low	1847	130	4

(d)

Fig. 9.15 Phenotypic plasticity in pea seedlings. (a) Seeds of the same strain were germinated under conditions of normal light intensity, numbers 1 and 2 numbering from the left, and low light intensity, numbers 3 and 4. Seedlings 1 and 3 grew in abundant water and 2 and 4 in a normal level of moisture. The numbers 1 to 4 correspond to those appearing in the graphs in (b) and (c). (b) Shoot growth in high and low light intensity. (c) Root growth in conditions of high and low water availability. (d) Root and shoot lengths (mm) 14 days after germination in conditions of different availabilities of light and water.

In general plants show greater phenotypic plasticity than animals. Increasing intensity or light tends to retard the rate of growth of shoots, and soil moisture affects root growth similarly (Fig. 9.15).

Phenotypic plasticity is also reflected in the differences in microstructure between leaves from the same plant which develop in full sun and leaves which develop in shade. Sun leaves generally have two or more layers of palisade cells, whereas shade leaves normally have only one. The single layer of palisade cells in the shade leaf can be regarded as a useful economy measure in that at lower light intensities a second deeper cell layer specialized for photosynthesis might be redundant.

In animals temperature is known to affect the expression of a number of genes. The Himalayan fur pattern in the rabbit is due to a mutant gene which forms part of a multiple allelic series (p. 147) on the C locus, which determines the degree of expression of coat colour. The gene for albinism is also a mutant at this locus. A rabbit homozygous for the Himalaya gene has a white body with dark fur

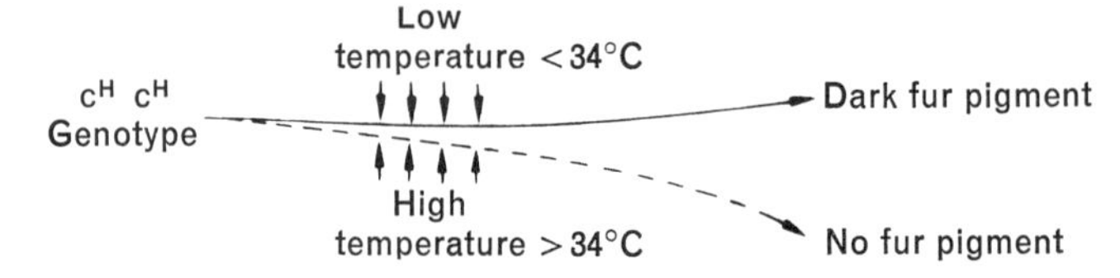

Fig. 9.16 The Himalaya breed of rabbit showing (above) the effect of temperature on coding for fur pigment and (below) the deposition of fur pigment at the cooler extremities.

on the tips of the ears, the nose, the tail and the feet. If fur is plucked from an area on the body and the animal is then kept in a cold room, the fur which grows in the plucked area is dark. Conversely, a young animal reared in a heated room is white, even at its extremities. This suggests that the Himalaya gene can only code for fur pigment at temperatures which are slightly lower than normal body temperature for the rabbit, hence the light body and dark extremities (Fig. 9.16).

A similar kind of phenotypic plasticity is present in Siamese cats, in which kittens raised in cool, out-of-door runs are darker in colour than siblings raised in-doors. In animals phenotypic plasticity probably reaches an exceptional extremity in the environmental determination of sex in *Bonellia* and *Crepidula* (p. 157).

Genotypic expression can also be affected by the internal environment. **Baldness** is more common in men than in women although the gene for baldness is not sex-linked. Its expression appears to be influenced by the levels of sex hormones in the body. Similar factors seem to concern the expression of **gout**. This is a condition in which deposits of uric acid appear in the joints which become inflamed and painful. It appears to be caused by an autosomal dominant gene whose expressivity is greater in men than in women.

In terms of evolutionary adaptability the role of phenotypic plasticity is clear. It provides a degree of flexibility and acts as a buffer between the genotype and the selective pressures which the environment is applying to the population. 'Most genes . . . do not produce stereotyped effects but merely determine the way in which the organism reacts to its environment' (Auerbach, 1962). Phenotypic plasticity enables populations to adapt immediately to temporary vicissitudes of the environment such as the occasional hot, dry summer in a temporate climate which is predominantly cool with a moderately heavy rainfall. Adaptations to long-term, permanent changes in the environment, such as consistent climatic trends, or the establishment of new patterns of drainage in a region following the erosion of new river beds, are achieved by progressive changes in population gene frequencies. Phenotypic plasticity can convey both **variability**, where similar genotypes in a population are subjected to varied environments and **uniformity** where slightly different genotypes exist in the same environment.

2. Sexual or Genetic Recombination

Sexual recombination is the most potent force for variation in populations. It provides for endless assortments of the gene pool and thereby confers upon sexually reproducing species the capacity for relatively swift adaptation to changing or to new environments by means of natural selection. Sexual recombination can be regarded as operating at two levels. The first is by **chromosomal recombination**. This takes place in the first division of meiosis and means that the separation of any given pair of homologous chromosomes is not determined by the manner of separation of any other pairs. The mechanism of chromosome segregation has been described on p. 86 and the principle of independent assortment of chromosomes and the genes they carry is discussed on p. 141. The second level, which is superimposed on chromosomal recombination, is **chromosomal restructuring**.

The strength of chromosomal recombination is related to the number of linkage groups, that is the number of kinds of chromosome, present in the nucleus. The number of kinds of gamete that an organism can produce, based exclusively on the principle of chromosomal recombination, is 2^n (where $n =$ number of linkage groups, or the number of homologous pairs of chromosomes in the diploid nucleus). Thus a *Drosophila* male, in which the diploid chromosomal complement is 8, can produce 2^4 kinds of sperm and a female likewise can produce 2^4 kinds of ova. Thus the number of assortments of chromosomes in the zygotes is 2^8 (Fig. 9.17).

Organism	Diploid chromosome number	No. of kinds of gamete per individual	No. of kinds of offspring per breeding pair
Drosophila	8	2^7	2^8
Cabbage	18	2^9	2^{18}
Mouse	40	2^{20}	2^{40}
Potato	48	2^{24}	2^{48}

Fig. 9.17 Table of the relationship between chromosome number and the number of kinds of offspring obtainable in theory. This degree of offspring variability is increased still further by changes in chromosome structure, such as crossing over.

There is a clear advantage, therefore, in having many chromosomes because this gives greater offspring variability. On the other hand linkage groups are themselves advantageous because of the ways that genes interact with their neighbours (see the Position Effect, p. 147). It would seem, therefore, that organisms have evolved the number of chromosomes that provides the best compromise between the conflicting advantages of increased chromosome number and increased linkage grouping.

3. Changes in Chromosome Structure

Crossing-over

Crossing-over is reciprocal translocation between homologous chromosomes (Fig. 9.18). Changes in chromosome structure due to crossing-over occur as a normal event in meiotic Prophase I. Points of cross-over, or chiasmata, are a regular feature of meiosis and usually occur at least once and often as much as five or six times in each tetrad (see Meiosis on p. 85 and Cross-over Value on p. 143). Since crossing-over is a common feature of meiosis the likelihood of a parental (P1) chromosome passing unchanged into an F2 organism is small. The effect of this, therefore, is greatly to increase the number of kinds of gamete produced by an individual and therefore the number of kinds of offspring genotypes capable of being produced

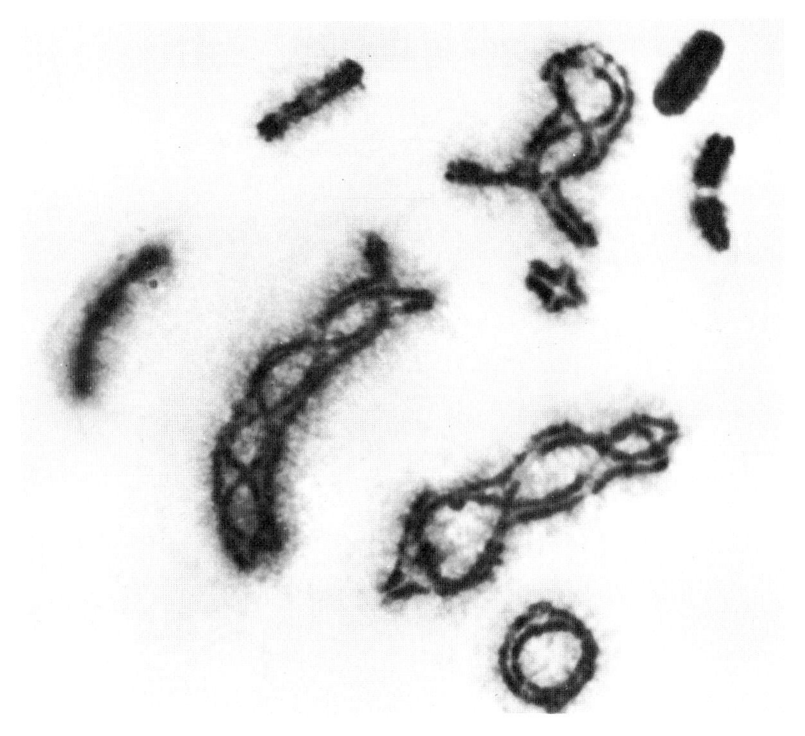

Fig. 9.18 Crossing over in the chromosomes of *Tradescantia* at the late diplotene stage of meiosis. In the longest tetrad there are five chiasmata. (Courtesy of Dr. K. R. Lewis.)

by each mating pair.

As a consequence of crossing-over new linkage groups are formed. The resultant gametic chromosomes, or **recombinants**, have an altered structure by virtue of their having exchanged a portion of their length with a homologous chromosome which originated in a different parent. In evolutionary terms this increases the genetic shuffling in the gametes and tends to randomize the gene pool as it passes to the next generation.

Crossing-over has been described as a common feature of meiosis and a normal mechanism for the disruption of linkage groups. Crossing-over differs from most of the other structural changes which occur to chromosomes in that the number of genes and the nature of the gene loci present in the recombinants remains unchanged; the only result of crossing over is that the genes find themselves with new neighbours. In the other kinds of structural changes in chromosomes **deletion** and **duplication** result in changes in the number of gene loci on a chromosome. **Inversion** and **translocation** result in changes in linkage groups, and alterations in the allelism of gene loci on the recombinants, respectively.

If changes in the structure of chromosomes are to have an evolutionary effect they must be heritable. In order to begin meiosis, and thus enter a gamete, an altered chromosome must satisfactorily undergo **synapsis** in the zygonema stage of Prophase I. Synapsis is the very exact juxtaposition of gene loci which occurs when homologous pairs of chromosomes come together. Therefore any structural change which leads to failure of synapsis would lead to an infertile gamete which would then be eliminated from the gene pool.

Deletion

Deletion is probably the simplest change in chromosomal structure and it involves the loss of a segment of a chromosome either from the end or, more commonly, internally (Fig. 9.19). If the deletion is **heterozygous**, that is it occurs in only one chromosome of a homologous pair, the alleles of the genes in the missing part which may originally have shown low penetrance when heterozygous, are now expressed in the phenotype and are said to be **hemizygous**.

In evolutionary terms deletions are injurious since they represent a reduction in the number of gene loci, and a loss of heterozygosity from a portion of the chromosome, together with the consequential loss of heterosis and balanced polymorphism. In effect there is a loss for the hemizygous region of those advantages conferred by diploidy (p. 89). **Homozygous deletion**, the loss of the same gene locus or

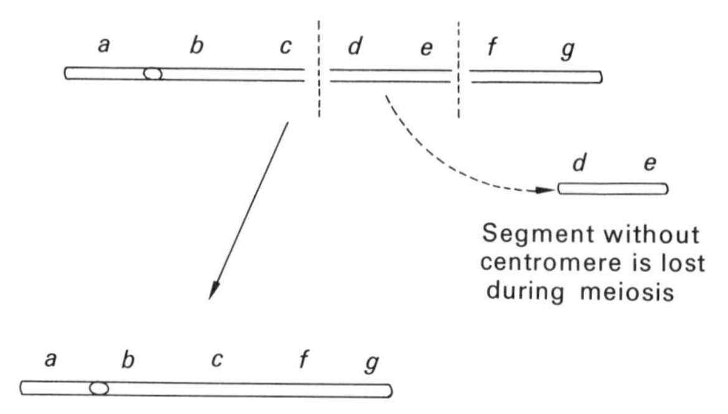

Fig. 9.19 Deletion of a middle portion of a chromosome.

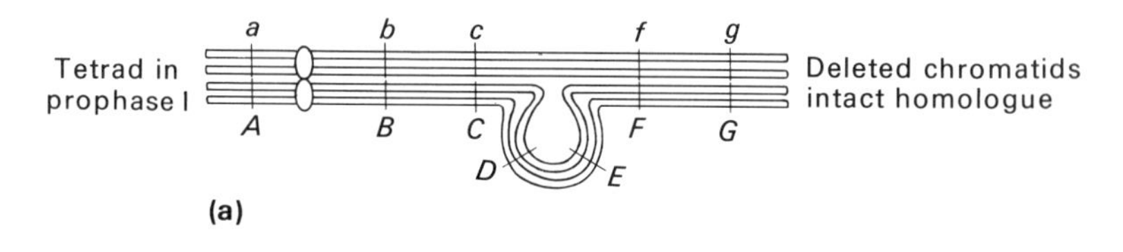

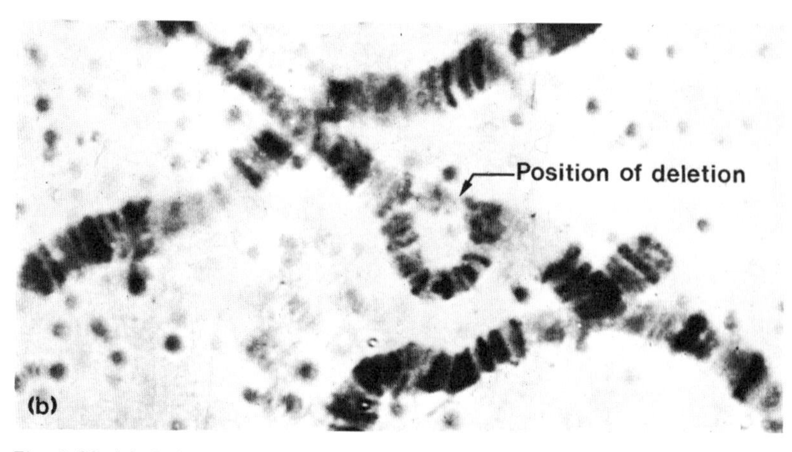

Fig. 9.20 (a) Deletion in a heterozygous organism synapsis of the chromosomes can still take place in meiosis by a looping of the segment of the unaffected chromosome. (b) A situation similar to that in meiosis can be observed in the giant salivary gland chromosomes of *Drosophila* in which the chromosomes synapse and are then multiplied many times to produce the polytene form that is visible in the interphase, or non-dividing cell. In this photograph a heterozygous deletion is clearly visible in chromosome III and the normal homologue forms a loop. (Courtesy Dr. O. G. Fahmy.)

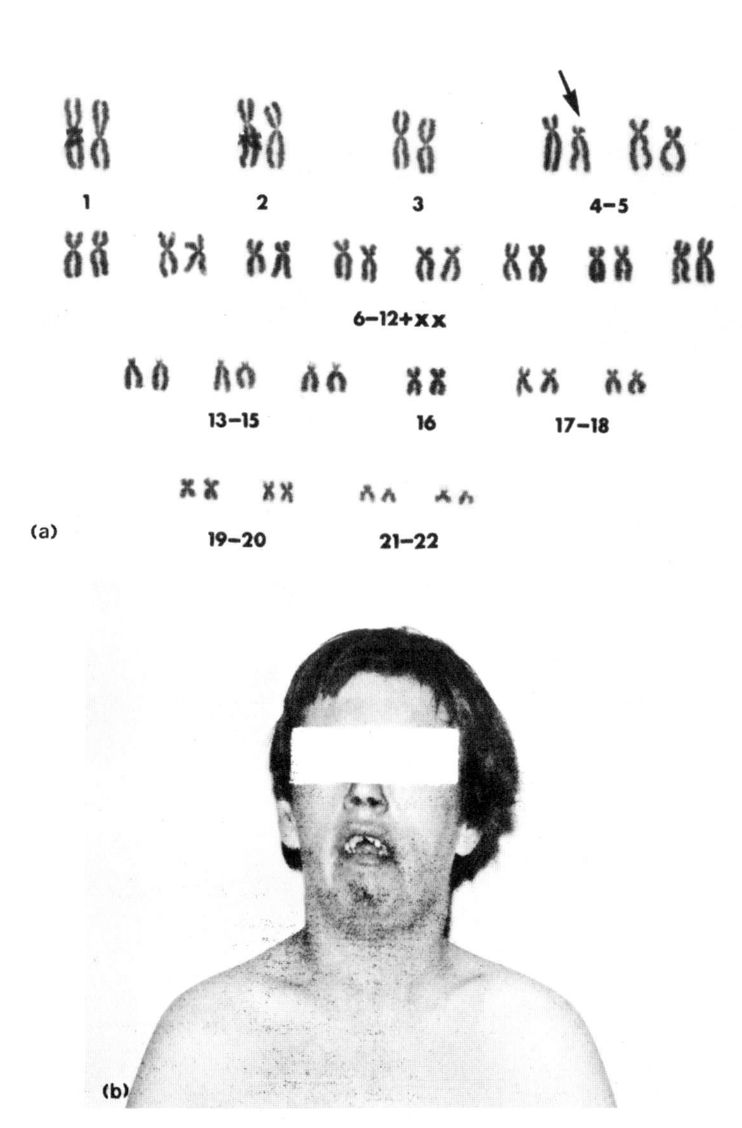

Fig. 9.21 In man deletions are usually lethal and therefore rare. (a) In this example a short-arm deletion of one of the larger chromosomes (arrowed in the karyotype) produces the 'Cri du chat' syndrome. (b) This female patient has severe mental abnormality (I.Q. untestable) and utters a characteristic high pitched mewing cry from which the name of the syndrome is derived. (Courtesy of the MRC Clinical Genetics Unit, Edinburgh.)

loci from both homologous chromosomes, is usually lethal, although homozygous deletion might account in part for the loss of redundant genes from populations which have evolved to occupy new habitats, such as parasitic species. Deletions induced by ultra-violet irradiation have been used extensively to relate linkage groups to cytologically identified chromosomes, particularly in maize (Singleton, 1962).

At meiotic prophase synapsis is achieved between the deleted chromosome and its normal homologue by a 'looping' of the latter (Figs. 9.20 and 9.21).

Duplication

Duplication is the opposite of deletion in that it represents an increase in the gene number by the addition of a portion of chromosome containing one or more gene loci. The additional gene or genes are duplicates of genes that are already present. The effect on the organism varies from slight to severe. In some plant species large duplications and duplications of whole chromosomes can occur without serious effect (see Trisomy, p. 195), but in other cases small duplications can have a serious effect, principally by altering genic balance. In general, the effects of duplications on viability are less severe than those of deletions. Both kinds of structural change can induce variability by altering the balance between genes and by the position effect. In addition, it has been suggested that a duplicated gene locus could be available for mutation to provide additional genes in the gene pool (Stebbins, 1966). Synapsis of a heterozygote duplication occurs by 'looping' in a parallel manner to that in deletion (see Fig. 9.20).

Cross-over analysis of the Bar-eye effect in *Drosophila* led Sturtevant and Morgan to the conclusion that this sex-linked mutant, which behaves like an incomplete dominant gene, was due to a duplicated portion of chromosome. This remarkable piece of work was subsequently confirmed after the discovery of the polytene salivary gland chromosomes (p. 151). By cytological methods Bridges (1936) was able to show that the Bar-eye condition, in which the eye is narrower and the number of ommatidia in it is reduced from about 200 to about 70 in the homozygous state, was due to duplication of a five-banded segment and that the Ultra-Bar, which was an extreme reduction to about 45 ommatidia, was due to a double duplication. In Bridge's words: '*The Bar-eye reduction is thus seen to be interpretable as the effect of increasing the action of certain genes by doubling or triplicating their number—a genic balance effect*' (Fig. 9.22).

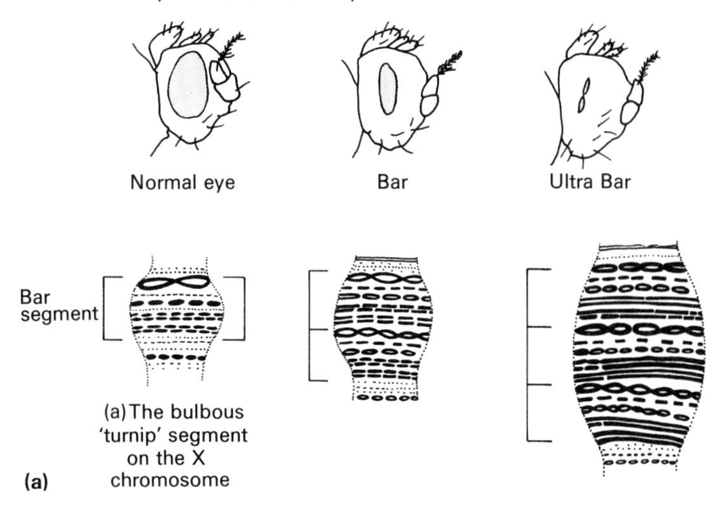

Fig. 9.22 (a) The Bar-eye duplication in *Drosophila*. (After Bridges, 1936.) (b) The Bar-eye in *Drosophila* and the normal eye for comparison. (c) The Bar duplication in the heterozygous state. (Courtesy of Dr. O. G. Fahmy.)

Inversion

Inversion occurs when a segment of a chromosome breaks and rejoins in an inverted sequence.

Inversions probably arise most frequently in the early prophase sequences of both meiosis and mitosis when the chromosomes adopt elongated, slender forms, are frequently looped and appear to be moving about in the nucleus. Chromosomal fragmentation at this stage can also give rise to deletions and duplications, although crossing-over in the pericentric form of inversion can also give rise to deletions and duplications. Synapsis between homologous chromosomes showing heterozygous inversion is achieved by a reverse loop (Fig. 9.23).

The consequences of inversion in terms of variation and evolution are two-fold. First, there is no change in genetic complement; unlike the cases of deletion and duplication, inversion does not result in an altered genotype. However, the sequence of genes is altered across the inverted region, which can give rise to phenotypic variation through the position effect. Second, the genes located in an inverted region, and their alleles on the normal homologue, are always inherited in a block because they cannot be separated by crossing-over. The reason that inversion 'fixes' a linkage group is that if crossing-over occurs between the chromatids in an inverted loop when the centromeres lie outside the loop (called a **paracentric inversion**) it results in one chromatid portion being devoid of a centromere (Fig. 9.24). This is therefore lost during subsequent

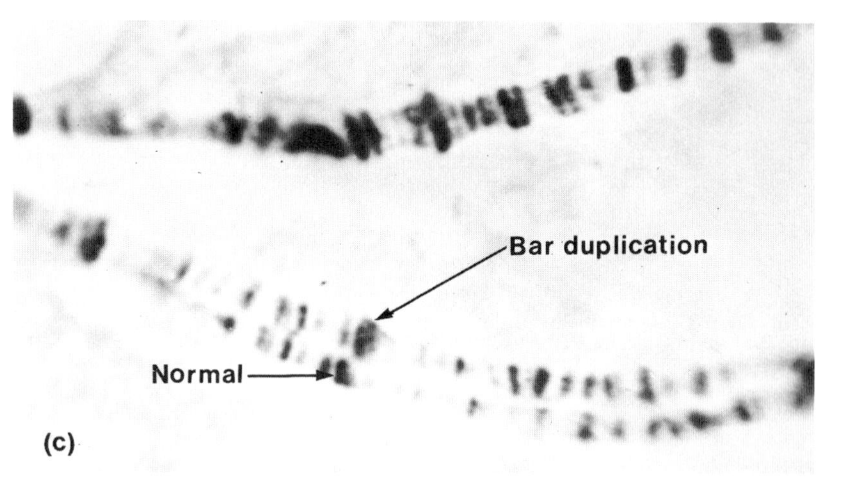

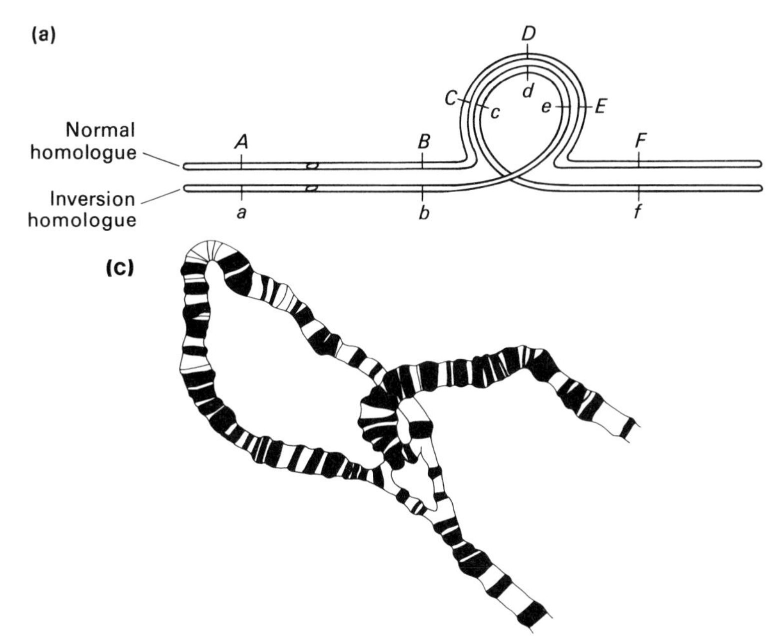

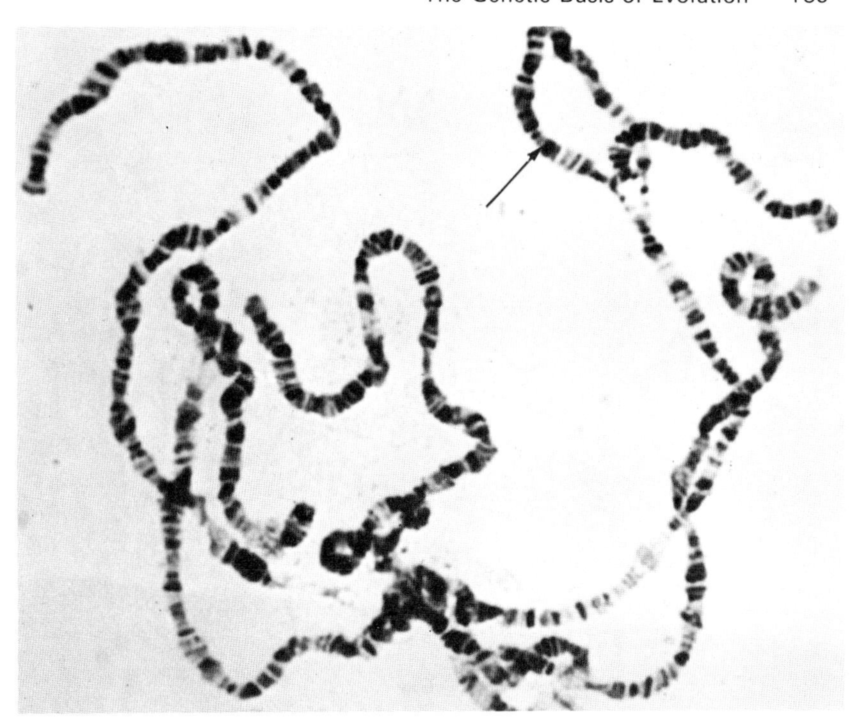

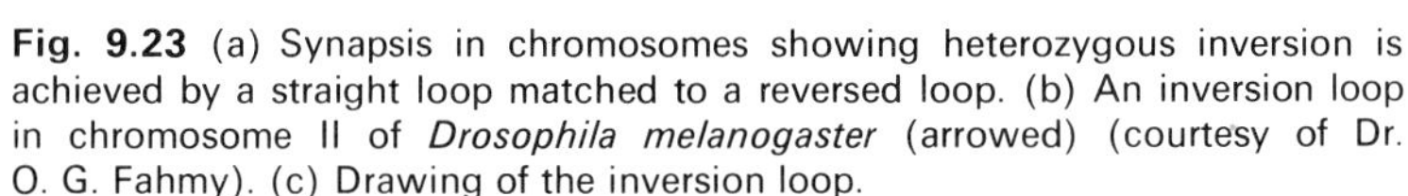

Fig. 9.23 (a) Synapsis in chromosomes showing heterozygous inversion is achieved by a straight loop matched to a reversed loop. (b) An inversion loop in chromosome II of *Drosophila melanogaster* (arrowed) (courtesy of Dr. O. G. Fahmy). (c) Drawing of the inversion loop.

divisions. There is also a second portion which is attached by two centromeres to dyads being pulled to opposite poles of the spindle. If this strand breaks at Anaphase I the daughter gametes which inherit the pieces are likely to be non-viable owing to shortage of genes. Inversion, therefore, can be a mechanism whereby advantageous gene combinations are held together.

When crossing-over occurs in an inversion loop, when the centromeres lie **inside** the loop (termed a **pericentric inversion**) the four gametic chromatids all have a centromere, but the recombinant pair show a deletion and a duplication (Fig. 9.25).

Drosophila has proved particularly valuable as a tool for the study of inversion owing to the presence of polytene chromosomes (p. 151). These giant structures can be prepared for microscopical inspection with relatively little difficulty; heterozygous inversions are identifiable immediately (see Fig. 9.23) and homozygous inversions can be located by examination of the sequence of banding. A number of types of inversion have been identified for Chromosome III of *Drosophila*, which have been given names such as Standard, Arrowhead, Pikes Peak and Chiricahua, and the frequencies of these inversions have been related to natural populations in the South Western United States (Dobzhansky and Epling, 1944). These and related studies reflect the different adaptive properties of the inversion types and demonstrate the evolutionary significance of fixed linkage groups (Fig. 9.26).

The adaptive advantages of inversions in natural populations have also been shown for the Standard and Chiricahua inversions in a population of *D. pseudo-obscura* from the Piñon Flats region of Southern California (Dobzhansky, 1947). During the spring the Chiricahua type appears more advantageous than the Standard type, but this trend is reversed at the height of summer (Fig. 9.27).

In population cages containing insects of both kinds of inversion at high density the competitive advantages of each inversion were related to the ambient temperature. In such mixed populations kept at 16.5°C the frequencies of the inversion types remained fairly constant, but at 25°C the frequency of the Standard inversion rose from approximately 10% to approximately 70% over a period of nine months and the frequency of the Chiricahua inversion declined accordingly. After this period the respective frequencies remained in equilibrium (Fig. 9.28).

These experiments by Dobzhansky and his co-workers suggest that the gene combinations held by the Standard inversion is of greater selective advantage at higher temperatures than that of the Chiricahua inversion. The higher temperature favours an equilibrium in the population of 70% Standard and 30% Chiricahua, whereas at lower temperatures no such selective advantage is shown. Laboratory experiments of this kind also show that evolutionary changes can take place fairly rapidly; they do not always require time spans of the order of the geological time scales that are sometimes associated with the fossil evidence for evolutionary change.

There is now quite good evidence that DDT affects the distribution of inversions in *Drosophila pseudo-obscura*. This insecticide is used in large quantities by farmers and fruit-growers of the fertile central valley of California and the prevailing westerly winds carry it across the Sierra Nevada and as far as Nevada, Arizona and New

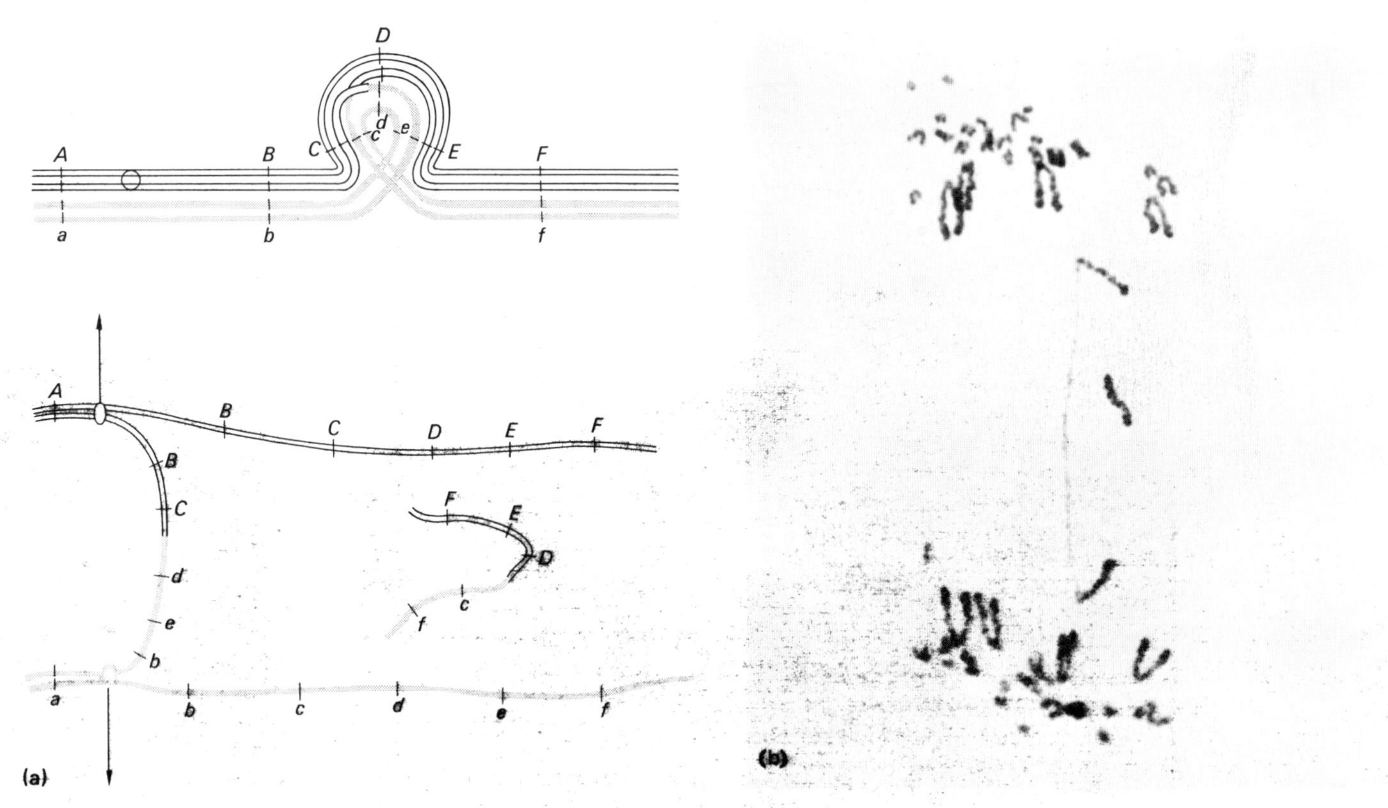

Fig. 9.24 (a) Crossing-over within the loop of a paracentric inversion and the resultant arrangement of the chromatids at Anaphase I. (b) Photomicrograph of failure of disjunction in *Agave stricta* following a paracentric inversion; the stage is Anaphase I of meiosis. The chromosomal bridge and fragment are clearly visible. (Courtesy of Dr. P. E. Brandham.)

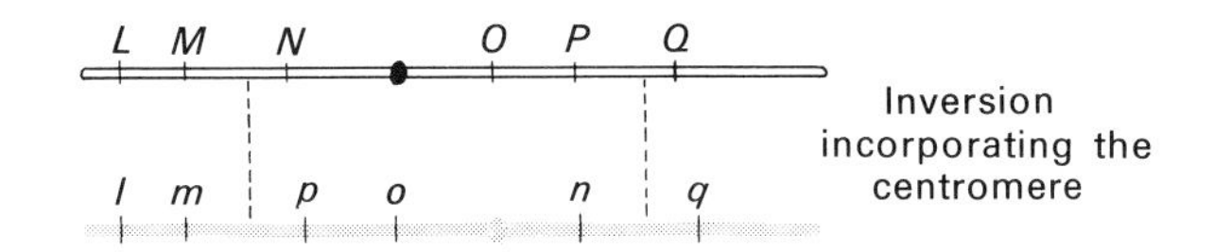

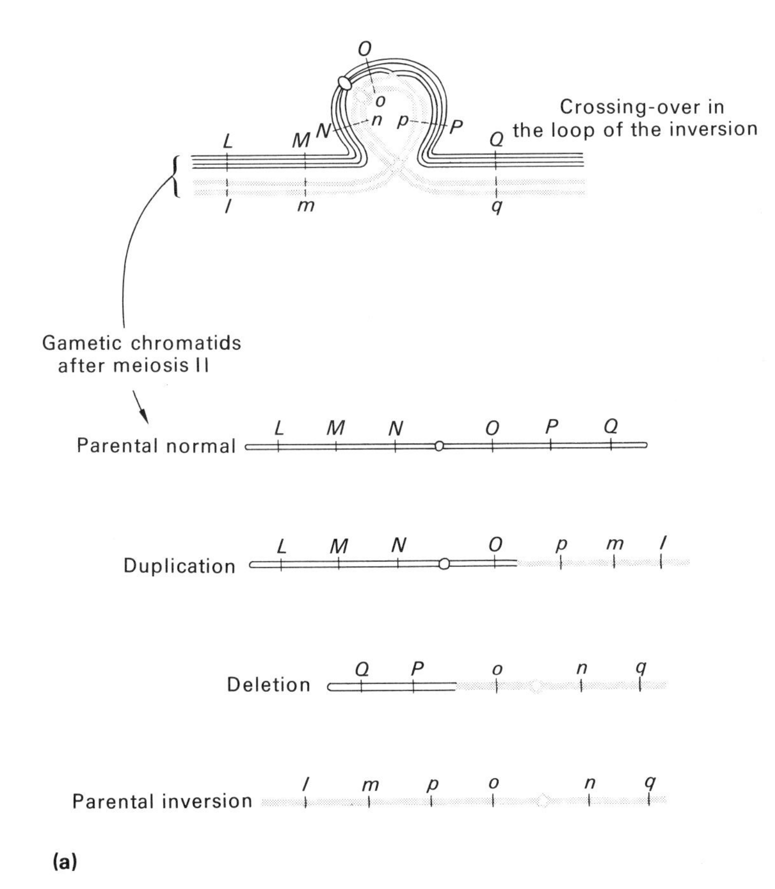

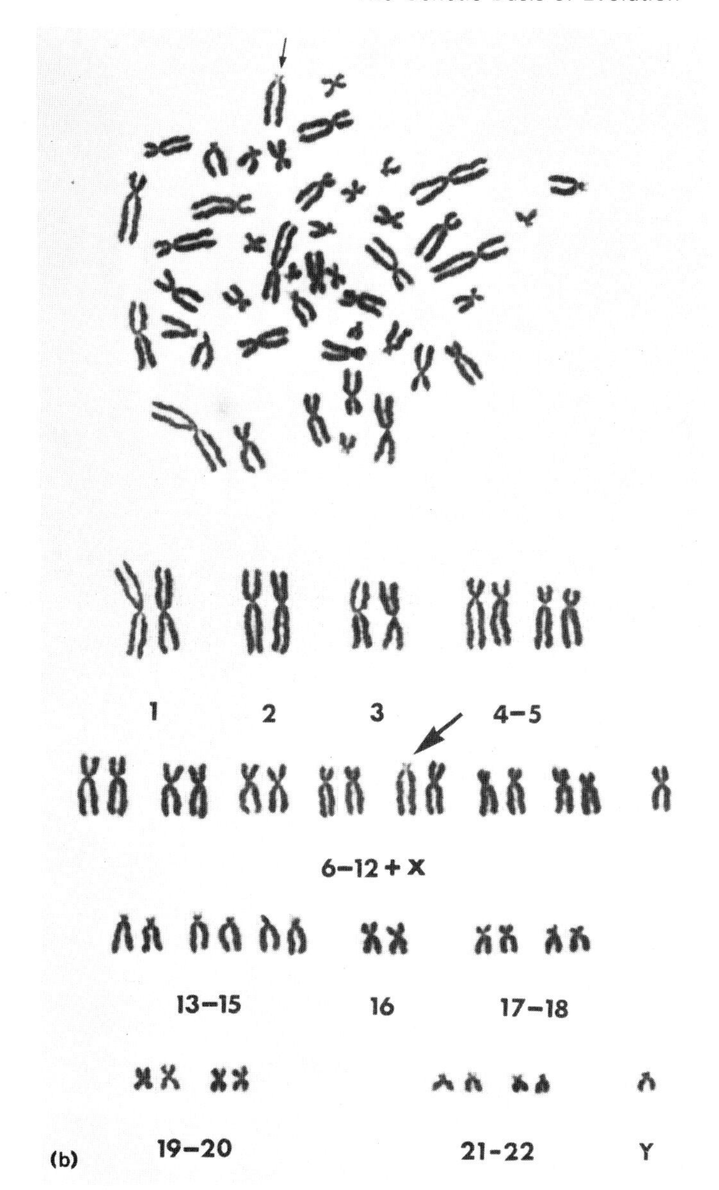

Fig. 9.25 (a) Crossing-over within the loop of a pericentric inversion, and the resultant gametic chromosomes. (b) A pericentric inversion in one of the 6–12 group chromosomes in man. In this case the phenotype was a normal male showing that genic interaction had not been affected to a detectable degree. (Courtesy of the MRC Clinical Genetics Unit, Edinburgh.)

Mexico. The distribution of DDT has been shown to coincide closely with shifts in the geographical distribution of several chromosomal varieties.

The gene combinations held together by an inversion add stability to the genome and their joint effect, a sort of 'summation' of co-adapted genes, may confer a powerful selective advantage on organisms that possess them. Such a group of genes has been termed a **supergene** (Darlington and Mather, 1949) and defined as 'a group

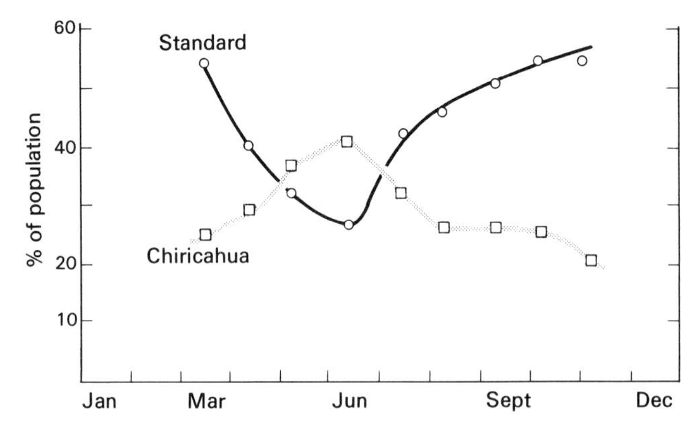

Fig. 9.27 Seasonal changes in the frequencies of the Standard and Chiricahua inversions in a natural population of *D. pseudoobscura*. (After Dobzhansky, 1947.)

of genes acting as a mechanical unit in particular allelic combinations'. A supergene tends to be inherited as a single Mendelian unit; only rarely are its components separated by crossing-over.

Ford (1965) has stressed the evolutionary significance of supergenes and points to their widespread occurrence and abundance in plants and animals. In most organisms they are less easily identified than in the polytene nuclei of the Diptera, but they can be detected by the presence of chromatid bridges (p. 186) in a wide range of organisms including man (Koller, 1937).

Spurway (1953) has argued that inversions may have had great significance in the formation of subspecies from races in the European newt, genus *Triturus*. Hybrids between subspecies showing different reciprocal inversions are either inviable or show F2 hybrid breakdown (p. 226) owing to chromosomal incompatibility. She suggests that inversion 'races' may have become reproductively isolated following the geographical isolation that almost certainly

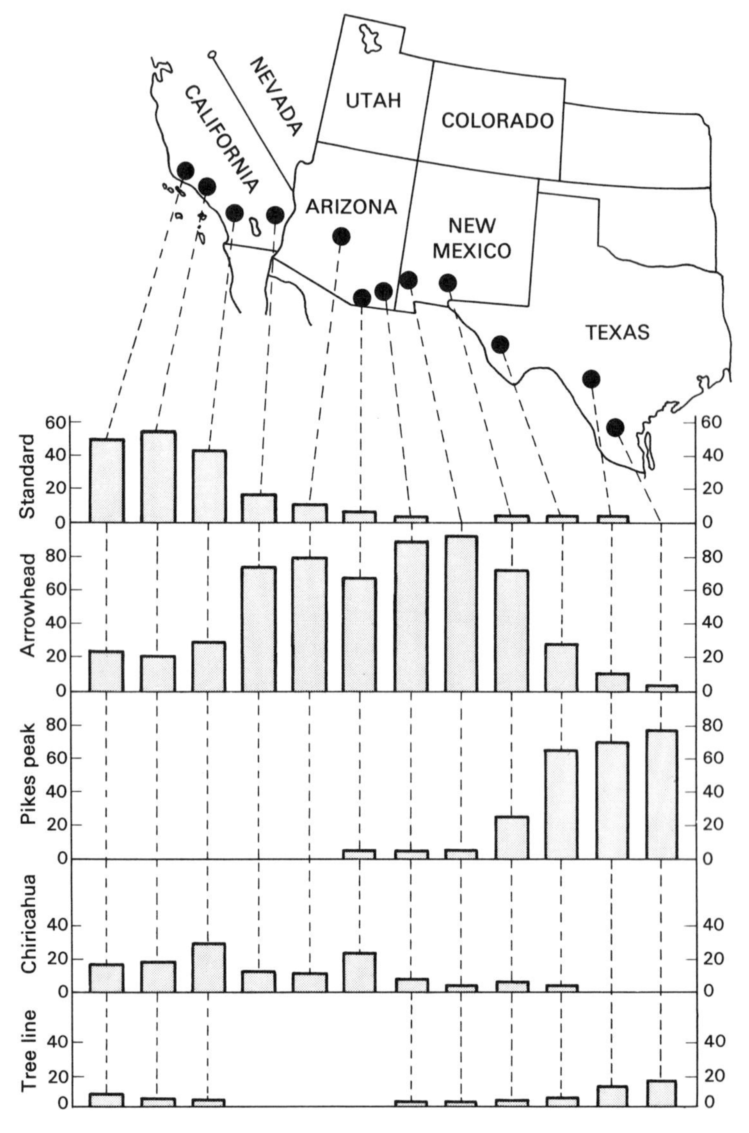

Fig. 9.26 Map showing distribution of inversion types in the third chromosome of *Drosophila pseudoobscura* along a transect through the Southwestern United States. (From Dobzhansky and Epling, Carnegie Institute of Washington, Publication No. 554, 1944.)

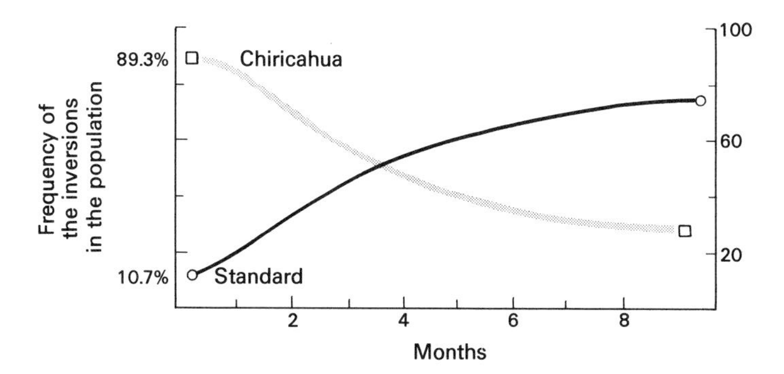

Fig. 9.28 Populations of *Drosophila* kept under competitive laboratory conditions at 25° C showed the selective advantage of the Standard inversion. (After Dobzhansky, 1947.)

occurred during the slow migration overland from pond to pond (p. 178).

Translocation

Translocation consists of a reciprocal interchange of a gene or a group of genes between **non-homologous** chromosomes. As in the case of inversion, translocation represents a change in the arrangement of the genes on their chromosomes rather than a change in the number or nature of the genes (Fig. 9.29).

In the case of translocation, synapsis at meiotic Prophase I is achieved by attraction between the homologous regions of both the affected pairs of chromosomes. This results in a cross-shaped configuration, in effect two conjugated bivalents (Fig. 9.30).

Instead of forming a bivalent, therefore, the translocation heterozygote forms a **quadrivalent**, attached by two pairs of centromeres to opposite poles of the spindle. This gives rise to a number of characteristic formations which are mainly either figure-of-eight forms or open-ring forms, although both may be modified if chiasmata are present in those regions not affected by the translocation (Fig. 9.31).

The gametes which result from the open-ring configurations show either deletions or duplications and as a result may be non-viable. Only the figure-of-eight form leads to 'correct' disjunction, whereas the open-ring forms lead to substantial portions of homologous chromosomes being drawn towards the same pole of the spindle, which contravenes the central dogma of meiosis.

In evolutionary terms the effects of translocation would seem to parallel those of inversion in that, in the heterozygous state, both would tend to hold together particular linkage groups which might be of selective advantage. Studies on *Drosophila* populations, of the kind already described, suggest that translocation occurs less frequently than inversion. Both inversion and translocation would be expected to induce phenotypic variation through the position effect and it could be argued that translocation can give rise to new orders of chromosomes which, in conjunction with duplication, could pro-

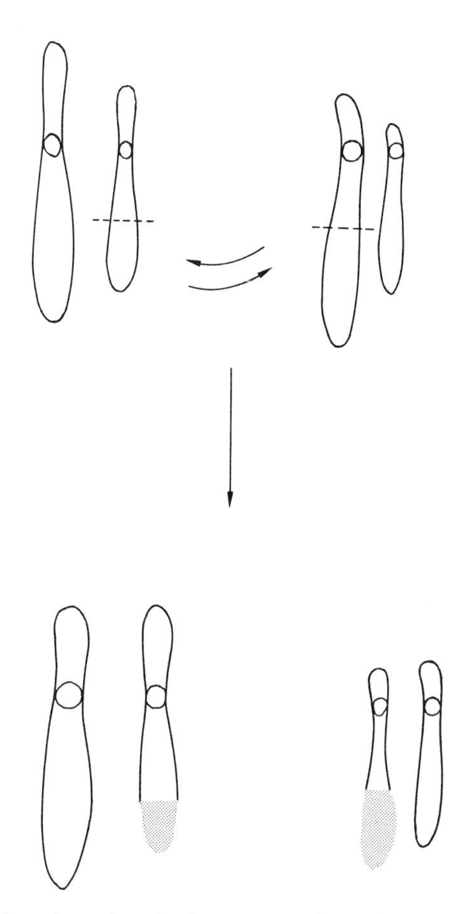

Fig. 9.29 Reciprocal translocation between non-homologous chromosomes results in two-homologous pairs of chromosomes.

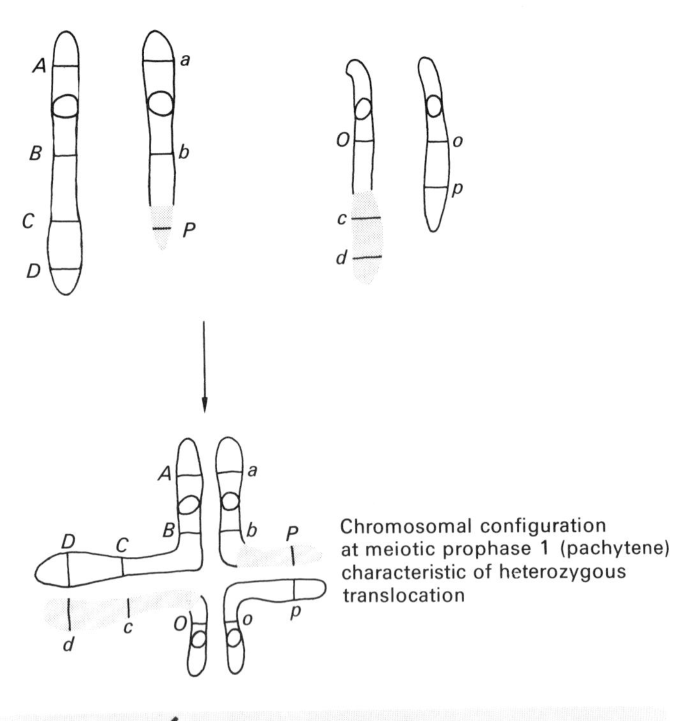

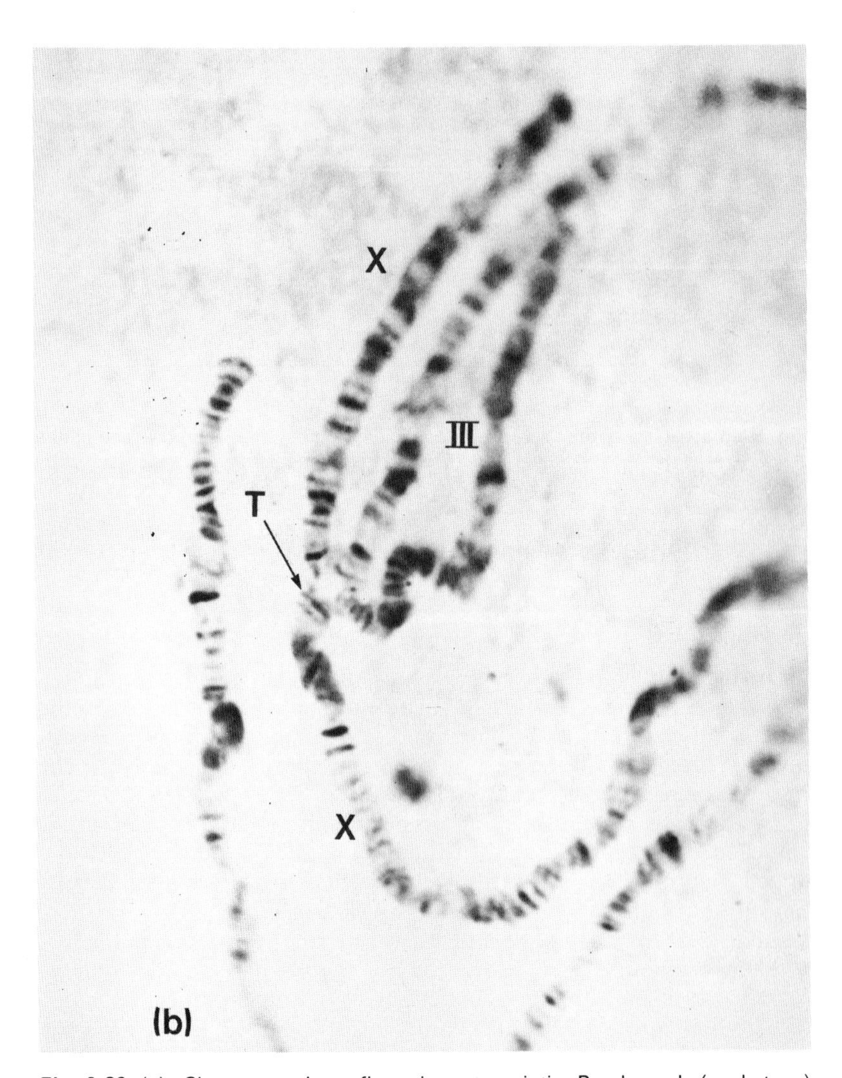

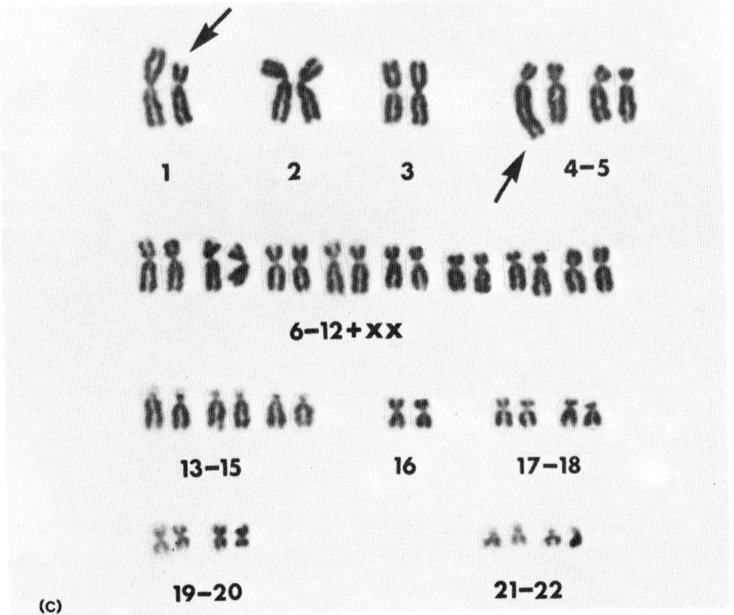

Fig. 9.30 (a) Chromosomal configuration at meiotic Prophase I (pachytene) characteristic of heterozygous translocation. (b) Translocation in *Drosophila melanogaster* showing the characteristic cross-shaped configuration (T). The translocation is between the X chromosome and chromosome III (see Fig. 9.20). How do you know that this is a female complement? (Courtesy of Dr. O. G. Fahmy.) (c) A balanced reciprocal translocation between chromosome 1 and one of the 4–5 group chromosomes in man. There is no net change in gene complement and in this case the phenotype was a normal female, showing that gene interaction, or gene balance, was unaffected. (Courtesy of the MRC Clinical Genetics Unit, Edinburgh.)

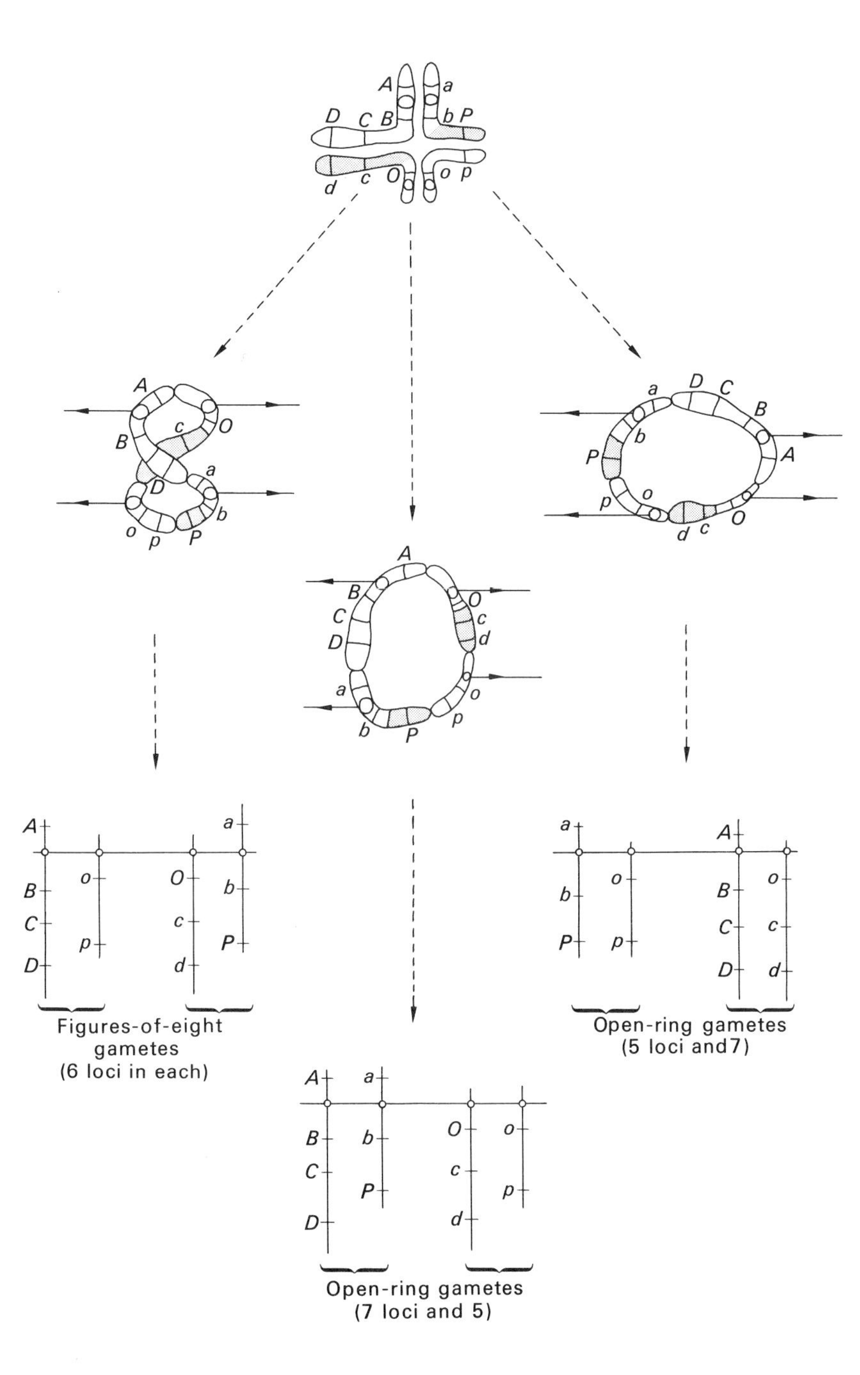

duce the sorts of fundamental changes in chromosomal architecture associated with the origins of new species.

4. Changes in Chromosome Number

Multiples of Whole Chromosomal Complements, or Polyploidy

The terms **haploidy** (n), in which each chromosome is represented singly in a nucleus, and **diploidy** (2n), in which each chromosome is represented by a homologous pair, have been discussed in connection with sexual reproduction. Other kinds of chromosomal number are known, however, particularly in plants. **Triploidy** (3n), **tetraploidy** (4n), **pentaploidy** (5n) and so on represent higher multiples and collectively are termed **polyploids**.

Broadly, polyploids can be derived in two ways. **Autopolyploidy** occurs when the chromosomal complement is multiplied in such a way that the additional homologues are derived from the same pure strain. This most commonly arises within the single organism when, for example, cytoplasmic cleavage fails to follow mitosis so that the diploid number of chromosomes is doubled in the daughter nucleus, forming a tetraploid. In such an autopolyploid, therefore, each member of a homologous pair of chromosomes is duplicated exactly.

In an **allopolyploid** nucleus, on the other hand, the chromosomes are derived from different organisms not of a pure strain, and the term usually refers to chromosomes derived from different species—an interspecific cross. It is, in effect, a polyploid hybrid. In most ordinary cases an interspecific hybrid is sterile owing to the incompatibility of chromosome pairs at meiotic synapsis, although genic incompatibility may also be a factor in hybrid sterility. By contrast, polyploids of hybrids can be fertile since each chromosome has a duplicate with which to synapse at the beginning of gametogenesis.

Autopolyploids are usually recognizably of the same species as the diploid plants from which they arose. Such polyploidy arises spontaneously in a wide variety of plant species. Brook (1964) reports that among new shoots arising from a diploid tomato plant which has been cut back some may be conspicuously larger and more robust. Such shoots, which produce larger flowers and larger, relatively seedless, fruits are probably tetraploid (4n = 48).

Polyploidy in animals is relatively rare. Polyploid individuals in

Fig. 9.31 There are three possible ways in which separation of the translocation can take place in Anaphase I. Only the 'figure-of-eight' situation (in which the spindle fibres are, as it were, attached to the centromeres that are diagonal) results in the proper distribution of genetic material to the gametes.

Drosophila have been identified but the faulty distribution of sex chromosomes in the formation of gametes makes a tetraploid strain impossible. In a tetraploid male *Drosophila* (4n = 2X + 2Y + 4 sets of autosomes) meiotic zygotene results in the synapsis of the X chromosomes together, and of the Y's together, instead of the X + Y synapsis that occurs in the normal diploid male. Thus the sperm carry an X and a Y chromosome which results in a sterile *intersex* offspring. Some tetraploid races of animals have been recorded, however, for example in species of *Artemia*, the brine shrimp, *Echinus*, the sea urchin, and *Ascaris*, the nematode worm (Herskowitz, 1962b) and polyploid nuclei have been demonstrated in cells of normal human liver. The polyploid status of the giant chromosomes from certain tissues in *Drosophila* and *Chironomus* have been discussed in section 8.4 **The Structure of Chromosomes**. Triploidy has also been described in a parthenogenetic species of the American whiptail lizard *Cnemidophorus* (Pennock, 1965, and Fig. 9.32).

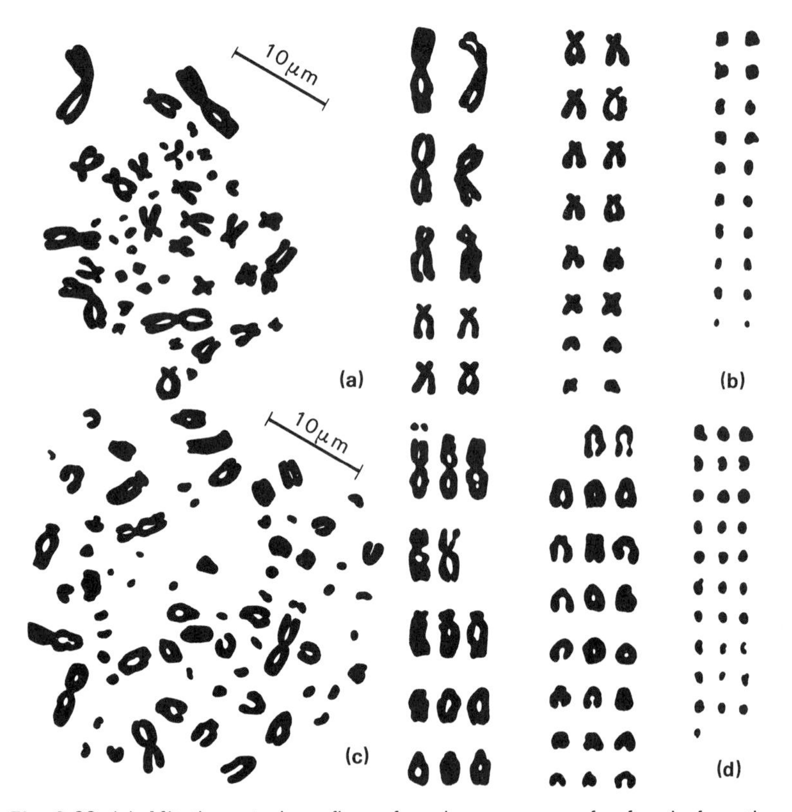

Fig. 9.32 (a) Mitotic metaphase figure from bone marrow of a female from the diploid bisexual subspecies *Cnemidophorus tigris septentrionalis*. (b) Karyotype prepared from (a) showing 46 chromosomes. (c) Mitotic metaphase figure from bone marrow of the pathenogenetic and presumably triploid species *Cnemidophorus velox*. (d) Karyotype prepared from (c) showing 68 chromosomes. (After Pennock, 1965.)

The drug **colchicine**, which can be extracted from the corm of the autumn crocus *Colchicum autumnalis* and has been used medicinally in the treatment of gout, can induce autopolyploidy. In concentrations of about 0.01% this substance inhibits spindle formation, thus preventing the separation of chromatids at mitotic anaphase (Fig. 9.33).

These results led to the widespread use of colchicine in agriculture in attempts to produce polyploid strains which would give higher crop yields. The increase in size which accompanies polyploidy is due primarily to the increase in cell size in the plant; polyploidy is accompanied by increase in the volume of cytoplasm, keeping the nucleo-cytoplasmic volume constant (see also p. 13). In natural populations the presence of autopolyploids undoubtedly could be a source of variation, but unless polyploid individuals are self-fertilizing or parthenogenetic, or unless they appeared simultaneously in sufficient numbers so that sexual reproduction can occur between them, the type is unlikely to be perpetuated. This is because the polyploid gamete, for example the diploid gamete of a tetraploid plant would, if combined with the normal haploid gamete of an ordinary neighbouring plant, give rise to a triploid embryo. Genic imbalance would probably prevent the development of this triploid individual, but even if it reached maturity it would be unable to produce satisfactory chromosomal configurations at meiosis and would be sterile.

Allopolyploidy is thought to have had a more important role in plant evolution because it serves to restore stable sexual reproduction to hybrids. Allopolyploid plants are usually conspicuously different from either parental species and provide what amounts to a source of instant variation. In the simplest case the somatic nucleus of an allopolyploid plant contains both full sets of parental chromosomes. During meiosis the parental homologous pairs undergo synapsis. Segregation in Anaphase I results in gametes containing chromosomes derived from both parental species. Such gametes do not give rise to viable progeny if they combine with the gametes produced by either of the parental species. The allopolyploid is therefore repro-

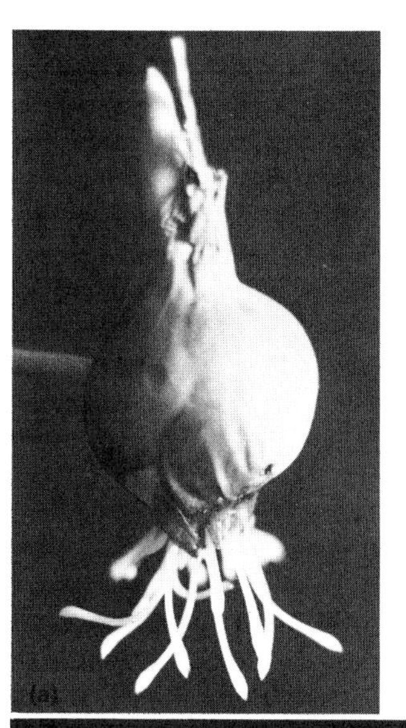

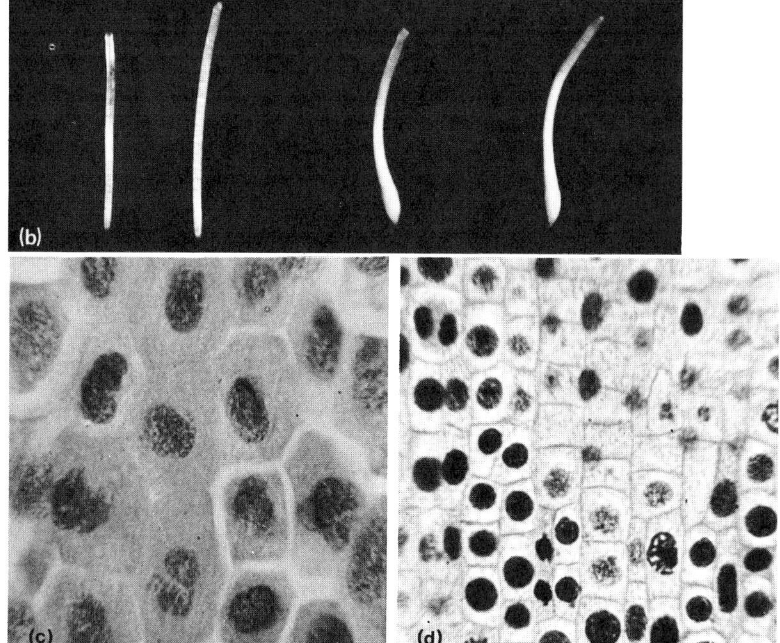

Fig. 9.33 Induced polyploidy in the root tip. (a) An onion bulb showing the swollen root tips characteristic of colchicine tumours. (b) Two polyploid root tips at the right compared with two normal root tips. (c) Sections of colchicine-treated root tips. (d) Normal root tip seen at the same magnification, showing the larger cell size of the polyploid.

ductively isolated from its parental species and by this criterion can be said to constitute a new species.

The classical example of an artificially produced allopolyploid is *Raphanobrassica* (Karpechenko, 1928). The radish plant *Raphanus sativus* (2n = 18) and the cabbage plant *Brassica oleracea* (2n = 18) were cross-pollinated in the laboratory and produced an F1 hybrid which was largely sterile. The sterility of this hybrid is due mainly to chromosomal incompatibility, as shown by the widespread failure of synapsis and the disorderly aggregation of chromosomes at meiotic metaphase. The polyploid of this hybrid was shown by Karpechenko to contain both parental sets of chromosomes duplicated, a situation which is termed **amphidiploidy**. Whereas the somatic cells of the sterile F1 hybrid contain 18 chromosomes (9 from *Raphanus* and 9 from *Brassica*), the amphidiploid *Raphanobrassica* contains 36

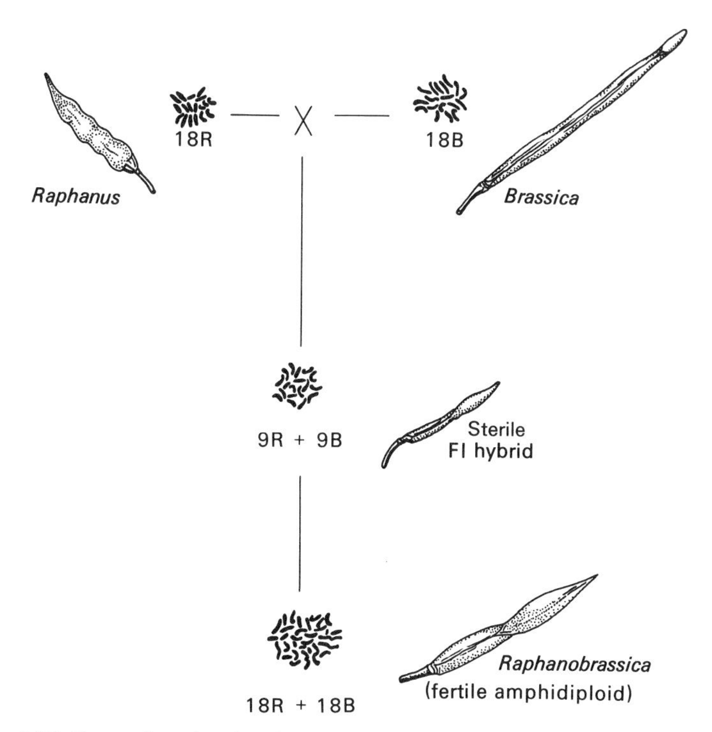

Fig. 9.34 Types of seed pod and somatic chromosomal complements of *Raphanobrassica* and its derivative forms. (After Karpechenko, 1928.)

chromosomes, namely 2 × 9 from *Raphanus* and 2 × 9 from *Brassica*. In meiosis the homologous chromosomes from each parental species form bivalents and the metaphase plate is well ordered. Fully fertile, diploid pollen and ovules are produced (Fig. 9.34).

An interesting example of a naturally occurring allopolyploid occurs in the salt-marsh grass *Spartina*. Historical records in the herbaria dating back to the early 17th century show that populations of *Spartina maritima* were indigenous in southern and eastern England, southern Europe and Africa. The species was restricted to the eastern side of the Atlantic. In about 1800 an American species *S. alterniflora* was introduced in Southampton Water, probably as seeds or rhizomes in ships' ballast. By 1805 this species was well established and co-existed with *S. maritima*. In 1870 an F1 interspecific hybrid was recorded in the area, and named *S.* × *townsendii*.

This plant shows signs of heterosis; it is more vigorous and often larger than either parental species and appears to compete equally with *S. alterniflora* and successfully against *S. maritima* (Fig. 9.35). Since the pollen and ovules were sterile, however, the spread of this

S. maritima —— × —— *S. alterniflora*
(2n = 60) (2n = 62)

(F1 hybrid) *S.* × *townsendii*
(2n = 62)

(amphidiploid) *S. anglica*
(2n = 120, 122 or 124)

hybrid was slow, by means of the rhizomes only. About 20 years later, in 1892, an allopolyploid was recorded at Lymington and this spread rapidly through the region, vegetatively and by seed formation. The allopolyploid was designated *S.* × *townsendii* (Allopolyploid) although the name '*Spartina anglica*' is being used increasingly, *S.* × *townsendii* being reserved for the F1 hybrid (Fig. 9.36, and see also the allopolyploid of *Primula kewensis*, Fig. 9.37).

The numbers of chromosomes are higher in all species of *Spartina* than in *Raphanus* and *Brassica* and they are correspondingly smaller in size. The apparent discrepancies in chromosome number arise from duplications (polysomy) and deletions of whole chromosomes in *S. alterniflora* and tolerance of small gametic chromosome number variations in *S. anglica*.

	2n	Mid-range of leaf size (mm)		Anther length (mm)
		length	width	
S. maritima (indigenous P1)	60	95	55	5.8
S. alterniflora (American P1)	62	300	90	6
S. × *townsendii* (hybrid F1)	62	210	85	5.4
'*S. anglica*' (allopolyploid)	124 (occ. 120 or 122)	305	105	10.5

Fig. 9.35 Chromosome numbers in *Spartina* species.

Polyploidy in the plant kingdom is not restricted to the Angiosperms. Manton (1953) has carried out chromosome counts in many

Fig. 9.36 *Spartina* × *townsendii*, the F1 hybrid, is sterile and the anthers are shrivelled and atrophied.

Fig. 9.37 *Primula kewensis* is a fertile hybrid which is an allopolyploid arising from a cross between *P. floribunda* and *P. verticillata*. This hybrid is a novel species, it produces fertile offspring only when crossed with its own species. (Courtesy of the Royal Botanic Gardens, Kew.)

species of fern and found a high frequency of polyploidy, from triploidy to hexaploidy. In tropical species, collected in Ceylon, the frequency of polyploidy was 60% compared with a frequency of 53% for species in the British Isles. These results conflict with the view that the frequency of polyploidy increases with increase of latitude and that polyploidy was an adaptation to survival under the rigorous climatic fluctuations that beset the temperate latitudes, such as the ice ages.

Manton concludes that the higher frequency of polyploidy in tropical species, together with the higher grades of ploidy (some species of 8n, 10n and even 12n have been described from Ceylon compared with a maximum of 6n in European ferns) is an indication that evolution is proceeding faster in the tropics than in temperate latitudes. She argues that the long dormant period in the highly seasonal annual growth cycle of the temperate species makes for a slower rate of evolution than tropical conditions. In the latter there are more habitats in a more densely populated and stratified vegetation, and the frequent landslides give more opportunity for re-colonization.

It should be borne in mind, however, that the rate of evolution is determined by environmental selection. A high frequency of polyploidy, or high grades of ploidy, do not themselves confer high rates of evolution any more than does a high frequency of point mutation (p. 208). Manton's conclusion that, as a general rule for all organisms, evolution is proceeding faster in the tropics than in temperate latitudes needs careful examination.

Multiples of Single Chromosomes, or Aneuploidy

The addition of a single somatic chromosome to the normal diploid complement is termed **trisomy**. The deficiency of a single chromosome from the normal complement is called **monosomy**. These conditions are relatively rare in animals, in which genetic balance seems of greater importance, and the effect on the phenotype is usually deleterious and severe. In plants, by contrast, monosomy and trisomy are more common and induce a variety of phenotypic variants. The classic work on trisomy in plants was carried out in the United States by Blakeslee and co-workers on *Datura stramonium*, the Jimson weed or Thorn Apple. This plant belongs to the Solanaceae family which includes the potato and a number of species having medicinal properties, such as Deadly Nightshade, from which the alkaloid atropine can be obtained, and Henbane, which yields the opium-like narcotics hyoscine and hyoscyamine. In Britain *Datura* is found occasionally in waste places and extracts were once used in the preparation of asthma powders. It has white trumpet-shaped flowers and a 4-partite, prickly capsule.

In what is regarded as one of the great experiments in genetics Blakeslee was able to correlate flower and seed capsule shape with trisomy of each of the twelve sets of chromosomes of *Datura* in turn (Fig. 9.38).

This work on *Datura* demonstrates that the effects of trisomy in plants are not necessarily particularly severe. The effects are distinctive, however, in that the twelve different trisomics are clearly distinguishable in terms of capsule size and form. The evolutionary significance of trisomy has not been defined. Although a character such as the form of the seed capsule may not confer an immediately obvious selective advantage, other structural or physiological characteristics appearing as variants in trisomics may do so. Unlike allopolyploids, however, trisomics are not true breeding. During meiosis the extra chromosome usually synapses with its homologous bivalent, forming a **trivalent**. This separates at Anaphase I, which results in half the gametes carrying a normal haploid complement for that chromosome and the remaining half of the gametes being diploid. Cross-pollination between trisomic plants, therefore, would result in only half the progeny being trisomic and the other half being tetrasomic and diploid (Fig. 9.39). The tetrasomic might prove non-viable, although some tetrasomics have been identified in *Datura*, but diploid plants would be a constant feature of a trisomic population. The subsequent reduction by 50% of the reproductive pressure of a trisomic population need not be disadvantageous if the trisomic variant is a sufficiently successful competitor, as Dobzhansky and Pavlovsky (1955) showed in the example of *Drosophila tropicalis*.

Trisomy is known in man only in respect of one of the chromosomes in the smallest set, Group G, probably Number 21. Trisomy 21 results in **Down's Syndrome**, or **Mongolism**, a complex set of symptoms which includes mental retardation, the typical moon-shaped face with a fold of skin over the inner corner of the eye, and characteristic palm prints (Fig. 9.40). Trisomy of the larger chromosomes, if it occurs at all, probably has such serious effects on metabolism that embryogenesis fails and the foetus aborts.

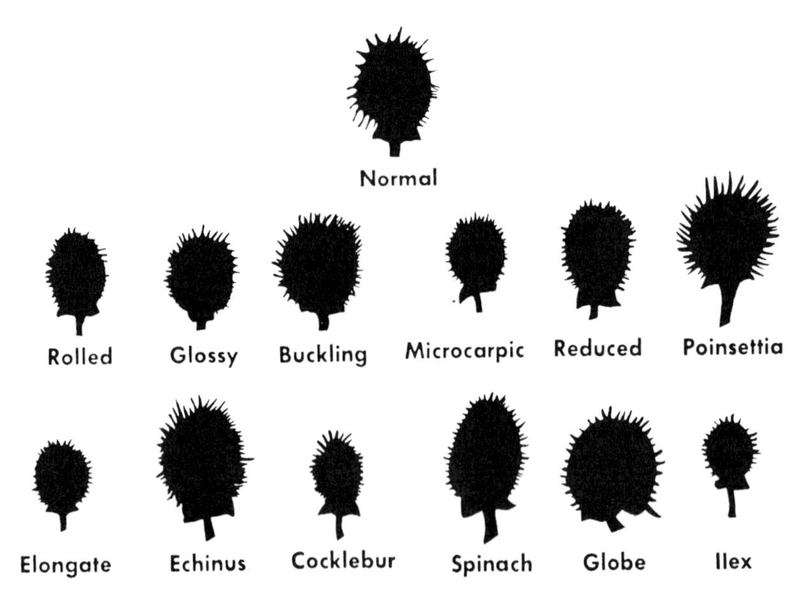

Fig. 9.38 Silhouette drawings of seed capsules in *Datura*. There are 12 chromosomes (2n = 24) and the effect of capsule size and shape caused by the trisomic state of each chromosome in turn can be seen. (After Blakeslee, 1934.)

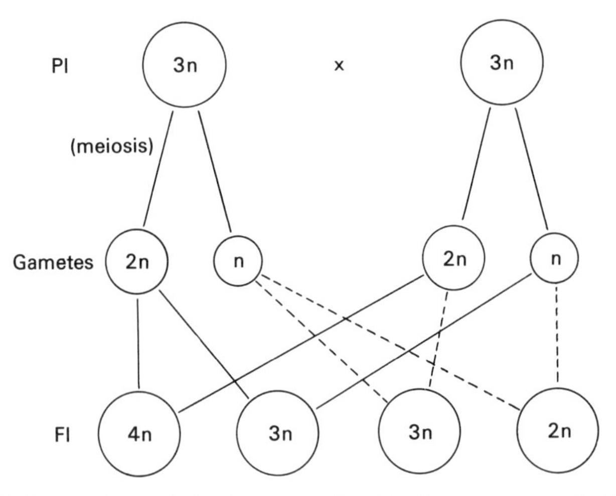

Fig. 9.39 A trisomic population is not true-breeding because during the formation of gametes the irregular separation, or disjunction, of the chromosomes results in widespread disorganization in chromosome number.

Trisomy probably occurs through non-disjunction in Anaphase I of meiosis. The daughter cells of this division would therefore have either two chromosome 21's or none. Fertilization of the resultant gametes with a normal haploid gamete would lead to trisomy and monosomy respectively. Monosomy 21, which might be expected to occur in man as frequently as trisomy 21, does not in fact do so and it must be concluded that monosomy 21, like trisomy of the larger chromosomes, is non-viable (Fig. 9.41).

In man there is a positive correlation between the incidence of trisomy 21, or mongolism, and the age of the mother. The probability of a 20-year old woman having a mongol child is about 0.0003; for a

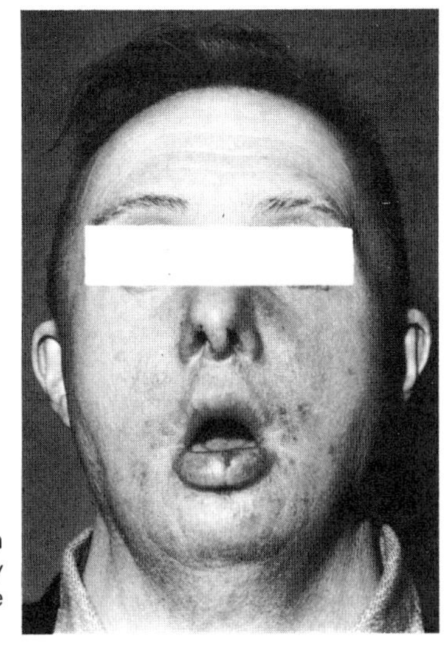

Fig. 9.40 Mongolism, the karyotype (right) shows an extra chromosome in the small 21–22 group, commonly believed to be trisomy of No. 21. (Courtesy of the MRC Clinical Genetics Unit, Edinburgh.)

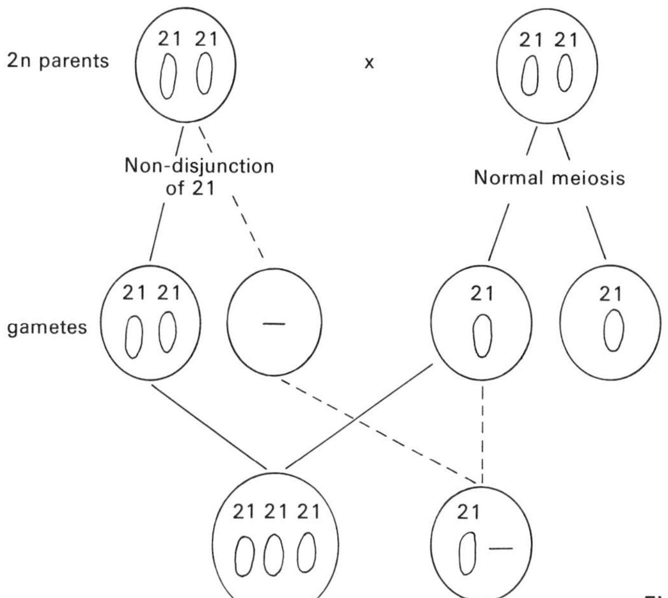

Fig. 9.41 Non-disjunction of chromosome during meiosis can result in a gamete which is diploid for that chromosome. The subsequent fertilization of such a gamete will result in a trisomic 21 zygote.

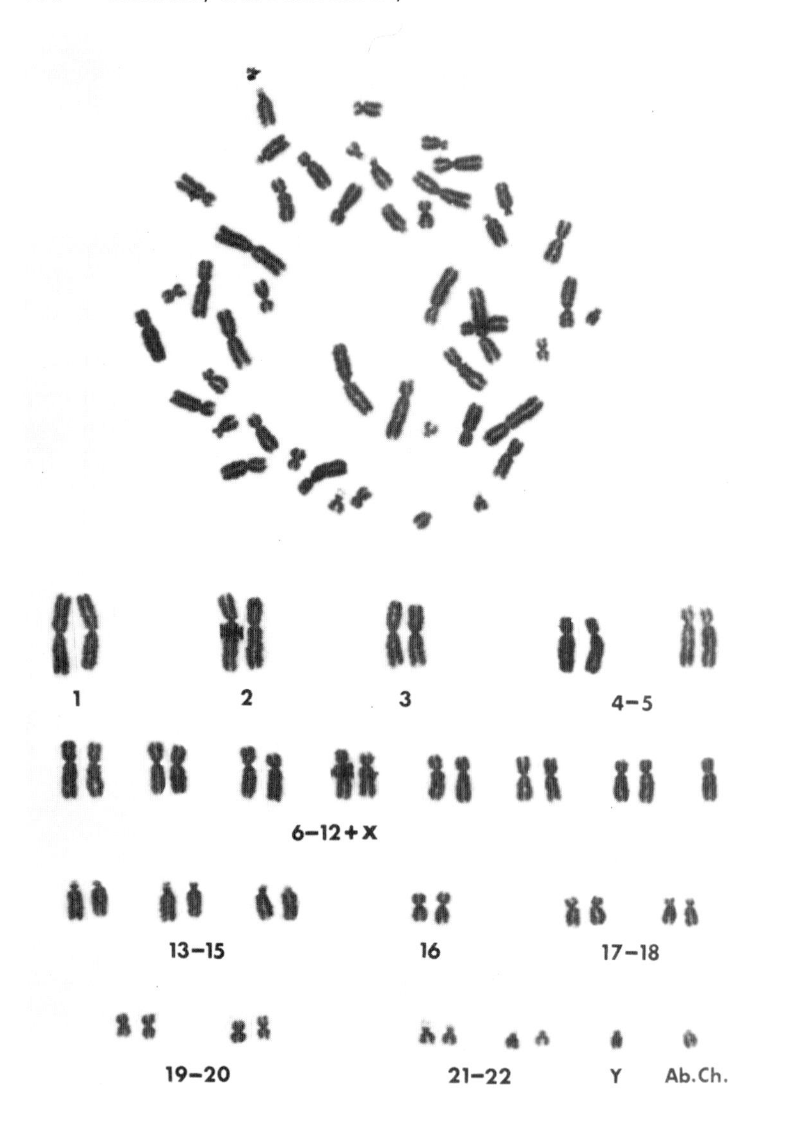

Fig. 9.42 (a) The karyotype of this patient shows an XY or male complement and an additional small chromosome or abnormal chromosome fragment (Ab·Ch·). The phenotypic effects of this additional chromosome included severe emaciation and contracture of the limb joints, and severe mental retardation, I.Q. untestable.

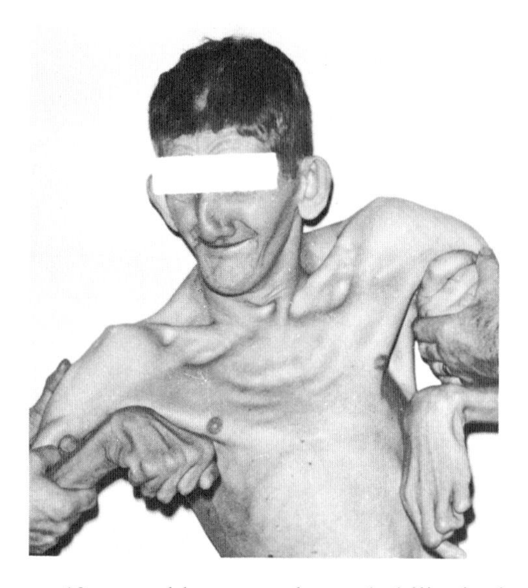

40-year old woman the probability is about 0.01 and for a 45-year old woman the probability is about 0.02. There is no correlation between the age of the father and the incidence of mongolism. The reason for the age correlation is not entirely clear but it is known that ovum formation begins in the embryonic ovary and that during oogenesis the development of all the egg cells that a woman will produce in her life time are suspended in Prophase I of meiosis (see p. 85). It might be that primordial follicles which come to maturity in older women have a higher non-disjunction rate owing to the chromosomes having been synapsed for a longer period. By contrast meiosis in spermatogenesis is a continuous process during the reproductive span of a man (Fig. 9.42).

5. Genic Interaction: Gene Balance, Position Effects, Epistasis and Pleiotropy

Quantitative changes in the chromosomal complement of an organism usually cause pronounced changes in the phenotype. Such changes are less marked in **euploidy**, that is when the whole chromosomal complement is multiplied, such as in tetraploidy, than in **aneuploidy**, which concerns multiples of less than the whole complement. The principal reason for this is that in euploidy the numerical balance between the genes is unchanged.

An example of the effect of gene balance has already been men-

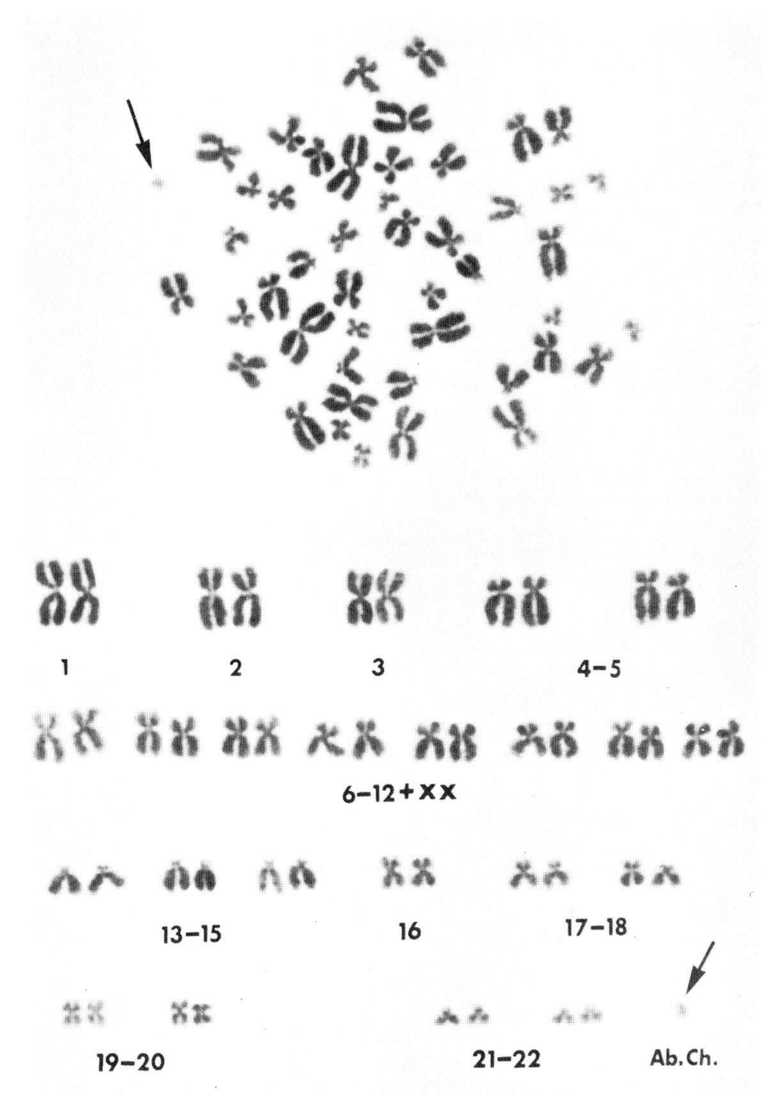

Fig. 9.42 (b) In a similar situation to Fig. 9.42 (a) an additional abnormal chromosome (arrowed in an XX complement) gave rise to a female patient with mental retardation, I.Q. approximately 39. Mental retardation is a widespread feature of chromosomal abnormalities, particularly where supernumeraries are concerned. (Courtesy of the MRC Clinical Genetics Unit, Edinburgh.)

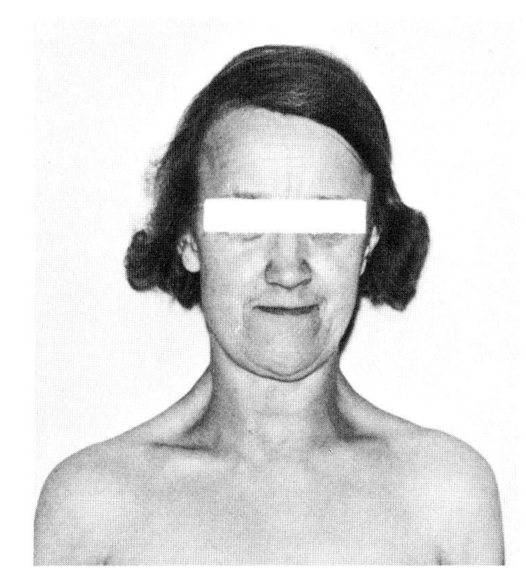

tioned in Sex Determination in *Drosophila*. In this insect the Y chromosome appears to carry no male promoting genes and sex is dictated by the balance between genes carried on autosomes which determine maleness and genes carried on the X chromosome which determine femaleness (Fig. 9.43).

X	Y	Autosomes (II, III and IV)	Ratio of X chromosome to sets of autosomes	Sex
1	1	2 sets	1:2	diploid ♂
2	–	2 sets	1:1	diploid ♀
3	–	3 sets	1:1	triploid ♀
2	–	3 sets	2:1	intersex (sterile)

Fig. 9.43 Table of sex determination in *Drosophila* showing that sex depends on the balance between the sex chromosomes and the autosomes.

On a smaller scale heterozygous duplications in chromosomes can provide a source of phenotypic variation by altering genic balance, which is exemplified by the Bar-eye situation in *Drosophila*. Genic imbalance also results from deletions, although in the case of heterozygous deletions the resultant variation is supplemented by the expression of previously recessive genes in that portion of the homologous chromosome opposite the deletion. The consequences of chromosomal inversions in terms of stabilizing linkage groups has been discussed on p. 184. The effects of inversion on the phenotype are generally less marked than those following duplication or deletion, since the genetic complement is unchanged. It is known, however, that the expressivity of genes is affected by neighbouring genes. Thus, in Figure 9.23, the activities of the genes *b*, *c*, *e* and *f* may be modified by the inversion; *b*, once next to *c*, is now adjacent to *e*, and so on.

The position effect which occurs in pseudoallelism has been described on p. 147. Recombination within the gene is known to show a position effect in the expression of the whole gene and this has been termed the **cis-trans** position effect. The garnet-eye locus in *Drosophila* shows a cis-trans position effect in crossover tests and analysis reveals that two mutants, g^1 and g^2 are involved; if both lie adjacent, in the **cis** position, normal eye colour is produced, but in the **trans** position garnet eye colour results (Fig. 9.44).

The cis-trans position effect can be explained by regarding the cistron as a complete coding unit. In the **cis** position the code is complete and produces normal enzymes for synthesizing red eye colour. In the trans position the translation of mRNA along the cistron is interrupted by the mutant **recon** and an incomplete enzyme pattern results.

It is less easy to account for the 'neighbourhood' position effect which occurs, for example, in inversion. There is no evidence that the transcription of mRNA on one cistron is altered by the characteristics of a neighbouring cistron except where these are in a related

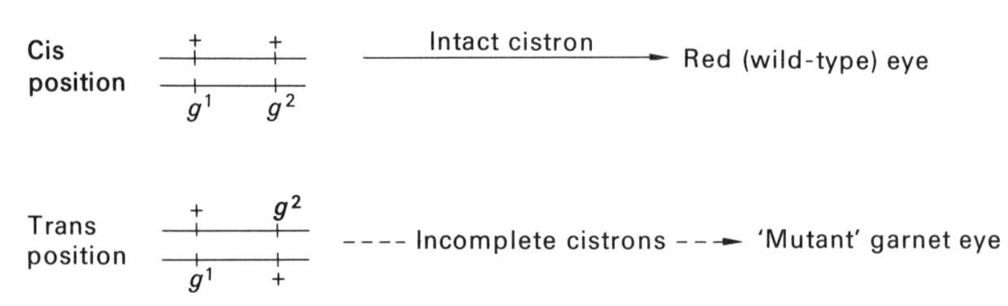

Fig. 9.44 The *cis-trans* position effect in *Drosophila*. In the *cis* position the coding sequence of the cistron is complete and therefore codes for normal enzyme for normal eye pigment. But crossing-over within the cistron (between the recons, or 'pseudoalleles') results in recombinants in which the coding sequence is incomplete and furnishes a relatively incompetent enzyme.

biochemical pathway (see Control of Gene Action, p. 135). On the other hand there is evidence that the change in position of a gene from its normal site to a site which is adjacent to a heterochromatic portion of a chromosome can alter its expressivity. For example, if an inversion moves the red-eye gene in *Drosophila* (the $+^w$ locus on the X chromosome) from its normal position in the euchromatin to a region of heterochromatin adjacent to the centromere, the eye is variegated in colour, containing speckles of white. In this case the addition to the genotype of an extra heterochromatin-rich chromosome, such as an extra Y chromosome, reduces the amount of variegation. The mechanism whereby this suppression of variegation is produced is unknown but it is additional evidence for the concept of the balanced interdependence of both forms of chromatin in the activity of the whole genotype (Commoner, 1964 and p. 160).

An example of a wide range of phenotypic variation resulting from genic interaction occurs in the expression of coat colour in mammals. A number of unlinked genes are known to interact in the expression of coat colour and mutants of these genes give rise to a variety of pattern and pigmentation (Fig. 9.45). At the ***A* locus** the dominant wild-type allele *A* is termed **agouti** (after the South American rodent of that name). This gene codes for a band of yellow close to the tip of the hair which improves the camouflage properties of the fur, giving the black fur a grey-brown appearance. The mutant allele *a* is called **non-agouti** and in an animal which is homozygous for *a* the agouti band is absent.

At the ***B* locus** the dominant wild-type allele *B* is **black** and its recessive, mutant allele *b* produces brown pigment. Various combinations of the *A* and *B* alleles can produce fur which is grey, black, brown or cinnamon.

At the ***C* locus** lies the gene for colour expression. The dominant wild-type allele *C* codes for colour expression, admitting the expression in the phenotype of the pattern coded for by the *A* and *B* loci. The recessive mutant allele *c* in the homozygous condition prohibits pigment formation. Such an animal is **albino** whatever the nature of the *A* and *B* loci. The *C* gene has a relatively high frequency of mutation in a wide range of chordate animals and albino individuals appear spontaneously in most natural populations of mammals as well as in the lower vertebrates, fish reptiles and birds.

The *C* locus is said to be **epistatic** (placed above). Bateson used this term to describe the 'dominant' role of one gene towards another that was not allelic to it. The genes whose actions are suppressed by an epistatic gene are said to be hypostatic.

A locus	*B* locus	*C* locus	Phenotype
A –	*B* –	*C* –	grey-brown
a a	*B* –	*C* –	black
A –	*b b*	*C* –	cinnamon
a a	*b b*	*C* –	brown
– –	– –	*c c*	albino

Fig. 9.45 The *A*, *B* and *C* genotypes and coat colours in the mouse.

In addition to the *A*, *B* and *C* series there are other genes which interact in the coat colour system. In the *D* series a recessive allele promotes a **dilution** of the pigment coded for by the *A* and *B* genes, producing a pale colouration in several dog breeds, such as the Great Dane, and giving a blue quality to the coat of some rabbit breeds like the Blue Bevern.

The ***S* locus** expresses solid coat colour and a number of alleles at this locus confer various kinds of spotting. In the beagle the most common form of spotting, piebald, consists of a dark saddle with tan points superimposed on a *B* – (black) or *b b* (liver) coat. A beagle pack exhibits an almost continuous variation in coat colour pattern (p. 210).

The mechanism of the cis-trans position effect has been postulated in terms of interruption of mRNA transcription at the cistron. Such a mechanism is less likely to operate in the case of the 'neighbourhood' position effect. In this case and in epistasis a more likely explanation of the mechanism can be found in terms of the interaction of gene products. The example of epistasis shown by the *C* locus in coat colour appears to be caused exclusively at the gene product level. An animal which is homozygous for the *c* allele is unable to produce the substrate, dopaquinone, from which the *B* locus produces the fur or skin pigment, melanin (Fig. 9.46).

The total effect of the discoveries of genic interaction and the balance between heterochromatin and euchromatin is to present a picture of the genotype as a compound unit rather than a complex of independently acting units. The significance of genic balance leads to the concept of the integral nature of gene action. A gene does not act in isolation, its action is modified by its genetic environment. In one genetic environment a gene may prove advantageous to an organism, whilst in another the same gene may be deleterious. One example of this is found in a gene concerned in camouflage in the fresh-water fish *Platypoecilus*. This gene codes for the formation of large melanocytes in the skin, giving the fish a black spotted appearance, but the transfer of this gene to a hybrid, in crosses between *Platypoecilus* and the sword-tail, *Xiphiphorus*, causes the production of lethal melanotic tumours (Kosswig, 1929).

A second example occurs in the work of Kettlewell (1965) with the Peppered Moth *Biston betularia*. In this European species the *carbonaria*, or black, form is a dominant condition. In breeding experiments in the laboratory, however, it has been shown that in

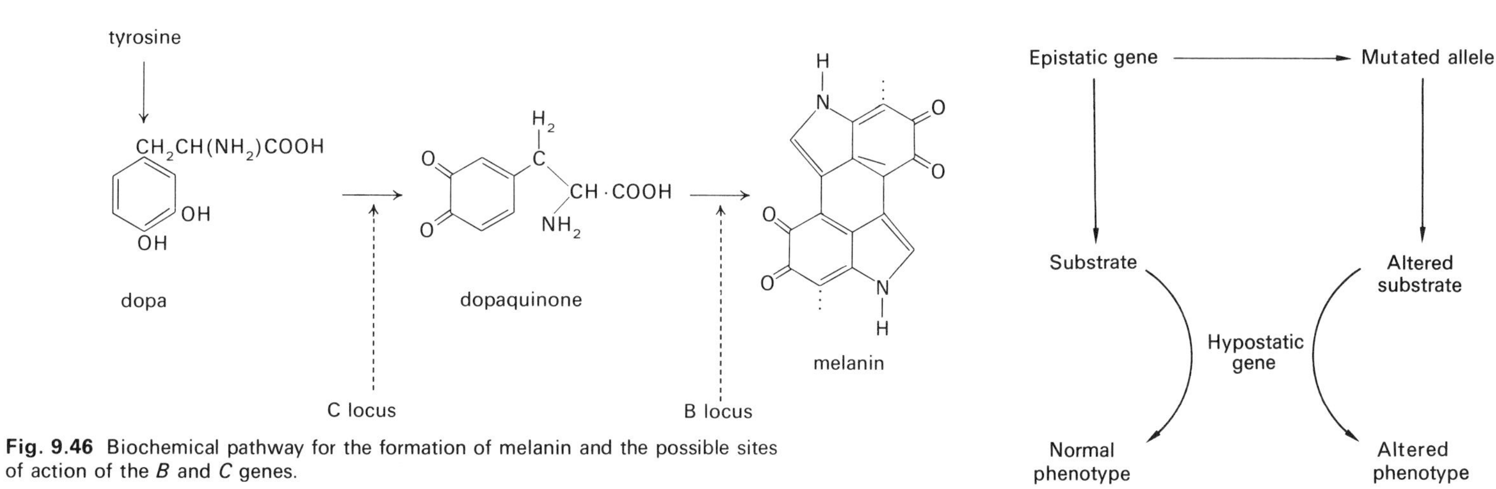

Fig. 9.46 Biochemical pathway for the formation of melanin and the possible sites of action of the *B* and *C* genes.

crosses between *B. betularia* and the related species *B. cognataria* of the northern United States the carbonaria gene ceased to have a dominant effect in the offspring.

This concept of the compound nature of the genome raises two important considerations. In the first place, organisms that survive successfully to the age of reproduction, at which samples of their genome can be contributed to the gene pool of the next generation, are unlikely to contain radically reorganized genotypes. Because the genome has evolved by natural selection of its phenotype any such radical variants are most likely to be deleterious and in a vigorous population with a high degree of internal competition would tend not to be passed on to the next generation. In this sense the integrated genome is highly **conservative** and the genetic sources of discontinuity discussed in this section of the book tend to be of a minor nature. The second consideration, which arises from this concept of the conservative nature of the genome, is the problem of major evolutionary advance. Inherent in the concept of speciation is the idea of the changing environment or occupation of a new environment which provides novel selective pressures and favours certain variations of the genotype over others. This establishes a directional trend in the gene frequencies of the population. However, many of the fundamental steps in evolution, such as the origins of triploblasty, the coelom, metamerism and the origins of phyla, appear to have taken place in a fairly constant marine environment over relatively short time spans. There is some difficulty, therefore, in reconciling the sorts of major changes in the genetic complement needed to accomplish these relatively sudden steps with the concept of the balanced, conservative genome.

An additional factor in this discussion is the phenomenon of **pleiotropy**. On analysis single genes are usually found to affect several apparently unrelated characters in the phenotype. In perhaps the most extreme case the gene **polymorph** in *Drosophila* is known to affect body proportions, wing size and vein pattern, eye colour, bristle pattern, gonad formation, growth and viability! In the pea, the gene for violet petal colour also produces grey seed coats. In man, multiple phenotypic effects accompany many lethal genes. The gene for sickle-cell anaemia, for example, produces not only sickling of the red cells but also heart malformation, kidney malfunction, growth abnormalities and skin lesions.

Dobzhansky and Holtz (1943) carried out detailed anatomical examinations of two strains of *Drosophila* which were essentially genetically identical except at one locus. One strain was homozygous for the white-eye gene (w) and the other strain was homozygous for its wild-type allele ($+^{w}$) which codes for red eye colour. One difference recorded was the shape of the spermatheca, the sperm storage organ in the female fly, which was found to be significantly different in the two strains. The conclusion from this, and other work, is that in general genes are pleiotropic (Herskowitz, 1962c).

The Evolution of Dominance, or Dominance-modification

The effects which a gene has on an organism, that is its expression in the phenotype, are not fixed but are capable of change. T. H. Morgan (1929) demonstrated the variability of the expression of the gene 'eyeless' in *Drosophila*, work which provided evidence that dominance and recessiveness might themselves be capable of evolutionary change.

The concept that selection operates on the expressivity of genes is attributed to Sir R. A. Fisher (1926) and its importance in evolution has been stressed by many workers, notably E. B. Ford (1965). The concept embraces the argument that dominance is not a property of genes themselves but of individual genetic characters, although this has been contested from the point of view of molecular biology (Crosby, 1963). It has already been shown that most single genes have multiple effects and variable degrees of expression; some may be dominant, others recessive and yet others of intermediate expression. Similarly a particular character is likely to be influenced by several genes which may or may not be structurally related, that is they may or may not be allelic or linked as a supergene (p. 188). In this sense a particular gene cannot be described as dominant or recessive since its expressivity may vary between the characters it affects.

If a character is strongly advantageous, selection will favour an increase in its dominance. Conversely a disadvantageous character will tend to become recessive. Some evidence for the evolution of dominance was gained from examination of entomological records and from breeding experiments on the Peppered Moth, *Biston betularia* (see Polymorphism in Insects, p. 213). Examination of the earliest recorded specimens of the *carbonaria* form taken in Manchester in 1848 reveals that the wings were marked with white scales and frequently with white lines. The wings of modern *carbonaria* moths, however, are completely black except for white spots at the bases of the wings and even these are disappearing in some areas (Kettlewell, 1965), which suggests a more intensive expression of the *carbonaria* gene.

Breeding experiments in which the black forms were crossed with pale forms collected from Cornwall, thus inserting the *carbonaria* gene into a gene complex to which it was not adjusted, yielded dark forms with white scales similar to the original melanic forms collected over a century earlier. The dominance of the *carbonaria* form was broken down still further in crosses with a related American species, *B. cognataria*.

These experiments point to the action of modifiers (p. 210) and serve to stress once again that the phenotype is the resultant of interactions within the whole gene complex; the qualities of a single genetic character are almost invariably the result of a compromise between the actions of several genes. From this point of view the concept of natural selection acting on the phenotype and inducing changes in gene frequencies in a population must, in reality, be orders of magnitude more complex than the hypothetical example of the pod length of a leguminous plant described in the section on the Origins of Race.

The principle of the evolution of dominance was summed up by Sheppard (1953) '...selection will cause the disadvantageous characters controlled by a gene to become recessive and the advantageous characters dominant.' This accounts for the observations that wild-type characters occurring in populations are almost always dominant to characters arising from mutant alleles. It is also notable that in all cases where industrial melanism in moths has been studied (p. 213) the dark form is dominant and although in many moths recessive dark forms do arise from time to time by mutation these have never reached high frequencies in populations.

6. *Mutation*

In Chapter 2 the process of protein synthesis has been outlined. Messenger RNA, produced as a complementary molecule to one strand of DNA in the nucleus, is known to move into the cytoplasm where it combines with ribosomes and acts as an instructor in the formation of proteins. The sequence of amino acids in the protein, as it is assembled at the ribosome, is determined by the sequence of bases in the mRNA, the so-called **genetic code**.

It has been established that three bases act as a coding unit defining a particular amino acid or coding for an ancillary piece of information, such as the beginning or end of a coding sequence. The genetic code is said to be degenerate, that is an amino acid may be coded for by two or three base triplets. In cases where two or more base triplets code for an amino acid the first two bases are common and the third is variable (Fig. 9.47).

Following the original postulation by Crick (1958) of the colinear relationship of nucleic acids and proteins a broad and intensive attack was launched to elucidate the nature of the genetic code (Crick *et al.*, 1961; Nirenberg *et al.*, 1965). Genetic evidence for the meaning of the triplet 'codewords' was amassed chiefly by experiments involving mutation in the DNA of bacteriophage viruses. In addition, valuable biochemical evidence was obtained from experiments in bacterial genetics, some of which have already been discussed (p. 149).

First position	Second position U	C	A	G	Third position
U	Phe	Ser	Tyr	Cys	U
	Phe	Ser	Tyr	Cys	C
	Leu	Ser	non.	non.	A
	Leu	Ser	non.	Trp	G
C	Leu	Pro	His	Arg	U
	Leu	Pro	His	Arg	C
	Leu	Pro	Gln	Arg	A
	Leu	Pro	Gln	Arg	G
A	Ile	Thr	Asn	Ser	U
	Ile	Thr	Asn	Ser	C
	Ile	Thr	Lys	Arg	A
	Met ('capital letter')	Thr	Lys	Arg	G
G	Val	Ala	Asp	Gly	U
	Val	Ala	Asp	Gly	C
	Val	Ala	Glu	Gly	A
	Val	Ala	Glu	Gly	G

Fig. 9.47 Table of the Genetic Code. The code is described as *degenerate* because more than one code may exist for a particular amino acid. The basis of the code is that three letters (representing three bases on a messenger RNA strand, the *codon*) code for one amino acid (see Chapter 2). (After Ambrose and Easty, 1970, courtesy of Thomas Nelson and Sons Ltd.)

The term **mutation** in its broadest sense encompasses all forms of genetic change including the changes in chromosomal structure, such as inversion and duplication, which have been described earlier in this section. However, the term is increasingly used for one particular form of mutation, that which concerns a single locus and was formerly called **point mutation**. The effect of such mutations is to increase the number of alleles that can occupy a locus, so increasing both heterozygosity and diversity of the gene pool in populations.

The disease **sickle cell anaemia**, the inheritable nature and symptoms of which have been described on p. 172, is known to be caused by a point mutation. Red cells in the blood of patients with this disease contain haemoglobin which differs from normal haemoglobin in several physical respects. For example, its mobility in electrophoresis is altered (Pauling *et al.*, 1950) and it is less soluble at low oxygen tensions (Perutz and Mitchison, 1950).

The haemoglobin molecule consists of four polypeptide chains (two α chains and two β chains) to each of which the iron-porphyrin prosthetic group **haem** is attached. The haem group carries a central ferrous iron atom (see Fig. 9.48) to which an oxygen molecule can be bonded loosely. Thus each haemoglobin molecule has the capacity for carrying four oxygen molecules. The oxygen absorption characteristics of the haemoglobin molecule, however, are determined not by the prosthetic haem groups but by the associated polypeptides. Different forms of haemoglobin, such as foetal and muscle haemoglobin, differ from normal adult haemoglobin only in respect of their polypeptides.

Fig. 9.48 A model of the molecule of haemoglobin comprising two α polypeptide chains and two β chains. Each chain has a haem group attached (see Fig. 2.17) giving the whole molecule the capacity for carrying four oxygen molecules. (Courtesy of Dr. J. C. Kendrew and Thames and Hudson Ltd.)

Analysis of the amino acid sequence in the polypeptide chains of sickle cell haemoglobin revealed the substitution of a single amino acid in each of the β chains; glutamic acid was replaced by valine (Fig. 9.49, and Ingram, 1956, 1961).

(a)

Haemoglobin type	Amino acid sequence at the amino-terminal of the β chain
Adult	······· Glu–Glu–Pro–Thr–Leu–His–Val–NH_2
Sickle cell	··· same ···–Val–········ same······–NH_2

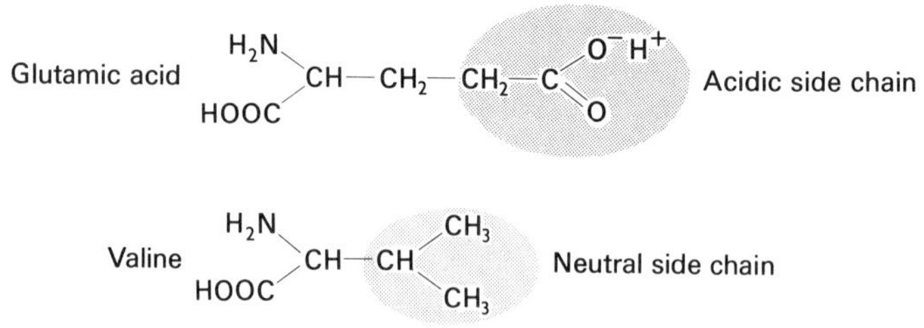

Glutamic acid: H_2N–CH(HOOC)–CH_2–CH_2–C(=O)–O^- H^+ — Acidic side chain

Valine: H_2N–CH(HOOC)–CH(CH_3)(CH_3) — Neutral side chain

(b)

Fig. 9.49 (a) Analysis of the amino acid sequences in normal adult and sickle-cell types of haemoglobin has revealed a single substitution near the amino terminal of the β chain. (b) Sickle-cell haemoglobin has a valine replacing a glutamic acid, in effect substituting a neutral side chain for an acidic one.

Thus the loss of a pair of acidic amino acid residues, with their charged side chains, and its substitution by an uncharged, neutral pair of amino acids distorts the whole molecule allosterically (p. 47) and changes its properties in the ways that have been described. (The total amino acid complement of the haemoglobin molecule is approximately 570 residues.)

From Fig. 9.50 it can be seen that the mRNA triplets which code for glutamic acid in normal haemoglobin are CAA and CAG. The triplets which code for valine are GUU, GUC, GUA and GUG. Thus any mutation in the DNA which codes for these bases resulting in a change from normal to sickle cell haemoglobin must involve at least two bases.

Glutamic acid	mRNA triplets DNA	CAA GTT	CAG GTC		
Valine	mRNA triplets DNA	GUU CAA	GUC CAG	GUA CAT	GUG CAC

Fig. 9.50 Codon changes required in the mutation giving rise to sickle-cell haemoglobin (refer to Fig. 9.47).

Confirmation of point mutation as a cause of sickle cell anaemia is given by analysis of **haemoglobin C** (HbC) which is found in patients with a different form of anaemia. HbC differs from normal haemoglobin in having a **lysine** residue in the place of the same glutamic acid residue that is substituted in the sickle cell haemoglobin, and both forms of anaemia are known to be alleles at the same locus.

This type of mutation then is the result of **substitution** of a few base pairs in the sequence of DNA (the cistron) which codes for the β polypeptides in the haemoglobin molecule. This leads to the formation of a complete, but altered, polypeptide. **Base deletion**, the loss of one or two base pairs, is likely to be more deleterious owing to the rest of coding sequence being altered, or put out of step. The loss of three base pairs would lead to a polypeptide being deficient in one amino acid. In the same way the addition of one or two bases, **base duplication**, alters the remainder of the code (Fig. 9.51).

Since metabolic pathways in the cell are regulated by enzymes which are proteins and which are coded for by DNA, mutations can lead to a variety of metabolic changes, most of which are deleterious (Fig. 9.52). In the eukaryotic organism most intensively studied at a genetical level, *Drosophila*, mutated genes are known for many loci (Fig. 8.14).

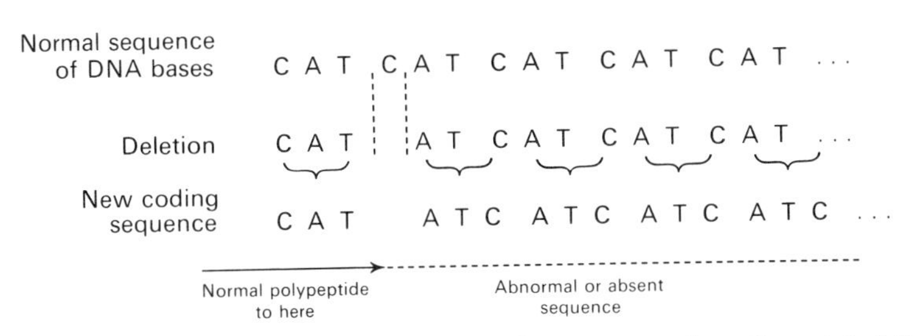

Fig. 9.51 Base deletion results in a faulty coding sequence from the point of the deletion unless there is a base duplication to follow which would put the code back in step. (After Paul, 1965.)

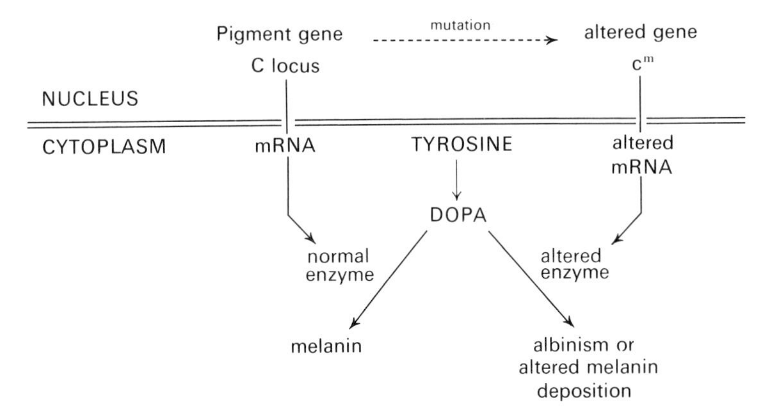

Fig. 9.52 Possible mode of action of mutation at the *C* locus giving rise to changes in pigment deposition in vertebrates (p. 200).

Rate of Mutation

The rates at which genes undergo spontaneous mutation, measured in terms of number of mutations per million sex cells, vary both between organisms and between gene loci in a single organism. The degree of variation, however, is less than might be expected. If mutation occurred randomly on a time basis one would expect wider differences. In *Drosophila*, for example, a two-week life span coincides with a mutation rate of about 30 per locus per million sex cells. Man, whose life span in terms of reproductive cycle is about twenty-five years and whose germ cells are therefore exposed to mutational forces for a period which is over six hundred times greater than that in *Drosophila*, might be expected to show a rate of mutation of about 19 000 per locus per million gametes. This is clearly not the case and it appears that mutation rate is adjusted to length of life cycle.

Mutation rates are easier to study in micro-organisms, which have shorter life cycles. In the fungus *Neurospora* a mutation which results in the reduced capacity of the organism to synthesize adenine occurs approximately six times in 10^8 spores. Similar biochemical mutants in bacteria have mutation rates of about half that figure.

(a)	Gene	Mutations per million gametes
Man (approximate values)	Muscular dystrophy (sex-linked, recessive)	50
	Haemophilia (sex-linked, recessive)	30
	Chondrodystrophy (autosomal, dominant)	70
	Albinism (autosomal, recessive)	28
	Phenylketonuria (autosomal, recessive)	25
Drosophila	White eye (sex-linked, recessive)	29
	Cut wing (sex-linked, recessive)	150
Maize	Waxy endosperm	0
	Sugary endosperm	2.4
	Yellow endosperm (dominant)	2.2
	Plant colour (R) (atypically high rate)	492
	Red aleurone colour	11

(b)

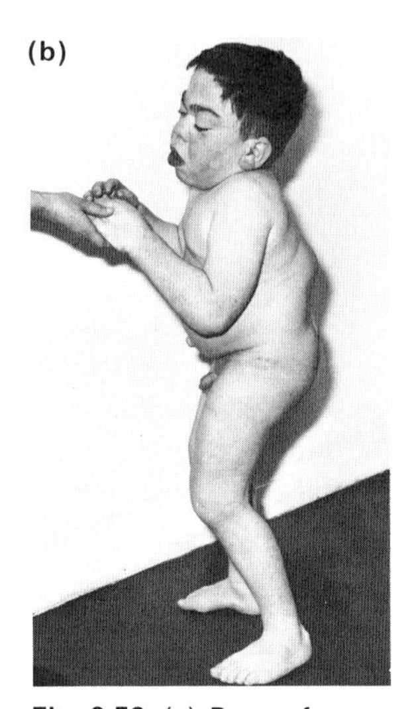

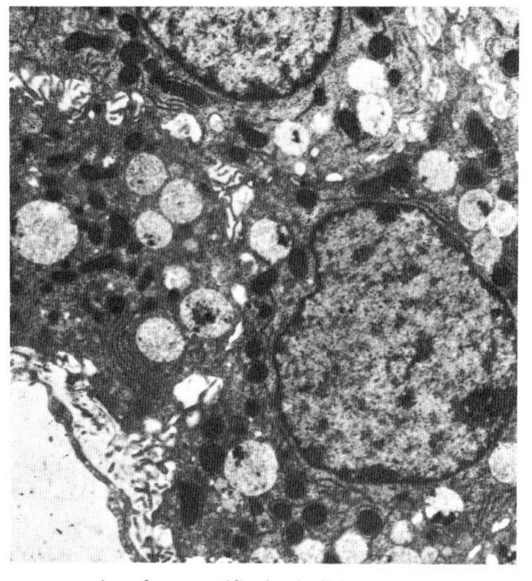

Fig. 9.53 (a) Rates of spontaneous mutation for specific loci. (b) A mutation in a gene regulating one stage in carbohydrate metabolism can cause Hurler's Syndrome, with widespread skeletal abnormalities (frequently lethal through respiratory infection in childhood) and mental retardation. (Right), an E.M. of liver from Hurler autopsy showing lipid inclusions which are absent in normal liver. (Courtesy of Dr. R. Lax, Kennedy-Galton Centre.)

The evolution of antibiotic-resistant strains of bacteria which, over successive generations, are able to tolerate increasing concentrations of the drugs occurs as a result of a 'stepwise' summation of mutant genes, each of which confers some additional tolerance.

Since mutation is of extreme importance in evolution one might expect the mutability of genes to be regulated. The overall rate of mutation is critical, since too high a rate would be damaging to the viability of too many individuals in the population, and too low a rate would lead to a reduction in diversity and adaptability. In a wide range of organisms mutation rates have been shown to be genetically controlled, both in respect of point mutations and chromosomal breakage.

In bacteria the existence of **mutator** genes can affect the rates at which other genes mutate in response to mutagens, such as X-rays. In maize the mutator gene **Dotted** (*Dt*) facilitates the mutation of the gene a_1 (colourless aleurone in the seed) to its allele A_1 (dark aleurone). In a plant with a *Dt* a_1 genotype the kernel shows many dark seeds corresponding to the mutated regions. In plants in which *Dt* is absent the a_1 gene is stable. This mutator gene **Dotted** does not affect the mutation rates of genes at other loci, or even of other alleles at the *a* locus (Fig. 9.54, and see Rhodes, 1938; Brink, 1954). Also present in maize are the loci **Dissociator** (*Ds*) and **Activator** (*Ac*) which regulate both point mutation rates at other loci and the frequency of chromosomal breakage and rearrangement (McClintock, 1950). It is interesting to note, in view of the previous discussion (p. 160) of the regulatory and quantitative role of that kind of chromatin, that the three loci which affect mutability in maize, *Dt*, *Ds* and *Ac*, are believed to lie in heterochromatic regions of chromosomes.

If the rate of mutation is critical with respect to the evolutionary success of a population in the ways described, it would seem that natural selection pressures would operate to optimize mutation rates, possibly by acting on mutator genes. Some evidence that this is the case has been gained from comparisons of mutation rates in two races of *Drosophila*. In the wild state one race inhabited a tropical environment and the second a temperate one. Spontaneous mutation rates were substantially the same in both wild populations, although it might have been expected that the race in the environment with the higher ambient temperature would have had a higher mutation rate. (Under laboratory conditions a 10C° rise in temperature increases the rate of spontaneous mutation in *Drosophila* populations between four and five times). However, when both races were grown in the laboratory under temperate conditions the race formerly in the tropical environment was found to have the lower mutation rate (Herskowitz, 1962d). It would seem, therefore, that genetic regulators of mutation maintained the rate at the same, possibly optimum, value in both races, suppressing the rate for the tropical race or increasing it for the temperate one.

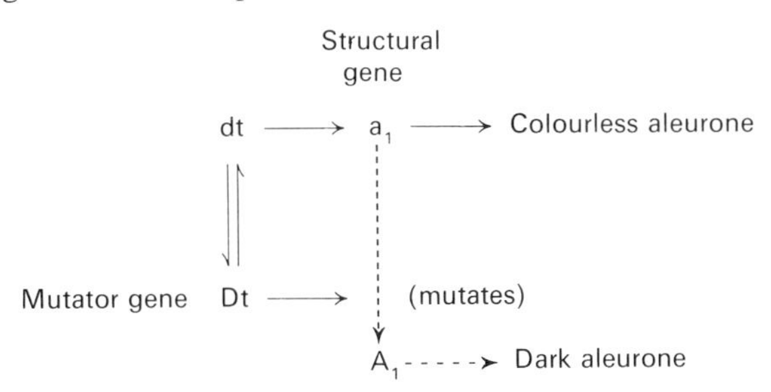

Fig. 9.54 In maize the presence of the Mutator gene facilitates the mutation of another gene from colourless aleurone (*a*1) to dark aleurone (*A*1).

Evolutionary Role of Mutation

Since mutation involves some change in DNA structure it is inherited by the descendents of the cell in which it occurs. **Somatic mutations**, that is mutations which occur in the cells of the body's tissues at large, have relatively little effect in evolutionary terms. The cells in which they occur are diploid, so the effect tends to be masked in the heterozygous state. Furthermore, unless they occur early in embryogenesis they are unlikely to affect a significant proportion of tissue cells. Somatic mutation in the meristematic tissue of the plant body can give rise to wider areas of affected tissue, thus the growth of an axillary bud in which a somatic mutation has occurred can give rise to aberrant branches, the so-called 'bud-sports'.

Although a somatic mutation may exercise some influence over the gene pool from which the next generation is drawn by, for example, reducing the viability of the organism in which it occurs, the mutated gene itself does not enter the gene pool. The mutant gene is lost from the population when the organism in which it occurred dies.

Mutations which have the greatest effect in evolutionary terms are **germinal mutations**, that is those occurring in cells which give rise to gametes. When a gamete carrying a mutated gene enters a zygote, subsequent mitosis ensures that the mutant gene is present in all the cells of the new organism. This represents a dramatic 'one-step' increase; the mutated gene is not present in the parental somatic cells but is present in all the cells of the offspring. In the parental organism in which the mutation occurred perhaps one gamete in a hundred thousand will carry the mutant gene. In the heterozygous offspring, however, half the gametes will contain the gene.

Most natural populations of organisms are in existence because they have survived in competitive conditions well adapted to relatively stable or slowly changing environments, over thousands or millions of generations. Although genes show low rates of mutation most loci will have mutated many times and the most advantageous mutants will have been incorporated in the gene pool at high frequencies, forming the 'wild-type' genes.

In this sense natural selection could be said to extend beyond the selection of 'gene frequencies and combinations' to the selection of the most suitable base sequencies within each cistron. At the ultimate level it is the survival of the DNA configuration best fitted to its environment, that confers the capacity for survival upon the organism. It follows, therefore, that in a well adapted population most mutations tend to be disadvantageous at the level of the individual organism. In the same way it can be seen that, because selection tends to make advantageous characters dominant, most mutations tend to have recessive effects.

The presence of mutated alleles is disadvantageous to the population only in respect of the likely loss of individuals possessing them in the homozygous state. In many cases there are considerable advantages gained through heterosis and genic interaction, which have been discussed in the previous section. In the wider sense mutation has provided the great variety of DNA configurations and combinations which has determined the wide range of organic life on this planet. Without mutation there can have been no evolution. At the population level gene mutations provide a relatively constant level of variation by replacement of alleles that have been eliminated from the gene pool.

The extent of the change resulting from a mutation is an important factor in evolution. Some mutations, for instance those involving the deletion of a single base pair early in a coding sequence, may have conspicuous and lethal effects. Their evolutionary significance is likely to be small, although such radical changes may have been involved in major evolutionary steps, as in those between phyla. Mutations with small effects, however, may be advantageous even in the short term by improving the population's adaptability to environmental changes. Stebbins (1966) has used a neat analogy, likening the organism in its habitat to a racing car tuned to perform on a particular circuit under specified conditions. Major alterations to the mechanics of the vehicle are likely to reduce its speed, or prevent it from running altogether, but minor adjustments, for example to the carburettor, can help to sustain the performance or even improve it if the weather conditions change.

Finally, it should be emphasized that, although mutation occupies a central position in the concept of the genetic basis of evolution, mutation itself is a random process. Mutation does not give direction to evolution; it cannot induce changes in gene frequencies in the population. It simply provides a source of genetic variability. Similarly, the rate of mutation does not affect the rate of evolution. Both the direction and the rate of evolution are determined by factors outside the organism. Mutation is rather like a passport. A passport allows a person to travel but in no way dictates the directions in which he should go or how quickly he should travel. Furthermore, a passport does not let a traveller go anywhere he wants; further modifications to the passport, namely visas, are usually necessary if he wants to travel into certain 'difficult environments'.

9.4 Continuous and Discontinuous Patterns of Variation

Genetic Sources of Continuous Variation

Most of the specimens, examples and problems selected for students of elementary genetics feature characters that are controlled by a single pair of genes. For example, in the mouse the *B* locus determining coat colour provides phenotypes which are either black (genotypes *BB* or *Bb*) or brown (genotype *bb*) and crosses between heterozygous animals can be used to demonstrate the expected phenotypic ratio of three black to one brown in the offspring. Mendel's experimental crosses with the garden pea involved the same genetic principle: his plants were either tall (*TT* or *Tt*) or dwarf (*tt*); either red-flowered (*CC* or *Cc*) or white flowered (*cc*); or had either round seeds (*RR* or *Rr*) or wrinkled seeds (*rr*). By the choice of such characters Mendel was able clearly to demonstrate the particulate theory of inheritance. By quantifying his results he was further able to show the nature of segregation and independent assortment.

A broader examination of phenotypes, especially in higher organisms, shows few such clear-cut divisions. Most physical and physiological characteristics show a gradation or spectrum effect across the population. In man, for example, height (Fig. 9.47) and weight show a frequency pattern that is approximately a normal distribution, as do I.Q. and a range of other characteristics, such as skin colour in mixed racial communities. Similar spectra of variation can be found in many domesticated animals and plants. Yields of milk or beef in cattle, and crop yields of grain or vegetable oil, show quantitative gradation between low-yield and high-yield varieties, selection of which forms the basis of the improvements given by systematic plant and animal breeding programmes.

The arguments in favour of the view that mutations having small effects are likely to be more useful in terms of the adaptability of the population than those having large effects can be extended to other genetic determiners of variation. A gradual spectrum of variation is likely to confer greater adaptability on a population than a small number of distinct types.

One genetic mechanism resulting in continuous variation occurs when a character is determined by several genes at different, unlinked loci, the **multiple gene** effect. Each locus may be occupied by a **contributory** gene or by its non-contributory, or **neutral**, allele. The phenotypic effect of the contributory alleles is cumulative; the phenotypic effect of the neutral genes is nil.

Consider a character such as pod length in a leguminous plant which, let us say, is controlled by two gene loci. If the shortest pod length is two centimetres and each contributory gene (*c*) adds one centimetre to the length (the neutral gene (*n*) adds nothing) one would have the following situation:

Genes at the pod-length loci	Pod length (cm)
All neutral genes (*nn nn*)	2
One contributory and three neutral (*cn nn*)	3
Two contributory and two neutral (*cc nn* or *cn cn*)	4
Three contributory and one neutral (*cc cn*)	5
All contributory genes (*cc cc*)	6

A cross between plants showing the shortest and longest pod lengths would give offspring plants of intermediate pod length:

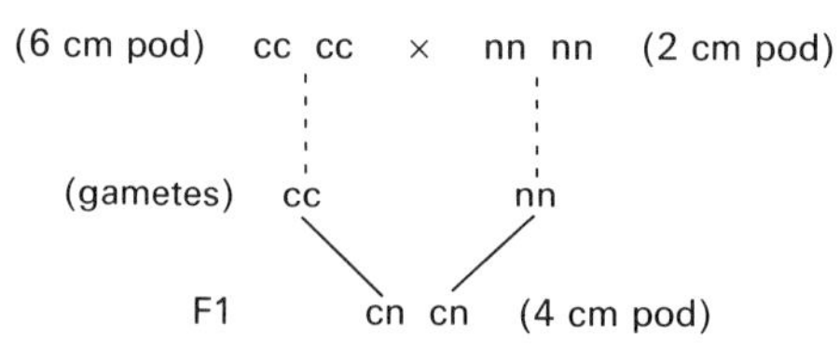

The offspring of a cross between the F1 plants, which have an intermediate pod length, is as follows:

F1 (4 cm) × F1   *cn cn* × *cn cn*

(gametes) $(\frac{1}{4}cc + \frac{1}{2}cn + \frac{1}{4}nn) \times (\frac{1}{4}cc + \frac{1}{2}cn + \frac{1}{4}nn)$

F2 $\frac{1}{16}cc\ cc + \frac{4}{16}cc\ cn + \frac{6}{16}cn\ cn + \frac{4}{16}cn\ nn + \frac{1}{16}nn\ nn$

Thus out of sixteen offspring the largest proportion, six, would be expected to show intermediate pod length. Four out of the sixteen would be expected to be shorter or longer than the intermediate length by one centimetre, and only one in sixteen would be expected to be shorter or longer by two centimetres.

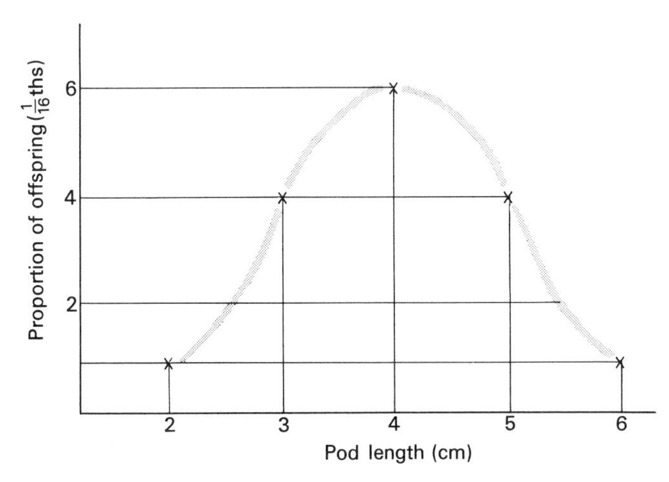

Where a greater number of loci is involved the distribution can be found by expanding the binomial $(c+n)^N$, where N = number of genes involved, that is the number of n genes plus the number of c genes. In the example quoted above the number of genes is four and the expansion is:

$$(c+n)^4 = c^4 + 4c^3n + 6c^2n^2 + 4cn^3 + n^4$$

in which the exponents refer to the number of genes in each category (c^3n = three contributory genes and one neutral gene) and the coefficients refer to the number of individuals possessing that genotype out of the total. Perhaps the easiest way to determine the coefficients is by means of Pascal's triangle (Fig. 9.55), in which each coefficient is obtained by adding together the two immediately above. Only alternate lines are relevant because the alleles in a diploid organism must be an even number.

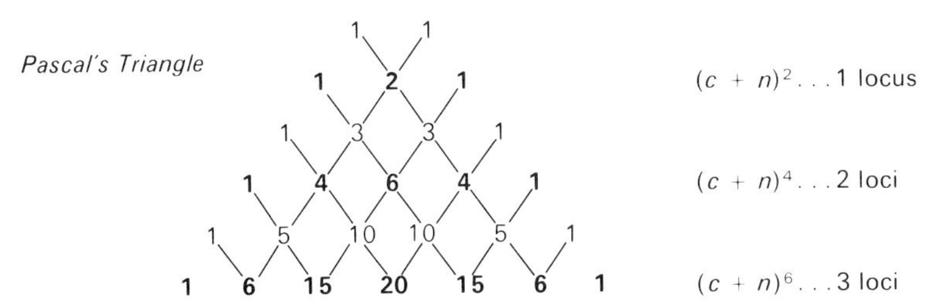

Fig. 9.55 Pascal's triangle can be used to determine the coefficients of the binomial expansion. Only the lines in bold are relevant to diploid organisms.

In terms of the distribution of variants in a population, however, one must proceed a stage further. The distribution of c and n genes in the offspring of the cross described above can only be applied to the whole population if c and n are in equal proportions in the gene pool, that is, if the proportion, or frequency, of c is $\frac{1}{2}$ and the frequency of n is $\frac{1}{2}$. Using the same hypothetical example of pod length in a legume, consider a case in which natural selection in a particular habitat had favoured greater pod length, so that the frequency of the contributory genes (termed p) was $\frac{3}{4}$ and the frequency of the neutral genes (termed q) was $\frac{1}{4}$, then

$$(p+q)^4 = p^4 + 4p^3q + 6p^2q^2 + 4pq^3 + q^4$$

and the distribution would be that shown in Fig. 9.56.

The distribution of variants, therefore, forms a bell-shaped curve when the frequencies are equal and a skew-shaped curve when they are unequal. In either event the advantage of such a pattern of variation is immediately apparent, namely that the largest proportion of the population is centred about the mean value of the character; in Galton's words, 'the majority are mediocre'. The system is conservative in that the number of individuals which vary widely from the mean, and are therefore less well adapted in terms of the evolutionary history of the population, is small.

The role of the extreme variants is probably limited to the larger and more radical steps in evolution (Mather, 1953); they have been described as 'hopeful monsters'. The fitness of a population, in the Darwinian sense, with respect to a particular environment, is the resultant of a number of factors (p. 169) among which is phenotypic variability. Extremes of phenotype with respect to one character, however, tend to lead to deletion, which can result in the loss from the gene pool of 'useful' combinations of genes associated with **other** characters. This hazard is reduced by mechanisms which increase genetic mixing and increase the randomness of inheritance from the gene pool (Fig. 9.7) but is promoted by factors which tend to fix linkage groups and increase genetic stability. In this sense, fitness is a compromise between genetic variability and genetic stability.

Modifying Genes

A second kind of genetic mechanism producing continuous variation is known to occur when a multiple gene system acts as a **modifier**.

Proportion of the population		$\times\frac{1}{256}$	Genotype	Pod length (cm)
$1\cdot(\frac{3}{4})^4$	$=(\frac{81}{256})$	81	c^4	6
$4\cdot(\frac{3}{4})^3(\frac{1}{4})$	$=4(\frac{27}{256})$	108	c^3n	5
$6\cdot(\frac{3}{4})^2(\frac{1}{4})^2$	$=6(\frac{9}{256})$	54	c^2n^2	4
$4\cdot(\frac{3}{4})(\frac{1}{4})^3$	$=4(\frac{3}{256})$	12	cn^3	3
$1\cdot(\frac{1}{4})^4$	$=(\frac{1}{256})$	1	n^4	2

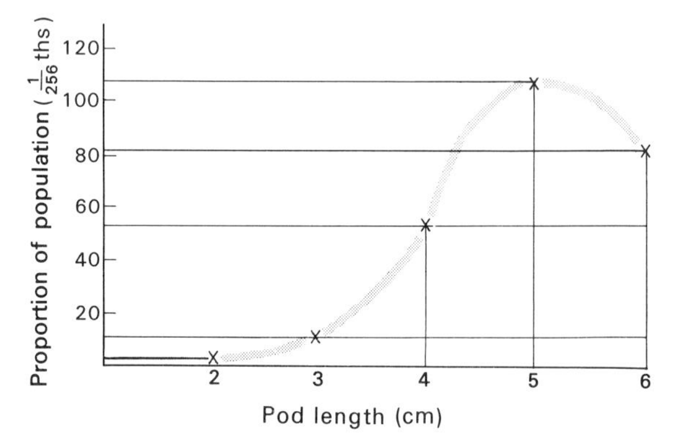

Fig. 9.56 The distribution of phenotypes in a population adopts a skew-shaped curve when the frequencies of the alleles are unequal. In this case the frequency of the *c* gene is $\frac{3}{4}$ and the *n* gene $\frac{1}{4}$.

Fig. 9.57 The effects of modifying genes can be seen in these hounds from the Eton College beagle pack. The expression of the piebald gene (s^p) is modified epistatically and shows variation from an almost completely dark animal, in the foreground, to a pale animal, in the background.

For example, in many mammals the so-called *S* locus determines the extent of the colour of the coat. An animal with a *S* – genotype has a solid coat colour, but the recessive homozygote *ss* has a spotted appearance.

In the beagle breed of dog there are several alleles at the *S* locus, of which the commonest is s^p, or **piebald spotting**. Dogs that are homozygous for the s^p allele are piebald, showing patches of black, tan and white. But in a beagle pack (Fig. 9.57) the pattern of piebald marking shows continuous variation from almost solid colour to almost complete white. This is known to be due to a number of genes at different loci which modify the action of the s^p gene in an epistatic (see p. 200) manner.

Genetic Sources of Discontinuous Variation: Genetic Polymorphism

A population which exhibits two or three (rarely more) distinct phenotypes with respect to a particular character is said to exhibit **polymorphism**. Genetic polymorphism has been defined by E. B. Ford (1940) as, 'the occurrence of two or more discontinuous forms, or phases, of a species in such proportions that the rarest of them cannot be maintained merely by recurrent mutation'. Thus a distinction can be drawn between the discontinuous pattern of variation arising as a result of genetic polymorphism, and either the continuous variation resulting from the concerted action of multiple genes, or the discontinuous variants that arise from time to time by mutation. Polymorphic variants arise either through a single pair of alleles having a large effect or a group of genes acting as a unit, or supergene (p. 188).

A population which is evolving, that is in which gene frequencies are changing consistently in response to environmental selection pressures, may exhibit a form of polymorphic variation which changes as evolution proceeds, thereby creating a temporary, or **transient polymorphism**. Perhaps the more interesting form of polymorphism, however, is that in which the phases are maintained in a relatively fixed state in the population, the so-called **balanced polymorphism**.

There are several genetic mechanisms which generate balanced polymorphism in populations. For example, in wild *Drosophila* populations chromosomal inversions (p. 188) are known to occur widely. The genes in the inversion tend to be held together because crossing-over in heterozygous flies is either inhibited or results in the production of non-viable gametes. Where such groups of genes, or supergenes, are of selective advantage there is clearly much to be gained by heterozygosity.

In these populations there exist phases which are homozygous for the normal chromosome, homozygous for the inversion, and heterozygous. The proportions of these phases in the population is determined partly by the relative frequencies of the normal and inverted chromosomes, except that in highly competitive conditions, such as overcrowded population cages (p. 189), the proportion of heterozygotes is higher than would be expected from a Hardy–Weinberg distribution. This suggests that there are additional advantages gained as a result of **heterosis**; inversion heterozygotes probably carry different alleles at many loci in the region of the inversion for which both forms of the homozygote will carry the same allele. In wild populations the distribution of the chromosomal phases also appears to be subject to strong selection pressures, in that they show fluctuations in response to seasonal changes (Dobzhansky, 1947) and differences between different habitats (p. 188).

Polymorphism due to single gene effects is, like the supergene effect, maintained by heterozygous advantage and also appears to reflect the presence of strong selective pressures. Probably the most striking example of polymorphism in man is **sickle cell anaemia**, the pattern of inheritance and symptoms of which have been described elsewhere in this book (p. 172). The polymorphic phases of sickle cell anaemia are: homozygous for the normal gene $(++)$, heterozygous $(+h^s)$ and homozygous for the recessive allele (h^sh^s).

The mutant allele h^s has lethal effects. Children who are born homozygous for this allele usually die before reaching reproductive age. When their blood is exposed to low oxygen tensions, such as exist in capillary beds in the tissues, the haemoglobin molecules in the red cells coagulate to form insoluble, linear aggregates. These distort the red cells, which then may clog blood vessels. Such 'sickled' red cells tend to disintegrate, which leads to anaemia and to jaundice due to the high concentration of the bile pigment bilirubin in the blood.

In temperate climates the frequency of the h^s gene in human populations is at the low level that one associates with a mutational source undergoing continuous deletion and consequently the disease is rare. In African Negroes in equatorial regions, however, infant mortality due to sickle cell anaemia is considerably higher, about 4%.

Thus, if the frequency of the sickle cell gene is termed q and the frequency of its normal allele is termed p, the distribution of the gene in the population would be expected to be given by the Hardy–Weinberg equation:

$$\underset{(++)}{p^2} + \underset{(+h^s)}{2pq} + \underset{(h^sh^s)}{q^2} = 1 \quad \text{(p. 175)}$$

Since $q^2 = 0.04$
then $q = 0.2$ or **20%** (the frequency of the h^s gene)
and since
$p+q = 1$, $p = 0.8$ or **80%** (the frequency of the normal allele)

The frequency of heterozygous individuals in the population would be expected to be given by $2pq = 2(0.8)(0.2)$, or **32%**.

In the next generation, however, the frequency of the sickle cell gene would be expected to be reduced owing to the failure of the

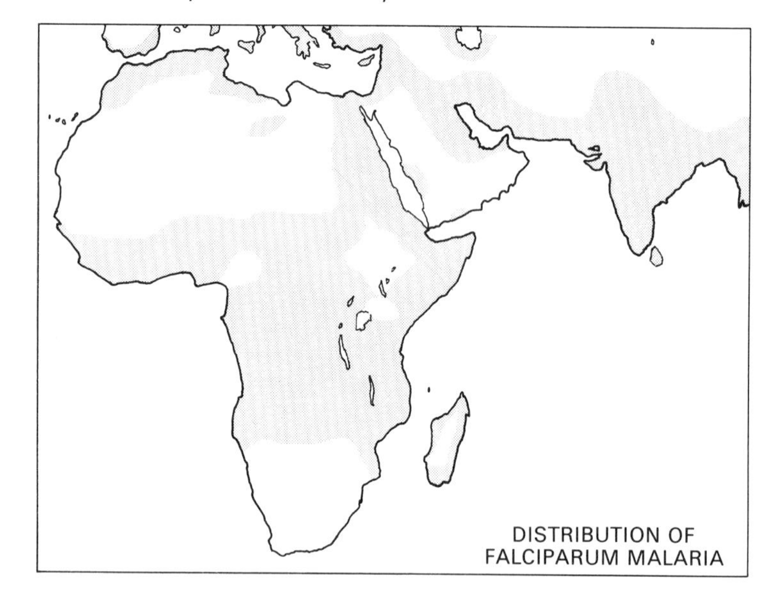

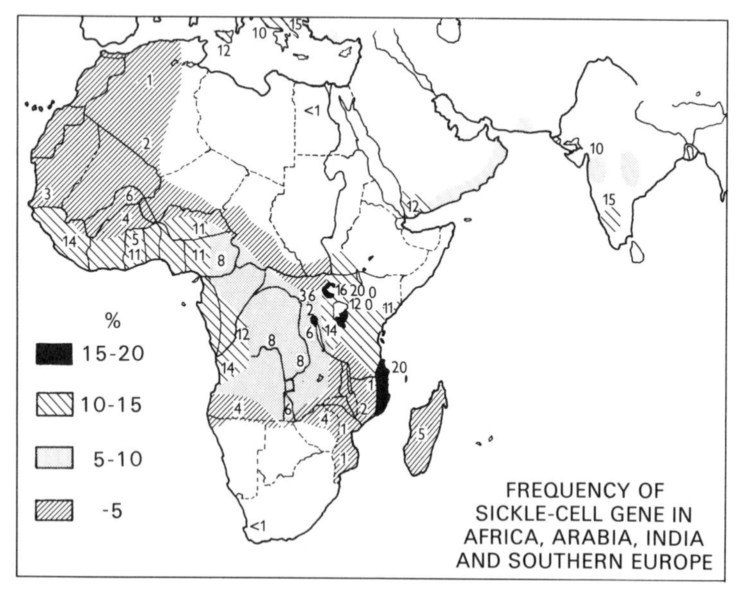

Fig. 9.58 Maps to compare the distribution of malignant tertian malaria and the frequency of the sickle-cell gene. (Courtesy of Dr. A. C. Allison.)

homozygotes (h^sh^s or q^2) to contribute their genes to the gene pool. The incidence of sickle cell anaemia among the newborn of the next generation would be expected to be given by:

$$\left(\frac{q}{1+q}\right)^2 = \left(\frac{0.2}{1.2}\right)^2 = 2.8\% \quad \text{(p. 176)}$$

Therefore the value of q, the frequency of the sickle cell gene, would fall to $\sqrt{0.028}$, or approximately 17%.

Thus in successive generations the frequency of the sickle cell gene and the incidence of sickle cell disease would be expected to decrease in accordance with the curve on p. 177. In tribes living in equatorial Africa, however, this does not happen. Allison (1954, 1956) has demonstrated the association between the high frequency of the lethal gene and the areas where malignant tertian malaria, which is caused by the sporozoan parasite *Plasmodium falciparum*, is transmitted by mosquitoes all the year round (Fig. 9.58).

The frequency of the h^s gene can only remain at the abnormally high value of 20% if the proportion of heterozygotes reproducing in the population is higher than would be expected from the Hardy–Weinberg equation. In other words, to sustain an incidence of sickle cell disease due to the homozygous recessives in the population (q^2) of 4% the value of 2pq must be raised each generation. This is because the heterozygous individuals are the only source of h^s genes for the gene pool from which the next generation is drawn. The value of 4% for the frequency of the homozygous recessive individuals arises from an expected one-quarter of the offspring from parents who are both heterozygous.

Thus
$$\tfrac{1}{4}(2pq)^2 = 0.04$$
$$2pq = \sqrt{0.16}$$
$$= 40\%$$

Therefore, to sustain the frequency of the h^s gene the proportion of the heterozygotes contributing to the next generation must rise from 32% to 40%. From these figures it can be argued that an environmental selective force is favouring the heterozygous individuals against the normal homozygous ones by a factor of twenty-five per cent (Fig. 9.59). (The heterozygote is 1.25 times better fitted to the environment than the homozygote.)

Allison has shown that the incidence of malaria among children is significantly lower for heterozygotes and he has produced evidence from clinical trials that the malarial parasite is the selective agent maintaining the polymorphism. In Negro tribes living in regions

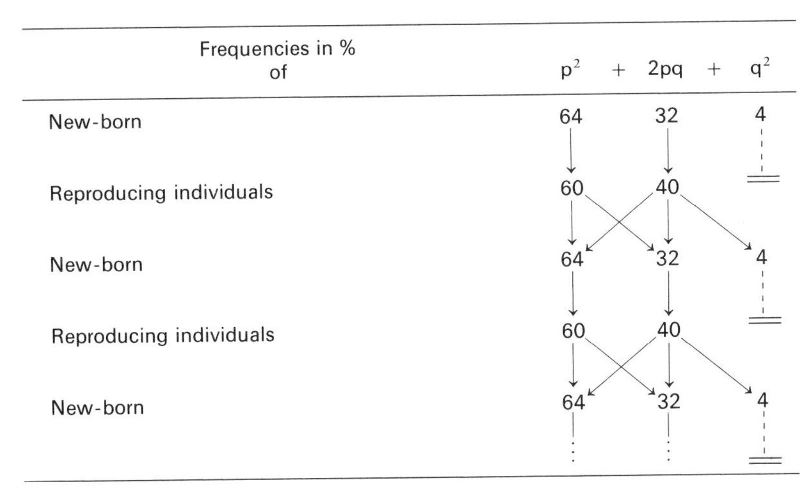

Frequencies in % of	p^2	+ $2pq$ +	q^2
New-born	64	32	4
Reproducing individuals	60	40	
New-born	64	32	4
Reproducing individuals	60	40	
New-born	64	32	4

Fig. 9.59 In a malarious region in which the human population has a 4% incidence of sickle-cell anaemia amongst the children (shown by q^2), it must be argued that the proportion of heterozygous individuals ($2pq$) must increase from 32% of the population at birth to 40% of the population at maturity. Likewise, the proportion of homozygotes (p^2) must fall from 64% to 60%.

adjacent to those described, in which malaria is less prevalent or benign, the sickle cell gene is rare. Here a balanced polymorphism does not exist; the frequency of the disease conforms approximately to the occurrence of the gene by mutation and its selective elimination. In malarial regions the polymorphic nature of sickle cell anaemia is clearly established (Fig. 9.60).

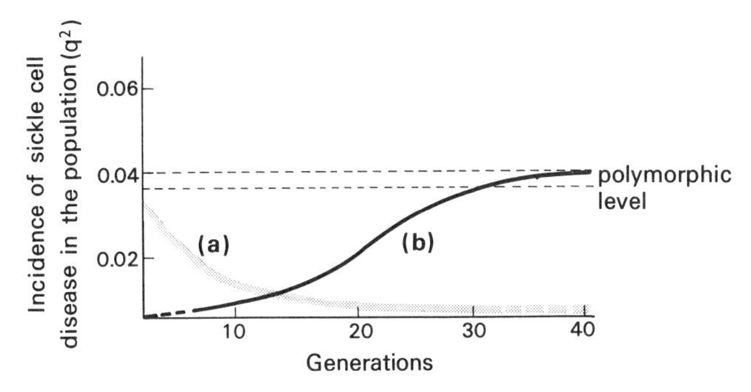

Fig. 9.60 Hypothetical changes in the incidence of sickle-cell anaemia in populations moving out of a malarial region (a) and entering a malarial region (b).

Polymorphism in Insects

Conditions of genetic polymorphism are known to exist in a wide variety of natural populations of plants and animals (Ford, 1957). The Lepidoptera have proved profitable for studies in polymorphism and evolution mainly because field trials are relatively easy to conduct and readily yield quantitative data which can be complemented by laboratory tests (Kettlewell, 1965).

The Scarlet Tiger Moth. In the scarlet tiger moth, *Panaxia dominula*, a colour polymorphism due to a mutation at a single locus is present. The normal form, *dominula*, shows disruptive colouration on its fore-wings. Its scarlet hind-wings give a flashing effect in flight which makes the line of flight difficult to follow. Its colouration is also 'warning' in that the moth appears to be distasteful to predators such as dragonflies and birds (Sheppard, 1953). The second form, in which the mutant gene is in the homozygous state, is termed *bimacula* and carries widespread dark areas on the fore- and hind-wings. No dominance is present and the heterozygote is distinguishable from either homozygote, being intermediate in colouration. It is termed *medionigra* (Fig. 9.61).

There are probably several selective factors maintaining the polymorphism in *Panaxia* but their precise roles are not fully understood. Laboratory breeding experiments have confirmed observations in the wild that mating is **disassortative**, or non-random, that is mating tends to take place between unlike genotypes (p. 223). Choice of mates is exercised by females, who usually mate only once. Males, however, have been observed to mate with up to ten females. When the pigment gene is at low frequencies, therefore, a heterozygous, or *medionigra*, male will be at an advantage since he is more likely to mate with more normal, or *dominula*, females than would a normal male. The *medionigra* gene therefore confers a selective, behavioural advantage on the insect.

On the other hand the *medionigra* males appear to have reduced fertility. Eggs of back-crosses between *dominula* and *medionigra* (+ + × + b) were released in a field trial near Oxford (Sheppard, 1953). The expected ratio of adult moths to emerge would be 1 *medionigra*:1 *dominula*, but when mature larvae were collected and allowed to pupate the ratio of adults emerging was 9 *medionigra*: 21 *dominula*. This highly significant discrepancy suggests that the *medionigra* gene is at a selective disadvantage in the part of the life cycle between fertilization and pupation. Indeed calculations show that the survival rate of *medionigra* from egg to imago is only 75% of

Fig. 9.61 The polymorphic phases of *Panaxia*, the Scarlet tiger moth. Top, the *dominula*, or normal male (left) and female forms, which are homozygous for the normal gene (+ +). Centre, the male and female *medionigra* form, which are heterozygous for the mutated allele (+b) and bottom, a single specimen of *bimacula*, the homozygote (bb), which is now very rare. (Courtesy of Professor E. B. Ford.)

that of the homozygous *dominula* (Sheppard and Cook, 1962).

The selective disadvantage of the *medionigra* gene appears to be increasing. Records of numbers of the three forms of the moth collected from a colony at Cothill, near Oxford, between 1939 and 1952 show a consistent and rapid decline in the frequency of the gene in the population, from approximately 10% to 3%. Over this period, therefore, the polymorphism was not balanced. The *medionigra* gene has been shown to affect mating behaviour, in which it is advantageous at low frequencies, and fertility, colour pattern, liability to predation and survival in the larval stage, in which it is generally disadvantageous. During this period Sheppard has calculated that the gene had an average disadvantage of about 8%, but has been unable clearly to demonstrate how the gene could have been of greater advantage prior to 1939, or what changes in the environment since then have increased the disadvantage of the gene.

The sampling of the colony of *Panaxia* at Cothill continued after 1952 and the frequency of the *medionigra* gene was found to decline more slowly. Projections of the curve would seem to suggest that the frequency of the gene is levelling off at about 1%. At this level the homozygous form of the insect, *bimacula*, would be expected to be very rare (one in ten thousand) and, indeed, in samples taken at Cothill it has not been found since 1959 (Fig. 9.62).

Polymorphism and Selection in the Peppered Moth. The Peppered Moth, *Biston betularia*, is a light coloured insect marked with fine dark lines and dots which give cryptic colouration on lichen and on the pale bark of trees such as birch. The insect is active at night and by day rests, with wings slightly outspread, fully exposed on tree trunks.

A melanized or dark form, *carbonaria*, was first recorded in Manchester in 1848 and Kettlewell (1955, 1956) has demonstrated the

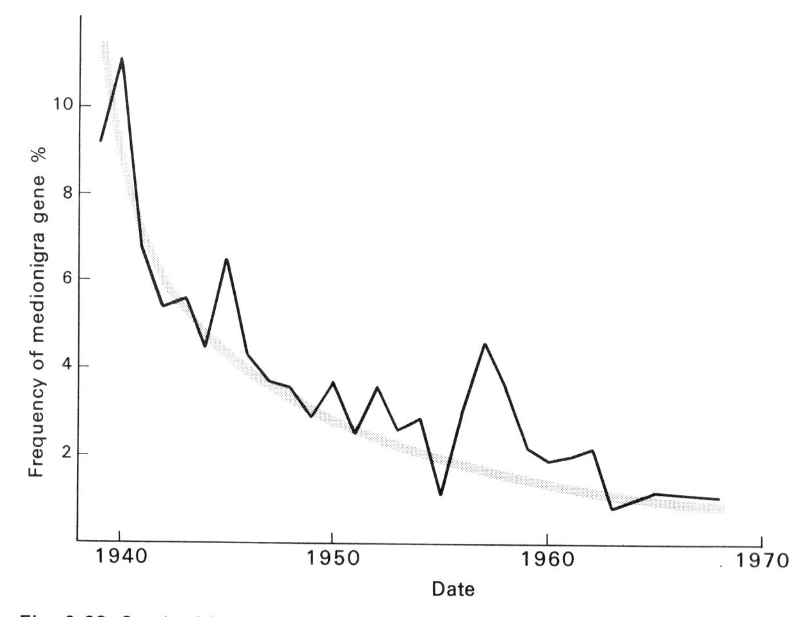

Fig. 9.62 Graph of frequency of the *medionigra* gene in the Cothill colony.

spread of this form through the industrialized areas of north-west England and the Midlands. He released a number of marked moths of both the normal and the melanized forms in a bird reserve near Birmingham where the trees were sooty with smoke pollution. Later he set a light trap and recorded the moths he recaptured, which were compared to the numbers released (Fig. 9.63).

	Number released	Number recaptured
Melanized form	416	119
Light form	168	22
% melanized	71	84
% light form	29	16

Fig. 9.63 Table of the numbers of light and dark forms of the Peppered Moth released in a field trial showing the high proportion of dark forms recaptured in a sooty area. (After Kettlewell, 1956.)

Of the 584 moths released, 71% were melanized and 29% were of the light form. By contrast, of the 141 moths recaptured 84% were of the melanized form and only 16% were of the light form.

Observations through binoculars by day showed that birds, particularly robins and hedgesparrows, preyed on the insects when they were resting on the trees. It is reasonable to conclude that in sooty areas the lighter, normal form is more conspicuous and actively selected against by the predators. The melanized form is cryptically coloured in these conditions.

The dark form of the wing is due to a single gene and is dominant to the normal, light form. Since the gene originated its frequency has increased rapidly and in industrialized areas has now stabilized at a frequency of 0.96 to 0.98. Like other forms of genetic polymorphism industrial melanism in the peppered moth appears to be maintained by heterozygous advantage. Back-crosses between heterozygous parents and parents homozygous for the recessive form (light-coloured) would be expected to show a 1:1 ratio of offspring between the dark and light forms. But when broods of such crosses were raised and fed on the sooty, polluted food of the industrialized areas the ratio was 108 dark-winged:65 light-winged forms (Kettlewell, 1957). From experiments of this kind pointing to the mode of heterozygous advantage, and from calculations of the gene frequencies it has been estimated that the polymorphic pattern shown by *Biston*

Fig. 9.64 *Biston betularia*, the peppered moth. Top, a pigmented, or *carbonaria*, form and a normal homozygote resting on lichen on a tree in Dorset, an area free from smoke pollution. Below, moths of the same types resting on a smoke-blackened (no lichen) tree trunk in the Birmingham area. The selective advantages of each polymorphic phase with respect to predation from birds is clear. (Courtesy of Professor E. B. Ford.)

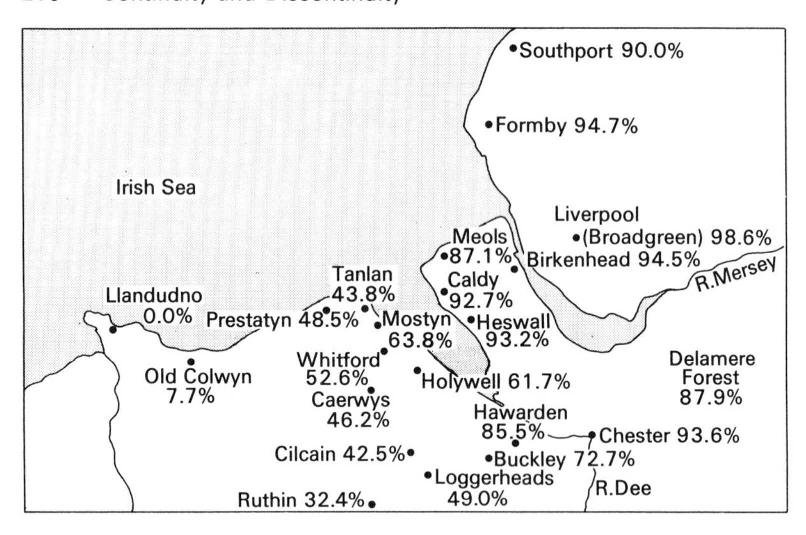

Fig. 9.65 Map showing the varying percentage of the black form of *Biston betularia* found on industrial Merseyside and in the countryside of the Wirral. (Courtesy of BBC Publications.)

betularia in an industrial environment is consistent with an approximate fitness of *Cc* of 100; *CC* of 90 and *cc* of 50 (Figs. 9.64 and 9.65. See also Haldane, 1956; Ford, 1965).

Blood Group Polymorphisms in Man

Sickle cell polymorphism in human populations in malarial regions has proved to be a remarkable study for several reasons. Most important was the ability to identify individuals who were heterozygous for the sickle cell gene. Their blood cells contain some S-haemoglobin and when a sample of the blood is exposed to oxygen tensions appreciably lower than those found in the tissues some sickling can be observed. This has been termed 'sickle cell trait'. In addition, the identification of the malarial parasite as the selective agent neatly demonstrated the role of heterozygous advantage in maintaining polymorphism.

Another polymorphism in man is that shown by the ABO blood groups. Although a considerable amount of data is available owing to the medical importance of blood grouping, the nature of the polymorphism and the factors sustaining it are less clear than in the case of sickle cell anaemia.

The blood groups of the ABO series, which have importance in relation to blood transfusions because of the accompanying serum antibodies, are determined by a single locus. Three principal alleles are present in the gene pool of most human populations, so in the diploid cells of an individual the locus may be occupied by any two out of the three. Two alleles (which some authorities term I^A and I^B—'*I*' standing for 'isoagglutinin' or antigen) are co-dominant. If they occur together in the genotype in a heterozygous state the red cells of that person carry both A and B antigens and the blood is termed Group AB. A person homozygous for the recessive allele *i* carries no antigens of this type and the blood is termed Group O.

The medical significance of the ABO blood group system is that the blood plasma contains complementary antibodies, or **agglutinins**, which coagulate red cells with the appropriate antigen. Thus a person with Group A blood carries A antigens on the red cells and anti-B agglutinins in the plasma. The transfusion into such a person of blood from a Group B individual can have serious consequences owing to the clumping of the donor red cells as they enter the recipient's blood system (Fig. 9.66).

Geographical and Racial Distribution of Blood Groups. The frequencies of the blood group alleles in human populations appear to be relatively constant, forming stable polymorphic patterns. The frequencies do vary, however, between populations native to different parts of the world, giving rise to fairly characteristic racial blood group frequencies.

A sample of a native English population showed a high frequency of the recessive allele *i*, 71%, compared with a frequency of the I^A allele of 24% and a frequency of the I^B allele of 6%. The frequency of the I^B allele increases eastwards across Europe, with the exception of the Basque and Lapp populations in whom it is very low or absent, and Asia until it reaches a value of approximately 30% in a sample of the native Siberian population. The same allele, however, was found

Genotype	Group terminology	Red cell antigen	Plasma agglutinin
ii	0	none	anti-A and anti-B
I^AI^A or I^Ai	A	A	anti-B
I^BI^B or I^Bi	B	B	anti-A
I^AI^B	AB	A and B	none

Fig. 9.66 Table of the ABO blood group alleles in man and the associated red cell and serum agglutinogens and agglutinins.

	Gene frequencies in %		
	I^A	I^B	i
Europe			
Scottish (highlands)	17.9	5.1	76.9
English	24.0	6.0	71.0
French	26.4	4.5	69.1
Swiss	27.0	5.8	67.1
Poles	32.7	14.4	52.9
Hungarians	35.0	15.4	49.6
Russians (west)	25.0	19.0	57.0
Asia			
Persians	22.5	15.7	61.9
Pakistanis (Punjab)	18.3	26.2	55.5
Indians (N. Cent. province)	18.6	29.7	51.8
Siberia (Irkutsk)	15.0	28.0	57.0
Chinese	22.0	20.0	59.0
Japanese	28.0	17.0	55.0
America			
Eskimos (Hudson's Bay)	25.6	1.5	72.9
Indians (British Columbia)	9.6	0.4	90.0
Indians (Chippewa)	6.4	0.0	93.6
Indians (Navajo)	13.3	0.0	86.8
World Total population*	21.5	16.2	62.3

frequency of i

frequency of I^B

Fig. 9.67 Table of approximate blood group frequencies. (Data from Mourant, 1954, and Boyd, 1953.)

to be absent in samples of American Indians of the Navajo and Blackfeet tribes, and in samples of Australian aborigines. (See Fig. 9.68b.)

In most of the populations sampled the frequency of the recessive allele i exceeded 50%, being the highest in the North American Chippewa Indians of Minnesota (93.6%) and the lowest in the Hungarians (49.6%).

Evolution of Blood Groups. The facts that the ABO blood groups are determined at one genetic locus (are unifactorial) and that the phases are maintained at relatively high frequencies indicate a genetic polymorphism. It has been suggested that the differences between various populations with respect to blood group frequencies can be attributed to genetic drift, that is the chance divergencies that can occur when populations are small and isolated, such as undoubtedly happened in man's early evolutionary history (Wright, 1940 and p. 177). Boyd (1940) and Wright have suggested that blood groups have neutral survival value and pointed to the absence of anatomical differences attributable to the ABO genes and the absence of any effect on the choice of mates.

* Estimated by McArthur and Penrose (1951).

Ford (1942, 1965), however, has argued consistently that no gene can have been devoid of selective advantage and that a polymorphic situation such as exists in human blood groups implies strong selective forces past or present. Evidence has been accumulating slowly which has suggested relationships between various diseases and specific blood groups. Aird *et al.* (1953) showed a positive correlation between cancer of the stomach and Group A and it was subsequently shown that duodenal ulcer is commoner in persons of Group O than of other groups. Admittedly these diseases generally affect people at a post-reproductive age, at any rate in contemporary society, so perhaps of greater significance in relation to natural selection were the demonstration that more Group A than Group O babies die of bronchopneumonia in the first two years of life (Struthers, 1951) and the evidence for higher viability in women of Group B than of other groups (Allan, 1954).

As already shown, analysis of blood groups reveal that the associated genes occur at specific and different frequencies in different races. Ford (1965) has argued that the differences are attributable directly to the forces of natural selection which have determined other racial differences, '. . . polymorphic genes stabilized at given frequencies in one population may be maintained with equal precision at quite different proportions in another. That situation is determined partly by an adjustment to the balanced gene-complex of each race and partly by the necessity to interact favourably with the environment.'

The maps (Figs. 9.68a, c, d) showing the distribution of the blood group genes in Europe, from the remarkable compilation and analysis of data by Mourant (1954) display a cline from west to east with respect to the i and I^A genes. The map in Fig. 9.68(a) shows a similar cline with respect to the rarer I^B gene outwards from a high of about 30% in central Asia. These patterns probably reflect gene flow as a result of the migrations of human populations over the last one or two thousand years. The fact that some populations which tend to be reproductively isolated by cultural barriers maintain blood group frequencies which differ markedly from those of surrounding populations, for example the Basques and the Lapps, which have among the lowest frequencies of Group B and the highest frequencies of

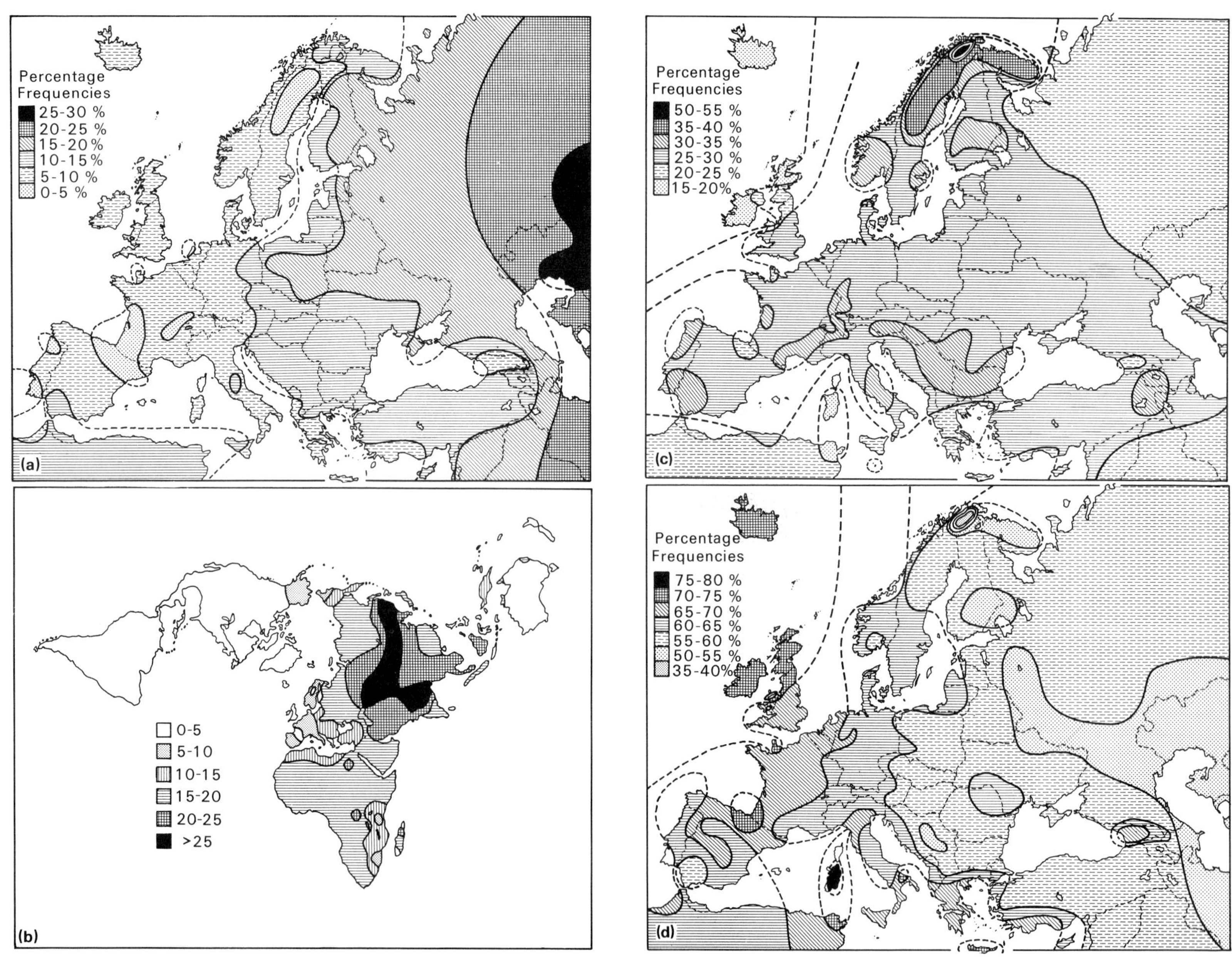

Fig. 9.68 Maps of blood group frequencies. (a) Blood group Gene *B* in Europe. (b) World map of Gene *B* frequencies. (c) Blood group Gene *A* in Europe. (d) Blood group Gene *O* in Europe. (After Mourant, 'The Distribution of Human Blood Groups', Blackwell Scientific Publications, Oxford.)

Rhesus negative in the world, and the Gypsies (which are essentially a Hindu race inhabiting western Europe) suggests that the ABO gene frequencies are primarily fitted to the racial gene complex. On the other hand, the high frequency of the I^B gene in Northern India and Central Asia is common to Caucasoid *and* Mongoloid peoples, which suggests that it had a selective advantage in that environment (Fig. 9.69).

	I^A	I^B	i	Rh^-
French (Toulouse)	29.6	5.8	64.6	43.9
Basques (San Sebastian)	24.0	0.0	76.0	53.2
Spanish (Galicia)	29.0	5.0	66.0	24.9

Fig. 9.69 Blood group samples from people in the Basque country of northern Spain differ markedly from the French to the north and the Spanish immediately to the south suggesting a separate origin and long isolation.

In considering the possible roles of blood group genes in human evolution, ethnology or anthropology, the distinction must be borne in mind between the patterns of continuous variation described in the previous section and the clines of polymorphic frequencies. The ABO blood groups are truly polymorphic in that all individuals must be Group A, B, AB or O and no intermediate groups are possible.

9.5 From Variation to Speciation

In this chapter we have discussed several sources of variation, both environmental and genetic, and the common modes of variation, both continuous and discontinuous, that exist in natural populations of plants and animals. It is now necessary to draw together the threads and attempt to relate these sources and modes of variation to the concept of evolution by consistent changes in gene frequency that was discussed at the beginning of the chapter.

One of the central pillars of contemporary evolutionary theory is the concept of **adaptive radiation**. A large population of a species spread over a wide geographical area is subject to two conditions which both promote and direct its further evolution. The first is that gene flow is reduced through populations which occupy large areas. An animal is more likely to choose a mate from the animals in its immediate vicinity. Similarly a flowering plant is more likely to be pollinated, especially by insects, but even by the wind, from near-by plants of its own species. Such populations, therefore, do not conform to the criteria of the Hardy–Weinberg Principle, since mating or pollination are not random and gene mixing does not occur throughout.

Second, a large geographic area is likely to contain more than one, probably many, different habitats. The species may therefore spread over many regions showing a variety of climates, altitudes, and so on, and cohabiting species. As the species spreads, or radiates, outwards and adapts to new environments, these two conditions, the break-down of gene flow and the exposure to different habitats, lead to the formation of races within the species.

The Race Concept

Races within a species differ from one another in respect of gene frequency. As already described, the conditions which lead to race formation are found when members of a widespread population with reduced gene flow move to occupy new regions which have different environmental selective pressures. Such selective pressures act by changing the frequencies of the genes in the gene pool (p. 170). They give rise to localized pockets in the population which differ from the bulk of the species and from the members of other 'pockets', or races, **quantitatively**, that is by virtue of the proportions of the genes which constitute the gene pool, rather than by a qualitative distinction arising from the presence of different genes.

In summary, therefore:

population spreading to new habitats
+
differential adaptive values of gene combinations
+
non-random mating

} = formation of new races

Many of the improvements in plant and animal breeding have consisted of the artificial selection of desired characters, resulting in many races of cattle, horses and dogs (termed 'breeds'), of wheat and other cereals (termed 'strains') and of flowers such as roses (termed 'varieties'). In 1895 agronomists at the University of Illinois began an experimental selection programme in maize which was to last 50 years. They selected and inbred for four characters; high and low oil content, and high and low protein content of the seed. The results were published in 1952 (Woodworth *et al.*, 1952). After 50 generations the high-oil race yielded 14% and the low-oil race yielded 1%. The high protein race yielded 21%; the low protein

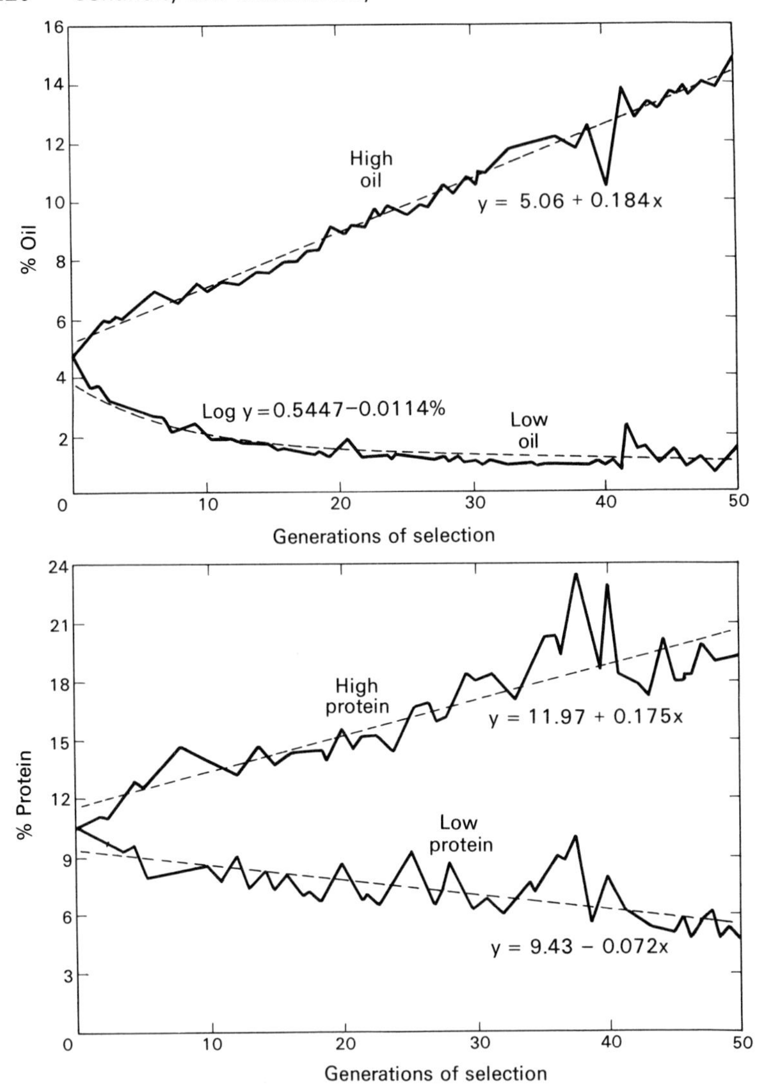

Fig. 9.70 Artificial selection for high and low content of oil and protein of the seed in maize, over fifty generations. With the exception of low oil the selection process seems capable of still further progress. That low oil reached a near plateau fairly soon suggests that the seeds of the original plants had evolved about the minimum amount of oil commensurate with survival. (After Stebbins, 1966, Processes of Organic Evolution, Prentice-Hall.)

race yielded 6% (Fig. 9.70).

These results demonstrate that even superficially rather simple phenotypic features are affected by many genes. After fifty generations of selective, non-random breeding there is no evidence that the gene combinations affecting high oil yield or high and low protein yields have been maximized. Their frequencies have not become fixed and the races are apparently still capable of further divergent, or adaptive, radiation with respect to these characters.

In natural populations racial boundaries are usually less distinct than in experimental populations. Consider a hypothetical natural population that is expanding to occupy a number of peripheral habitats. Each of these habitats will exert selective pressures on the local gene pool and, since most characters appear to be controlled by multiple genes, the frequencies of 'contributory and neutral' genes will change. Extending the argument of p. 208 concerning a population of legumes to this situation one can visualize the main body of a population having intermediate pod length expanding into two peripheral habitats. In one habitat longer pod length is advantageous and in the other shorter pod length is advantageous. In the former the frequency of contributory genes (*c*), termed **p**, will therefore increase and in the latter the frequency of the *c* gene will decrease. Assuming that the value of **p** in the first habitat rises to $\frac{3}{4}$ and in the second habitat falls to $\frac{1}{4}$, the frequencies in the gene pool of the whole population could be seen as:

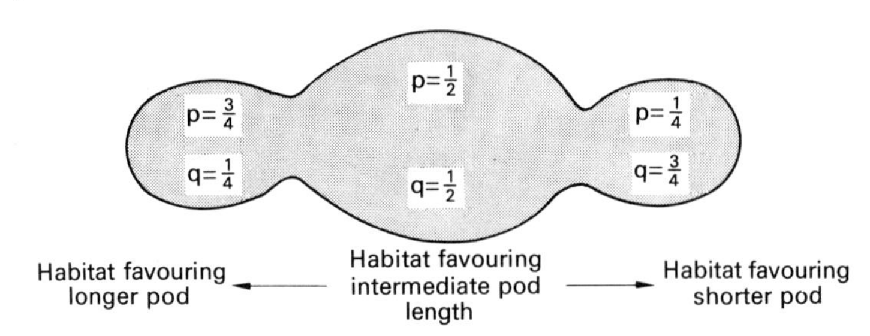

This is illustrated in greater detail in Fig. 9.71.

It is possible to think of the origins of race at a geographical level in terms of organisms at the periphery of a population's distribution encroaching on a different environment for which their genotypes are relatively not well equipped. Initially adaptation to the new environment may occur through phenotypic plasticity; the organisms being able to satisfy the selective requirements for survival in the new environment by physical changes which do not have a direct genetic

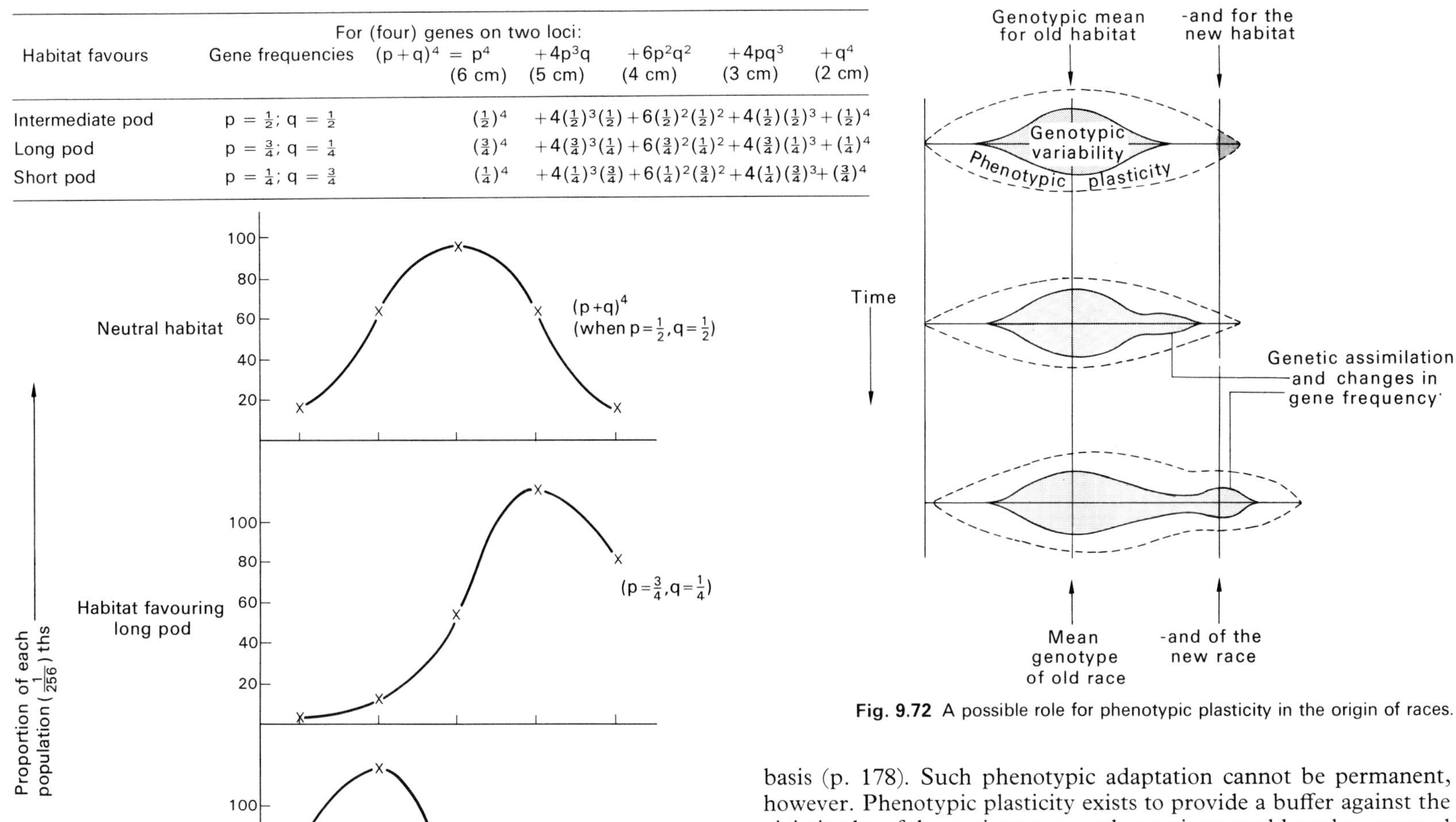

Habitat favours	Gene frequencies	For (four) genes on two loci: $(p+q)^4 = p^4$ (6 cm)	$+4p^3q$ (5 cm)	$+6p^2q^2$ (4 cm)	$+4pq^3$ (3 cm)	$+q^4$ (2 cm)
Intermediate pod	$p=\frac{1}{2}$; $q=\frac{1}{2}$	$(\frac{1}{2})^4$	$+4(\frac{1}{2})^3(\frac{1}{2})$	$+6(\frac{1}{2})^2(\frac{1}{2})^2$	$+4(\frac{1}{2})(\frac{1}{2})^3$	$+(\frac{1}{2})^4$
Long pod	$p=\frac{3}{4}$; $q=\frac{1}{4}$	$(\frac{3}{4})^4$	$+4(\frac{3}{4})^3(\frac{1}{4})$	$+6(\frac{3}{4})^2(\frac{1}{4})^2$	$+4(\frac{3}{4})(\frac{1}{4})^3$	$+(\frac{1}{4})^4$
Short pod	$p=\frac{1}{4}$; $q=\frac{3}{4}$	$(\frac{1}{4})^4$	$+4(\frac{1}{4})^3(\frac{3}{4})$	$+6(\frac{1}{4})^2(\frac{3}{4})^2$	$+4(\frac{1}{4})(\frac{3}{4})^3$	$+(\frac{3}{4})^4$

Fig. 9.71 Distribution of contributory and non-contributory genes (whose frequencies are represented by p and q respectively) in populations in three habitats.

Fig. 9.72 A possible role for phenotypic plasticity in the origin of races.

basis (p. 178). Such phenotypic adaptation cannot be permanent, however. Phenotypic plasticity exists to provide a buffer against the vicissitudes of the environment and organisms could not be expected to remain in a viable and competitive state, capable of further adaptation, if they are at the limits of their phenotypic plasticity.

A second phase might be genetic assimilation, the mechanism whereby the genetic capacity to respond to characters initially induced by the environment is incorporated in the gene pool (p. 170). Thereafter consistent changes in the frequencies of adaptive genes would shift the 'genotypic mean', thus constituting a true, or genetic, race.

Such a scheme for racial formation consisting of first, a phenotypic adaptation, or 'phenotypic race', followed by genotypic adaptation would allow populations to enter new environments which represented a 'step-wise' change in selective pressures. If adaptation were to depend solely upon genotypic adaptation new races could only arise in new environments whose selective pressures represented a relatively gradual change (Fig. 9.72).

Reproductive Isolation

The breeding experiments in maize described in the previous section resulted in races giving high or low yields in the fat and the protein content of the seed. The formation of such races was made possible only because the experimenters ensured that pollination was not random; for example, only pollen from high oil yield plants was used to fertilize the ovules of the high oil yield strain.

Crosses between races produce **inter-racial hybrids** which generally have characteristics intermediate between the parental types. In turn, the offspring of crosses between such interracial hybrids show a wide range of characteristics and destroy racial distinctions based on differences in gene frequencies. To return to the example of the leguminous plant: cross-pollination between long-pod race (5 cm) and short-pod race (3 cm) plants would be expected to produce offspring of which about half would be of intermediate form. Crosses between these intermediate plants in turn would be expected to produce the full range of variants:

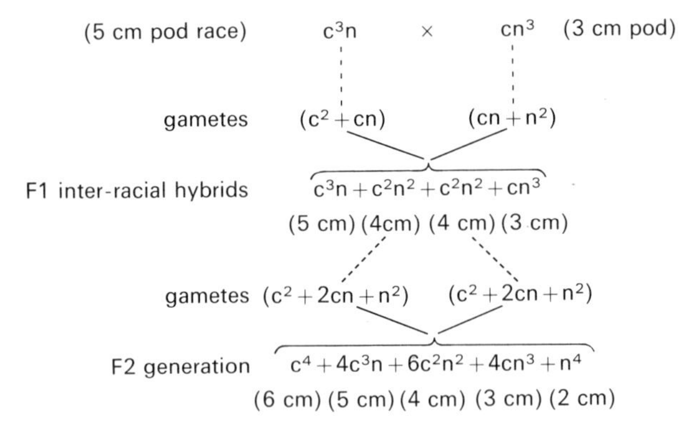

It follows, therefore, that racial differences can only be sustained if gene flow between adjacent races of a species is reduced or stopped. In the early stages of race formation, when the genetic constitution of the new race is not very different from the ancestral population from which it arose, the isolating mechanisms must be entirely **prezygotic**, that is they must operate to prevent fertilization. Any one mechanism may be sufficient to isolate a race and prevent the influx of genes from adjacent races of the same species, but in natural populations racial separation probably occurs as a result of the simultaneous, concerted action of several mechanisms. The principal prezygotic isolating mechanisms are:

1. *Spatial*

Spatial or geographic isolation is probably a relatively weak isolating mechanism. It is likely to be effective only when the distances between the racial epicentres are such that the rate of selection is greater than the rate of gene flow between the races. Man-made disruptions of the environment, however, can create isolated pockets in which races can originate, such as might occur when woodlands are intersected by agricultural reclamation or when open ground is divided by urban extensions and interurban ribbon development. For example, Sheppard (1953) reports that a mile of agricultural land is an effective isolation barrier in the case of the colonial, non-migratory Scarlet Tiger Moth. Spatial isolation is probably more significant in plant than in animal evolution because the mobility of animals aids genetic mixing. In many mammals and birds seasonal migration tends to negate the effects of localized spatial isolation.

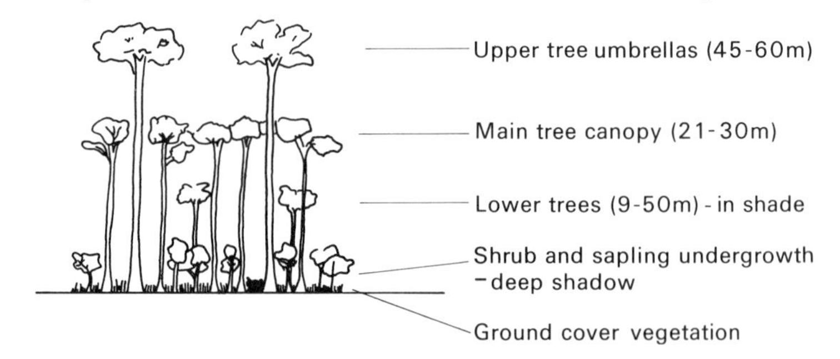

Fig. 9.73 A stable climactic forest can provide distinct habitats vertically because of its characteristic stratification.

Associated with spatial isolation is **habitat** isolation in which the races are separated not by horizontal distance but by ecological factors; the races occupy the same region, but different habitats within it. In this manner spatial isolation can be vertical as well as horizontal. Tropical forest, for example, is clearly stratified into upper and lower tree umbrellas, shrub layers, and surface growth, thus offering relatively distinct ecological habitats (Fig. 9.73).

2. *Temporal and Behavioural*

The optimum times of spore production and flowering in plants, or copulation in animals, probably differ in different habitats and this may be a powerful isolating mechanism. The genetic basis of reproductive behaviour has been demonstrated in the fowl (Wood–Gush, 1960) and in *Drosophila* (Manning, 1961). Other aspects of behaviour, which may have a bearing on reproductive behaviour, have been shown to have a genetic basis in the rat, in which strains were selected for high and low 'spontaneous activity' in a running wheel (Rundquist, 1933).

Manning's experiment consisted of the repeated selection in *Drosophila* for high and low mating speed. After eighteen generations the speed of mating was significantly different in the high and low races, both from each other and from the unselected controls which represented the ancestral race. It was found, in comparisons of samples of fifty pairs from each race, that about 80% of the 'fast-mating' race had mated before the first pair of the 'slow-mating' race began (Fig. 9.74).

There is evidence, however, that in some species behavioural characteristics may have quite the opposite effect, namely to favour interracial hybridization. In laboratory breeding experiments Sheppard (1952) noted that there was a strong tendency for mating to take place between *unlike* genotypes in the Scarlet Tiger Moth, *Panaxia dominula* (see p. 213). This non-random, or **disassortative** mating behaviour is known also to occur in *Drosophila* (Rendel, 1951), but it is probably a relatively uncommon phenomenon.

In man, **cultural barriers**, rather than behavioural ones, frequently serve to isolate races. The distinction between the patterns of blood group gene frequencies in the population of Europe at large and those of the Basques and Lapps has been discussed previously (p. 219). The blood group clines in the European population are partly a reflection of the mobility of its peoples. The turbulent history of Europe in the last thousand years or so has seen widespread shifting of national boundaries and the compounding of nations into empires, and their dissolution. National characteristics tend to be as blurred at the borders between most countries as are their respective languages.

The Basques, by contrast, have jealously preserved their culture. Their language is unrelated to either Spanish or French and strong social pressures have operated for many hundreds of years to encourage Basques to marry only Basques. The evidence from analysis of a large number of blood group antigens suggest that the Basques are an ancient and long isolated population (Mourant, 1954).

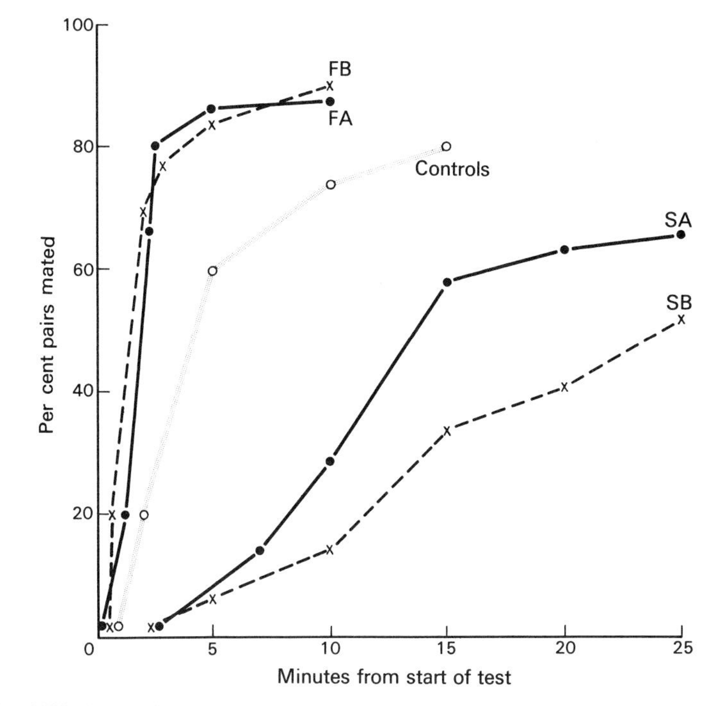

Fig. 9.74 The mating speed of groups of 50 pairs of *Drosophila melanogaster* from two lines selected for fast mating, FA and FB, and two selected for slow mating, SA and SB, compared with unselected controls. These samples are from the 18th selected generation. Some 80% of the fast lines have mated before the first of the slow lines begin. (After Manning, 1961, *Anim. Behav.*, **9**, 82.)

The Swedish Lapps also show signs of long isolation from the European population. The frequency of the I^B gene in the Lapps is almost as low as in the Basques and in addition they have a conspicuously low frequency of the Rhesus negative gene complement. It is interesting to reflect that the Basques and the Lapps are the only people in Western Europe who speak a language the roots of which do not appear to be Indo-European in origin (Fig. 9.75).

Racial features are preserved by cultural barriers in many expatriot races. The Chinese in New York City, many of whose families have lived in the United States for many generations, retain the blood

	Approximate gene frequencies (%)			
	I^A	I^B	i	Rh^-
European	25–30	5–10	60–70	40
Lapp	50–55	4	55–60	4
Basque	15–20	0	70–75	53
		In New York City	Chinese	7
			Negroes	30
			Whites	36

Fig. 9.75 Table of some blood group gene frequencies showing the effects of cultural isolation. The Lapps and Basques differ significantly from other Europeans and the same is true of the Chinese of New York City who have been established for many generations.

group gene frequencies characteristic of the Southern Chinese, such as the low frequency of the genes for Rhesus negative. By contrast, the blood group genes of American Negroes show the admixture of European gene patterns and a study of a wide range of blood group antigens enabled Glass and Li (1935) to calculate the rate of inter-racial hybridization.

3. *Structural*

Racial differences which result in differences in the structure of reproductive organs, particularly with respect to the anatomy of copulatory organs in animals and pollinating organs in plants, could serve to reduce interracial hybridization. In animals copulatory mechanisms tend to be relatively uniform, however, and the evidence that anatomical differences between races can provide an effective isolation barrier is not good. Such differences are not marked even at the species level, where interspecific hybridization in mammals is well known. In flowering plants, by contrast, pollinating mechanisms which use insects as pollen vectors are highly refined. The insect pollinators tend to be selective in the size, shape, odour and colour of the flowers they visit, so racial differences which incorporate differences in flower form can lead to the establishment of a reproductive barrier. Levin (102) has shown that both bees and butterflies tend to visit particular flower forms in mixed populations of the loosestrife, *Lythrum* spp.

	Gene frequency (%)		
	I^A	I^B	i
United Kingdom mean	26	6	68
Scottish Highlands	16.7	6.8	76.5

Fig. 9.76 The native highlanders of Scotland have been isolated by economic factors for many generations and their blood group gene frequencies differ significantly from the U.K. mean.

4. *Unilateral Gene Flow*

In the case of wind-pollinated species in regions where prevailing winds have great constancy of direction, such as the southern parts of South America and on islands in the Trade Wind belts, gene flow is largely unidirectional and the parts of the population that are 'up-wind' tend to be reproductively isolated.

In man economic factors can lead to unidirectional gene flow. The Scottish Highlands has suffered steady emigration of its population with little immigration and as such as been almost free from genetic influx for several hundred years. The frequency of the I^A blood group gene is lower and the frequency of the i gene is higher in Scottish Highlanders than in corresponding samples of an English population, and in this respect differ significantly from the western European mean values (Fig. 9.76).

From Races to Species

A race originates when a group of organisms is isolated from the gene pool of the whole population and is subjected to different environmental selection pressures than those acting on the rest of the population. If reproductive isolating mechanisms are in operation those organisms reproduce only with members of their own race and contribute to a separate, racial gene pool. A continuation of the selection pressures leads to continued genetic response and increasing genetic divergence between the gene pool of the new race and that of the 'ancestral' population.

As has been described, the early stages of race formation are characterized by the ability of interracial hybrids to arise, thus destroying divergence. The new race still shares the capability of dipping into the gene pool of the whole population. Dobzhansky (1962) has described such a situation as a **genetically open system**. In the later stages of evolutionary divergence the constitution of the new gene pool may become so different, in both a quantitative and a qualitative sense, that hybridization with the ancestral gene pool becomes impossible owing to **genic disharmony**. Qualitative differences are differences arising from the presence of new genes, by mutation, and new genic and chromosomal arrangements by inversion, translocation, and so on (p. 182). At such a level the genetically open systems (races) become **genetically closed systems**, or **species**.

Definition of the term 'species' has proved controversial for over a century and terms such as variety, race, sub-species, species and genus long antedate the application of genetics to evolutionary theory. In *Origin of Species* Darwin summarized the confusion that existed in plant and animal taxonomy in a section entitled *Doubtful Species . . . 'as representative species . . . are distinguished from each other by a greater amount of difference than that between the local forms and sub-species, they are almost universally ranked by naturalists as true species. Nevertheless, no certain criterion can possibly be given by which variable forms, local forms, sub-species, and representative species can be recognized . . . there is no possible test but individual opinion to determine which of them shall be considered as species and which as varieties.'*

It is certainly true that this taxonomic confusion arose in part from errors of communication, such as the problems of subjectivity in description, difficulties in translation, and the necessity for drawing rather than photographing specimens, but it mainly arose from the lack of clear physical distinctions between representative organisms, particularly between those of species having an evolutionary relationship.

In the determination of speciation, phenotypic plasticity is inclined to make physical features less reliable than genetic features. Even so, the evolutionary divergence of races to species, from genetically open systems to genetically closed ones, does not exhibit a threshold effect and there are degrees of genic disharmony. Therefore, although genetic differences represent a much finer method of determining taxonomic differences between species, the boundaries still tend to be blurred. While interracial hybrids are viable and fertile, interspecific hybrids tend to vary from viable but largely sterile to non-viable according to the degree of evolutionary divergence. In this context Dobzhansky (1951) states '. . . although the process of divergence is a gradual one, the speciation in the strict sense, i.e. the development of reproductive isolation, is a crisis which is passed relatively rapidly.'

Reproductive barriers existing between species comprise those prezygotic mechanisms we have discussed already as existing between races, together with a number of genetic, or **post-zygotic** mechanisms:

5. *Hybrid Sterility*

In cases where species are closely related, that is where the degree of evolutionary divergence is relatively small, hybridization may result in viable offspring. In higher animals a combination of detailed courtship patterns and behavioural 'imprinting', that is the early attachment of an offspring to a 'mother figure', which affects its choice of mate when mature, results in prezygotic barriers to inter-specific hybridization. But in captivity or domestication a number of inter-specific hybrids are known; for example the mule (horse × ass), the zebron (zebra × horse) and the tigon (tiger × lion).

In spite of its hardy constitution the mule is sterile, probably in both sexes, although some fertility in female mules has been reported. Sterility of hybrids is usual and occurs either as a result of gonadal malformation or, if the gonads develop normally, gametic malformation. In the mule the gonads appear to develop normally and sexual behaviour and mating have been known to take place. Gametes are incompetent, however, mainly because the chromosomes in the hybrid gamete mother cells do not synapse sufficiently closely at Meiotic Prophase I (p. 85). (In the mule the chromosomal complement is 63; horse $2n = 60$ and ass $2n = 66$). The failure of synapsis and the consequential failure of gametogenesis has been discussed in the section on Changes in Chromosome Structure.

Some crosses involving interspecific hybrids in *Drosophila* produce sterile male offspring but fertile female offspring in the F1 generation. But the offspring of crosses involving those females, such as back-crosses with either parental species, are generally non-viable or weakly and sterile, a phenomenon which has been termed F2 hybrid breakdown. Haldane noted the relationship between interspecific hybrid sterility and sex, and pointed out that in those instances where one sex is fertile or partly so it is generally the homogametic sex (Haldane's Rule of reduced F1 fertility, Huxley, 1942). In higher plants hybrid infertility resulting from chromosomal incompatibility and synaptic failure in meiosis can be overcome by polyploidy. Doubling of the chromosomal complement of the hybrid provides a mechanism for chromosomal pairing and normal gametogenesis, providing there are no sex chromosomes (see Allopolyploidy, p. 191).

Thus hybrid infertility is an effective barrier to gene flow between species. The interspecific allopolyploid is not an exception since it is itself unable to produce fertile offspring in crosses with either parental species. In that sense an allopolyploid is a new species comprising an isolated gene pool, a genetically closed system.

6. *Hybrid inviability*

Hybridization between species that are still further divergent involves a still greater degree of genic disharmony and is more likely to result in abnormal chromosome numbers (see Aneuploidy, p. 195). This is most likely to affect the F1 embryo at early stages in development, when the balanced genome begins to direct embryogenesis. In flowering plants this may result in a seed containing a malformed embryo or a seed wholly malformed through accompanying endosperm failure, and in mammals in an embryo which either aborts or is absorbed *in utero*.

Moore (1941) obtained hybrids between several species of frog of the genus *Rana*, in the United States. Such interspecific hybrids showed different degress of inviability. Some hybrid eggs developed into larval stages but most failed at gastrulation. Some hybrids between more distant species failed to undergo normal cleavage. In general, Moore found that the degree of hybrid inviability was approximately proportional to the difference in latitude of the habitats of the parental species, which in turn is probably a reflection of their evolutionary divergence.

In the European newt, genus *Triturus*, interspecific hybridization produces several degrees of inviability in a similar manner to *Rana*. Hybrids between *T. cristatus* and *T. marmoratus* and between *T. carnifex* and *T. karelinii* usually die at the most critical and complex stages of embryogenesis, the 'ontological crises' (Lantz, 1947; Spurway, 1953). Spurway reports that in the hybrid offspring of some crosses, such as *T. cristatus* × *T. carnifex*, inviability is associated with the transfer from nutrition on yolk to feeding on crustaceae. Hybrids between 'subspecies', such as *T. cristatus carnifex* × *T. cristatus karelinii*, showed a 90% survival rate through metamorphosis. The survivors showed hybrid vigour, but on reaching sexual maturity exhibited F2 hybrid breakdown.

Spurway attributed this breakdown primarily to chromosomal incompatibility and pointed to the presence of different reciprocal inversions (p. 188) in their chromosomes (Callan and Spurway, 1951). She argued that these taxonomic distinctions 'must have involved periods when the effective population size was drastically reduced, probably during migration into a previously inhospitable geographical region'.

In the natural state several isolation barriers may act in a concerted manner to prevent hybridization. Conversely, the presence of some

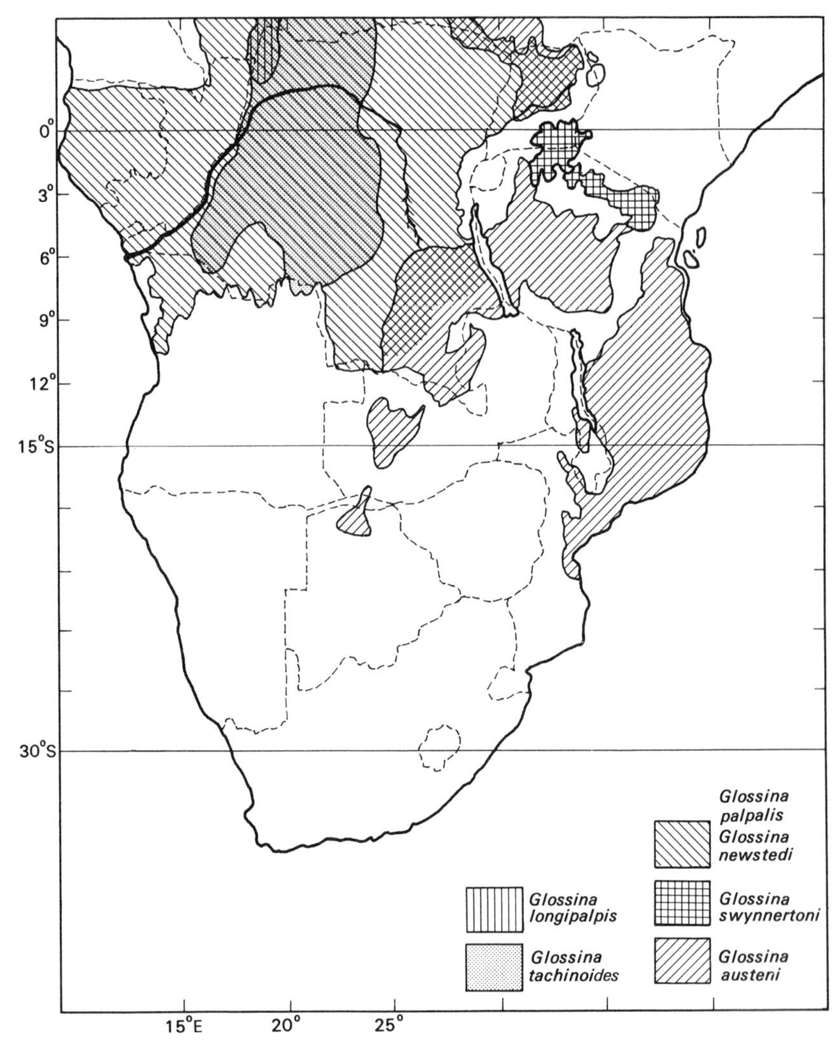

Fig. 9.77 Distribution of some Tsetse species in central and southern Africa. (Data from W. H. Potts and Messrs. Edward Stanford Ltd., 1954.)

barriers, for example the postzygotic barriers which operate between species (that is populations whose gene pools are so divergent that the level of genic or chromosomal disharmony in the hybrid results in sterility or death), obviates the necessity for other barriers such as geographical isolation. There are many examples of related species overlapping in their distribution ranges.

The tsetse fly, *Glossina* spp. is an insect vector of the flagellate protozoan *Trypanosoma* which is parasitic in the blood plasma of a wide range of vertebrate hosts. In man, a host to which *Trypanosoma* is not well adapted, the parasite causes the debilitating disease Trypanosomiasis, or sleeping sickness. W. H. Potts (1954), when chief entomologist of the Tsetse and Trypanosomiasis Research Organization, collated data on the geographical distribution of a number of species of *Glossina*, some of which are summarized in Fig. 9.77. Broadly, the eastern part of the African continent is dominated by *Glossina austeni* and the western and central areas by *G. palpalis* and *G. newstedi*. Over most of their ranges these species are geographically separate, or **allopatric**, although they overlap to the east of Lake Tanganyika and the north of Lake Victoria. In these areas the species are said to be **sympatric**.

Glossina swynnertoni is largely an isolated species with a relatively small distribution to the south-east of Lake Victoria along the border between Kenya and Tanzania. It is probably sympatric at the southern part of its range with *G. austeni*. One might speculate, therefore, that *G. swynnertoni* is a relatively new species which is largely isolated from, but may have its ancestral origins in, the much larger population of *G. austeni* to the south.

Much of Gabon and the Congo lies in the distribution of *G. palpalis* and *G. newstedi*. These species are sympatric with *G. tachinoides* in the central Congo over an area of some three hundred thousand square miles, which suggests that the speciation in this area is relatively ancient and well isolated by effective reproductive barriers other than geographical isolation (Fig. 9.78).

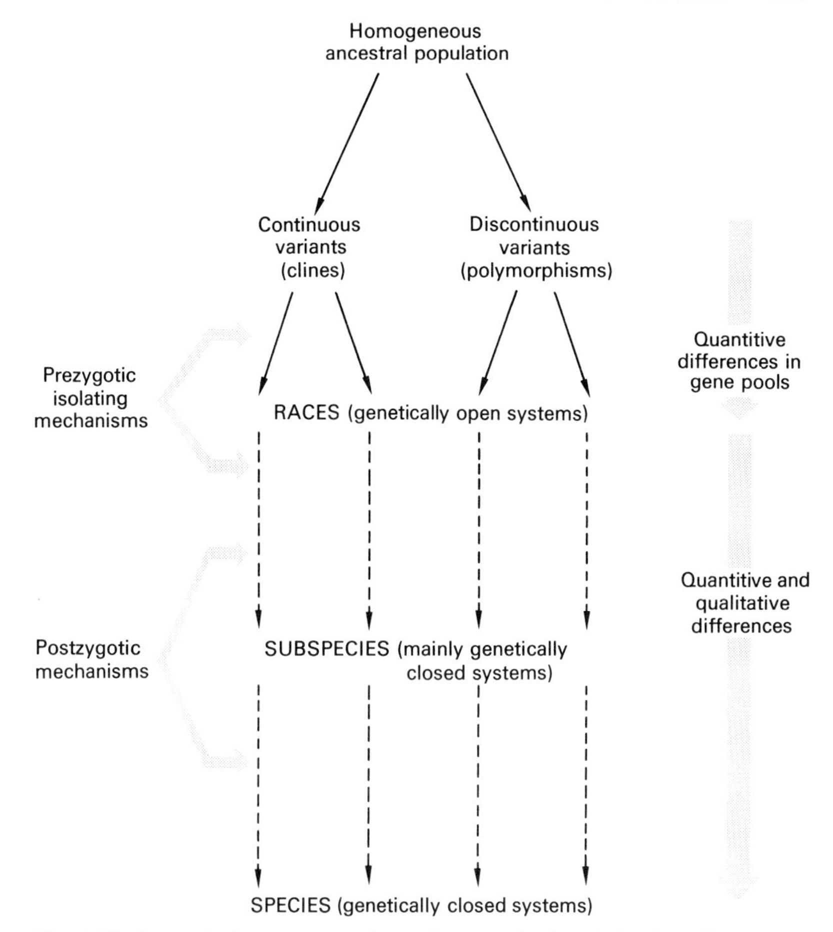

Fig. 9.78 A speculative sequence of genetic events in the origin of species.

10 Further Implications of Evolutionary Theory

This final chapter in *Patterns in Biology* is intended to be speculative in some measure and also to provoke discussion. Through an outline of the principles of ecology it seeks to relate evolution to competition between organisms and to the way energy flows through the biosphere (see the diagram on p. 2 of the Preface). Evolution, after all, is not simply the genetic changes that take place in a species with respect to time but the changes that occur in all the complex, interrelated organisms that comprise the biosphere of the planet Earth. Evolutionary changes arise out of mutations or from novel combinations or groupings of genes which change in frequency in a population over successive generations. Advantageous combinations would be expected to increase in frequency. 'Advantageous', however, means conferring a competitive advantage in an ecological sense. Therefore evolutionary changes in one species will affect neighbouring species in an ecosystem. To coin a phrase of Donne's, 'No species is an island'.

From this broader view of evolution the author goes on to speculate on what he has termed **collectivism** and **separatism**. The former comprises the blending or merging of organic units or organisms to produce more complex systems which more successfully exploit the energy available at their trophic level. The latter, which is the antithesis of collectivism, takes account of the appearance in evolution of mechanisms such as phagocytosis and later, the immune response. These, it is argued, act to maintain the genetic identity of the individual organism. They operate to ensure that an organism's phenotype, the features upon which natural selection acts, is a consequence only of its inherited genotype.

10.1 The Inter-related Nature of Genetics, Ecology and Behaviour

Evolutionary genetics is rapidly extending into the fields of behaviour and ecology. A genetic basis for some aspects of behaviour has been established and the significance of behaviour in terms both of adaptation and of isolation in animal populations has been recognized and incorporated into modern evolutionary theory. The successful development of evolutionary theory in genetical terms and the reinforcing data accumulated from laboratory tests has led to considerable pressure to demonstrate more clearly its influence in natural populations.

But a natural population cannot be treated in isolation because it is part of its own ecological landscape. During its evolution, a population is exposed to changes in biotic factors caused by surrounding populations, whose development is influenced in turn by the population that is evolving.

The principles referred to as 'Genetic Drift', and the 'Founder', or 'Adam and Eve Principles', whereby the isolation of a few animals, or perhaps a gravid female, gives rise to a new race by virtue of the fact that the genotypes of the isolates differs significantly, on a random basis, from the mean of the parental population, cannot be regarded as having had a major role in evolution. The evidence for the operation of such mechanisms outside the laboratory is poor. In nature, race formation is more likely to occur and to have occurred when the fringes of expanding populations were pressed into new and different habitats to which they adapted in the ways that have been described.

An expanding population is, in biological terms, a **successful** one. Biological success can be measured in a number of ways. A group of plants or animals can be said to be successful if it shows such great diversity and adaptability that it extends its geographical distribution in competition with other groups with which it does not reproduce. Some formulae of body organization seem particularly to lend themselves to expressions of great genetic versatility, such as that shown by the Class Insecta. A second measure of biological success is increase in **biomass**, that is the actual mass of living material which constitutes the reproducing group, or the source of the gene pool. On the other hand, groups of organisms which have neither greatly extended their distribution nor increased their biomass, but have remained relatively unchanged over millions of years, exhibiting an almost stoichal disregard for evolutionary change, such as many Lamellibranch molluscs, should perhaps also be credited with some measure of biological success.

Increase in the biomass of a particular group of organisms, the prime measure of biological success, means the carving out of a greater share of the total biomass for the particular economic level in the ecology of the habitat. The total economy of the planetary biosphere, which is reflected in the ecological structure of each ecosystem, can be examined in terms of energy flow.

The origin of all the energy used by biological systems is the radiation emitted from the **hydrogen fusion → helium** reaction in the sun. This radiation excites chlorophyll molecules in the cells of green plants and permits the production of 'high energy' electrons by which the plant cell carries out endergonic reactions. A variety of carbohydrate, lipid and proteinaceous end products are formed which the plant body uses for structural purposes, for food reserves in spores and seeds, or for overwintering. The end products of photosynthesis also form respiratory substrates which provide a source of energy at times of the day or season, or in parts of the plant body, where photophosphorylation does not occur.

The **net** end products of photosynthesis, that is the proportion of the total end products which are incorporated into the structure of the plant body or its storage organs, in turn become the origin of a chemical food chain which represents the flow of energy through the organisms in the habitat.

The change of energy from one state to another, such as from plant biomass to herbivore biomass is accompanied by 'wastage' or loss of heat, as would be expected from the Second Law of Thermodynamics. While the gross energy turnover in animals is probably more efficient than in most engines the net amount transferred into new biomass at a new trophic level probably does not exceed 10%. For this reason food chains are usually short.

Food chains operate usually at three, occasionally four, trophic levels. The first level, the **primary producers**, are autotrophs. They make organic molecules by photosynthesis and they comprise the green plants. The remaining levels in the food chains are heterotrophs. The second level constitutes the **primary consumers**, or herbivorous animals, and the third level is made up of **secondary consumers**, or carnivorous animals. A fourth trophic level, the **tertiary consumers**, comprises a number of parasitic forms and super-carnivores, which are carnivores that feed on other carnivores. One such example is the polar bear which feeds on seals (Fig. 10.1).

It follows, therefore, that in any given ecological system or **ecosystem** (a functional ecological unit of the biosphere, self-contained in an energetic and organic sense, such as a rock-pool, prairie or coral reef) the available energy at each trophic level is limited. It is not rationed, but is 'up for grabs'. Plants compete for light, herbivorous animals for the best grazing or browsing, and carnivores for prey.

In plants competition is both inter- and intraspecific since it

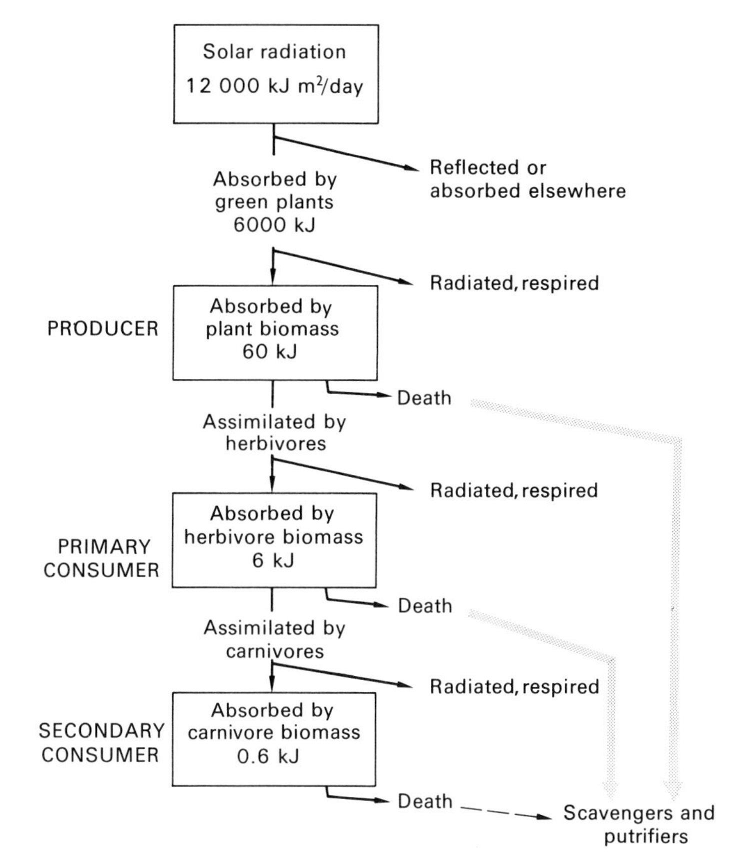

Fig. 10.1 The general sequence of energy flow through the biosphere. (Data from Odum, 1963.)

revolves primarily around the ubiquitous light and water. In animals, by contrast, competition is mainly intraspecific since the spectrum of food is inevitably narrower. This contributes in part to intraspecific aggression which is most marked in mammals and is reflected in territoriality (Lorenz, 1963).

A successful population which is increasing its biomass must do so either by increasing its density in the ecosystem, at the expense of competing species at the same trophic level, or by expanding into neighbouring, hitherto unoccupied, areas, which is the basis of succession. Both are associated with the formation of new races in the ways described in the previous chapter. Without a clear understanding of the ecological relationships of the population under examination and, in the case of animals, the relevant aspects of behaviour, a complete analysis of the causes of evolutionary change in genetic terms is unlikely to be achieved. The work of Sheppard and his colleagues on the Scarlet Tiger Moth *Panaxia dominula*, described in the previous chapter under Polymorphism in Insects, provides a striking example of the rewarding results achieved by integrating behavioural and ecological data with changes in gene frequencies in a natural population.

The Role of Collectivism in Evolution

Most theories about the origin of life on this planet involve some kind of aggregation. Oparin's (1957) concept of the coacervate, for example, has proved to be a central pillar in these theories. It seems likely that the first two billion years of the Earth's existence saw the accumulation, from a strong reducing atmosphere at about 1000° C, of a variety of simple saturated and unsaturated hydrocarbons.

Under these conditions carbon would most likely be in the form of acetylene C_2H_2, ethylene C_2H_4 and methane CH_4 and nitrogen in the form of ammonia, NH_3, etc. Unsaturated hydrocarbons can react with superheated steam to form a variety of hydroxy derivatives, such as acetylene to acetaldehyde.

The cooling of the planet resulted in the condensation of water to form oceans, in which the organic molecules would accumulate in a colloidal state to form a hot, dilute soup. One of the features of colloids is that under specific conditions of concentration separation of the colloidal particles from the equilibrium medium can occur in a way which is termed **coacervation**. A coacervate is an aggregate of hydrophilic particles, each particle surrounded by a shell of water molecules parts of which are shared with the shells of adjacent particles. A **mixed coacervate** is an aggregate consisting of different kinds of colloidal particles (Fig. 10.2).

The constituents of a coacervate would be expected to associate randomly in proportions determined essentially by the properties and nature of the colloidal soup. At the same time the sizes and constitution of coacervates might be expected to show considerable variation due to differences in colloidal concentrations and dispersions, and to localized effects of temperature and tidal turbulence. The step from coacervate to living unit is clearly a large one. It involves the association of amino acids to form useful proteins and

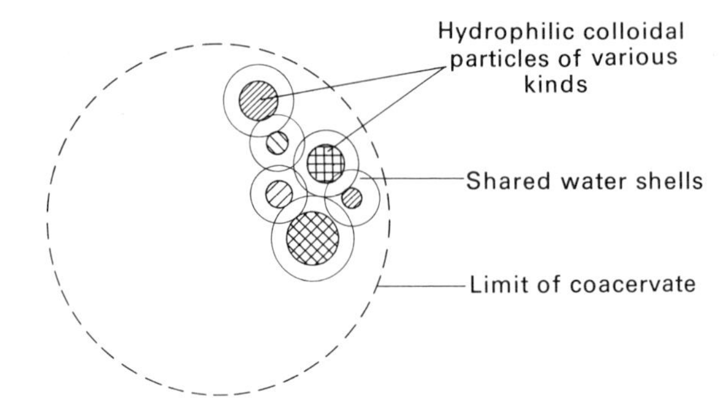

Fig. 10.2 Mixed coacervate. An aggregation of colloidal particles of various compounds, held together by shared water shells, forms a mixed coacervate. Such a system, it has been postulated, might have been a significant stage in the origin of life.

the formation of a boundary layer or 'membrane' which would allow the composition of the internal ingredients to remain distinct from the medium outside.

The origin of feeding, or the ingestion of neighbouring coacervates, would require the development of enzymes whose role would be to alter the nature of the ingested material to conform to that of the eater whose integrity would then be maintained. The development of such enzymes must have been hazardous since they would be potentially as capable of digesting the coacervates which developed them as those which were ingested.

These developments, what might be called **chemical evolution**, together with the origin of nucleic acids as organizers of enzyme synthesis, represent major stages in the origin of life. The changes wrought from simple coacervate to some form of simple prokaryotic life form probably represents the greatest step in organic evolution on this planet. In this sense it can be argued that the differences between the 'pre-living' complex coacervate of the kind we have described and the simplest form of protozoan or protophytan organism are orders of magnitude greater than the differences between the same protozoan or protophytan life forms and those of the highest evolved animals and plants. It seems hardly surprising, therefore, that about half of the evolutionary time span of the planet should have been taken up in the accomplishment of this single step (Fig. 10.3).

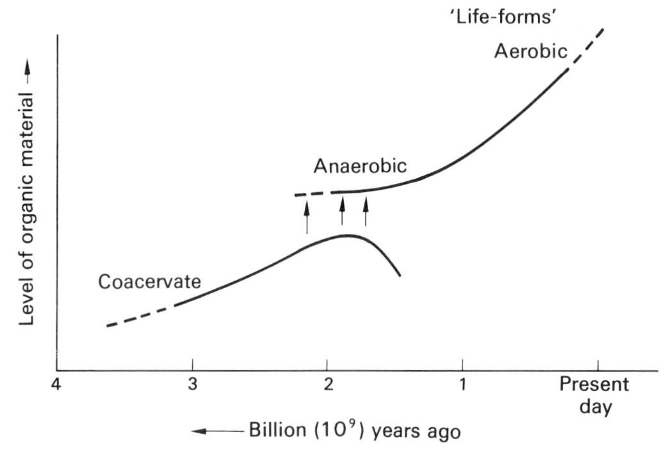

Fig. 10.3 A possible sequence of events in the origin of life (see text). (After Keosian, 1965.)

The Origin of Cell Organelles

There is now very good evidence that two other aggregations, apart from coacervation, of major importance occurred in organic evolution, namely the incorporation of mitochondria and chloroplasts into primitive life forms. Recent work (reviewed by Goodenough and Levine, 1970) strongly suggests that mitochondria and chloroplasts were originally independent life forms which took up a symbiotic or **mutualistic** relationship with other life forms, broadly in two ways. The first, in which both mitochondria and chloroplasts were present, gave rise to the autotrophic protophyta and consequently to the plant kingdom. In the second mitochondria only became established, giving rise to the heterotrophic protozoans and thereby to the animal kingdom. Since many cellular activities, a high proportion of which require aerobic respiration for successful completion, are common

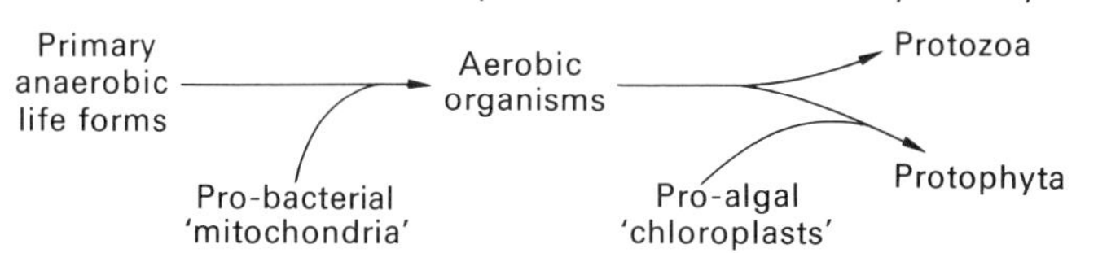

Fig. 10.4 A hypothetical sequence leading to the origins of plants and animals by successive symbioses.

to both plants and animals, it could be argued that the assimilation of mitochondria occurred before that of chloroplasts (Fig. 10.4).

Bacteria, chloroplasts and mitochondria have many features in common which suggest they have a shared evolutionary origin. There are a number of characteristic ways in which the contents of all three differ from the contents of eukaryotic cells. The first suggestion that mitochondria might have a bacterial origin was made over 80 years ago, based upon observations of their shape. Their mode of movement in the cell and their division were also reminiscent of bacteria. The view that they were bacterial in origin, however, remained unconvincing until the discovery of mitochondrial DNA. This form of DNA is arranged in loose fibrils, is free from histones (p. 120) and in the mitochondria of higher animals is arranged in a circular manner (Fig. 10.5). In these respects, and in view of the difference between its density and that of the DNA in the host cell nucleus, mitochondrial

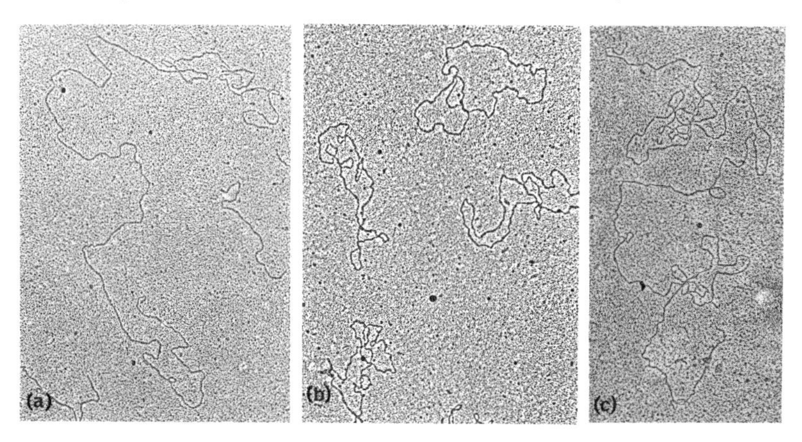

Fig. 10.5 Types of DNA strands extracted from cell organelles. (a) Linear DNA from the nucleus of a mouse cell. (b) Circular DNA molecules, about 4 μm in circumference, extracted from mouse mitochondria (compare Fig. 8.19). (c) DNA from chloroplasts is linear. (Courtesy of Dr. M. M. K. Nass.)

DNA closely resembles bacterial DNA.

In chloroplasts the DNA is also 'naked' and although its circular nature has not been described in other respects it resembles the prokaryotic DNA characteristic of, for example, bacteria and blue-green algae. Chloroplast DNA also differs from DNA in the host cell in respect of the proportions of base pairs. For example, the proportion of adenine + thymine may be significantly different in the two types of DNA.

Perhaps even more striking evidence relating chloroplasts and mitochondria to the prokaryotic level of organization has been gained from a study of the protein synthesizing apparatus in the organelles (Williamson, 1970). Ribosomes from the cytoplasm of the cells of higher organisms have a sedimentation rate of 80 S (Svendberg units) in which the RNA subunits (p. 39) have sedimentation rates of about 28 S and 18 S. Ribosomes extracted from mitochondria and chloroplasts, on the other hand, are smaller and more closely resemble the ribosomes of prokaryotic organisms. Their respective sedimentation rates are 70 S (whole ribosome), 23 S and 16 S (subunits). In both mitochondria and chloroplasts unique types of transfer-RNA are also present (Figs. 10.6 and 10.7).

The presence of unique kinds of DNA, ribosomes and transfer-RNA inside mitochondria and chloroplasts suggests an autonomous protein synthesizing mechanism, although there is good evidence that the mechanism is under the direct control of the genes in the host cell nucleus. The amounts of DNA in mitochondria and chloroplasts are relatively small, certainly not sufficient to code for all the proteins in the organelle. Mitochondrial DNA, for example, amounts to only about 0.5% of that present in a bacterial cell. It can be argued, therefore, that during evolution these cell inclusions must have lost many of the genes that were duplicated in the host cell nucleus.

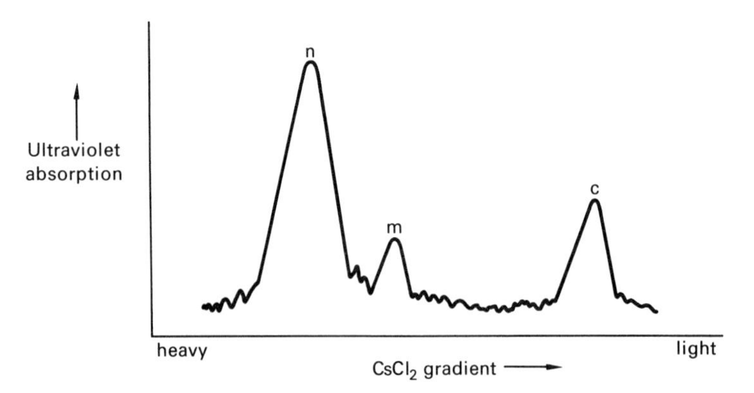

Fig. 10.6 The U.V. absorption characteristics of types of DNA extracted from the alga *Chlamydomonas* showing DNA which is presumed to be from the nucleus (n), the mitochondria (m) and the chloroplast (c). (After Goodenough and Levine, 1970)

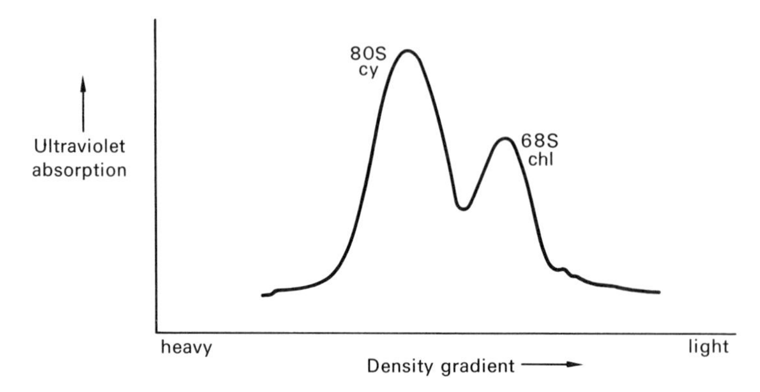

Fig. 10.7 The U.V. absorption properties of ribosomes from *Chlamydomonas* separated on a density gradient, showing 80 S particles from the cytoplasm of the cell and 68 S particles from the chloroplast. (After Goodenough and Levine, 1970.)

The results of molecular hybridization experiments, by which it is possible to identify nucleic acids which have complementary base sequences, indicate that the organelle DNA appears to be responsible for the coding of the organelle ribosomal-RNA and possibly codes for some proteins concerned in the structure of the organelle, for example in its membranes. Many of the enzymes contained in the organelles, however, seem to be coded for by DNA in the host cell nucleus. Cytochrome *c*, although found exclusively in mitochondria, is known to be synthesized on ribosomes in the cytoplasm, from whence it is presumably transported into the mitochondria. Similar 'deficiencies' in gene content have been reported for chloroplasts. The chloroplast membranes, or **grana** (p. 58), are known to be synthesized partly by enzymes which are made on the chloroplast's own 70 S ribosomes, together with chlorophyll, and partly by enzymes made on the 80 S ribosomes in the cytoplasm (Hoober *et al.*, 1969).

In many electron microscopy studies of cells mitochondria are frequently observed as a dumb-bell shape which suggests that they are dividing, that is undergoing a semi-independent reproduction of a fission kind such as is found in many micro-organisms (p. 72). Chloroplasts can similarly be observed with constrictions (Fig. 10.8).

In the moss *Splanchnum*, under certain growth conditions, both chloroplasts and mitochondria can be seen with bud-like structures similar to those found in budding yeast cells (Fig. 10.9). These have been interpreted as reproductive buds which break off and grow to form new organelles (Fig. 10.10, and see Maltzahn and Muhlethaler, 1962a and b).

More evidence that chloroplasts and mitochondria are able to divide by some fission or budding process has been provided by electron micrographs which show DNA fibrils as single entities in whole organelles and as paired units in organelles which are constricted (Fig. 10.11, and see Yotsuyanagi and Guerrier, 1965).

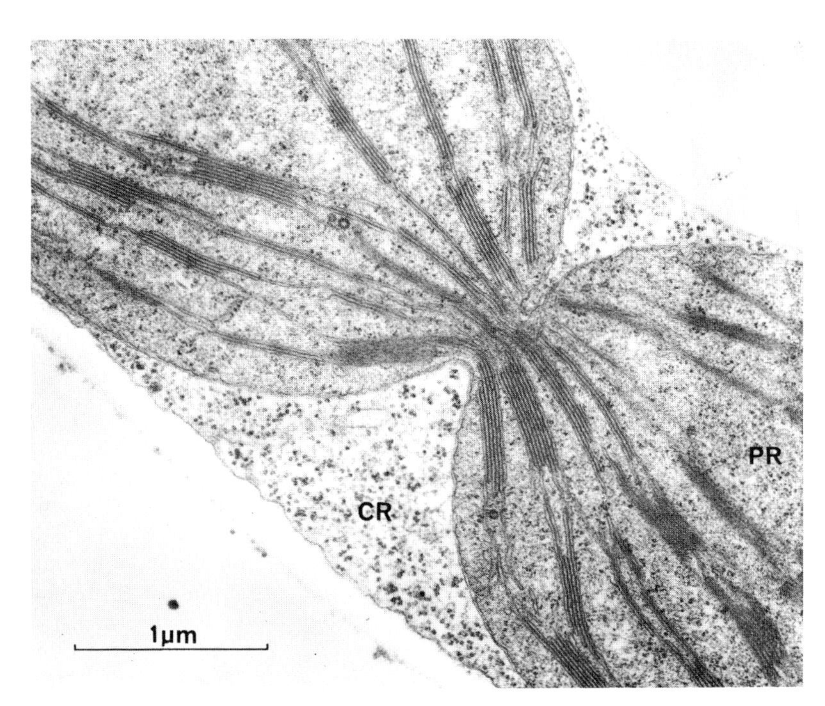

Fig. 10.8 Electron micrograph of a section of chloroplast from spinach leaf. The constriction suggests that it is dividing. Note also the difference in size between the plastid ribosomes (PR) and the cytoplasmic ribosomes (CR). (By courtesy of A. D. Greenwood.)

Genes which are present in organelles are not arranged on chromosomes during cell division or gametogenesis. Their inheritance is not Mendelian. In most cases very little cytoplasm accompanies the male gamete in fertilization, either in spermatozoa or pollen tube nuclei, so that the inheritance of organelles must be regarded as primarily maternal.

If indeed mitochondria and chloroplasts were originally free-living forms analogous to bacteria and algae, which became established symbiotically in certain cells, then their roles in increasing the efficiency of ATP production (p. 50) and in creating endergonic pathways have proved of immense significance in evolution. Other organelles may have a similar origin. Centrioles, for example, have a conspicuously flagellum-like structure and may represent the vestiges of some flagellate or ciliate ancestor incorporated into the 'collective' (Fig. 10.12).

There are numerous examples of what might be termed intermediate levels of interdependence between bacteria and other cells which lend credibility to the bacterial theory of the origin of cell organelles. In some cases a clear endosymbiotic relationship can be identified, such as the nitrogen fixing bacteria in the root cortex of leguminous plants. In other cases the contributory role of the bacteria is harder to ascertain because they cannot be cultured outside the host cells, but it seems likely that they may help in a variety of chemical syntheses in much the same way that some of the intes-

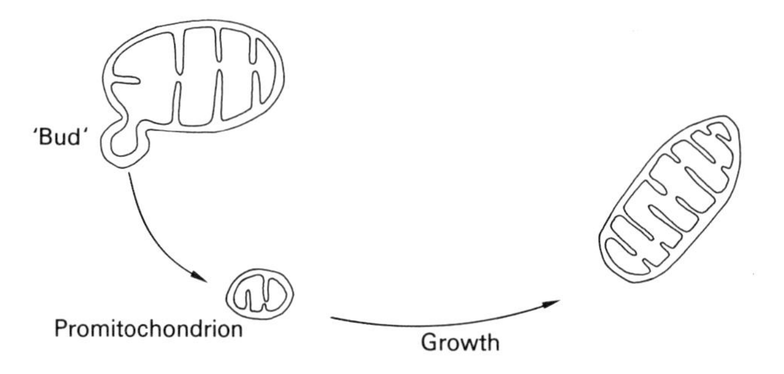

Fig. 10.9 A 'budding' process postulated for the observed increase in numbers of mitochondria during cell growth (after Maltzahn and Muhlethaler, 1962).

tinal bacteria in man provide him with a source of nutrients, like vitamin K.

Endosymbionts resembling Gram-negative bacteria inhabit both the cytoplasm and macronucleus of the protozoan ciliate *Paramoecium*. Sonneborn (1938) identified the presence of what were termed Kappa particles in the cytoplasm of certain strains of *Paramoecium*. In the presence of these particles the ciliate could release a toxic material (at that time called 'paramecin') into the water which killed other strains in which the Kappa particles were not present. The value of the endosymbiont in conferring a competitive advantage on the host cell is clear. What is also interesting to note is that the capacity of the killer strain of *Paramoecium* to support the endosymbionts was found to be dependent upon the presence of a specific gene K in the host cell nucleus.

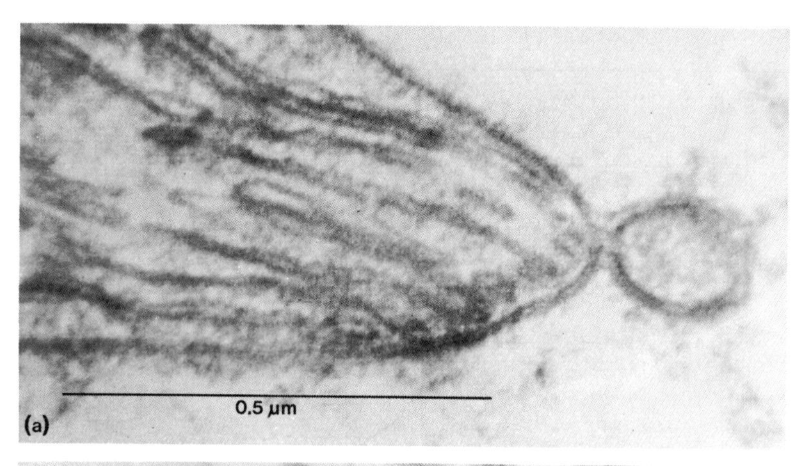

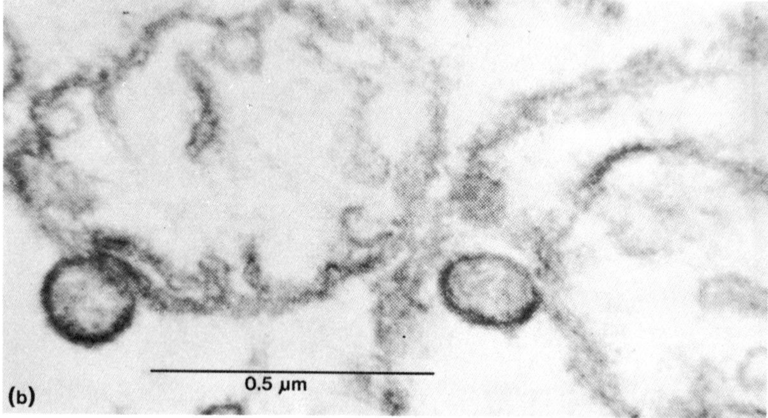

Fig. 10.10 (Upper) An electron micrograph of a section of dedifferentiating cells from the moss *Splanchnum* showing the peripheral region of a chloroplast with a conspicuous 'bud' which may be a way in which chloroplast number is increased. (Lower) An E.M. from the same source showing two mitochondria with similar buds. (By courtesy of K. F. von Maltzahn.)

Further work revealed that a heterogeneous assortment of microorganisms comprise the endosymbionts of *Paramoecium*, which have been termed Alpha, Nu, Delta, Mu, Lambda, and so on varying in length from 0.5 µm to 15 µm (Fig. 10.13, and see Beale *et al.*, 1969).

The Role of Separatism in Evolution

The cases of endosymbiosis previously discussed occur between protozoa or protophyta and bacteria. Examples from higher phyla are less common. In the coelenterate *Hydra viridis* green algae are symbiotic in the ectodermal musculo-epithelial cells where, presumably, they supply their host with oxygen and possibly various substrates during photosynthesis and in return gain a relatively constant environment, carbon dioxide and nitrogenous substrates for photosynthesis.

In the free-living platyhelminth worm *Convoluta roscoffensis* a green or blue-green alga is endosymbiotic in the cells of the epidermis exhibiting an interdependence similar to that of the green *Hydra*. In higher animals and plants, however, endosymbiosis is conspicuously absent. It is as though the pattern of evolutionary development, having favoured collectivism and interdependence at an intracellular level during the early stages, suddenly changed policy. The entity, or one-ness, of the individual organism became thematic in evolution. Cohabitation of organisms still occurred widely, but only rarely in a mutualistic or symbiotic manner. The cohabitation generally became of one-sided benefit, namely parasitism.

Perhaps the most refined form of parasitism at an intracellular level occurs in the viruses, whose structure and habits in the host cell have been described previously (p. 39). Viruses represent the ultimate level of parasitic degeneration. Just as intestinal parasites, such as tapeworms, which inhabit a relatively constant environment, show losses in the genetic capacity to form locomotory, digestive and sense organs, so viruses show loss of those genes which are duplicated in the host cell nucleus. Viruses have no capacity for autonomous protein synthesis. They cannot make enzymes and thus have no

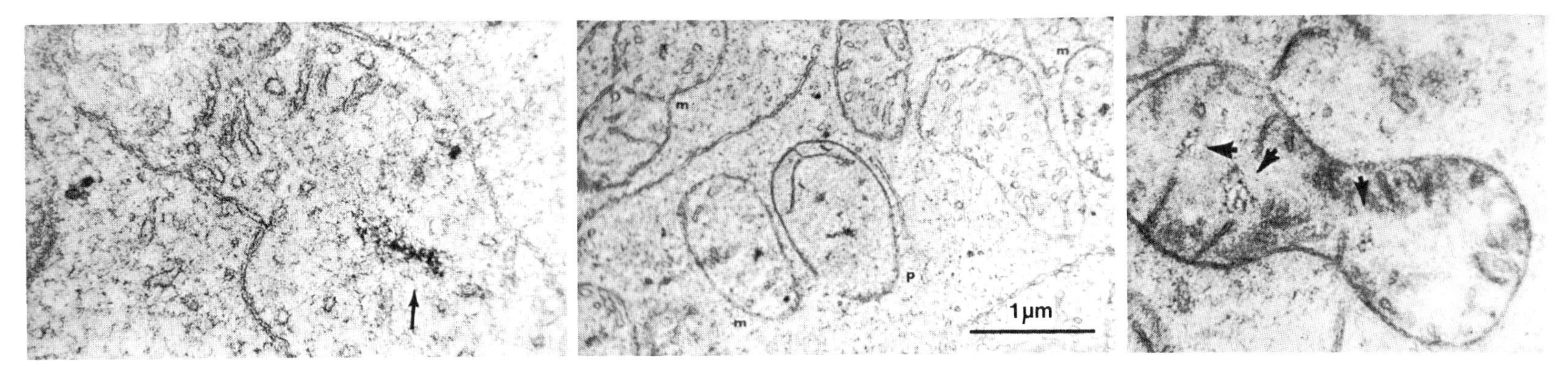

Fig. 10.11 Electron micrographs of young cells from the onion *Allium cepa* stained with ribonuclease uranyl acetate. (Left) A mitochondrion containing an electron dense mass thought to be DNA (arrowed). (Centre) Similar DNA masses in mitochondria (m) and a proplastic (p). (Right) A mitochondrion with a constriction, possible dividing, in which the aggregations of DNA (arrowed) suggest separation as a prelude to division. (By courtesy of Mlle Claudie Guerrier.)

independent metabolic processes such as respiration, growth and excretion. Their genetic data seems to be capable solely of instructing the protein synthesizing and nucleic acid replicating mechanisms of the host cell for the construction of more virus. There is no evidence that even the passive, or non-virulent viruses, such as the herpes-like viruses which probably inhabit all human cells, have a beneficial effect on the host organism.

Homothermic animals, that is birds and mammals, have evolved a constant body temperature so giving their metabolism a greater degree of independence from the vicissitudes of the environment. These animals might be expected to provide a prime organic structure for endosymbiosis. That they do not do so is attributable to a number of barriers (summarized in Harrison, 1970) amongst which is the immune response.

The Immune Response

The immune response is the production of **antibodies** in response to the invasion of the tissues by foreign protein or **antigen**. Antibodies combine with antigen, usually in a highly specific manner, coagulating it and, if the antigen is an invasive organism, generally rendering it biologically inert. The immune response appears to be present in all vertebrates but is particularly well developed in birds and mammals.

In mammals several classes of antibody are known, which have different modes of action. One class is secreted in mucus in, for example, saliva and in sweat and clearly represents an early barrier to recurrent infection. Other classes are found in the serum fraction of blood.

Sites of Antibody Production. Antibodies are formed by cells in the reticulo-endothelial system. These cells are dispersed in several organs of the body but the main centres of antibody production are in bone marrow, thymus, spleen and lymph nodes. The thymus gland lies in the thorax, ventral and anterior to the heart. In young animals the thymus gland is relatively large and it diminishes in size as the animal matures.

If the thymus gland is removed from newborn mice these animals are subsequently found to be unable to respond to certain kinds of antigen. Thymectomized mice, for example, seem to be unable to produce antibodies to sheep red cells, although their capacity to produce antibodies to other foreign proteins may be unimpaired.

In an elegant experiment Mitchison (1967) has shown that the immune response may require the cooperation of more than one type of cell. He selected certain substances of relatively low molecular weights compared with proteins, termed **haptens**, which have no antigenicity, that is they elicit no immune response when injected into animals. One such hapten is **NIP** or 4 hydroxy-3 iodo-5 nitro-

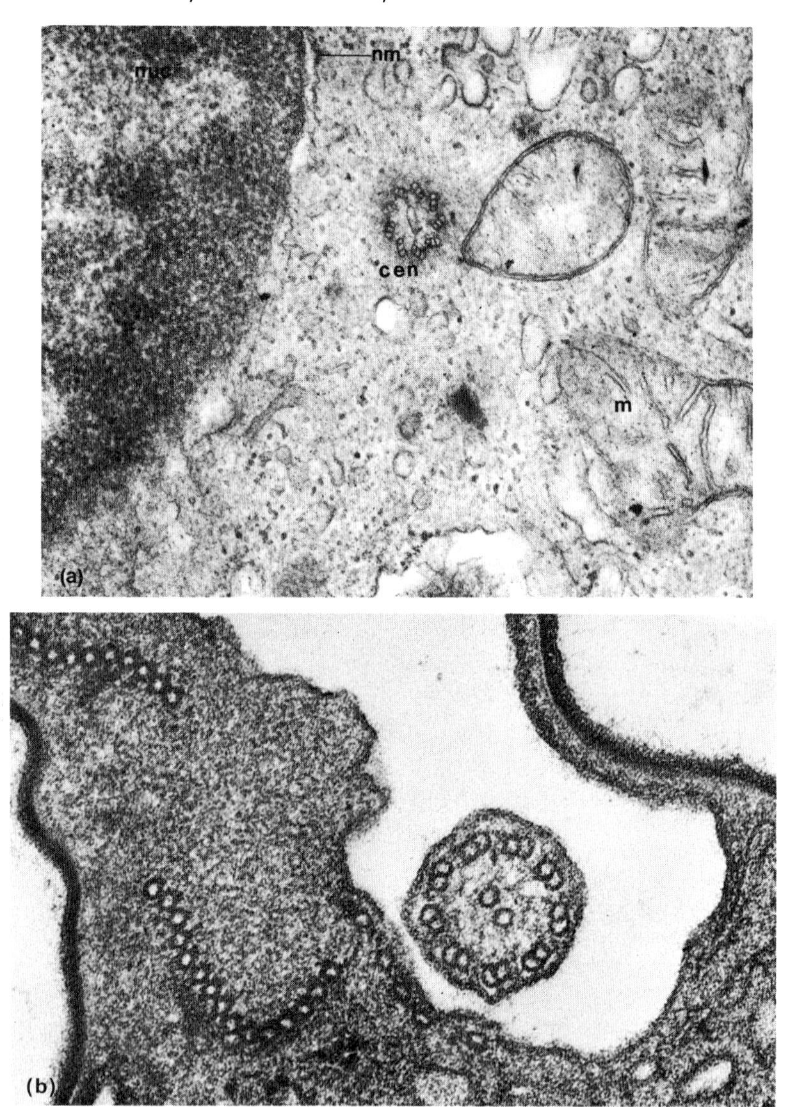

Fig. 10.12 (a) Electron micrograph of a section of a lymphocyte showing the centriole (cen) in transverse section, the nuclear membrane (nm) and mitochondria (m). (By courtesy of Alan Ross.) (b) Section of the flagellum of the protozoan blood parasite, *Trypanosoma* (by courtesy of Dr. K. Vickerman). Compare this structure with the centriole, above, and the sperm tail section in Figure 6.12.

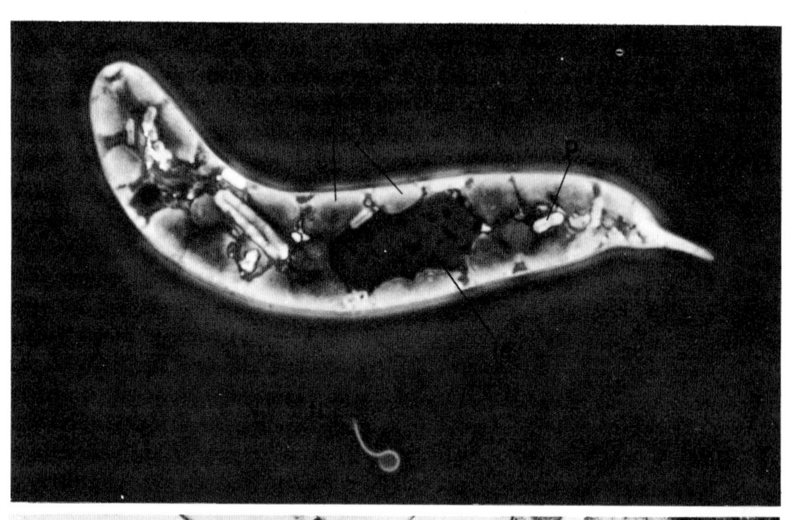

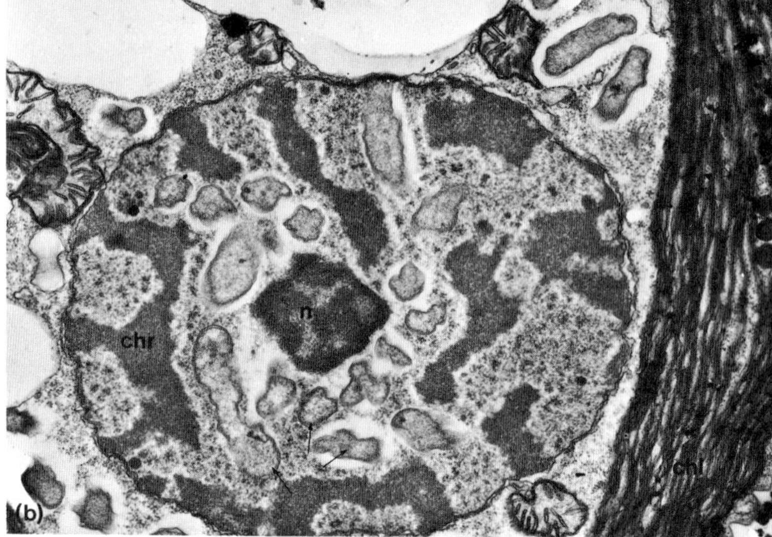

Fig. 10.13 (a) Phase contrast photomicrograph of *Euglena spirogyra* showing the flagellum (fl), frequently discarded in microscopic preparations, paramylum grains (p) for carbohydrate storage, the chloroplasts (c) and the conspicuous intranuclear endosymbionts (ie). (b) E.M. of the perinuclear region of the euglenoid flagellate, *Trachelomonas oblonga* showing part of a chloroplast (chl), nucleolus (n) and chromatin (chr). Bacteria-like endosymbionts (arrowed) can be seen inside the nucleus and in the cytoplasm. (Courtesy of Dr. G. F. Leedale.)

phlynyl acetic acid. Haptens do elicit a specific antibody response, however, when they are conjugated with a carrier protein, such as ovalbumin (OA) or bovine serum albumin (BSA).

In this experiment one group of mice was injected with NIP-OA, that is ovalbumin to which the hapten NIP was conjugated. A second group of mice was injected with BSA, or bovine serum albumin, alone. Later, when the two groups of mice were judged to have responded to their respective antigens, the animals were killed. Their spleen cells were extracted and injected into three further sets of mice as shown in Fig. 10.14. These three sets of mice had been irradiated with X-Rays, a treatment which destroys the antibody producing capacity of animals. The injected spleen cells, therefore, took up residence in their new hosts where they formed the sole antibody producing tissue.

The three sets of mice were injected with NIP-BSA, that is the same hapten, but this time conjugated with bovine serum albumin.

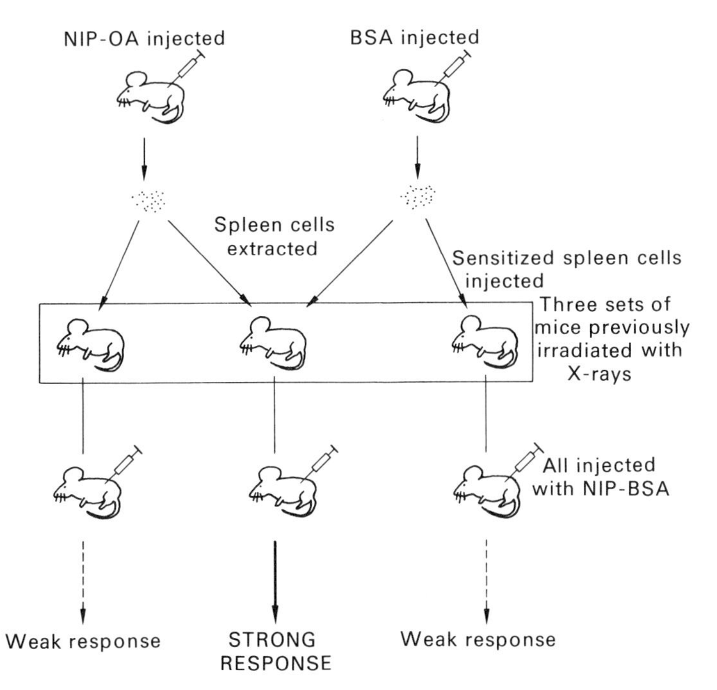

Fig. 10.14 In this experiment (see account in text) the strong response suggests that there has been some cell interaction in the production of antibody to NIP-BSA.

The immune response that followed was found to be relatively small for the two sets of mice which had received spleen cells from either the NIP-OA or the BSA sensitized animals alone. But the response was large from the third set of mice which had received a mixture of spleen cells. It seems, therefore, that the two kinds of cells, one sensitized to NIP and the other to BSA, cooperated in the immune response (Fig. 10.14).

The Structure of Antibody. The structure of one class of antibody, that of a gamma-globulin in blood serum, has now been described and has produced considerable speculation about its mode of production. γ-globulins are protein molecules having molecular weights of approximately 150 000. The sequence of amino acids in a human γ-globulin has been determined together with the positions of disulphide bonds (p. 32) which sustain the tertiary structure (Edelman *et al.*, 1969).

The molecule consists of two long or **heavy** chains of amino acids and two shorter or **light** chains. In the molecule described by Edelman and his co-workers the long chains each comprised 446 amino acids and the short chains each had 214 amino acids. This protein is thus the largest yet for which the amino acid sequence has been determined (Fig. 10.15).

The most striking feature of this molecule is that the linear sequence

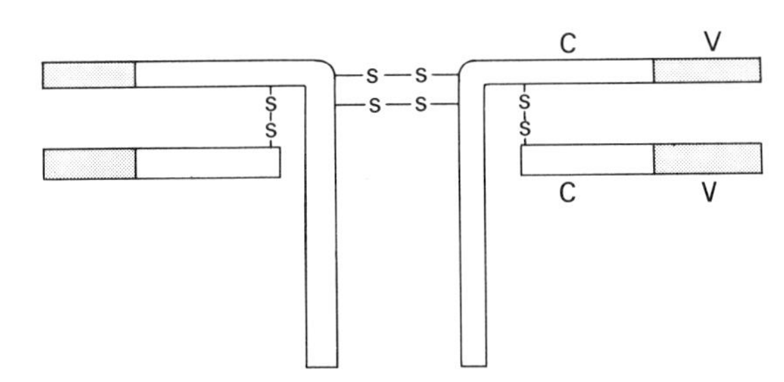

Fig. 10.15 Proposed structure of an antibody molecule comprising two heavy and two light chains, each with a variable region (shaded) and a constant region (unshaded).

in each chain is clearly demarcated into a *constant* or *C* portion, in which the sequence of amino acids is relatively fixed for a wide range of immunoglobulins, and a **variable** or **V** portion. It is this latter portion which is believed to give the molecule the high degree of specificity which characterizes the antibody-antigen reaction.

The Genetic Basis of Antibody Formation. The antibody molecule is a protein comprising, in the case of the immunoglobulins, four polypeptide chains held together by disulphide bridges. The fact that each chain consists of two regions separate in character and probably in function raises the question of whether they are coded for by two genes or cistrons (p. 148), in which case the concept of 'one gene one polypeptide' may have to be revised. Indeed, there is now good evidence (reviewed by Pink and Marrack, 1969) that the constant and variable regions are in fact coded for by separate cistrons, or DNA coding units.

The C sequence of each chain appears to be quite stable and its gene, in a relatively uniform state, is common to various animal species. It seems likely that its role in the immune response is of a generalized or mechanical nature, such as interaction with complement (a substance which heightens the antibody-antigen reaction) rather than acting in the specific sense of antigen recognition.

The V site poses more problems in terms of genic control because of the many kinds of antibody molecule that can be produced by an animal. If the potential number of antibody types is, say, a million one has to postulate that there are a million genes capable of coding for the V region. It has been shown that any given cell in the antibody forming tissues produces only one antibody (Nossal *et al.*, 1964). Thus in an animal which has been injected with two quite distinct antigens, the respective antibodies are produced by two cell colonies and no one cell produces both antibodies.

This evidence is accommodated by the **clonal selection theory** (Burnet, 1961). This theory holds that during embryogenesis the antibody producing cells undergo diversification and differentiation. Their multitude of V genes are randomly and permanently 'switched-off' (p. 121) except for one, or perhaps a few, that remain permanently active. Thus the full spectrum of antibody production is maintained across the tissue as a whole, but only a few cells are capable of making any one antibody.

In the clonal selection theory the role of antigen is seen as a trigger, promoting the proliferation, by repeated mitosis, of the cell or few cells which produce the corresponding antibody. The theory neatly accommodates many observations concerning the immune reaction and its strong feature is that it makes the formation of antibody proteins conform to the mechanism whereby cells produce other proteins (p. 34). Thus instructions for antibody protein synthesis reside in the genetic data of the cell and this avoids the rather untidy earlier 'instructive' theories whereby antigen itself entered the cell and formed some kind of 'template' for the formation of complementary antibody.

Phenomena such as **immunological tolerance**, whereby an animal's antibody synthesizing mechanism distinguishes between its own proteins and those from a foreign source, can be incorporated by postulating that those cells which would produce antibodies to 'self' proteins are eliminated as they arise during embryogenesis. However, a concept requiring the presence of a large number of V genes of which all but one must be permanently switched-off is uneconomic and biologically unsatisfactory. Burnet therefore proposed that during early embryogenesis those cells that would give rise to the antibody forming tissue would undergo rapid somatic mutation at the V locus, probably a thousand times more rapidly than normal mutation rates. In this way diversity could be acquired without the use of a large pool of redundant genes.

Fig. 10.16 Sir McFarlane Burnet, KBE, FRS born in Australia in 1899. He is a world recognised authority on virus diseases and on the immune response. He was awarded the Nobel Prize for medicine in 1960. (By courtesy of Universal Pictorial Press.)

The principle of variation at the V locus has stimulated many hypotheses about possible mechanisms, or 'generators of diversity'. Brenner and Milstein (1966) proposed the action of a 'cleaving enzyme' which, commencing at a recognition site on the V gene, would break down the DNA in the cistron. During its repair random errors could arise in the nucleotide sequence and thus in the genetic code. Excision and repair systems for DNA are known (p. 27) but this hypothesis requires an additional 'mutagenic' polymerase enzyme operating at only one locus and permitting errors in strand construction in a manner which is in direct conflict with the semi-conservative principle of DNA replication. Dreyer and Bennett (1965) have proposed the involvement of viruses or episomes and Whitehouse (1967) has devised a model for the generator based on intrachromosomal crossing-over. In the latter scheme antibody variability would arise by 'somatic recombination' similar to that in meiotic recombination (p. 181) but involving looping of the same chromosome.

The clonal theory of antibody production, then, goes a long way both to accommodate many of the observed phenomena about the immune response and to bring its operation into line with what is known about other activities of the cell. It does not, however, readily incorporate the findings that several cells may be involved in a cooperative manner in inducing the antibody response (possibly macrophages and lymphocytes derived from thymus and bone marrow) and it may require further modification as more evidence comes to light.

In addition to the immune response which affords a barrier to foreign proteins there occurs in birds and mammals the **interferon** response. Interferon is complementary to the action of antibodies but differs from it in two ways. First, interferon is produced by cells in response to foreign nucleic acids, and second, it is non-specific (Isaacs, 1963). Interferon does not agglutinate foreign material, nor does it appear to prevent the entry of invasive organisms. Its activity, in so far as virus particles are concerned, appears to lie in inhibiting the replication of the foreign nucleic acid.

The immune response appears to be limited to one animal phylum, the Chordata. In lower phyla the capacity to resist invasion by 'foreign protein' appears to rest solely with macrophages. These cells conform to the principle of genic identity in that they can distinguish between 'self' and 'non-self' proteins but they are not **adaptive** in an immunological sense. If an insect larva, for example, successfully rejects a parasite this in no way improves the larva's capacity for resisting a further invasion by the parasite.

Earlier it was stated that the immune response is one of the barriers to endosymbiosis, but its evolutionary origin occurred long after that of other barriers of proven effectiveness and other reasons for its evolutionary adoption are possible. Its origin in the Phylum Chordata coincides with the evolutionary origin of continuously proliferating tissues, such as epithelia. Such continuous proliferation throughout the life of the organism, past its normal growth phase, raises the hazard of somatic mutation giving rise to ontogenesis, or cancer, and Burnet and others have suggested that the immune response may have evolved for the recognition of, and resistance to, this event.

It appears, then, that in all animal phyla except the Protozoa, with isolated exceptions in some lower phyla such as the Coelenterata and Platyhelminthes, various mechanisms exist to prevent collectivism and to enforce separatism. It is noteworthy that a parallel situation exists in the plant kingdom although the mechanisms by which plants maintain their genetic identity, or separateness, are as yet unknown.

The underlying reasons for the switch, at an early level in evolutionary history, from collectivism to separatism are not difficult to postulate. Natural selection, as has already been described in the previous chapter, acts on the phenotype. Those phenotypes in a population which are best fitted to the selection pressures exerted by the environment are likely to be the most successful in contributing their genomes to the gene pool from which the next generation is drawn. It is axiomatic, therefore, that the phenotype should truly represent the manifestation or expression of the inherited genome.

Collectivism operates against this principle. The complex nature of genic interaction (p. 174) ensures a fine degree of phenotypic variability, which would be destroyed by the admixture of foreign, cytoplasmic, genomes, Medawar (1957) coined the term 'uniqueness of the individual' which describes this principle. Collectivism would introduce a source of phenotypic variation other than that conferred by the genetic data inherited from the gene pool. Phenotypic plasticity would become extensive and the process whereby progressive changes in gene frequencies lead to race formation and speciation would be blurred.

The genetic integrity of the individual organism is central to any theory of evolution considered in genetic terms. That all but the most primitive organisms reflect this 'uniqueness' lends credence to those concepts embraced by the genetic basis of evolution outlined in this book.

Journal Title Abbreviations

A. Rev. Biochem.	Annual Review of Biochemistry
Adv. Enzymol.	Advances in Enzymology
Adv. Protein Chem.	Advances in Protein Chemistry
Am. J. Bot.	American Journal of Botany
Am. J. phys. Anthrop.	American Journal of Physical Anthropology
Am. Nat.	American Naturalist
An. Estac. exp. Aula Dei	Anales de la Estación experimental de Aula Dei
Anat. Anz.	Anatomischer Anzeiger
Anim. Behav.	Animal Behaviour
Annls Sci. Nat. (Bot.)	Annales des sciences naturelles
Ann. Eugen.	Annals of Eugenics
Ann. N.Y. Acad. Sci.	Annal of the New York Academy of Sciences
Biblphia genet.	Bibliographia genetica
Biochem. J.	Biochemical Journal
Biochem. Soc. Symp.	Biochemical Society Symposia
Biochim. biophys. Acta	Biochimica et biophysica acta
Biol. Bull. mar. biol. Lab., Woods Hole	Biological Bulletin. Marine Biological Laboratory, Woods Hole, Mass.
Biol. Rev.	Biological Review
Bot. Rev.	Botanical Review
Br. J. prev. soc. Med.	British Journal of Preventive and Social Medicine
Br. med. J.	British Medical Journal
Bull. Johns Hopkins Hosp.	Bulletin of the Johns Hopkins Hospital
Can. med. Ass. J.	Canadian Medical Association Journal
Can. J. Genet. Cytol.	Canadian Journal of Genetics and Cytology
Chromosoma	Chromosoma
Circulation	Circulation
Cold Spring Harb. Symp. quant. Biol.	Cold Spring Harbor Symposia on Quantitative Biology
C. r. hebd. Séanc. Acad. Sci. Paris	Comptes rendus hebdomadaire des seances de l' Academie des Sciences
C. r. Trav. Lab. Carlsberg, (Chem.)	Comptes rendus des travaux du Laboratoire Carlsberg, (Chemistry)
Endocrinology	Endocrinology
Entomologist	Entomologist
Evolution, Lancaster, Pa.	Evolution
Experientia	Experientia
Expl Cell Res., Suppl.	Experimental Cell Research, Supplement
Fedn Proc. Fedn Am. Socs exp. Biol.	Federation proceedings. Federation of American Societies for Experimental Biology
4th. Int. Conf. Electron Micr.	Fourth International Conference on Electron Microscopy
Genetics, Princeton	Genetics
Heredity, Lond.	Heredity
J. biol. Chem.	Journal of Biological Chemistry
J. biophys. biochem. Cytol.	Journal of Biophysical and Biochemical Cytology
J. Cell Biol.	Journal of Cell Biology
J. Cell. Sci.	Journal of Cell Science
J. Comp. Psychol.	Journal of Comparative Psychology
J. exp. Med.	Journal of Experimental Medicine
J. exp. Zool.	Journal of Experimental Zoology
J. gen. Microbiol	Journal of General Microbiology
J. gen. Physiol.	Journal of General Physiology
J. gen. Virology	Journal of General Virology
J. Genet.	Journal of Genetics
J. Hered.	Journal of Heredity
J. Membrane Sci.	Journal of Membrane Science
J. molec. Biol.	Journal of Molecular Biology
J. Linn. Soc. (Bot.)	Journal of the Linnean Society
J. theor. Biol	Journal of Theoretical Biology
Jh. Ver. vaterl. Naturk. Württ.	Jahresheft des Vereins für vaterländische Naturkunde in Württemberg
Lancet	Lancet
Nature, Lond.	Nature
Naturwissenschaften	Naturwissenschaften
New·Phytol.	New Phytologist
New Scient.	New Scientist
Obstet. Gynec., Tokyo	Obstetrics and Gynecology
Öst. bot. Z.	Österreichische botanische Zeitschrift
Planta	Planta
Proc. Am. phil. Soc.	Proceedings of the American Philosophical Society
Proc. natn. Acad. Sci. U.S.A.	Proceedings of the National Academy of Sciences of the United States of America
Proc. R. Soc.	Proceedings of the Royal Society
Proc. R. Soc. Edinb.	Proceedings of the Royal Society of Edinburgh
Proc. zool. Soc. Lond.	Proceedings of the Zoological Society of London
Publs Carnegie Instn	Publications. Carnegie Institution of Washington
Sch. Sci. Rev.	School Science Review
Sci. J. Lond.	Science Journal
Science, N.Y.	Science
Scient. Am.	Scientific American
Symp. Soc. exp. Biol.	Symposia of the Society for Experimental Biology
Univ. Calif. Publs Agnc. Sci	University of California Publications in Agricultural Science
Z. indukt. Abstamm.-u. VererbLehre	Zeitschrift für induktive Abstammungs-u. Vererbungslehre

References

AIRD, I., BENTALL, H. H. and ROBERTS, J. A. F. (1953). A relationship between cancer of the stomach and the ABO blood groups. *Br. med. J.*, **1**, 799–801.

ALFERT, M. and NIRMAL DAS. (1969). Evidence for control of the rate of nuclear DNA synthesis by the nuclear membrane of eukaryotic cells. *Proc. natn. Acad. Sci. U.S.A.*, **63**, 123–8.

ALLAN, T. M. (1954). Fitness, fertility and blood groups. *Br. med. J.*, **2**, 1486.

ALLISON, A. C. (1954). Protection afforded by the sickle cell trait against subtertian malarial infection. *Br. med. J.*, **1**, 290–2.

ALLISON, A. C. (1956). Sickle cells and evolution. *Scient. Am.*, **195**, 87–94.

ALLISON, A. C. (1970). Viruses, immunology and cancer. *New Scient.*, **47**, 714, 330–1.

AMOROSO, E. C. (1970). Development of the early embryo. *Sci. J. Lond.*, **6**, 58–64.

ARNON, D. I. (1960). The role of light in photosynthesis. *Scient. Am.*, **203**, 5, 104–18.

ASTBURY, W. T. (1947). X-ray studies of nucleic acids. *Symp. Soc. exp. Biol.*, **I**, Nucleic Acid, 66–76.

AUERBACH, C. (1962). *The science of genetics.* Hutchinson & Co. Ltd. London.

BANKS, B. E. C. (1970). A misapplication of chemistry in biology. *Sch. Sci. Rev.*, **52**, 179, 286–97.

BARON, W. M. M. (1963). Investigation of the catalytic properties of extracted chloroplasts (the Hill Reaction). Appendix A in *Organisation in Plants*, Edward Arnold, London.

BARR, M. L. and BERTRAM, E. G. (1949). A morphological distinction between neurones of the male and female, and the behaviour of the nucleolar satellite during accelerated nucleoprotein synthesis. *Nature, Lond.*, **163**, 676–77.

BARR, M. L., CARR, D. H., POZSONYI, J., WILSON, R. A., DUNN, H. G., JACOBSON, T. S. and MILLER, J. R. (1962). The XXXXY sex chromosome abnormality. *Can. med. Ass. J.*, **87**, 891–901.

BARR, M. L., CARR, D. H., SOLTAN, H. C., WEINS, R. G. and PLUNKETT, E. R. (1964). The XXYY variant of Klinefelter's syndrome. *Can. med. Ass. J.*, **90**, 575–80.

BASSHAM, J. A. (1962). The path of carbon in photosynthesis. *Scient. Am.*, **206**, 6, 89–100.

BASSHAM, J. A. and CALVIN, M. (1957). *The Path of carbon in photosynthesis.* Prentice-Hall, Inc. N.J.

BATESON, W. (1894). *Materials for the study of variation treated with especial regard to discontinuity in the origin of species.* Macmillan and Co. (London).

BEALE, G. H., JURAND, A. and PREER, J. R. (1969). The classes of endosymbiont of *Paramecium aurelia. J. Cell Sci.*, **5**, 65.

BEERMAN, W. and CLEVER, U. (1964). Chromosome puffs. *Scient. Am.*, **210**, 4, 50–65.

BELL, E. (1969). I-DNA: its packaging into I-somes and its relation to protein synthesis during differentiation. *Nature, Lond.*, **224**, 326–8.

BENZER, S. (1962). The fine structure of the gene. *Scient. Am.*, **206**, 1, 70–84.

BERG, P. (1961). Specificity in protein synthesis. *A. Rev. Biochem.*, **30**, 293.

BEUTLER, E., YEH, M. and FAIRBANKS, V. F. (1962). The normal human female as a mosaic of X-chromosome activity: studies using the gene for G–6–Pd deficiency as a marker. *Proc. natn. Acad. Sci. U.S.A.*, **48**, 9–16.

BIRNSTEIL, M. L., CHIPCHASE, M. I. H. and HYDE, B. B. (1963). The nucleolus, a source of ribosomes. *Biochim. biophys. Acta*, **76**, 454.

BLACKWOOD, M. (1956). The inheritance of B chromosomes in *Zea mays. Heredity, Lond.*, **10**, 353.

BLAKESLEE, A. F. (1934). New Jimson weeds from old chromosomes. *J. Hered.*, **25**, 80–108.

BLOBEL, G. and SABATINI, D. D. (1970). Controlled proteolysis of nascent polypeptides in rat liver fractions. I. Location of polypeptides within ribosomes. *J. Cell Biol.*, **45**, 130–57.

BOYD, W. C. (1940). Critique of methods of classifying mankind. *Am. J. phys. Anthrop.*, **27**, 333–64.

BRAIN, P. (1952). Sickle cell anaemia in Africa. *Br. med. J.*, **2**, 880.

BRENNER, S., JACOB, F. and MESELSON, M. (1961). An unstable intermediate carrying information from genes to ribosomes for protein synthesis. *Nature, Lond.*, **190**, 576–81.

BRENNER, S. and MILSTEIN, C. (1966). Origin of antibody variability. *Nature, Lond.*, **211**, 242–3.

BRIDGES, C. B. (1925). Sex in relation to chromosomes and genes. *Am. Nat.*, **59**, 127.

BRIDGES, C. B. (1936). The Bar gene, a duplication. *Science, N.Y.*, **83**, 210–1 and reproduced in *Classic Papers in Genetics*, ed. J. A. PETERS. (1959). Prentice-Hall, Inc.

VAN BRINK, J. M. (1959). L'expression morphologique de la digamétie chez les Sauropsides et les Monotrèmes. *Chromosoma*, **10**, 1–72.

BRINK, R. A. (1954). Very light variegated pericarp in maize. *Genetics, Princeton.*, **39**, 724–40.

BROOK, A. J. (1964). *The Living Plant.* Edinburgh University Press.

BROWN, D. D. and GURDON, J. B. (1964). Absence of ribosomal RNA synthesis in the anucleolar mutant of *Xenopus laevis. Proc. natn. Acad. Sci. U.S.A.*, **51**, 139–46.

BURGESS, R. G., TRAVERS, A. A., DUNN, J. J. and BANTZ, E. K. F. (1969). Factor stimulating transcription by RNA-polymerase. *Nature, Lond.*, **221**, 43–6.

BURNET, SIR MACFARLANE. (1961). The mechanism of immunity. *Scient. Am.*, **204**, 1, 58–67.

BUTLER, J. A. V. (1968). *Gene Control in the Living Cell.* Allen and Unwin, London.

BUVAT, R. (1958). Recherches sur les infrastructures du cytoplasm d'Elodea canadensis. *Annls. Sci. nat. (Bot.)*, **19**, 121.

CALLAN, H. G. and SPURWAY, H. (1951). A study of meiosis in inter-racial hybrids of the newt, *Triturus cristatus. J. Genet.*, **50**, 235–49.

CANELLAKIS, E. S. and HERBERT, E. (1960). Studies on s-RNA synthesis; 1. Purification and general characteristics of the RNA-enzyme complex. *Proc. Natn. Acad. Sci.*, **46**, 170.

CARO, L. G. and PALADE, G. E. (1964). Protein synthesis and discharge from the pancreatic exocrine cell. *J. Cell Biol.*, **20**, 473–96.

CARR, D. H. (1965). Chromosome studies in spontaneous abortions. *Obstet. Gynec. Tokyo*, **26**, 308–26.

CASPARI, E. (1950). On the selective value of the alleles *Rt* and *rt* in *Ephestia kuhniella. Am. Nat.*, **84**, 367.

CHARGAFF, E. (1950). Chemical specificity of nucleic acids and mechanism of their enzymic degradation. *Experientia*, **6**, 201–9.

CLAPHAM, A. R., TUTIN, T. G. and WARBURG, E. F. (1962). *Flora of the British Isles.* 2nd edition. Cambridge University Press, London.

CLARK, B. F. C. and MARCKER, K. A. (1966). The role of N-formyl methionine-sRNA in protein biosynthesis. *J. molec. Biol.*, **17**, 394–406.

CLELAND, R. E. (1936). Some aspects of cytogenetics of *Oenothera. Bot. Rev.*, **2**, 316–48.

CLOWES, F. A. L. and JUNIPER, B. E. (1968). *Plant Cells*. Blackwell's Scientific Publications, Oxford and London.

COHEN, J. (1969). Is sexual reproduction wasteful? *New Scient.*, **44**, 674, 282–4.

COMMONER, B. (1964). Roles of deoxyribonucleic acid in inheritance. *Nature, Lond.*, **202**, 960–8.

CRICK, F. H. C. (1958). On protein synthesis. *Symp. Soc. exp. Biol.*, **12**, 138–63.

CRICK, F. H. C., BARNETT, L., BRENNER, S. and WATTS-TOBIN, R. J. (1961). General nature of the genetic code for proteins. *Nature, Lond.*, **192**, 1227–32.

CROSBY, J. L. (1963). The evolution and nature of dominance. *J. theor. Biol.*, **5**, 35–51.

DA CUNHA, A. B., DOBZHANSKY, T., PAVLOVSKY, O. and SPASSKY, B. (1959). Genetics of natural populations. XXVIII. Supplementary data on the chromosomal polymorphism in *Drosophila Willistoni* in its relation to the environment. *Evolution*, **13**, 389–404.

DARLINGTON, C. D. and MATHER, K. (1949). *The Elements of Genetics*. Allen and Unwin, Ltd., London.

DARWIN, C. (1859). *The origin of species by means of natural selection or the preservation of favoured races in the struggle for life*. John Murray, London.

DAVIDSON, E. H. (1968). *Gene Activity in Early Development*. Academic Press, Inc. New York and London.

DAVSON, H. and DANIELLI, J. F. (1943). *The Permeability of Natural Membranes*. Cambridge University Press, London.

DOBZHANSKY, T. (1947). The adaptive changes induced by natural selection in wild populations of *Drosophila. Evolution, Lancaster, Pa.*, **1**, 1–16.

DOBZHANSKY, T. (1951). *Genetics and the Origin of Species*. 3rd edition. Columbia University Press, N.Y.

DOBZHANSKY, T. (1959). Variation and evolution. *Proc. Am. phil. Soc.*, **103**, 252–63.

DOBZHANSKY, T. (1962). Reproductive Isolation, in *Mankind Evolving*, Yale University Press, New Haven, Connecticut.

DOBZHANSKY, T. (1970). *Genetics of the Evolutionary Process*. Columbia University Press, N.Y.

DOBZHANSKY, T. and EPLING, C. (1944). *Contributions to the genetics, taxonomy and ecology of Drosophila pseudo-obscura and its relatives. Publs. Carnegie Instn.*, **554**.

DOBZHANSKY, T. and HOLTZ, A. M. (1943). A re-examination of manifold effects of genes in *Drosophila melanogaster. Genetics, Princeton*, **28**, 295–303.

DOBZHANSKY, T. and PAVLOVSKY, O. (1955). An extreme case of heterosis in a central American population of *Drosophila tropicalis. Proc. natn. Acad. Sci.*, **41**, 289.

DOURMASHKIN, R. R. and TYRRELL, D. A. J. (1970). Attachment of two myxoviruses to ciliated epithelial cells. *J. gen. Virology*, **9**, 77–88.

DREYER, W. J. and BENNETT, J. C. (1965). The molecular basis of antibody production, a paradox. *Proc. natn. Acad. Sci.*, **54**, 864–9.

DU PRAW, E. J. (1968). *Cell and Molecular Biology*. Academic Press, N.Y.

DURHAM, A. (1971). How a virus assembles itself. *New Scient.*, **49**, 736, 200–3.

EAST, E. M. and MANGELSDORF, A. J. (1925). A new interpretation of the hereditary behaviour of self-sterile plants. *Proc. natn. Acad. Sci.*, **11**, 2, 166–71.

EDELMAN, G. M., CUNNINGHAM, B. A., GALL, W. E., GOTTLEIB, P. D., RUTISHAUSER, U. and WAXDAL, M. J. (1969). The covalent structure of an entire γ-immunoglobulin molecule. *Proc. natn. Acad. Sci.*, **63**, 1, 78–85.

ERRERA, M., HELL, A. and PERRY, R. P. (1961). The role of the nucleolus in RNA and protein synthesis. *Biochim. biophys. Acta*, **49**, 58–63.

FAWCETT, D. W., ITO, S. and SLAUTTERBACK, D. (1959). The occurrence of intercellular bridges in groups of cells exhibiting synchronous differentiation. *J. biophys. biochem. Cytol.*, **5**, 453–60.

FERNANDEZ-MORAN, H. (1962). Low temperature electron microscopy and X-ray diffraction studies of lipoprotein components in lamellar systems. *Circulation*, **26**, 1039–65.

FISHER, SIR R. A. (1931). The Evolution of Dominance. *Biol. Rev.*, **6**, 345.

FORBES, T. R. (1961). Endocrinology of reproduction in cold-blooded vertebrates. In *Sex and Internal Secretions*, Ed. W. C. YOUNG, Baillière, London.

FORD, C. E. and HAMERTON, J. L. (1956). The chromosomes of man. *Nature, Lond.*, **178**, 1020–3.

FORD, C. E., JONES, K. W., POLANI, P. E., DE ALMEIDA, J. C. and BRIGGS, J. H. (1959). A sex chromosome anomaly in a case of gonadal dysgenesis (Turner's syndrome). *Lancet*, **1**, 711–3.

FORD, E. B. (1940). Polymorphism and Taxonomy. In *The New Systematics*, Ed. J. HUXLEY, Clarendon Press, Oxford.

FORD, E. B. (1942). Anthropology and Blood Groups. In *Genetics for Medical Students*, Methuen, London.

FORD, E. B. (1957). Polymorphism in plants, animals and man. *Nature, Lond.*, **180**, 1315–9.

FORD, E. B. (1965). Genetic Polymorphism. *All Souls Studies V*, Faber and Faber, London.

FORD, E. B. and SHEPPARD, P. M. (1969). The *medionigra* polymorphism of *Panaxia dominula. Heredity, Lond.*, **24**, 561–9.

FOWLER, L. R. and RICHARDSON, S. H. (1963). Studies on the electron transfer system. I. On the mechanism of reconstitution of the mitochondrial electron transfer system. *J. biol. Chem.*, **238**, 456–63.

FREESE, E. (1958). The arrangement of DNA in the chromosome. *Cold Spring Harb. Symp. quant. Biol.*, **23**, 13–8.

FROMSON, D. and NEMER, M. (1970). Cytoplasmic extraction: polyribosomes and heterogeneous ribonuclear proteins without associated DNA. *Science, N.Y.*, **168**, 266.

GAULDEN, M. E. and PERRY, R. P. (1958). Influence of the nucleolus on mitosis as revealed by ultraviolet microbeam irradiation. *Proc. natn. Acad. Sci.*, **44**, 553.

GEIDUSCHEK, E. P., NAKAMOTO, T. and WEISS, S. B. (1961). The enzymatic synthesis of RNA: complementary interaction with DNA. *Proc. natn. Acad. Sci.*, **47**, 1405–15.

GLASS, B. and LI, C. C. (1935). The dynamics of racial admixture: an analysis based on the American negro. *Am. J. hum. Genet.*, **5**, 1–20.

GOODENOUGH, U. W. and LEVINE, R. P. (1970). The genetic activity of mitochondria and chloroplasts. *Scient. Am.*, **223**, 5, 22–30.

GROSS, J. (1961). Collagen. *Scient. Am.*, **204**, 5, 120–34.

HALDANE, J. B. S. (1956). The theory of selection for melanism in Lepidoptera. *Proc. R. Soc. B.*, **145**, 303–6.

HARDY, G. H. (1908). Mendelian proportions in a mixed population. *Science, N.Y.*, **28**, 49–50 and in *Classic Papers in Genetics*, Ed. J. A. PETERS. (1959). Prentice-Hall, Inc. N.J.

HARRISON, D. (1970). Simplified linkage map of *Drosophila*. Appendix III in *Problems in Genetics*, Addison-Wesley, Ltd., London.

HARRISON, D. (1971). Defence against infection. In Advanced Biology Notes, 2nd. edition., Macmillan, London.

HARRISON, K. (1965). A Guide Book to Biochemistry. 2nd. edition, Cambridge University Press, London.

HERSKOWITZ, I. H. (1962a). Recombination in bacteriophage II. In *Genetics*, Little, Brown and Co., Boston.

HERSKOWITZ, I. H. (1962b). Changes involving genomes and chromosomes. In *Genetics*, Little, Brown and Co., Boston.

HERSKOWITZ, I. H. (1962c). Pleiotropism, penetrance and expressivity. In *Genetics*, Little, Brown and Co., Boston.

HERSKOWITZ, I. H. (1962d). Races and the Origin of Species. In *Genetics*, Little, Brown and Co., Boston.

HILL, R. and WHITTINGHAM, C. P. (1955). Photosynthesis. Methuen and Co., London.

HOKIN, L. E. and HOKIN, M. R. (1959). Evidence for phosphatidic acid as the sodium carrier. *Nature, Lond.*, **184**, 1068–9.

HOLLEY, R. W., APGAR, J., EVERETT, G. A., MADIAN, J. Y., MARQUISEE, M., MERRILL, S. H., PENSWICK, J. R. and ZAMIR, A. (1965). Structure of a ribonucleic acid. *Science, N.Y.*, **147**, 1462–5.

HOLTER, H. and MARSHALL, J. M. (1954). Studies on pinocytosis in the amoeba *Chaos chaos*. *C.r. Trav. Lab. Carlsberg*, (Chem.), **29**, 7–26.

HOOBER, K. SIEKEVITZ, P. and PALADE, G. E. (1969). Formation of chloroplast membranes in *Chlamydomonas reinhardi* y-l. *J. biol. Chem.*, **244**, 2621–31.

HOOKER, C. W. (1960). Reproduction in the Male. In *Medical Physiology and Biophysics*, Ed. T. C. RUCH and J. F. FULTON, Saunders, Philadelphia.

HUXLEY, J. S. (1942). *Evolution, the Modern Synthesis*. George Allan and Unwin Ltd., London.

HUXLEY, J. S. (1949). *Soviet Genetics and World Science*. Chatto and Windus, London.

HUXLEY, SIR J. and KETTLEWELL, H. B. D. (1965). *Charles Darwin and his World*. Viking Press, Inc. N.Y. and Thames and Hudson, London.

INGRAM, V. M. (1956). A specific chemical difference between the globins of normal human and sickle-cell anaemia haemoglobins. *Nature, Lond.*, **178**, 792–4.

INGRAM, V. M. (1961). Gene evolution and the haemoglobins. *Nature, Lond.*, **189**, 704–8.

ISAACS, A. (1963). Foreign nucleic acids. *Scient. Am.*, **209**, 4, 46–50.

JACOB, F. and MONOD, J. (1961). Genetic regulatory mechanisms in the synthesis of protein. *J. molec. Biol.*, **3**, 318–56.

JACOBS, P. A. and STRONG, J. A. (1959). A case of human intersexuality having a possible XXY sex-determining mechanism. *Nature, Lond.*, **183**, 302–3.

JENKINS, R. A. (1970). The fine structure of a nuclear envelope associated endosymbiont of *Paramecium*. *J. gen. Microbiol.*, **61**, 355–9.

JENNINGS, H. S. (1929). Genetics of the protozoa. *Biblphia genet.*, **5**, 108.

JONES, D. F. (1925). *Genetics in Plant and Animal Improvement*. John Wiley and Sons, Inc., N.Y.

JUNIPER, B. E. and BARLOW, P. W. (1969). The distribution of plasmodesmata in the root tip of maize. *Planta*, **89**, 352–60.

KAEMPFER, R., MESELSON, M. and RASKAS, H. (1968). Cyclic dissociation into stable subunits and reformation of ribosomes during bacterial growth. *J. molec. Biol.*, **31**, 277.

KAMEN, M. D. (1958). A universal molecule of living matter. *Scient. Am.*, **199**, 2, 77–82.

KARPECHENKO, G. D. (1928). Polyploid hybrids of *Raphanus sativus* L. X *Brassica Z. indukt. Abstamm. u. VererbLehre*, **48**, 1–85.

KASHNIG, D. and KASPER, C. (1969). Isolation, morphology and composition of the nuclear membrane from rat liver. *J. biol. Chem.*, **244**, 3786.

KELLENBERGER, E. (1960). The physical state of the bacterial nucleus. In *Microbial Genetics*, Cambridge University Press, London.

KEMP, R. (1970). Cell Division and Heredity. Institute of Biology, Studies in Biology, No. 21. Edward Arnold Ltd., London.

KENDREW, J. C. (1961). The three dimensional structure of a protein molecule. *Scient. Am.*, **205**, 6, 96–111.

KENDREW, J. C., BOBO, G., DINTZIS, H. M., PARRISH, R. C., WYCKOFF, H. and PHILLIPS, D. C. (1958). A three dimensional model of the myoglobin molecule obtained by X-ray analysis. *Nature, Lond.*, **181**, 662–6.

KENDREW, J. C., DICKERSON, R. E., STRANDBERG, B. E., HART, R. G., DAVIES, D. R., PHILLIPS, D. C. and SHORE, V. C. (1960). Structure of myoglobin: a three dimensional Fourier synthesis at 2Å resolution. *Nature, Lond.*, **185**, 422–7.

KEOSIAN, J. (1965). The Origin of Life. In *Selected Topics in Modern Biology*, Reinhold Publing. Corporation and Chapman and Hall Ltd., London.

KERR, W. E. and WRIGHT, S. (1954). Experimental studies of the gene frequencies in very small populations of *Drosophila melanogaster*. *Evolution, Lancaster, Pa.*, **8**, 172.

KETTLEWELL, H. B. D. (1955). Selection experiments on industrial melanism in the Lepidoptera. *Heredity, Lond.*, **9**, 323–42.

KETTLEWELL, H. B. D. (1956). Further selection experiments on industrial melanism in the Lepidoptera. *Heredity, Lond.*, **10**, 287–30.

KETTLEWELL, H. B. D. (1957). Problems in industrial melanism. *Entomologist*, **90**, 98–105.

KETTLEWELL, H. B. D. (1965). Insect survival and selection for pattern. *Science, N.Y.*, **148**, 1290–6.

KING, R. (1970). Reception of a sex hormone. *New Scient.*, **46**, 704, 472–3.

KLINEFELTER, H. F., REIFENSTEIN, E. C. and ALBRIGHT, F. (1942). Syndrome chracterised by gynecomastia, aspermatogenesis without aleydigism and increased excretion of follicle stimulating hormone. *J. clin. Endocr. Metab.*, **2**, 615–27.

KOBEL, H. R. (1962). Heterochromosomen bei *Vipera berus*. *Experientia*, **18**, 173–4.

KOLLER, P. C. (1937). The genetical and mechanical properties of the sex chromosomes. III. Man. *Proc. R. Soc. Edinb.*, **57**, 194–214.

KORNBERG, A., LEHRMAN, I. R. and SIMMS, E. S., 1956. Polydesoxyribonucleic synthesis by enzymes from *Escherichia coli*. *Fedn. Proc. Am. Socs exp. Biol.*, **15**, 291–2.

KOSSWIG, C. (1929). Ueber die veranderten Wirkung von Farbgenen des *Platypoecilus*. *Z. indukt. Abstamm.-u. VeterbLehre*, **50**, 63.

KYLE, H. M. (1926). *The Biology of Fishes*. Sidgwick and Jackson, London.

LAKE, J. A. and BEEMAN, W. W. (1968). On the conformation of yeast transfer-RNA. *J. molec. Biol.*, **31**, 115.

LANTZ, L. A. (1947). Hybrids between *Triturus cristatus* and *Triturus marmoratus*. *Proc. zool. Soc. Lond.*, **117**, 247–58.

LEEDALE, G. F. (1969). Observations on endonuclear bacteria in euglenoid flagellates. *Öst. bot. Z.*, **116**, 279.

LEVIN, D. A. (1963). Natural hybridization between *Phlox maculata* and *P. glaberrima* and its evolutionary significance. *Am. J. Bot.*, **50**, 714–20.

LEVINTHAL, C., KEYMAN, A. and HIGA, A. (1962). Messenger RNA turnover and protein synthesis in *B. subtilis* inhibited by Actinomycin D. *Proc. natn. Acad. Sci. U.S.A.*, **48**, 1631–8.

LEWIS, E. B. (1952). The pseudoallelism of White and Apricot in *Drosophila melanogaster*. *Proc. Natn. Acad. Sci.*, **38**, 953–61.

LEWIS, W. H. (1931). Pinocytosis. *Bull. Johns Hopkins Hosp.*, **49**, 17.

LIPMANN, F. (1941). Metabolic generation and utilization of phosphate bond energy. *Adv. Enzymol.*, **1**, 99–162.

LITTLEFIELD, J. W., KELLER, E. B., GROSS, J. and ZAMECNIK, P. C. (1955). Studies on cytoplasmic ribonucleoprotein particles from the liver of the rat. *J. biol. Chem.*, **217**, 111–24.

LONGUET-HIGGINS, H. C. and ZIMM, B. H. (1960). Calculation of the rate of uncoiling of the DNA molecule. *J. molec. Biol.*, **2**, 1.

LORENZ, K. (1963). *On Aggression.* University Paperback Editions (1967) Methuen, London.

DE LUCIA, P. and CAIRNS, J. (1969). Isolation of an *E. coli* strain with a mutation affecting DNA polymerase. *Nature, Lond.*, **224**, 1164–6.

MCARTHUR, N. and PENROSE, L. S. (1951). World frequencies of the O, A and B blood-group genes. *Ann. Eugen.*, **15**, 302–5.

MCCLINTOCK, B. (1944). The relation of homozygous deficiencies to mutations and allelic series in maize. *Genetics, Princeton.*, **29**, 478–502.

MCCLINTOCK, B. (1950). The origin and behaviour of mutable loci in maize. *Proc. natn. Acad. Sci. U.S.A.*, **36**, 344–55.

MCCLUNG, C. E. (1901). Notes on the accessory chromosome. *Anat. Anz.*, **20**, 220–6.

MALTZAHN, K. VON and MUHLETHALER, K. (1962a). Observations on division of mitochondria in differentiating cells of *Splanchnum ampullaceum* L. *Experientia*, **18**, 315–6.

MALTZAHN, K. VON and MUHLETHALER, K. (1962b). Observations on chloroplast division in dedifferentiating cells of *Splanchnum ampullaceum* L. *Naturwissenschaften*, **49**, 308–9.

MANNING, A. (1961). The effects of artificial selection for mating speed in *Drosophila melanogaster*. *Anim. Behav.*, **9**, 82–92.

MANTON, I. (1953). The cytological evolution of the fern flora of Ceylon. *Symp. Soc. exp. Biol.*, VII, 174–85.

MARCHANT, C. J. (1967). Evolution in *Spartina* (Graminae). I. The history and morphology of the genus in Britain. *J. Linn. Soc. (Bot.)*, **60**, 381, 1.

MARCHANT, C. J. (1968). Evolution in *Spartina* (Graminae). II. Chromosomes, basic relationships and the problem of *S. x townsendii agg.* *J. Linn. Soc. (Bot.)*., **60**, 383, 381–409.

MARSDEN, J. C. (1970). Why do cells pump ions? *New Scient.*, **45**, 152–5.

MARSH, R. E., COREY, R. B. and PAULING, L. (1955). An investigation of the structure of silk fibroin. *Biochim. biophys. Acta*, **16**, 1–14.

MATHER, K. (1953). The genetical structure of populations. *Symp. Soc. exp. Biol.*, VII (Evolution), 66–95.

MATTHEY, R. (1949). *Les Chromosomes des Vertébrés.* Rouge, Lausanne.

MAYNARD SMITH, J. (1958). *The Theory of Evolution.* Pelican Biology Series, Penguin Ltd., London.

MAYR, E. (1954). Change of Genetic Environment and Evolution. In *Evolution as a Process*, Ed. J. HUXLEY, George Allen and Unwin, London.

MAZIA, D. (1960). The analysis of cell reproduction. *Ann. N.Y. Acad. Sci.*, **90**, 455.

MEDAWAR, P. B. (1957). The Uniqueness of the Individual. Methuen and Co., London.

MESELSON, M. and STAHL, F. W. (1958). The replication of DNA in *Escherichia coli*. *Proc. natn. Acad. Sci. U.S.A.*, **44**, 671–82.

MEYER, H. (1938). Investigations concerning the reproductive behaviour of *Mollienisia formosa*. *J. Genet.*, **36**, 329–66.

MIESCHER, F. (1871). On the chemical composition of pus cells. *Hoppe-Seyler's medizinische-chemische Untersuchungen*, **4**, 441–60 and reprinted in *Great Experiments in Biology*, Ed. M. L. GABRIEL and S. FOGEL, Prentice-Hall, Inc., N.J.

MITCHELL, P. (1963). Molecule group and electron translocation through natural membranes. *Biochem. Soc. Symp.*, **22**, 142–68.

MITCHISON, N. A. (1967). Antigen recognition responsible for the induction *in vitro* of the secondary response. *Cold Spring Harb. Symp. quant. Biol.*, **32**, 431–9.

MITTWOCH, U. (1967). *Sex Chromosomes.* Academic Press, N.Y.

MITTWOCH, U., KALMUS, H. and WEBSTER, W. S. (1966). Deoxyribonucleic acid values in dividing and non-dividing cells of male and female larvae of the honey bee. *Nature, Lond.*, **210**, 264–6.

MONTGOMERY, T. H. (1904). Some observations and considerations upon the maturation phenomena of the germ cells. *Biol. Bull. mar. biol. Lab.*, *Woods Hole*, **6**, 137–58.

MOORE, J. A. (1941). Developmental rate of hybrid frogs. *J. exp. Zool.*, **86**, 405.

MORGAN, T. H. (1929). The variability of eyeless. *Publs. Carnegie Instn.*, **399**, 139–68.

MORGAN, T. H. and BRIDGES, C. B. (1919). The origin of gynandromorphs. *Publs. Carnegie Instn.*, **278**, 1–122.

MORGAN, T. H. and GOODALE, H. D. (1912). Sex-linked inheritance in poultry. *Ann. N.Y. Acad. Sci.*, **22**, 113–33.

MOSELEY, B. E. B. (1969). Keeping DNA in good repair. *New Scient.*, **41**, 641, 626–8.

MOURANT, A. E. (1954). *The Distribution of the Human Blood Groups.* Blackwell Scientific Publications, Oxford.

MOURANT, A. E., KOPEC, A. C. and DOMANIEWSKA-SOBCZAK, K. (1971). Blood groups and blood clotting. *Lancet*, **7**, 692, 223–7.

NAVASHIN, M. (1931). Chromatin mass and cell volume in related species. *Univ. Calif. Publs. Agric. Sci.*, **6**, 207.

NICHOLSON, D. E. (1970). *Metabolic Pathways Chart and Booklet.* Reviewed annually and published by Koch-Light Laboratories, Colnbrook, England.

NIRENBERG, M., LEDER, P., BERNFIELD, M., BRIMACOMBE, R. TRUPIN, J., ROLTMAN, F. and O'NEIL, C. (1965). RNA codewords and protein synthesis. VII. On the general nature of the RNA code. *Proc. natn. Acad. Sci. U.S.A.*, **53**, 1161–8.

NOMURA, M. and GUTHRIE, C. (1968). Initiation of protein synthesis: a critical test of the 30S subunit model. *Nature, Lond.*, **219**, 232.

NOSSAL, G. J. V., SZENBERG, A., ADA, G. L. and AUSTIN, C. M. (1964). Single cell studies on 19S antibody production. *J. exp. Med.*, **119**, 485.

ODUM, E. P. (1963). *Ecology.* Modern Biology Series, Holt, Rinehart and Winston, Inc., N.Y.

OPARIN, A. I. (1957). The Origin of Life on the Earth. Academic Press, N.Y.

OWEN, J. J. T. (1965). Karyotype studies on *Gallus domesticus*. *Chromosoma*, **16**, 601–8.

PAUL, J. (1965). *Cell Biology.* Heinemann Educational Books Ltd., London.

PAUL, J. and FOTTRELL, P. F. (1963). Mechansim of D-glutamyltransferase repression in mammalian cells. *Biochim. biophys. Acta*, **67**, 334–6.

PAULING, L. and COREY, R. B. (1951). Atomic coordinate and structure factors for two helical configurations of polypeptide chains. *Proc. natn. Acad. Sci. U.S.A.*, **37**, 235–40.

PEACOCKE, A. R. and DRYSDALE, R. B. (1965). *The Molecular Basis of Heredity.* Butterworth and Co., London.

PENNOCK, L. A. (1965). Triploidy in parthogenetic species of the Teiid lizard, Genus Cnemidophorus. *Science, N.Y.*, **149**, 3683, 539–40.

PERUTZ, M. F. and MITCHISON, J. M. (1950). State of haemoglobin in sickle cell anaemia. *Nature, Lond.*, **166**, 677–9.

PINK, R. and MARRACK, P. (1969). Genes, cells and antibodies. *New Scient.*, **42**, 653, 573–5.

POTTS, W. H. (1954). *Distribution map of the Tsetse species in Africa, Sheet 3.* Directorate of Colonial Surveys, Publications, Edward Stanford, London.

RENDEL, J. M. (1951). Mating of ebony vestigial and wild type *Drosophila melanogaster* in light and dark. *Evolution, Lancaster, Pa.*, **5**, 226–30.

REZNIKOFF, W. S., MILLER, J. H., SCAIFE, J. G. and BECKWITH, J. R. (1969). A mechanism for repressor action. *J. molec. Biol.*, **43**, 201.

RHODES, M. M. (1941). The genetic control of mutability in maize. *Cold Spring Harb. Symp. quant. Biol.*, **9**, 138–55.

RIS, H. (1957). The Chemical Basis of Heredity. Ed. W. D. MCELROY and B. GLASS, Johns Hopkins Press, Baltimore.

RIS, H. (1960). Fine structure of the nucleus during spermiogenesis. *4th. Int. Conf. Electron Micr.*, **2**, 199.

RIS, H. (1961). Ultrastructure and molecular organisation of genetic systems. *Can. J. Genet. Cytol.*, **3**, 95–120.

RIS, H. and CHANDLER, B. L. (1963). The ultrastructure of genetic systems in prokaryotes and eukaryotes. *Cold Spring Harb. Symp. quant. Biol.*, **28**, 1–8.

ROBERTS, J. W. (1969a). Promoter mutation *in vitro*. *Nature, Lond.*, **223**, 480.

ROBERTS, J. W. (1969b). Termination factor for RNA synthesis. *Nature, Lond.*, **224**, 1168–74.

RUNDQUIST, E. A. (1933). The inheritance of spontaneous behaviour in rats. *J. comp. Psychol.*, **16**, 415–38.

SANGER, F. (1952). The arrangement of amino acids in proteins. *Adv. Protein Chem.*, **7**, 1–67.

SEARS, E. R. (1944). Cytogenetic studies with polyploid species of wheat. III. Additional chromosomal aberrations in *Triticum vulgare*. *Genetics, Princeton*, **29**, 232–46.

SELMAN, G. and PERRY, M. (1970). How cells cleave. *New Scient.*, **46**, 695, 12–14.

SHAVER, E. L. (1962). The chromosomes of the opossum, *Didelphis virginiana*. *Can. J. Genet. Cytol.*, **41**, 62–8.

SHEPPARD, P. M. (1952). A note on non-random mating in the moth *Panaxia dominula* L. *Heredity, Lond.*, **6**, 239–41.

SHEPPARD, P. M. (1953). Polymorphism and population studies. *Symp. Soc. exp. Biol.* VII (Evolution), 274–89.

SHEPPARD, P. M. and COOK, L. M. (1962). The manifold effects of the *medionigra* gene of the moth *Panaxia dominula* and the maintenance of a polymorphism. *Heredity, Lond.*, **17**, 415–26.

SIBATANI, A., DE KLOET, S. R., ALLFREY, V. G. and MIRSKY, A. E. (1962). Isolation of a nuclear RNA fraction resembling DNA in its base composition. *Proc. natn. Acad. Sci. U.S.A.*, **48**, 471–7.

SINGLETON, W. R. (1962). Architectural changes in chromosomes. In *Elementary Genetics*. Van Nostrand, Inc., N.Y.

SINNOTT, E. W., DUNN, L. C. and DOBZHANSKY, T. (1958). Genetics of Race Formation. In *Principles of Genetics*. McGraw-Hill, N.Y.

SINSHEIMER, R. L. (1962). Single stranded DNA. *Scient. Am.*, **207**, 1, 109–19.

SIRLIN, J. L. (1962). The Nucleolus. In *Progress in Biophysics and Biophysical Chemistry*, **12**, Eds. J. A. V. BUTLER, R. E. ZIRKLE and H. E. HUXLEY, Pergamon Press, Oxford.

SKELDING, A. D. and WINTERBOTHAM, J. (1939). The structure and development of the hydathodes of *S. townsendii*. *New Phytol.*, **38**, 69–79.

SMITH, J. D. and DUNN, D. B. (1959). The occurrence of methylated guanines in ribonucleic acids from several sources. *Biochem. J.*, **72**, 294.

SONNEBORN, T. M. (1938). Mating types, toxic interactions and heredity in *Paramecium aurelia*. *Proc. Am. phil. Soc.*, **79**, 411–34.

SPENCER, M., FULLER, W., WILKINS, M. H. F. and BROWN, G. L. (1962). Determination of the helical configuration of ribonucleic acid molecules by X-ray diffraction study of crystalline amino-acid-transfer ribonucleic acid. *Nature, Lond.*, **194**, 1014–20.

SPIRIN, A. S. (1966). *Current Topics in Developmental Biology*. Eds. A. MONROY and A. A. MOSCONA, Academic Press, N.Y.

SPURWAY, H. (1953). Genetics of specific and subspecific differences in European newts. *Symp. Soc. exp. Biol.* VII (Evolution), 200–37.

STEBBINS, G. L. (1966). *Processes of Organic Evolution*. Prentice-Hall, Inc., N.J.

STEVENSON, G. (1967). *The Biology of Fungi, Bacteria and Viruses*. Contemporary Biology Series, Edward Arnold Ltd., London.

STEWARD, F. C. (1959). *Plant physiology: Vol II. Plants in Relation to Water and Solutes.* (Editor). Academic Press Inc., N.Y.

STEWART, J. S. S. (1960). Genetic mechanisms in human intersexes. *Lancet*, **1**, 825–6.

STOECKENIUS, W. (1963). Some observations on negatively stained mitochondria. *J. Cell Biol.*, **17**, 443–54.

STRUTHERS, D. (1951). ABO groups of infants and children. *Br. J. prev. soc. Med.*, **5**, 223–8.

STURTEVANT, A. H. (1913). The linear arrangement of six sex-linked factors in *Drosophila*, as shown by their mode of association. *J. exp. Zool.*, **14**, 43–59.

STURTEVANT, A. H. and MORGAN, T. H. (1923). Reverse mutation of the Bar gene correlated with crossing-over. *Science, N.Y.*, **57**, 746–7.

TAYLOR, J. H., WOODS, P. S. and HUGHES, W. L. (1957). The organisation and duplication of chromosomes as revealed by auto-radiographic studies using tritium-labelled thymidine. *Proc. natn. Acad. Sci. U.S.A.*, **43**, 122–8.

TJIO, J. H. and LEVAN, A. (1956). Note on the sex chromosomes of the rat during male meiosis. *An. Estac. exp. Aula Dei*, **4**, 173–84.

TOMKINS, G. M., GELEHRTER, T. D. and GRANNER, D. (1969). Control of specific gene expression in higher organisms. *Science, N.Y.*, **166**, 1474–80.

TRAUTNER, T. A., SCHWARTZ, M. N. and KORNBERG, A. (1962). Enzymatic synthesis of deoxyribonucleic acid, X. Influence of bromouracil substitutions on replication. *Proc. natn. Acad. Sci. U.S.A.*, **48**, 449.

TURNER, H. H. (1938). A syndrome of infantilism, congenital webbed neck and cubitus valgus. *Endocrinology*, **23**, 566–574.

VOLKIN, E. and ASTRACHAN, L. (1957). RNA metabolism of T-2 infected *Escherichia coli*. In *The Clinical Basis of Heredity*, Eds. W. D. MCELROY and B. GLASS, Johns Hopkins Press, Baltimore.

WADDINGTON, C. H. (1953a). The 'Baldwin effect', 'genetic assimilation' and 'homeostasis'. *Evolution, Lancaster, Pa.*, **7**, 386–7.

WADDINGTON, C. H. (1953b). Genetic assimilation of an acquired character. *Evolution, Lancaster, Pa.*, **7**, 118–26.

WADDINGTON, C. H. (1953c). Epigenetics and Evolution. *Symp. Soc. exp. Biol.*, VII (Evolution), 186–99.

WALLACE, A. R. (1889). *Darwinism.* Macmillan and Co., London.

WARMKE, H. E. (1946). Sex determination and sex balance in *Melandrium. Am. J. Bot.*, **33**, 648–60.

WATSON, J. D. and CRICK, F. H. C. (1953). A structure for deoxyribose nucleic acid. *Nature, Lond.*, **171**, 737–8.

WEILER, C. and OHNO, S. (1962). Cytological confirmation of female heterogamety in the African water frog, *Xenopus laevis. Cytogenetics* (Basel), **1**, 217–23.

WEINBERG, W. (1908). Über den Nachweiss des Vererbung beim Menschen. *Jh. Ver. vaterl. Naturk. Württ.*, **64**, 368–82.

WEISS, P. (1960). Molecular reorientation as unifying principle underlying cellular selectivity. *Proc. natn. Acad. Sci. U.S.A.*, **46**, 993.

WEITZMAN, D. (1969). Self control for enzymes. *New Scient.*, **43**, 666, 514–5.

WENT, H. A. and MAZIA, D. (1959). Immunological study of the origin of the mitotic apparatus. *Expl. Cell Res., suppl.*, **7**, 200–18.

WHITEHOUSE, H. L. K. (1967). Crossover model of antibody variability. *Nature, Lond.*, **215**, 371–4.

WILKINS, M. H. F., STOKES, A. R. and WILSON, H. R. (1953). Molecular structure of deoxypentose nucleic acids. *Nature, Lond.*, **171**, 738–40.

WILSON, G. B., SPARROW, A. H. and POND, V. (1959). Sub-chromatid rearrangements in *Trillium erectum. Am. J. Bot.*, **46**, 309.

WILLIAMSON, D. (1970). Where did mitochondria come from? *New Scient.*, **47**, 720, 624–6.

WINGE, O. (1922). One-sided masculine and sex-linked inheritance in *Lebistes reticulatis. J. Genet.*, **12**, 145–62.

WINGE, O. (1923a). On sex chromosomes, sex determination and preponderance of females in soke dioecious plants. *C. r. Trav. Lab. Carlsberg*, **15**, 1–26.

WINGE, O. (1923b). Crossing-over between the X- and Y-carrying chromosomes in *Lebistes. J. Genet.*, **13**, 201–17.

WITSCHI, E. (1929). Studies on sex differentiation and sex determination in amphibians. III. Rudimentary hermaphroditism and Y-chromosome in *Rana temporaria. J. exp. Zool.*, **54**, 157–223.

WOODBURY, J. W. (1960). The Cell Membrane: Ionic Gradients and Active Transport. In *Medical Physiology and Biophysics*, Eds. T. C. RUCH and J. F. FULTON, W. B. Saunders, Philadelphia.

WOOD-GUSH, D. G. M. (1960). A study of sex-drive of two strains of cockerel through three generations. *Anim. Behav.*, **8**, 43–53.

WOODWORTH, C. M., LENZ, E. R. and JUGENHEIMER, R. W. (1952). Fifty generations of selection for Protein and Oil in Corn. *Agron. J.*, **44**, 60–65.

WRIGHT, S. (1931). Evolution in Mendelian populations. *Genetics, Princeton*, **16**, 97–159.

WRIGHT, S. (1940). The statistical consequences of Mendelian heredity to speciation. In *The New Systematics*, Ed. J. HUXLEY, Oxford University Press, London.

WRIGHT, S. (1948). On the roles of directed and random changes in gene frequency in the genetics of populations. *Evolution, Lancaster, Pa.*, **2**, 279–94.

WYATT, G. R. (1952). The nucleic acids of some insect viruses. *J. gen. Physiol.*, **36**, 201.

YCAS, M. and VINCENT, W. S. (1960). An RNA fraction in yeast related in composition to DNA. *Proc. natn. Acad. Sci. U.S.A.*, **46**, 804.

YOTSUYANAGI, Y. and GUERRIER, C. (1965). Mise en évidence par les techniques cytochimiques et la microscopie électronique d'acide désoxyribonucléique dnas les mitochondries et les protoplastes d'*Allium cepa. C. r. hebd. Séanc. Acad. Sci., Paris*, **260**, 13, 2344–7.

Index